Y

ations

D1071640

CONTRIBUTORS TO VOLUME 4

Walter H. Bauer
James T. Bergen
A. Bondi
Edward A. Collins
H. L. Goldsmith
Julian F. Johnson
H. F. Mark
Hershel Markovitz
S. G. Mason
Roger S. Porter
Arie Ram
M. Reiner
G. W. Scott Blair

RHEOLOGY
Theory and Applications

Edited by
FREDERICK R. EIRICH
Polytechnic Institute of Brooklyn

Brooklyn, New York

VOLUME 4

1967
ACADEMIC PRESS, NEW YORK AND LONDON

ACADEMIC PRESS, INC.
111 Fifth Avenue, New York, New York 10003

United Kingdom Edition published by
ACADEMIC PRESS, INC. (LONDON) LTD.
24/28 Oval Road, London NW1 7DD

LIBRARY OF CONGRESS CATALOG CARD NUMBER: 56-11131

Second Printing, 1972

PRINTED IN THE UNITED STATES OF AMERICA

LIST OF CONTRIBUTORS

Numbers in parentheses refer to the pages on which the authors' contributions begin.

WALTER H. BAUER, School of Science, Rensselaer Polytechnic Institute, Troy, New York (423)

JAMES T. BERGEN, Research and Development Center, Armstrong Cork Company, Lancaster, Pennsylvania (285)

A. BONDI, Shell Development Company, Emeryville, California (1)

EDWARD A. COLLINS, B. F. Goodrich Chemical Company, Development Center, Avon Lake, Ohio (423)

H. L. GOLDSMITH, Pulp and Paper Research Institute of Canada, Montreal, Canada (86)

JULIAN F. JOHNSON, Chevron Research Company, Richmond, California (317)

H. F. MARK, Polytechnic Institute of Brooklyn, Brooklyn, New York (411)

HERSHEL MARKOVITZ, Mellon Institute, Pittsburgh, Pennsylvania (347)

S. G. MASON, Pulp and Paper Research Institute of Canada, Montreal, Canada (86)

ROGER S. PORTER, Polymer Science and Engineering Program, University of Massachusetts, Amherst, Massachusetts (317)

ARIE RAM, Department of Chemical Engineering, Technion, Haifa, Israel (251)

M. REINER, Department of Mechanics, Strength of Materials Division, Technion, Haifa, Israel (461)

G. W. SCOTT BLAIR, The National Institute for Research in Dairying, University of Reading, Shinfield, Reading, England (461)

PREFACE

When the exigencies of work and interest led this editor more and more into rheological studies during the early nineteen fifties, there was very little comprehensive literature available for self-study short of an extensive undertaking. Fortunately, there was a growing awareness of this need in many places and the editor found a ready and constructive understanding among his colleagues who responded with the chapters which now form Volumes 1–3 of "Rheology." These chapters presented not only valuable introductions and up-to-date reviews, but often, by the logic and necessity to broach new territory, broke new ground, provided much needed cross-correlation, and thus became minor classics which even ten years after their appearance are among the most quoted rheological literature.

Rheology has developed in the last decade at a still faster pace under the influence of the increasing interdisciplinary nature of modern technology in the midst of which rheology stands as a veritable meeting place for ideas and methodologies ranging from plasma physics to muscle contraction. In industry too, increasing complexity of demands and pressure of competition necessitates greater sophistication of processing procedures. All fabrication means forming and forming means one type of deformation or another, so that a great deal of stimulus has been provided to apply rheological theory to the most diverse applications and to wed continuum mechanics, theories of the solid and liquid states, colloid science, and polymer chemistry with various industrial machine operations. Thus, not only has rheology become broader still than it ever was, but new classes of demands on rheological understanding have arisen.

Consequently, Volume 4 and the forthcoming Volume 5, which were thought of as one volume, but once again had to be divided because of the accumulated material, do not contain revisions or updatings but cover new areas which just became visible at the time of publication of the earlier volumes. There are the important complementary contributions by Bondi, by Goldsmith and Mason as well as by Mark, and also the chapters that cover newer areas of interest such as those by Markovitz, Ram, Bergen, and Porter and Johnson. The contributions have been divided in such a way that Volume 4 contains the discussions pertaining to materials of more fluid nature, whereas the chapters collected in Volume 5 will comprise subjects related to viscoelasticity and more extensive deformation in solids.

One of the main, and justified, criticisms leveled at the preceding volumes was that of the Babylonian stage of nomenclature and of the symbols used. This editor tried to improve the situation, but found it unavoidable as it reflected the wide variety of backgrounds of the contributors and the great difficulties of finding a throughgoing rationale which would justify the usage of given terms in most diverse fields. Moreover, it was obvious that a single editorial enterprise could not succeed where many major authors as well as learned societies and international committees had failed. A notable exception is the nomenclature referring to viscoelasticity which under the leadership of prominent members of the American Society of Rheology became practically unified and is being used today almost exclusively as pioneered and introduced for instance by H. Leaderman in his chapter in "Rheology," Volume 2.

However, the pressures and needs for a unified nomenclature continued and are growing every day. The editor has therefore solicited an effort by two rheologists with an extremely broad acquaintance of rheological problems in communication, to compile a rheological dictionary and to explain the history and reasons for their selections. Even now in Volumes 4 and 5 it has not been quite possible to present a unified rheological language, largely because of the magnitude of the undertaking, but also because the chapter by Reiner and Scott Blair was not completed before nearly the end of the preparations, yet it is hoped that their contribution and dictionary will assist in further agreement on meaning and uses of terms.

It might also be pointed out that the indexes of Volumes 1–3 of "Rheology," and more so of this volume and the forthcoming volume, were arranged with the preknowledge of the coming efforts and with the aim to tie the many fields into a semblance of unity, thus enabling the readers of many backgrounds to find not only references to what they were looking for, but also references to what was related to their interests but unknown to them because it was either clothed in unfamiliar verbiage or dealt with in unfamiliar literature. The dictionary now presented and prepared independently by two competent authors recommends substantially the same usages as the indexes of "Rheology," which should further assist the readers. The dictionary included in this volume will be available also as a separate booklet to facilitate and encourage a further dissemination of a coherent nomenclature. It is very much hoped that this dictionary as well as the other chapters will continue to serve the discipline of rheology.

Brooklyn, New York   FREDERICK R. EIRICH
September, 1967

CONTENTS

CONTENTS OF PREVIOUS VOLUMES

RHEOLOGY

Theory and Applications

VOLUME 4

Chapter 1

VISCOSITY AND MOLECULAR STRUCTURE

A. Bondi

I. Purpose and Scope

The efforts of organic chemists at synthesizing lubricants for extravagant operating conditions, the data needs of engineers in process plants, and the curiosity of physical chemists have in the aggregate provided a sufficient volume of data to challenge the physicists' ability to generalize the accumulated information. Various attempts have been made to relate the viscosity of liquids to the structure of the molecules of which they are composed.[1-5]

[1] R. Kremann, "Physikalische Eigenschaften und Chemische Konstitution," p. 205. Steinkopff, Dresden, 1937.
[2] E. C. Bingham and L. W. Spooner, *J. Rheol.* **3**, 221 (1932).
[3] A. Bondi, *J. Chem. Phys.* **14**, 591 (1946).
[4] A. Bondi, *Ann. N.Y. Acad. Sci.* **53**, 870 (1951).
[5] L. Grunberg, *Kolloid-Z.* **126**, 87 (1952).

The justification for adding yet another article to the voluminous literature is this reviewer's impression that enough progress has recently been made in the molecular theory of liquids to encourage a new look at the viscosity even of those liquids which are composed of rather large and complicated molecules.

The effect of molecular structure on viscosity can be understood more easily if one is familiar with the effect of molecular structure on well-defined motions of molecules in liquids. The first part of this article is therefore devoted to a presentation of various manifestations of rigidity and flexibility of molecules in condensed phases and also of data on the effects of molecular structure on translational and rotational motions of molecules in liquids.

The subsequent discussion of viscosity consists of three parts. First, a review of a recent advance in the theory of transport phenomena brings this reviewer's article[5a] up to date. Next are presented a discussion of viscosity at high temperatures (reduced) and, finally, a discussion of viscosity at low temperatures or, rather, at high packing densities. The separation into these two regimes is connected with the fact that the viscosity of liquids at high reduced temperatures, i.e., above about 0.6 of the absolute normal boiling point, is well, and nearly quantitatively, represented as a comparatively simple function of molecular geometry and intermolecular forces obtained from independent measurements. Viscous flow at lower temperatures, or at high packing densities, depends on rather subtle features of molecular geometry, and its formal representation is therefore a rather more complicated function of structural and state variables. The accent in the present work is on the correlation of viscosity with molecular structure from a particular point of view, no attempt being made to present a comprehensive review of the literature.

II. Molecular Motions in Liquids

A qualitative understanding of the relations between the viscosity of liquids and the structure of its constituent molecules is more easily obtained if one can start with some information about the motions of molecules in liquids. Details of these motions are accessible to us through two pathways: direct observation, i.e., through the measurement of the relaxation times of nuclear magnetic resonance absorption and of electric dipole orientation, and analysis of equilibrium properties, such as density and heat capacity, by means of a molecular theory. In a short review of the results obtained up to the present time we shall start with the interpretation of equilibrium properties.

[5a] A. Bondi, in "Rheology: Theory and Applications" (F. R. Eirich, ed.), Vol. 1, p. 321. Academic Press, New York, 1956.

1. Evidence from Equilibrium Properties

a. Density

Analysis of the density of liquids composed of largish molecules, made by means of a theory first proposed by Prigogine and his collaborators[6,7] and somewhat modified for daily use by this writer,[8] permits one to estimate the number of external degrees of freedom ($3c$), including those due to internal rotation, excited in liquids over a certain fairly broad temperature range. In simple liquids this range extends from the melting point to the atmospheric boiling point. Significant deviations are obtained at packing densities of $\rho^* > 0.64$ in the case of glass transitions. With high polymers many degrees of freedom begin to freeze in at $\rho^* > 0.6$. A crude estimate yields, for the medium-temperature regime, $3c \doteq 0.391E°/RT(1.8)$, where $E°$ is the standard energy of vaporization and $T(1.8)$ is the temperature at which $V/V_w = 1.80$, V and V_w being the molal volume and van der Waals' volume,[9] respectively.

Rigid nonlinear polyatomic polar and nonpolar molecules exhibit $3c = 6 \pm 0.5$, as expected; flexible molecules, $3c > 6$. The data of Table I convey the rather good correlation between the flexibility ($3c - 6$) and the magnitude of the barriers to internal rotation for slightly flexible molecules. It should be noted that a good correlation is found only when the rotating molecule segments have more than a critical size. This size is determined by the simple consideration that molecular motion in liquids involves internal rotation only when the barrier to external rotation of a segment is larger than the barrier to its internal rotation. The barrier to external rotation of a molecule (segment) is of the order of $0.2E°$ of the molecule (segment). This makes it understandable why small alkane molecules act effectively as rigid molecules.

The reason for the relation between density and molecular flexibility is more clearly understood if one estimates the temperature $T(1.8) = 0.391E°/3cR$ for limiting values of $3c$. For the completely rigid molecule $3c = 6$, but for the completely flexible molecule $3c = 2n + 6$, where n is the number of freely moving segments per molecule. Hence, a liquid composed of molecules with ten freely moving segments is expanded to $V^* = 1.80$ at $0.23T(1.8)$ of a liquid composed of rigid molecules with the same value of $E°$. Conversely, a rigid molecule is always at a lower reduced temperature than is a flexible molecule. Pure tobacco mosaic virus for instance, is always near $T^* \approx 0$, regardless of how high the absolute temperature. Since

[6] I. Prigogine, "Molecular Theory of Solutions." Wiley (Interscience), New York, 1957.

[7] I. Prigogine, A. Bellemans, and C. Naar-Colin, *J. Chem. Phys.* **26,** 751 (1957).

[8] A. Bondi and D. J. Simkin, *A.I.Ch.E. J.* **6,** 191 (1960).

[9] A. Bondi, *J. Phys. Chem.* **68,** 441 (1964).

TABLE I

EFFECTS OF MOLECULAR STRUCTURE AND OF BARRIERS TO INTERNAL ROTATION ($V°$)
ON MOLECULAR FLEXIBILITY ($3c - 6$)
From density and vapor pressure data

Structure variable	Substance	$3c - 6$	$V°$ (kcal/mole) of linkages*
Effect of branch-crowding	n-C_9H_{20}	2.1	2.8 (8)
	$(C_2H_5)_4 \cdot C$	0.3	2.8 (4), 4.9 (4)
	2,2,4,4-Tetramethylpentane	0.5	4.9 (8)
	n-$C_{14}H_{30}$	4.2	2.8 (13)
	2,2,3,3,5,6,6-Heptamethylheptane	0.3	4.9 all
Effect of double bonds	n-C_6H_{14}	0.84	2.8 (5)
	n-Hexene-1	0.85	2.8 (3), 2.0 (1)
	1,5-Hexadiene	1.0_5	2.8 (1), 2.0 (2)
	1,3,5-Hexatriene, cis	> 1.7	1.2 (2)
	2,3,5-Hexatriene, trans	≈ 1.0	1.2 (2)
Rigidity of ring systems	Benzene, naphthalene, etc.	≈ 0	∞
	Cyclopentane, cyclohexane, decalins	≈ 0–0.5	10 (1, 2)†
	Perhydrophenanthrene ($N_c = 14$)	0.5	10 (3)†
	Perhydrodibenzofluorene ($N_c = 19$)	1.5	10 (4)†
Mobility of joined rings	Diphenyl, p-terphenyl	0.5, 0.9	4.5 (1), 4.5 (2)
	Dicyclohexyl, 1,3-dicyclohexyl cyclohexane	0.3_5, 2.1	3.6 (1), 3.6 (2)§
Crowding of rings on alkane axis	1,1- vs. 1,2-Diphenylethane	1.0, 1.5	> 0 (2), 2.8 (1), ~ 0 (2), 2.8 (1)
	1,1- vs. 1,2-Didecalylethane	1.2, 2.6	≈ 4 all, 3.5 (2), 2.8 (1)
	Tricyclopentylmethane	0.95	≈ 5 all
	Tricyclohexylmethane	1.4	≈ 5 all
	Tri(2-cyclohexylethyl)methane	5.1	≈ 3.6 (4), 2.8 (3)
Effect of monovalent functional groups‖	1-n-Alkyl bromides, iodides	+ 0.4, + 0.9	
	α,ω-Alkane dibromides, iodides	1.3, 1.4	
	1-n-Alkyl cyanides	≈ 0	
	α,ω-Alkane dicyanides	− 0.3	
	1-n-Nitroalkanes	≈ 0	

* The first figures give $V°$ in kilocalories per mole, and the figure in parentheses is the number of such barriers per molecule.

† Cycloparaffins carry out motions which are equivalent to internal rotation. In the case of cyclopentane it is called pseudorotation, and in the case of cyclohexane and its many derivatives it involves the boat–chain transition, for which the barriers are given.

§ The barriers to rotation around cyclohexyl ring junctions have been taken as equal to those around the diisopropyl bond.

‖ Here the $3c - 6$ column contains the difference $3c$ (for the polar compound) minus either $3c$ for the hydrocarbon homomorph or $3c$ for the hydrocarbon parent molecule, the latter in case of the alkane halides.

"completely flexible" molecules are hardly encountered in practice, experimental observations are restricted to a somewhat narrower range of effects. Nevertheless, the data of Table II show a rather wide range of values for $3c$ per segment in chain molecules of differing degrees of flexibility.

TABLE II

COMPARISON OF THE NUMBER OF EXTERNAL DEGREES OF FREEDOM ($3c$)
PER MOLECULE SEGMENT WITH THE BARRIER TO INTERNAL ROTATION ($V°$)

Repeating unit	$3c$	Probable compn. deg. of freedom*	$V°$, kcal/mole*
$-CH_2-$	0.44	0.44	2.8
$-CH_2-CH-$ Me	1.15	0.43 (2), 0.30 (1)	2.8 (2), 3.9 (1)
$-CH_2-C-$ Me ⟍ Me	1.0	0.25 (4)	4.9 (4)
$-CH_2-C=C_H-CH_2$(cis 1,4) Me	2.22	0.43 (2), 0.68 (2)	2.8 (2), 2.0 (2)
$-CH_2-CH-$ φ	2.80	0.43 (2), 1.94 (1)	2.8 (2), 0 (1)
$-CH_2-O-$	1.0	0.5 (2)	1.6 (2)
$-CH_2-CH^-$ O$-$C$-$Me ‖ O	2.46	0.43 (3), 0.57 (2)	2.8 (2), 2.5 (2), 0.5 (1)
$-CH_2-C-$ Me ⟍ C$-$O$-$Me ‖ O	1.89	0.43 (2), 0.28 (1), 0.25 (3)	4.3 (3), 2.8 (2), 0.7 (2)

* Number in parentheses is the number of bonds of a given kind per repeating unit.

b. Heat Capacity

Independent evidence of the nature of molecular motions in liquids can be derived from the configurational heat capacity[10,11] at constant volume $C_v^c = C_v(liq) - C_v(gas) + 3R$. A few typical examples for such data are presented on Figs. 1 and 2 for liquids composed of rigid and flexible molecules, respectively. The data of Fig. 1 point up a phenomenon that did not

[10] For a detailed discussion of configurational heat capacity and entropy in liquids see Refs. 6 and 11 and A. Bondi, *Ind. Eng. Chem., Fundamentals* 5, 442 (1966).

[11] J. S. Rowlinson, "Liquids and Liquid Mixtures." Academic Press, New York, 1959.

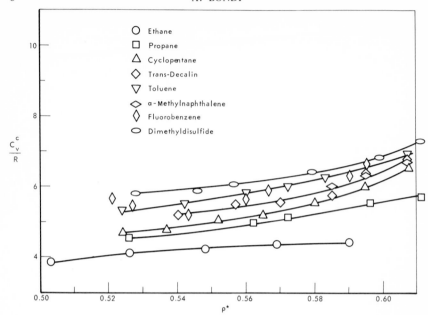

FIG. 1. Configurational heat capacity of various rigid molecules in the liquid state as a function of packing density.

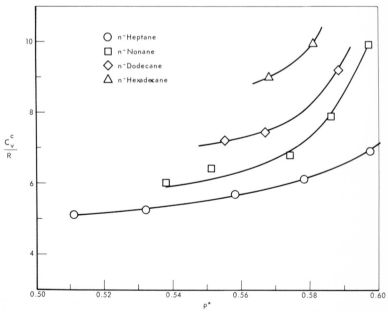

FIG. 2. Configurational heat capacity C_v^c of flexible molecules in the liquid state as a function of packing density.

appear in the density considerations; the hindered external rotation of rigid molecules in liquid composed of nonassociating molecules. The rise of C_v^c with increasing packing density ρ^* (that is with decreasing reduced temperature) is the result of the work that must be done when a molecule is rotated past the force field of its nearest neighbors. The effect can be calculated in much the same way as for hindered internal rotation, except that the barrier to external rotation is now a function of E° and of the number of nearest neighbors encountered during one rotation around the shortest axis of the molecule. The configurational heat capacity of liquids composed of flexible molecules results from a superposition, of the effects of the barrier to external rotation on those of the difference in barrier height to internal rotation between liquid and gas [contained in $C_v(\text{gas})$].

c. Vapor Pressure

Hindered external rotation of molecules is observed particularly easily in a comparison of the entropy of vaporization of the liquid under consideration with that of a "perfect liquid" at the same vapor volume, as proposed by Hildebrand.[12,12a] Typical values for this excess entropy of vaporization

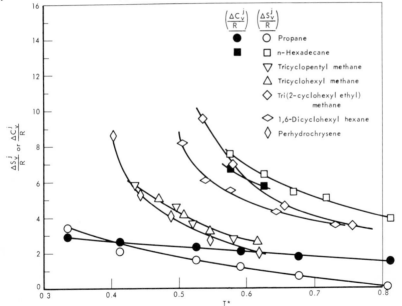

FIG. 3. Excess entropy of vaporization ΔS_v^j of various hydrocarbon liquids vs. their reduced temperature T^*. A few configurational heat capacity data have been entered for comparison.

[12] J. H. Hildebrand and R. L. Scott, "Solubility of Non-Electrolytes." Reinhold, New York, 1951.

[12a] A. Bondi, *J. Phys. Chem.* **58**, 929 (1954).

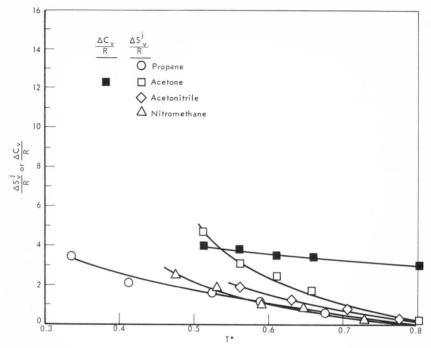

FIG. 4. Excess entropy of vaporization $\Delta S_v{}^j$ of several polar compounds as a function of their reduced temperature T^*.

$\Delta S_v{}^j$ are presented in Figs. 3 and 4, as a function of reduced density, for nonpolar and polar substances, respectively. In the case of nonpolar molecules $\Delta S_v{}^j$ could be related to the ratio between the volume required for free rotation in the liquid and the volume actually available for free motion in the liquid shown in Fig. 5. In the case of polar substances $S_v{}^j$ is a function both of dipole-orientation energy and of the volume requirements for free rotation.

2. Evidence from Transport Processes

Certain transport processes—the relaxation times of electric dipoles and of nuclear magnetic resonance absorption[13]—give rather direct information about specific molecular motions. Others, such as diffusion and spectral line-broadening give less clearly defined information about molecular motions. Yet the effects of molecular structure on all of these not only is instructive but also provides background information that will be valuable in interpreting viscosity data.

[13] For a detailed discussion of rotational motions in liquids see the series of papers by W. A. Steele, *J. Chem. Phys.* **38**, 2404, 2411, and 2418 (1963).

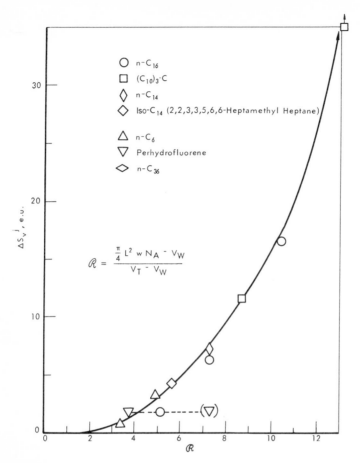

FIG. 5. Relation between degree of restriction $\mathscr{R}$ and the excess entropy of vaporization of various saturated hydrocarbons. Calculated from data in part from R. W. Schiessler *et al.*, AP1 Project 42, synthesis and properties of higher molecular weight hydrocarbons.

a. Relaxation Time of Electric Dipoles

The average rotation rate of polyatomic molecules in a medium-pressure gas according to kinetic theory is $(2\pi kT/I_i)^{1/2}$, where I_i is the appropriate moment of inertia. Comparison of the rotation rate of rigid molecules in the liquid, deduced from dipole relaxation time τ_e, with their gas kinetic rotation rate $\tau_e(2\pi kT/I_i)^{1/2}$ gives an idea of the extent of hindered rotation in the liquid. Typical retardation ratios $\tau_e(kT/I)^{1/2}$, shown in Fig. 6, increase so rapidly with decreasing temperature that most molecules in liquids much below their atmospheric boiling point should be considered more nearly

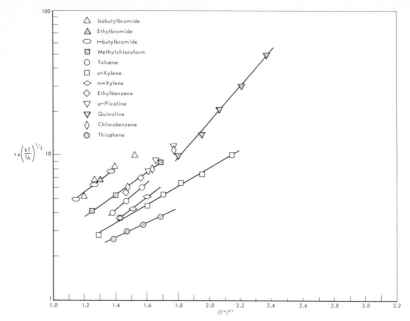

FIG. 6. Electric dipole relaxation time of rigid molecules as a measure of hindered external rotation in the liquid phase.[13a]

torsional oscillators than free rotors. This result seems to be in disagreement with Steele's indications[13] from NMR relaxation time data that rotational motions in liquids of the type shown are "almost free."

The oscillation frequency of a torsional oscillator in the liquid should be of the order of $(2\pi)^{-1}(aE^\circ/I_i)^{1/2} \exp(b/T^*)$, where the constant a is generally of the order of unity. Comparison of $\tau_i^* = \tau_e(E^\circ/I_i)^{1/2}$ vs. inverse reduced temperature[13a] for many different compounds on Fig. 7 separates specific from nonspecific effects of molecular structure. Certain groups are discernible: The plane five-membered rings exhibit the lowest reduced relaxation time and the bulky branched alkane compounds the highest, the benzenoid ring systems occupying the well-populated middle region.

When interaction of the electric field with the dipole permits rotation around more than one axis, there will obviously be more than one relaxation time. Because these differ rarely by large factors, resolution of the observed relaxation time (distribution) is of questionable value, even when the measurements extend into the millimeter wavelength region. Detailed analysis can establish in individual cases the preferred axis of rotation.[14]

[13a] A. Bondi, J. Am. Chem. Soc. **88**, 2131 (1966).

[14] K. Higasi, K. Bergmann, and C. P. Smyth, J. Phys. Chem. **64**, 880 (1960).

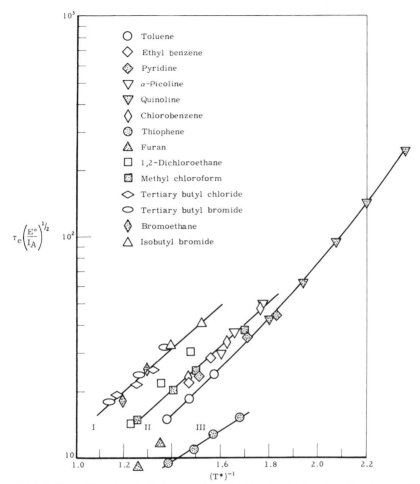

FIG. 7. Generalized electric dipole relaxation of rigid molecules (as reduced torsional oscillation period): I, bulky aliphatic halides; II, aromatic six-membered ring systems; III, heteroaromatic five-membered ring systems.[13a]

Rather large differences in the longest and shortest relaxation time per molecule are found when one deals with flexible molecules. The simplest case is the rigid (aromatic) ring with a small flexible substituent. Here the resolution into relaxation times characteristic of the whole molecule and that of the rotating side group is executed by calculation with the methods proposed by Budo[15] and others,[16,16a] improved by consideration of the moment

[15] A. Budo, *Physik.-Z.* **39,** 706 (1938); *J. Chem. Phys.* **17,** 686 (1949).

[16] A. Aihara and M. Davies, *J. Colloid. Sci.* **11,** 671 (1956).

[16a] H. Kramer, *Z. Naturforsch.* **15a,** 66 (1960).

of inertia. The results for various typical molecules are shown in Table III to indicate the relative mobility of the entire molecule and of its individual components. Noteworthy is the difference between large and small rotating links.

Solute (all at 293°C)	Whole molecule		Smaller segment rotating		
	τ, psec $\tau^* \overline{\text{ABC}}$		Segment	τ, psec	τ_R^*
Aniline	7.2	33	$-NH_2$	0.3 to 0.5‡	9 to 15
Phenol	11.4	54	$-OH$	1.1	42§
Anisol	9	36	$-OMe$	0.9	8
2-Methoxynaphthalene	19.7	56	$-OMe$	1.1	10

† Data from G. Klages and P. Knobloch, Z. Naturforsch. **20a,** 580 (1965). $\tau_R^* \equiv \tau(V°/I_{red})^{1/2}$, where $V°$ is the appropriate potential energy barrier to internal rotation and I_{red} the appropriate reduced moment of inertia. The magnitude of τ_{R^*} indicates that the "rotation" is more nearly like a torsional oscillation.

‡ The lower segment relaxation time from M. Stockhausen, Z. Naturforsch. **19a,** 1317 (1964).

§ This large value suggests appreciable retardation of rotation by H-bonding to neighboring molecules.

The rotatory motion of flexible n-alkane monohalides can be resolved into a short relaxation time τ_1 and a long relaxation time τ_2. The short relaxation time τ_1 is only weakly temperature dependent and can be related to the moment of inertia characteristic of rotation around the long axis of the molecule. Quite unexpectedly, τ_2 can be associated with the tumbling motion of the entirely stiff molecule (in the thermodynamically favored gauche-conformation of the polar end) up to $N_c = 14$ and to 55°C. The limit of available data[13a] is shown in Fig. 8. It is most likely that this correlation breaks down at higher (absolute) temperature and at larger values of N_c.

b. Relaxation Time of Nuclear Magnets

The relaxation of nuclear magnetic resonance absorption offers two specific advantages over electric-dipole relaxation: it can probe the motions of nonpolar molecules and, in some instances, permits the direct observation of the independent motion of two segments of a molecule, if they differ in the type of nuclear magnet or even in the "type" of proton attached to them.

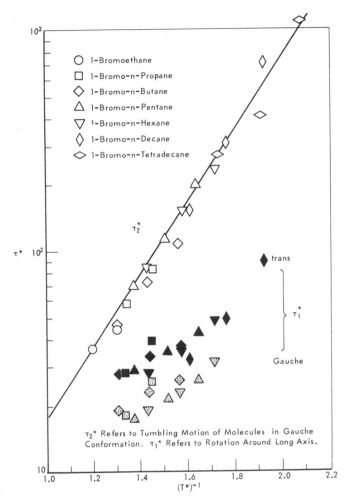

FIG. 8. Resolved generalized dipole relaxation times of 1-bromo-n-alkanes.[13a]

A typical example is toluene, the ring protons and methyl protons of which can be clearly differentiated. Hence, the rate of the rotation of ring and methyl group have been measured separately. A somewhat intuitive analysis of the motions of toluene and of other molecules with different proton signals, made by Powles,[17] yielded the relaxation times for several different rotations of the molecules; see Fig. 9. The interpretation of these data is far from simple.

[17] J. G. Powles, *Ber. Bunsenges. Physik. Chem.* **67**, 328 (1963).

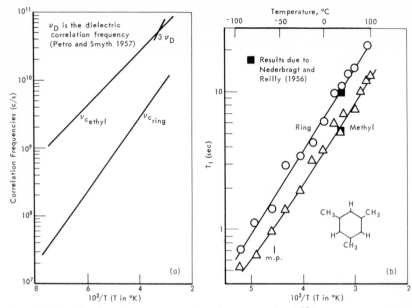

FIG. 9. Relaxation times for different portions of the same molecule[19] by NMR measurements. (a) Correlation frequencies for ethyl benzene deduced from the proton magnetic relaxation time. (b) Proton spin–lattice relaxation times for liquid mesitylene.

For many, but not all, rigid molecules the NMR relaxation τ_c is $\frac{1}{3}$ of τ_e as demanded by theory, if both measure the same motions in systems with a single relaxation time. When, as is often the case, there is a relaxation time distribution, the two methods can differ very appreciably in their results, because they weigh the different parts of a relaxation time spectrum very differently.

The small moment of inertia of methyl and ethyl groups leads to the expectation that they should rotate faster than the entire molecule. This expectation is borne out by experiment in the case of toluene,[17] ethyl benzene,[17] and isobutyl bromide.[18] The same is not true of paraxylene and mesitylene, whose methyl relaxation appears to be about as slow as that of the ring,[19] which is difficult to understand. Hence, although there is qualitative evidence of independent motion of molecule segments, its quantitative interpretation is still further away than one might have wished. More importantly, however, these data are well suited to investigate the coupling, if any, between molecular or segment motion and viscosity—a point to be taken up at the end of this section.

[18] K. Lusczynski and J. G. Powles, *Proc. Phys. Soc.* **74**, 408 (1959).
[19] J. G. Powles and D. J. Neale, *Proc. Phys. Soc.* **78**, 377 (1961).

c. Reorientation Time of Molecules by Optical Spectroscopy

The line width of Raman scattered light and of infrared absorption bands is an additional source of information about rotatory motion of molecules in liquids and crystals. An excellent summary of this somewhat unfamiliar method has been presented by Rakov.[19a] Specifically, the line width δ_i of a depolarized Raman scattering band (i) (or of an infrared band) at a given temperature (T) can be described as

$$\delta_i(T) = \delta_{0i} + \Delta v(T) = \delta_{0i} + (\pi c \tau(T))^{-1} = \delta_{0i} + [\tau_0 \exp(\Delta E_{\ddagger}/RT)]^{-1},$$

where δ_{0i} is a temperature independent component of the line width, τ is the reorientation time of a rotating molecule in the liquid (or crystal), τ_0 is the inverse of the equivalent torsional oscillation frequency at $0°K$, and $\Delta E_{\ddagger}$ is the activation energy for the rotatory motion. In all of the liquids studied so far $\Delta E_{\ddagger}$ (spectroscopic) $\approx \Delta E_{\ddagger}$ (viscosity).

Careful analysis of the Raman (or infrared) frequency, its assignment, depolarization and its electrical characteristics permit at least tentative—and in crystals firm—assignment of the prevalent axis of rotation of the molecule. In the few cases studied so far there seems to be agreement between conclusions from nuclear magnetic resonance and those from the Raman line width data. A fundamental limitation of the method is the inverse relation between line width and viscosity, which makes it experimentally inaccessible for the analysis of molecular motions in viscous liquids ($\eta > 0.5P$).

The virtual identity of the activation energy for viscous flow with that of molecule reorientation, among the simple liquids for which the line width method can be used, strongly suggests that molecule rotation is an important step in the motion of molecules past each other which underlies viscous flow in liquids composed of polyatomic molecules.

III. A Recent Advance in the Theory of Transport Phenomena

Molecular theory of liquids has made slow but steady advance since completion of the simple review presented in Volume I of this series.[5a] Much of this progress is due to the work of Rice and co-workers,[20] who continued Kirkwood's attack by expressing transport phenomena in terms of the radial distribution function of molecules $g(r)$, the potential function $U(r)$ determining the forces between molecules, and the correlation functions expressing the occurrence of energy dissipation in terms of the lack of

[19a] A. V. Rakov, in "Research in Molecular Spectroscopy" (D. V. Skobeltsyn, ed.). Consultants Bureau, New York, 1965.

[20] S. A. Rice, address on accepting the ACS Pure Chemistry Award, Los Angeles, 1963. I am indebted to Professor Rice for a prepublication copy of his address, published in Proc. Symp. Liquids, Structure, Properties and Solid Interactions, 1963. Elsevier, Amsterdam, 1965, where the references to the earlier work by J. G. Kirkwood can be found.

correlation of motion of molecules before and after strong interaction. Within the framework of interest to the chemist only the role of molecular parameters in the theory (a review of reviews on theories of viscosity by Brush[21] should be consulted for details on the current status of the subject) will be examined here.

As had been shown by Kirkwood (what follows are the results of a more recent treatment by Rice),[20] the viscosity of liquids is the sum of three terms of very unequal magnitude:

$$\eta = \eta_K + \eta_V(\sigma) + \eta_V(R > \sigma) \tag{1}$$

where η_K, the gas kinetic component, contributes less than 10% of the total, even at high temperatures. The second term, representing the contribution due to rigid-core encounters, rises with increasing temperature, like the gas kinetic viscosity, and is quite important, especially between T_b and T_c. Its relation to state and structure variables is apparent from the equation

$$\eta_V(\sigma) = \frac{5kT}{8\Omega^{(2,2)}} \left[\frac{2}{15} \left(\frac{2\pi\sigma^3}{v} \right) + 0.761 \frac{2\pi\sigma^3}{v} g(\sigma) \right]$$

$$\times \frac{1}{g(\sigma)} \frac{4\Omega^{(2,2)}}{\{8\Omega^{(2,2)} + 5\zeta_s v/m[g(\sigma)]^2\}} \left(1 + \frac{4\Omega^{(2,2)} g(\sigma)}{4\Omega^{(2,2)} + 5\zeta_s v/mg(\sigma)} \right) \tag{2}$$

where $\Omega^{(2,2)} = (4\pi kT/m)^{1/2}\sigma^2$ and m and σ are mass and rigid-core diameters of a molecule, v is the fluid volume per molecule, $g(\sigma)$ is the pair correlation function evaluated at $R = \sigma$, and ζ_s is the friction constant determined by the intermolecular potential. It is essentially a dense gas viscosity.

The most important term, especially at low temperatures, is $\eta_V(R > \sigma)$, the intermolecular force contribution to the shear viscosity of liquids:

$$\eta_V(R > \sigma) = \frac{\pi \zeta_s \rho^2}{15kT} \int_\sigma^\infty R^3 \frac{du}{dR} g_2^\circ(R) \psi_2(R) \, dR \tag{3}$$

where ρ is the number density of the fluid and u is the pair potential, and the integral is closely related to

$$\frac{2\pi\rho^2}{3} \int_0^\infty R^3 \frac{du}{dR} g_2^\circ(R) \, dR = \rho kT - P$$

obtainable from the pressure–volume–temperature properties of the fluid. The characteristic feature of $\eta_V(R > \sigma)$ is the function $\psi_2(R)$, the deformation factor for the pair correlation function $g_2(R)$ due to the flow field, $g_2^\circ(R)$ being

[21] S. G. Brush, Chem. Rev. 63, 513 (1963).

the unperturbed function. While the form of $\psi_2(R)$ is complicated, it contains no molecular property not already given. Hence, the mass m per molecule, its dimensions σ, and the intermolecular force constant ε, are the only molecular properties of immediate concern. The form of the functions $U(R)$, $g_2(R)$, and $\psi_2(R)$ is identical for all spherical molecules when expressed in the appropriate dimensionless coordinates. Specificity of molecular structure enters first through changes in the form of $g_2(R)$.

The form of such dimensionless viscosity functions,

$$\eta^* \equiv \frac{\eta\sigma^2}{(m\varepsilon)^{1/2}} = f(T^*, P^*)$$

where $T^* \equiv kT/\varepsilon$ and $P^* \equiv P\sigma^3/\varepsilon$ of simple substances, is shown in Fig. 10.

Experiments with isotopic substitution in simple molecules yield the expected constancy of η^*, which means that at constant reduced temperature

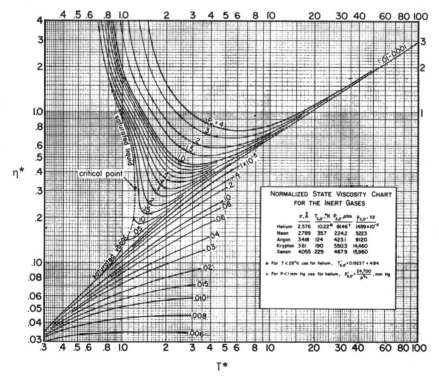

FIG. 10. Reduced viscosity correlation for the inert gases from H. Shimota and G. Thodos, *A.I.Ch.E. J.* **4**, 261 (1958).

$\eta \approx m^{1/2}$.[22] The same holds true for the substitution of heavy central atoms in well-shielded symmetrical compounds, as shown in a previous article.[22a] However, for more complicated molecules one should expect[22,23] and indeed finds $\eta \approx I_i^{1/2}$, where I_i is one of the principal moments of inertia. Then the reduced viscosity should be formulated $\eta_I^* = \eta\sigma^3/(I\varepsilon)^{1/2}$. There is no good a priori choice of the appropriate moment of inertia. In a few cases comparison of band shapes in gas phase and liquid phase infrared spectra suggests that the most hindered rotation is that normal to the axis of highest symmetry.[24] There are indications that this rotation is also retarding viscous flow the most.

The shape of the curves $\eta^* \sim \exp[(1/T^*)]$ is due largely to the form of the term $g(R)$ of equation (3) at comparatively high temperatures (where $r \geqslant \frac{3}{2}r_0$), namely $g(R) = V^{-2} \exp[-\phi(r)/kT]$.   Similarly,   the   proportionality $\eta^* \approx \exp(P^*/T^*)$ is related to the effect of pressure on the radial distribution function through $g(r) \approx V^{-2} \exp\{-[\phi(r_1) + Pr_1^3/Z]/kT\}$.[24a] At low temperatures, (at higher packing densities) $g(R)$ becomes rather complicated and the viscosity–temperature relation steeper than the simple exponential. The preexponential factor of the viscosity equation, dominated by the density fluctuation correlation term, also increases with decreasing temperature, further steepening the viscosity–temperature curve in the low-temperature region. However, simple substances freeze at such high reduced temperatures that this part of their viscosity curve is not observable.

One of the most important aspects left out of this discussion is the origin of irreversibility. Kirkwood and Rice[20] reinterpreted the role of probability, the classical source of irreversibility in particle mechanics, in terms of the time required for particle configurations and motions to lose correlation with those they had before their encounter (collision) with another particle. In the case of monatomic fluids this "forgetting time" τ_F turned out to be expressible as a function of their equilibrium properties. Hence, there was no need to consider τ_F explicitly in the dimensionless representation of the viscosity–temperature–pressure p functions for monatomic fluids. The same simplicity cannot be expected to hold for τ_F of polyatomic molecules with many internal degrees of freedom, especially if their equivalent frequency is at $<300\,\mathrm{cm}^{-1}$, that is, if they are excited at low temperatures. Since a quantitative formulation of τ_F for fluids composed of polyatomic molecules has not yet been produced, a correlation of τ_F with molecular structure is

[22] E. McLaughlin, *Physica* **26**, 650 (1960); J. P. Boon and G. Thomaes, *ibid.* **28**, 1197 (1962).

[22a] A. Bondi, *in* "Rheology: Theory and Applications" (F. R. Eirich, ed.), Vol. 1, p. 338. Academic Press, New York, 1956.

[23] J. Pople, *Physica* **19**, 668 (1953).

[24] H. E. Hallam and T. C. Ray, *Trans. Faraday Soc.* **59**, 1983 (1963).

[24a] See Ref. 22a, p. 333.

quite impossible. There may turn out to be several different forgetting times in such systems.

Hence, at this juncture one can only draw the qualitative conclusion that the viscosity of fluids composed of complicated polyatomic molecules can probably not be represented in dimensionless form as a universal function without an additional parameter. The following attempts at a dimensionless viscosity correlation should therefore be considered only as zeroth approximation until the extra parameter can be formulated.

IV. Semiempirical Correlation Schemes of Viscosity–Temperature–Pressure Relations

Plausible reasons can be advanced for dividing the temperature regime into three regions: a near-critical region, between the atmospheric boiling point and the critical temperature ($1.0 < T^* < 1.3$), a "high temperature" region ($0.6 < T^* < 1.0$), and a low-temperature region ($T^* < 0.5$). A peculiar feature of the near-critical region is the importance of the gas kinetic contributions, the first two terms on the right-hand side of equation (1). Here only the residual term $\eta - \eta_K$ can be correlated with T^*. This is essentially the meaning of the correlation of the residual reduced viscosity proposed by Thodos and co-workers[25] and shown in Fig. 11. The related correlation for more complicated compounds is shown on Fig. 12. Molecular structure enters into these correlations only in a very nonspecific way: through the effect on the mass per molecule, the average molecule dimensions, and the average molecular force field. A comparatively narrow range in the absolute magnitude of viscosity is covered in the near-critical region, especially when expressed as kinematic viscosity, as is shown by the graph of Fig. 13. The critical kinematic viscosity of most substances is of the order of 2 ± 0.5 mstokes.[26] Moreover, few complicated molecules survive chemically in the critical temperature region. This temperature regime is therefore of little interest in the present investigation.

The second regime $0.6 < T^* < 1.0$ coincides approximately with that of the simple exponential relation mentioned above. Many polyatomic substances with fairly high molecular weight are chemically stable at these temperatures. Their molecular geometry cannot be characterized as a sphere, thus precluding the use of the clearly defined parameters σ and ε; these must be replaced with other descriptions. The energy parameter is replaced with

[25] G. Thodos et al., Ind. Eng. Chem. **50**, 1095 (1958); A.I.Ch.E. J. **8**, 59 (1962); **10**, 275 (1964).
[26] E. L. Lederer, Kolloid–Beih. **34**, 270 (1931). It is apparent from his graph and from the figure given here that the kinematic viscosity, the property needed in most engineering calculations, is far more convenient to handle in the critical-temperature region than is the dynamic viscosity.

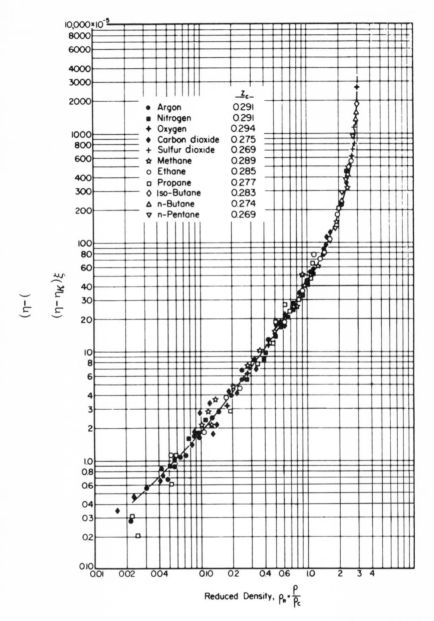

FIG. 11. Relationship between reduced residual viscosity $(\eta - \eta_K)\xi$ and reduced density ρ_{r} for nonpolar substances.[25]

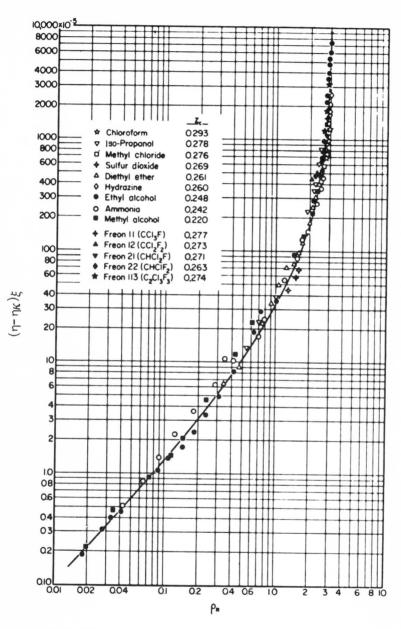

FIG. 12. Relationship between $(\eta - \eta_K)\xi$ and ρ_R for polar substances.[25]

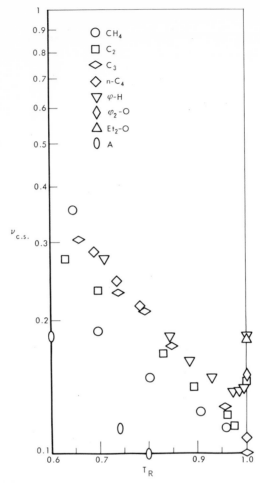

FIG. 13. Kinematic viscosity of various liquids in the critical temperature region. (The upswing at T_c is probably real.)

$2E°/Z$, where $E°$ is the standard energy of vaporization and Z is the number of nearest neighbors of a molecule or a molecule segment. Unless stated otherwise, we assume $Z = \text{constant} = 10$ in this temperature regime. The energy $E°$ can be estimated by adding group increments, available in Ref. 26a. The reduced temperature of systems composed of larger polyatomic molecules, $T^* = ZcRT/2E°$, contains also the constant c, which as $3c$ gives the total number of external degrees of freedom discussed earlier. Finally,

[26a] A. Bondi, "Physical Properties of Molecular Crystals, Liquids, and Glasses." Wiley, New York, 1967.

as a description of molecular geometry, we use the van der Waals volume V_w, which is easily calculated from molecular geometry data, the closely related surface area per mole of molecules, A_w, and a characteristic average molecule width $\bar{d}_w$, which is also obtained from well-known atom and atom group dimensions.

V. Effect of Molecular Structure on Viscosity of Nonassociated Liquids

1. At High Temperatures

The combining of geometry and energy parameters for large polyatomic molecules into a dimensionless viscosity can be guided by a few simple model considerations. At fixed and moderately high values of T^* ($0.6 < T^* < 1.0$)[27] one should expect $\eta \sim m^{1/2}$, as mentioned in Section III. Eyring's model for molecular motion in flow suggests that the area per molecule and a length related to the average jump length should be considered.[28] A reducing parameter for the viscosity of rigid molecules that meets all these requirements is $\eta_s' = (mE^\circ/A_w)^{1/2}/\bar{d}_w$. Since in a given family of compounds $E^\circ/A_w \approx$ constant, then $\eta_s' \approx m^{1/2}$, as required. Hence, we surmise that a plot of $\ln(\eta/\eta_s') \equiv \ln \eta_s^* = f(1/T^*)$ might be linear and should, ideally, be a "universal" curve for liquids composed of rigid molecules. Rigidity is here defined in terms of the constant $3c$. For rigid molecules $3c < 7$.

a. Rigid Nonpolar Molecules

There are three kinds of rigid nonpolar molecules: isoparaffins with so many closely spaced branches per molecule that the barriers to internal rotation become very high, polymethylene ring systems composed of connected or condensed naphthenic rings with less than 8 carbon atoms per ring, and aromatic ring systems. Examples of the viscosity–temperature relations of all three are presented in Figs. 14, 15, and 16.

Isoparaffins are rigid molecules, with $3c < 8$, if there are at least as many methyl or ethyl groups as carbon atoms on the backbone chain. Even then the barrier to internal rotation is at most 5 kcal/mole per C—C linkage, so that there is a finite amount of internal mobility left in the molecule. The randomly combined torsional oscillations add to the equivalent of one free internal rotation in a backbone chain length of about 12 carbon atoms for the most densely branched isoparaffins on which data are available.

[27] A. Bondi, *Ind. Eng. Chem., Fundamentals* **2,** 95 (1963).
[28] S. Glasstone, K. L. Laidler, and H. Eyring, "Theory of Rate Processes." McGraw-Hill, New York, 1941.

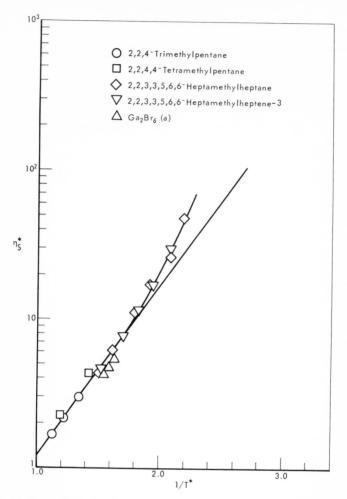

FIG. 14. Generalized viscosity-temperature relations for rigid isoparaffins.[27] (a) N. N. Greenwood and I. J. Worrale, *J. Inorg. Nucl. Chem.* **6**, 34 (1958).

The overall shape of certain inorganic molecules, such as

$$
\begin{array}{ccc}
\text{Br} & \text{Br} & \text{Br} \\
\diagdown & \diagup\diagdown & \diagup \\
& \text{Ga} \quad \text{Ga} & \\
\diagup & \diagdown\diagup & \diagdown \\
\text{Br} & \text{Br} & \text{Br}
\end{array}
$$

is somewhere between that of the tightly branched isoparaffins and the cycloparaffins. Accordingly, one finds that their viscosity–temperature curves closely coincide with those of the stiff isoparaffins. The absence of

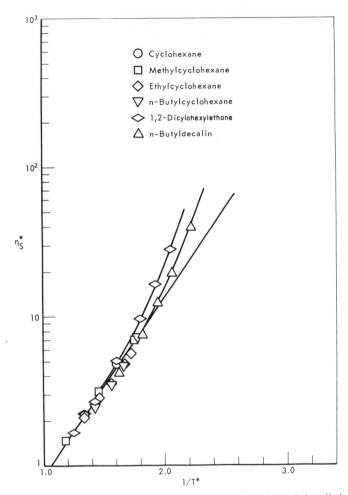

FIG. 15. Generalized viscosity-temperature relations for short-chain alkyl naphthenes $(N_B \leqslant 4)$.[27]

data and of reliable van der Waals' radii for metals handicaps the integration of inorganic compounds into this general molecular-structure correlation.

The work of the American Petroleum Institute Project 42 at Pennsylvania State University has made data available for a very large number of naphthenic hydrocarbons having up to 31 carbon atoms in branchless ring structures.[29] Their high-temperature viscosity data are surprisingly well

[29] A. E. Miller, *World Petrol Congr., Proc., 4th, Rome, 1955* Vol. 5, p. 127. Carlo Colombo, Pbl. 1955; R. W. Schiessler *et al., Ind. Eng. Chem.* **47,** 1660 (1955).

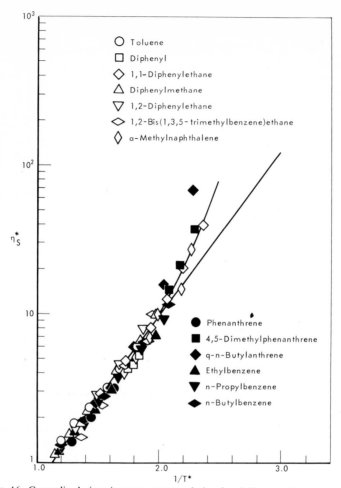

FIG. 16. Generalized viscosity-temperature relation for rigid aromatic hydrocarbons.[27]

represented by following the same rules for calculating the molecule width parameter as those used with isoparaffins.

This similarity in behavior is probably associated with the fact that the potential-energy barrier for transformation from the boat to the chair form of naphthenic rings is with 10 kcal/mole[30] of the same order as that for coordinated rotation of two branches around quarternary carbon atoms. Details of the spatial structure of alkylated naphthene hydrocarbons must be considered if the flexibility of the alkyl side or interring chains is to be

[30] J. E. Piercy, *J. Acoust. Soc. Am.* **33**, 198 (1961).

predicted. Unbranched side chains or interring chains more than 5 carbon atoms in length will, in general, raise $3c > 8$ and thus shift the substances toward the family of flexible molecules. The most interesting feature of rigid naphthenic hydrocarbons, their very high viscosity at low temperatures, will be discussed in Section V,2.

The most rigid hydrocarbon molecules are those of the aromatic species, especially the condensed cyclic polyaromatic ones. Their inconveniently high melting points have prevented the determination of the viscosity of many unalkylated condensed cyclic compounds. However, the presence of few short alkyl chains hardly increases molecular flexibility; hence, the data on many methylated and other short-chain alkylated aromatics (for all of which $3c < 8$) have been assembled on a single graph (Fig. 16) together with the few data on unalkylated ring systems. No particular sensitivity to specific structural features is noted in the high-temperature (straight line) region of the graph.

The close proximity of experimental points to a single curve ($\log \eta^* = (B/T^*) - A$) on Figs. 14 to 16 is only apparent. Closer examination of the data shows a systematic trend of A and B with molal volume, so that, more accurately,

$$\log \eta_s^* = [(a + bV_w)/T^*] - (g + dV_w)$$

where the empirical constants $a, b, d,$ and g would be "universal," if viscosity were describable by a single corresponding-states correlation. Actually, they can be grouped by families of compounds, as shown by the numbers of Table IV.

Were viscosity a function of volume only, B should be independent of molecule size. The inference from the observed size dependence of B is that there exists a significant isochoric temperature coefficient of viscosity $\Delta E_{\ddagger}^j$ (discussed in Section V,1,e) which rises with molecule size more rapidly than does E°. The absence of a molecule size dependence of A and B for rigid naphthene hydrocarbons may be due to the absence of high-temperature viscosity data for condensed ring naphthenes with $N_c \geqslant 18$, so that only the narrow size range $6 \leqslant N_c < 18$ is covered. The term "semirigid" for ring systems applies to such molecules as tricyclohexyl methane and 1,1-dicyclohexylethane.

Noteworthy specificities have been observed with the high-temperature viscosity of phenyl ring systems containing two or more methyl groups per ring (Fig. 17). Since the slope of the curves is affected, no change in the viscosity shift factor can bring these curves to coincidence. This could be accomplished only by changes in the characteristic temperature $E^\circ/5cR$, for instance, by changing the number of nearest neighbors from the arbitrary 10. The accuracy of the density and vapor pressure data, upon which $E^\circ/5cR$

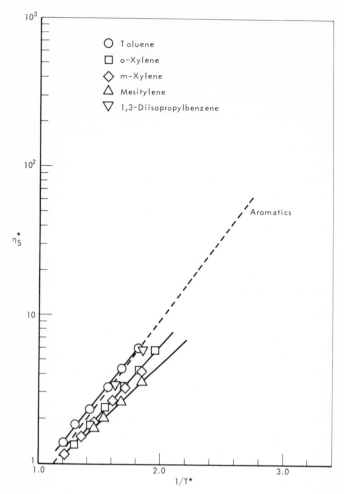

FIG. 17. Generalized viscosity of polymethylbenzene.[27]

are based, has to be tested before the apparent peculiarities of Fig. 17 are examined any further.

b. Rigid Polar Molecules

The most significant feature of liquids composed of polar molecules is the nonuniformity of the force field in which the molecules move past each other. The resulting hindrance to external rotation is quite different from that opposing the rotation of a highly anisometric molecule in a nearly uniform force field. Although the detailed consequences of this effect cannot

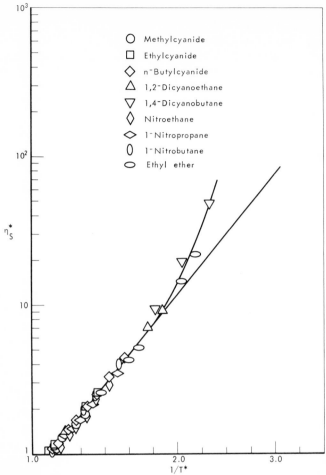

FIG. 18. Generalized viscosity-temperature relation for a few short-chain polar aliphatic compounds.[27]

be predicted from existing theory, we are warned that the uniformity of the force field implied by using the total energy of vaporization as energy parameter in the calculation of $\eta_R{}'$ might not yield a good representation of viscosity data.

Actual tests of the correlation with the very polar alkyl cyanides and nitroaliphatic compounds with $3c < 7$ yielded an unexpectedly good representation of the data, as shown in Fig. 18, neither the position nor the slope of the curve differing widely from that of hydrocarbons. Somewhat less concordant curves are obtained with cyclic compounds (Fig. 19), where a subdivision into families can be discerned.

TABLE IV

NUMERICAL VALUE OF CONSTANTS a, b, d, g, FOR VARIOUS GROUPS OF
LIQUIDS COMPOSED OF RIGID MOLECULES

Substances	a	b	g	d
Rigid isoalkanes	0.84	0.37×10^{-2}	0.53	0.62×10^{-2}
Rigid nonpolar aromatic ring systems	0.83	0.34×10^{-2}	0.93	0.382×10^{-2}
Rigid nonpolar naphthenic* ring systems		1.18		1.12
Semirigid nonpolar naphthenic* ring systems	0.65	0.58×10^{-2}	0.496	0.684×10^{-2}
Rigid polar aliphatic compounds	0.84	0.37×10^{-2}	0.80	0.80×10^{-2}
Rigid polar aromatic ring systems	0.79	0.36×10^{-2}	0.70	0.77×10^{-2}
Cyclopoly(dimethylsiloxanes)	1.17	0.284×10^{-2}		1.65

* Exclusive of cyclopentyl derivatives.

The steepness of the size dependence of the intercept A (shown by the magnitude of constant d in Table IV) and also of the other coefficients for polar molecules might not be real because of the paucity of data and of the wide scatter of coefficients.

c. Large-Ring Compounds

Once their peculiar packing-density relations have been taken into consideration, the reduced viscosity of cycloparaffins from cyclopentane through cyclododecane falls with little scatter around a single curve, shown in Fig. 20. The reduced viscosity of other large-ring compounds, however, deviates increasingly from the generalized curve, as the ring structure becomes more complex. Typical examples are the cyclic phosphazene fluorides and dimethylsiloxane polymers, the generalized viscosity–temperature curves of which are also shown in Fig. 20.

At equal reduced temperature one finds for cyclic phosphazenes $\eta \sim M$ and for the cyclic siloxanes $\eta \sim M^{1.7}$.

The increase in molecular-weight dependence of viscosity with increasing molecular flexibility of the large rings suggests that nearest-neighbor molecules are entangled in the ring and dragged over some distance before being released and replaced by others. The exponent on M can be expected to rise with increasing molecular weight and extent of entanglement. Improved theory in this area is clearly needed.

The steep molecular-weight dependence of the viscosity of large flexible ring compounds could also be accounted for by invoking the effect of the

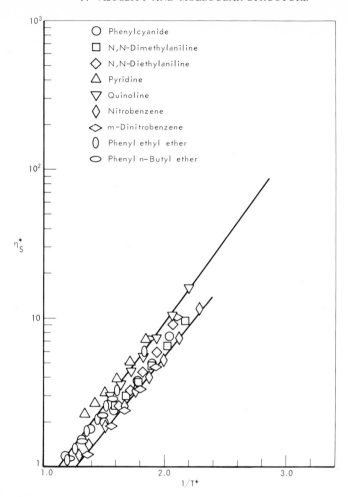

FIG. 19. Generalized viscosity-temperature relation for a few polar aromatic compounds.

moment of inertia, mentioned in Section III. For simple molecules of nearly spherical symmetry it is easy to see that the reducing parameters are $(\varepsilon m)^{1/2}/\sigma \sim (\varepsilon I_i)^{1/2}/\sigma^3$, because $I_i \sim m\sigma^2$. However, for anisometric molecules I_i can vary rather more steeply. Specifically, in the case of a *planar* ring the diameter varies in proportion to the number of ring segments; hence, the largest moment is $I_A \sim m^3$, with $E^\circ \sim m$. The effective-length dimension can range between the extremes $V_w^{1/3}$ and the ring thickness $\overline{d_w}$. In the former case $(E^\circ I_A)^{1/2}/V_w \sim m$ and in the latter case $(E^\circ I_A)^{1/2}/d_w \sim m^2$, all at constant T^*. Of the very distorted cycloparaffins I_A varies more nearly as m^2, and E°

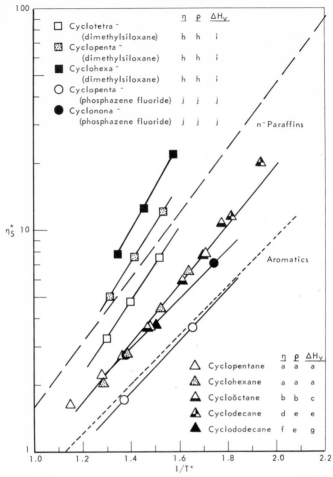

FIG. 20. Generalized viscosity vs. temperature curves for large ring compounds. (a) API Research Project 44, Selected Properties of Hydrocarbons, 1953. (b) Ref. 67. (c) W. L. Fink *et al.*, *J. Am. Chem. Soc.* **78,** 5469 (1956). (d) M. P. Doss, "Physical Constants of the Principal Hydrocarbons." Texaco, New York, 1943. (e) G. Egloff, "Physical Properties of Hydrocarbons," Vol. II. Reinhold, New York. (f) M. N. Gollis *et al.*, *J. Chem. Eng. Data* **7,** 311 (1962). (g) By correlation. (h) C. B: Hurd, *J. Am. Chem. Soc.* **68,** 364 (1946). (i) J. C. Osthoff and W. T. Grubb, *J. Am. Chem. Soc.* **76,** 399 (1954); D. F. Wilcock, *ibid.* **68,** 691 (1946). (j) A. C. Chapman *et al.*, *J. Chem. Soc.* **1960,** 3608.

varies more slowly than $\sim m$. Hence, their acceptable representation by $\eta \sim m^{1/2}$ is understandable. Whether this description is more valid than the "drag effect," suggested earlier with the reference to the phosphazene and siloxane rings, could be tested by isotope substitution effects.

d. Effects of Electrostatic Repulsion

Interactions between molecules of the type CX_4 or C_nX_{2n+2} with large permanent electric-dipole moments in the C—X bonds differ from those between other types of molecules, because electrostatic repulsion reduces the London force attraction during the very frequent C—X $\cdots$ X—C contacts. Hence, in such compounds the energy of vaporization contribution per —X group is very low (Table V). A systematic analysis of the manifestations of this repulsion effect in physical properties has yet to be made. A critical evalua- tion of the few available reliable property measurements over long ranges of temperature and molecular weight has yet to be undertaken.

TABLE V

COMPARISON OF STANDARD ENERGY OF VAPORIZATION INCREMENT OF ELECTROSTATICALLY REPULSIVE TIGHTLY COVERED "PERCOMPOUNDS" WITH THAT OF ORDINARY COMPOUNDS (ALL IN kcal/mole)

		Compounds*			
Group	E_v° of "percompounds"	Ordinary nonpolar		Polar	
—F	0.60	(F_2)	0.80	(R–F)	2.4
—Cl	2.05	(Cl_2)	2.6	(R–Cl)	3.5
—Br	2.80	(Br_2)	3.7	(R–Br)	4.0
—I	4.60	(I_2)	4.9	(R–I)	5.4
—NO_2	2.41	$(NO_2)_2$	4.57	(R–NO_2)	6.1
—C≡N	(3.)†	$(C≡N)_2$	3.26	(R–C≡N)	5.2

* Comparison compound in parentheses.
† Rough estimate from sublimation data.

However, even within the framework of the only moderately consistent data one finds important differences between the viscosity characteristics of the present group of compounds and the ordinary polar and nonpolar substances. The striking feature of the assembled data in Fig. 21 is the impossibility of bringing the curves to coincidence with each other as well as with the guide curve of the similarly shaped rigid isoparaffins. Their slope factor B increases from 1.30 for CCl_4, through 1.45 for $C(NO_2)_4$ to about 2.0 for the higher perfluorocarbon compounds. This distinct variation in behavior from that of most other compounds means, primarily, that for these compounds the contour of the potential-energy wells differs sub- stantially from substance to substance and certainly from the "average (i.e. geometrically similar) contour," which must be responsible for the great similarity in behavior among most "normal substances."

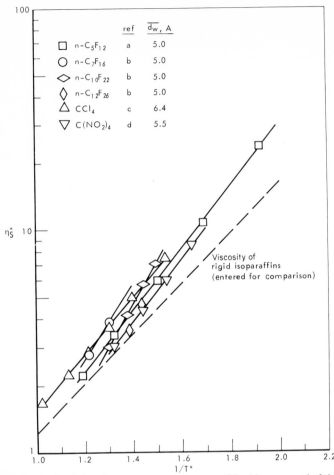

FIG. 21. Generalized viscosity vs. temperature curves of liquids composed of rigid molecules with significant electrostatic repulsion. (a) G. H. Cady *et al., J. Am. Chem. Soc.* **73,** 4243 (1951); *J. Phys. Chem.* **60,** 504 (1956). (b) R. N. Haszeldine and F. Smith, *J. Chem. Soc.* **1951,** 606. (c) J. Timmermans, "Physico-Chemical Constants of Pure Organic Compounds." Elsevier, Amsterdam, 1950. (d) E. Lucatu and G. Palade, *Acad. Rep. Populare Romine, Bul. Stiint., A. Ser. Mat. Fiz. Chim.* **1,** 125 (1949).

e. Viscosity Isochores of Rigid Molecules

One of the oldest notions regarding the viscosity of liquids is that it should be a function of the packing density.[31] The data in Fig. 22 show that this guess, while intuitively plausible, is at best only a very crude approximation to the facts, even in the case of the simplest liquids. The observed exponential

[31] A. Batschinsky, *Z. Physik. Chem.* **84,** 643 (1913).

decrease of viscosity along the isochore at high packing densities is in accord
with the molecular theory mentioned earlier. However, the onset of gaslike
behavior at $\rho < 2\rho_c$, first noted by Zhdanova,[32] is at a somewhat higher
viscosity level than is that predicted by theory. The observation can be
understood intuitively, however, if one uses the rather primitive picture of a
regular array of molecules. Molecules should be able to pass each other quite
freely, when their average distance equals their diameter. This happens when
the packing density ρ^* is 0.26. Assuming the hard-core diameter σ, one finds

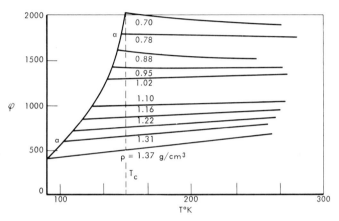

FIG. 22a. Temperature dependence of fluidity of liquid argon in equilibrium with its vapor
pressure at various densities. $\alpha - \alpha$ = saturation line.[32]

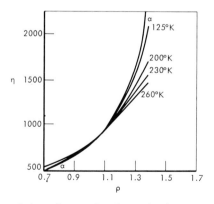

FIG. 22b. Isothermal plots of argon viscosity vs. density. $\alpha - \alpha$ = saturation line.

[32] N. F. Zhdanova, *Soviet Phys.—JETP (English Transl.)* **4**, 19 and 749 (1957).

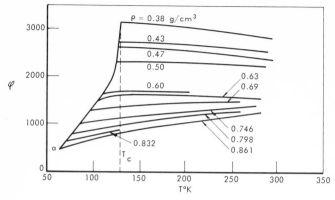

FIG. 22c. Temperature dependence of viscosity in liquid nitrogen.[32]

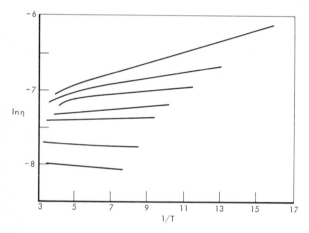

FIG. 22d. Dependence of $\ln \eta$ on $1/T$ for liquid nitrogen at constant density. Sequence of curves exact inverse of Fig. 22c.

that $\rho^* \approx 0.31$ at $2\rho_c$, which is only slightly higher than what is required for unimpeded motion.

The packing density ρ^* at which $(\partial \eta / \partial T)_v \to 0$ is rather higher (near $3\rho_c$) with nonspherical molecules, but no relation has been found so far between this inflection density and molecular structure. The data suggest that in general $(\partial \ln \eta / \partial 1/T)_v < \frac{1}{3}(\partial \ln \eta / \partial 1/T)_P$ at $\rho^* < 0.55$ and tends toward zero at $\rho^* = 0.45$, which is just above the atmospheric boiling point.

The isochoric temperature coefficient of viscosity may be considered a measure of the potential-energy barrier to the motion of molecules in the force field of their neighbors and is usually defined as $\Delta E_{\ddagger}{}^{j} = R(\partial \ln \eta / \partial 1/T)_v$. The magnitude of $\Delta E_{\ddagger}{}^{j}$ is two to three times smaller than necessary to account for the configurational heat capacity C_v^c discussed in Section II,1,b, if one ascribes the nonideal component of C_v^c to hindered external rotation in the liquid state. Thus, viscous flow involves either rotation only around a single axis or very limited translation in the sense of the old Eyring flow model. Rather good, although not perfect, is the old correlation of $\Delta E_{\ddagger}^{j}$ with $T \Delta S_v^{j}$,[3] where ΔS_v^{j} is the excess entropy of vaporization discussed in Section II,1,c as another measure of hindered rotation in the liquid state. But here, too, $\Delta E_{\ddagger}^{j}$ is much too small to account for the magnitude of ΔS_v^{j}, if the latter is ascribed to hindered external rotation. Hence, the exact nature of the motion for which $\Delta E_{\ddagger}^{j}$ is the potential-energy barrier remains obscure for the present.

Far too few data are available to relate $\Delta E_{\ddagger}^{j}$ to molecular structure. The curves for the truly rigid molecules in Fig. 23 indicate a smooth increase of $\Delta E_{\ddagger}^{j}$ with packing density ρ^*. Moreover, for these molecules it appears that $(\partial \Delta E_{\ddagger}^{j}/\partial T)_v = 0$. In these respects short-chain aliphatic compounds (as usual) behave like rigid molecules. However, the introduction of any

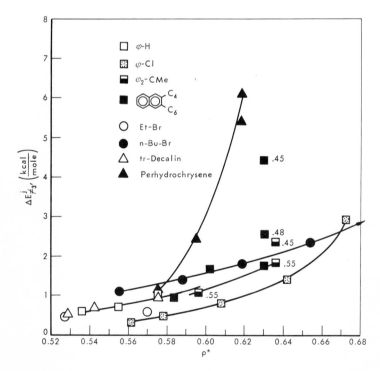

FIG. 23. Isochoric temperature coefficient $\Delta E_{\ddagger}^{j}$ vs. packing density ρ^* for liquids composed of rigid molecules. Parameter on some of the curves: T^*.

substantial amount of hindered internal mobility, as in 1,1-diphenylethane or in perhydrochrysene, leads to two simultaneous changes in behavior, which will be discussed in detail in Section V,2,c, $(\partial \Delta E_{\ddagger}^{j}/\partial T)_v < 0$ and a sharp, almost discontinuous, increase in $\Delta E_{\ddagger}^{j}$ as $\rho^* \to 0.58$.

f. Effect of Pressure

The effect of pressure on viscosity is primarily through its effect on the (packing) density of the liquid. Just as the density and temperature are

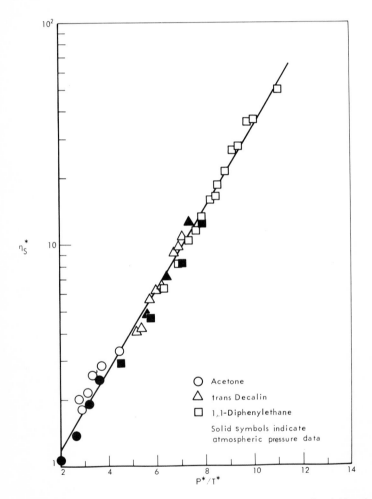

FIG. 24. Generalized viscosity vs. pressure and temperature curve for rigid molecules.

required to specify the viscosity, so pressure and temperature must be specified together to constitute a meaningful variable for viscosity. The appropriate unit of pressure for liquids is the sum of external pressure P and internal pressure $P_i \equiv (\partial E/\partial V)_T$. Hence, a reduced pressure is defined as $P^* \equiv 5V_w(P + P_i)/E^\circ$. A "natural" variable for viscosity might then be $P^*/T^* = V_w(P + P_i)/cRT$. The plots of $\ln \eta_s^*$ vs. P^* and vs. P^*/T^* in Fig. 24 indicate that this choice of variables permits a fairly good, although not very precise, description of viscosity–pressure–temperature relations. In the reduced-temperature range under consideration in this section one finds comparatively small molecular-structure sensitivity for the slope of these η_s^* vs. P^*/T^* curves, indicative of the close relationship between temperature and pressure coefficient of viscosity. Some of the variability in reduced viscosity level may be due to imperfect knowledge of P_i over wide temperature ranges (for the sake of simplicity, the value of P_i at atmosphere pressure was used throughout).

g. Flexible Molecules

"Flexible molecules," in the sense in which the term is used here, are those for which $3c \geqslant 10$. In the case of molecules with freely rotating segments this would correspond to a minimum of three segments per molecule. In the case of real molecules of usual hindered internal rotation many more segments often are required to attain $3c \geqslant 10$. A common feature of flexible chain molecules is their mode of motion which, from diffusion data, appears to be dominated by the cross section of the chain rather than by its length.[33] Such a mode of flow leads to the reducing parameter $\eta_F' = (ME^\circ)^{1/2}/N_A \overline{d_w^2}$. This form of η_F' also meets the requirement of experimental evidence that at a fixed value of T^* one should expect $\eta \sim M$, since $E^\circ \sim M$. Once the chain length exceeds that at which mutual chain entanglement becomes important, the experimental relation abruptly changes to $\eta \sim M^{3.4}$,[34] and the present treatment ceases to apply.

A detailed discussion of this critical chain length Z_c for the onset of entanglement is given by Porter and Johnson in Chapter 5 of this volume. In the present context only its relation to molecular structure is important. The not always concordant data have been collected in Table VI. One sees that Z_c increases with chain rigidity, with the size of the side chains, if any, and with the extent of polar interaction among chain segments. The increase in Z_c with polar interaction (in the case of the polyamides) is somewhat unexpected, since intuition would connect an early entanglement limit with stronger chain–chain interaction.

[33] D. J. Trevoy and H. G. Drickamer, *J. Chem. Phys.* **17**, 1120 (1949).

[34] F. Bueche, "Physical Properties of High Polymers." Wiley (Interscience), New York, 1962.

TABLE VI
ENTANGLEMENT CHAIN LENGTH Z_c

Polymer	Z_c*	Z_c†
Polyethylene	275	286
Aliphatic linear polyesters	285 ± 5	285 ± 5
Polyisobutylene	460	607
Poly(vinyl acetate)	570	570
Polystyrene	600	673
Poly(neopentyl succinate)	—	810
Polycaprolactam	324	
Poly(methyl methacrylate)	210	? > 210
Poly(oxypropylene glycol)	400	—
Poly(dimethylsiloxane)	630	784
Polyacrylonitrile	—	(50)‡
Poly(vinyl alcohol)	—	220‡§

* T. G Fox and V. R. Allen, *J. Chem. Phys.* **41**, 344 (1964).
† R. S. Porter and J. F. Johnson, *Chem. Rev.* **66**, 1 (1966).
‡ Evaluated from solution data.
§ Y. Oyanagi and M. Matsumoto, *J. Colloid Sci.* **17**, 426 (1962).

h. Flexible Hydrocarbon Molecules

The high-temperature end of the generalized viscosity–temperature relations of normal paraffins with $N_c \geqslant 12$ and for most alkylated cyclic hydrocarbons with sufficiently long side chains is represented by the generalized $\ln \eta_F^*$ vs. $(T^*)^{-1}$ curves shown in Fig. 25. The slope constant B of the straight-line portion is, with 1.47, rather larger than the 1.14 of the rigid molecules. The physical meaning of B may be that the number of degrees of freedom active in flow is $3c/B$ and thus smaller than in thermal expansion and thermal conductivity, reflecting the requirement of larger amplitudes for molecular displacements in bulk flow than for thermal expansion.

Seeing that $3dc/dN_c = 0.44$ for the n-paraffins and $3dc/dN_c = 2$ for the freely rotating chain link, a segment of, on the average, 4.5 methylene groups in length is equivalent to a freely rotating chain link in thermal expansion, and the freely moving segment in viscous flow of n-paraffins is 1.47/1.14 times that, or about 6 methylene groups long. This explains why the viscosity of n-paraffins with less than 12 carbon atoms is not representable by a single curve. Very similar conclusions had been reached from an inspection of the activation energies of the viscosity of n-paraffins, made by Powell and Eyring many years ago.[35]

[35] R. E. Powell and H. Eyring, *Advan. Colloid Sci.* **1**, 183 (1942).

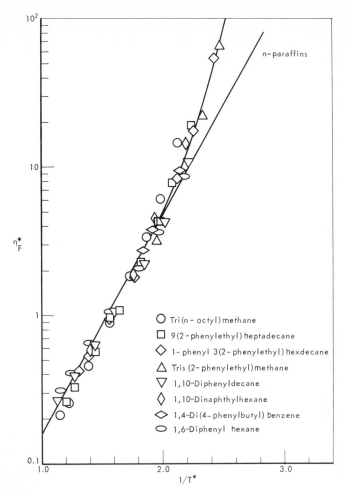

FIG. 25. Generalized viscosity-temperature relation for flexible alkyl aromatic hydro-carbons.[27]

Except for the slightly greater scatter of the alkylnaphthene than of the alkylaromatics data, there is no significant effect of molecular structure in the high-temperature linear region of the curves. However, the reduced temperature at which the viscosity curve bends upward increases slowly with increasing molecular weight, as is especially apparent from a comparison of the curves for polymer melts in Fig. 26 with those in Fig. 25. Owing to the absence of melt data at sufficiently high temperatures, their coincidence in the linear region can only be guessed at present.

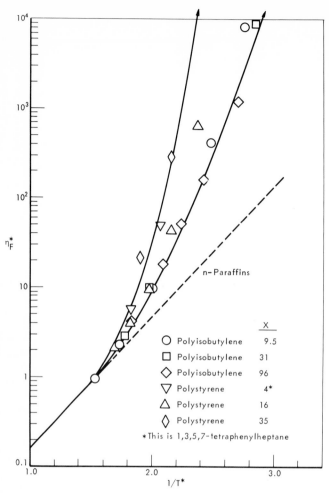

FIG. 26. Generalized viscosity of polyisobutylene and polystyrene of low molecular weight.[27]

i. Flexible Polar Molecules

A survey of the many viscosity data accumulated for the class of flexible polar molecules as a result of world-wide searches for synthetic lubricants[36] brings to light several common features. The slope of their generalized $\ln \eta_F^*$ vs. $1/T^*$ curves in Figs. 27a, b, c is comparatively independent of chemical composition.

[36] For an excellent summary of data see R. C. Gunderson and A. W. Hart (eds.), "Synthetic Lubricants." Reinhold, New York, 1962.

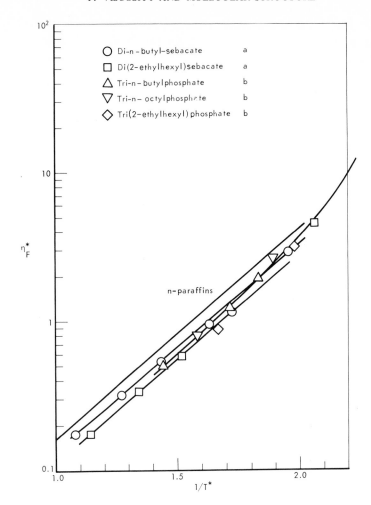

FIG. 27a. Generalized viscosity curve of alkane esters. (a) C. M. Murphy *et al., Trans. Am. Soc. Mech. Engrs.* **71**, 561 (1949). (b) R. E. Hatton in Ref. (38).

The slope constant B_F of the high-temperature viscosity equation of the liquids composed of flexible molecules,

$$\log \eta_F^* = B_F/T^* - A_F$$

depends somewhat on molecule size in a manner that differs from that encountered with rigid molecules. The reason is that B_F must become size-independent as $M \to \infty$, so that

$$B_F^{-1} = a_F + b_F N_s^{-1}$$

where N_s is the number of skeletal atoms in the backbone chain. The constant A_F may be represented as function of B_F, and the high-temperature viscosity relation for liquids composed of flexible molecules becomes

$$\log \eta_F^* = \frac{(a_F + b_F N_s^{-1})^{-1}}{T^*} + (g_F B_F^{1/2} - d_F)$$

Numerical values of constants a_F, b_F, g_F and d_F for different series of flexible molecules have been assembled in Table VII. The primary barrier to the external rotation of nonpolar flexible molecules in the liquid is the

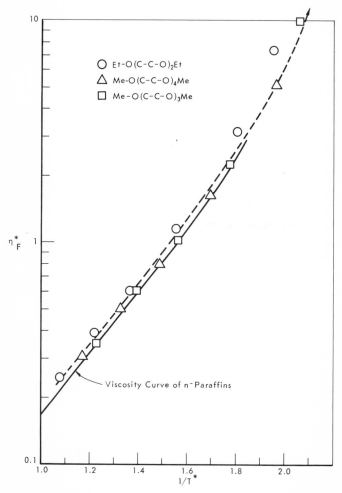

FIG. 27b. Generalized viscosity vs. temperature curves of alkane polyethers.

TABLE VII

Numerical Values of Constants a_F, b_F, d_F, g_F for Various Groups of
Liquids Composed of Flexible Molecules

Substances	a_F	b_F	g_F	d_F
n-Alkanes ($N_c > 8$)	0.565	2.56	3.26	1.70
Flexible isoalkanes	0.440	2.56*	3.17	1.60
Flexible alkylaromatics	$\varphi_R B_R + (1 - \varphi_R)B_b$†		3.26	1.70
Aliphatic polyethers	$B_p + 0.1$‡		3.26	1.70
m^n Polyphenylethers	0.450	0.985§	5.46	4.20

*Add a term $+1.10 \times 10^{-2}S_b$, where S_b is the number of carbon atoms per straight chain per branch point; at $S_b \geqslant 12$ set this condition equal to zero.

† The symbol φ_R is the (volume) fraction of carbon atoms in aromatic rings; B_b is to be calculated in the same manner as B_F for isoalkanes with "branch point" equal to ring-alkyl branch point; set equivalent branch length of phenyl group as $N_c = 4$ and equivalent branch length of cyclohexyl group as $N_c = 6$.

‡ $B_p = B_F$ of n-alkanes calculate for $N_c = N_s$.

§ Multiplier for b_F is N_φ^{-1}, where N_φ is the number of phenylene groups per molecule.

barrier to internal rotation of the chain segments relative to each other. Since the constant c in the characteristic temperature $\theta_L \equiv E^\circ/5cR$ is generally considered a measure of that chain flexibility, one might have expected B_F to be sensibly constant. The absolute magnitude of B_F and its rise with the chain length to an asymptotic limit may be interpreted as meaning that B_F is a measure of the ratio of the effective length of a freely rotating chain segment in viscous flow as compared with its effective length in equilibrium properties, and that this effective length reaches an asymptotic limit valid for the particular high polymer.

In an earlier discussion[27] the effective freely rotating length L_s of n-paraffins in liquids at equilibrium properties had been given as 4 to 5 methylene groups. For the more complete rotation required in shear flow the segment length might be approximately $B_F L_s$. In infinitely long methylene chains this argument yields an average segment length of 8 to 10 methylene groups. This number is somewhat smaller than, but of the same order as, the persistence length defined by Kratky and Porod.[37]

The trend of the constants of Table VII with molecular structure is in basic agreement with this picture of their physical meaning. The presence of branches increases B_F beyond the value expected for the given molecule size, because the increased bulk raises the height of the barrier to external rotation. Likewise, the small constant increment in B_F demanded for the polyethers can be attributed to the increased magnitude of the barrier to external rotation caused by the dipole-dipole orientation interaction between the ether

[37] G. Porod, *Monatsh. Chem.* **80**, 251 (1949).

oxygens. The method of counting segments in polyphenyl ethers is so different from that used for polymethylene that a direct comparison is not possible.

One polyether, poly(dimethylsiloxane), cannot be compared at all.
Oligomers up to the hexamer (the size limit for good data) behave reasonably
like the corresponding alkanes, as shown by the data in Fig. 27d. However,
somewhere between $x = 6$ and $x = 26$, the next set of available data, there
is a significant change in properties, as is evident from the fact that not only
is B constant at $T^* > 0.55$ but the sign of dB/dT is positive, in contrast to
the behavior of all "normal" liquids for which, at $T^* > 0.55$, $dB/dT \approx 0$.

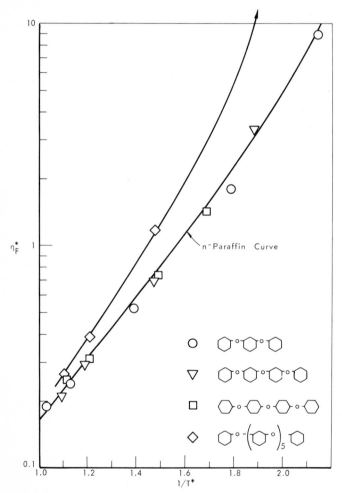

Fig. 27c. Generalized viscosity vs. temperature curves of various polyphenylethers. From
data by C. L. Mahoney and E. R. Barnum, in Ref. 38.

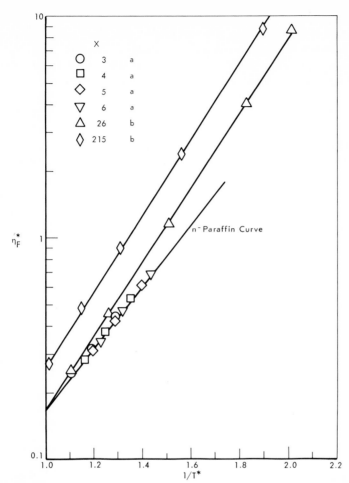

FIG. 27d. Generalized viscosity vs. temperature curves of long chain poly(dimethyl)siloxanes. (a) C. B. Hurd, *J. Am. Chem. Soc.* **68,** 364 (1946). (b) Ref. (a) of Fig. 27a.

This peculiar behavior of the siloxane polymers may have to do with the often-postulated formation of helices at low temperatures which, of course, must disappear at elevated temperatures.

Another item of interest is the group of "mixed" compounds, in which flexible chains are attached to rigid aromatic or naphthenic ring systems. Here B_F appears to be intermediate between the value for the flexible chain and that for the rigid ring at the same reduced temperature. A linear mixing rule has been developed to estimate B_F from B of the ring structures and B_F of the flexible chain. The latter, of course, must be calculated for the entire molecule. A somewhat arbitrary "equivalent" segment length is thus assigned to the ring systems. As it is well known that there is no significant difference between the viscosity of such a "chemical mixture" and that of the equivalent "physical mixture," the mixture rule found for the components of a single molecule should also be used to estimate the viscosity of mixtures involving ring structures and flexible chains.

The viscosity curves of many flexible polar compounds are parallel to, but do not coincide with, the n-paraffin curve. Only the constant $\overline{d_w}$ in the viscosity reducing parameter $(ME°)^{1/2}/N_A\overline{d_w}^2$ may be considered sufficiently disposable to serve as a shift factor in bringing the curves to coincidence. The values of $\overline{d_w}$ required have been assembled in Table VIII. Their physical meaning is clear in some cases. For instance, the ether data show that the ether oxygen is so buried in the molecule that its effective collision diameter is nearly zero, especially when the adjacent groups are large. The effective bulkiness of the ester group, on the other hand, is not easily intelligible.

TABLE VIII
EFFECTIVE VISCOSITY DIAMETER $d_w(\eta)$ OF CERTAIN FUNCTIONAL GROUPS

Group	$d_w(\eta)$, Å*	$d_w(\rho^*)$, Å†	Nominal d_w, Å‡
—O— in aliphatic polyethers	2.8	2.5	3.0
—O— in poly(dimethylsiloxanes)§	0	3.0	3.0
—O— in poly(phenyl ethers)	0	2.7	3.0
—C—O— in aliphatic esters (with =O)	7 ± 1		3.2 or 3.8‖
O=P(O—)₃ in aliphatic esters	4.8 ± 0.5**	2.9	3.2 or 4.3₅‖

* The diameter required to fit the generalized viscosity curve.
† The diameter required to fit the generalized ρ^*–T^* curve.
‡ This is $2r_w$.
§ Applies only to samples with $M \leqslant 2000$.
‖ The first number is just the conventional average $\overline{d_w}$: the second number is $(6V_w/\pi N_A)^{1/3}$.
** This is not a very meaningful number, because the slope of the phosphate ester in the $\ln \eta^*$ vs. $1/T^*$ curve is steeper than that of the n-paraffin reference curve.

j. Viscosity Isochores of Flexible Molecules

The few viscosity isochore data available in the low packing-density region ($\rho^* < 0.58$) of flexible molecules are similar to those of the rigid molecules at least with respect to the slope $(\partial\Delta E_\ddagger{}^j/\partial\rho^*)_{T^*}$. However, at $N_c > 6$ one finds with n-paraffins that $(\partial\Delta E_\ddagger{}^j/\partial T)_v < 0$. Consideration of the effect of molecular structure is thus confined to comparisons at constant

ρ^* and T^*, a restriction which begins to become quite important at $N_c \geq 16$, as is evident from the curves for *n*-hexadecane in Fig. 28. A cross-plot of the hydrocarbon data (at constant ρ^* and T^*) shows a regular rise of $\Delta E_\ddagger^j$ with increasing molecular weight. This trend suggests that $\Delta E_\ddagger^j$ is a barrier to rotational motion of the entire molecule. At some high molecular weight, however, $\Delta E_\ddagger^j$ must become constant and representative only of the extent to which hindered internal rotation must be excited for flow to occur. With slightly branched or cyclized alkanes, or both, this upper limit does not appear to have been reached at C_{25}, the highest carbon number for which low-density data are available.

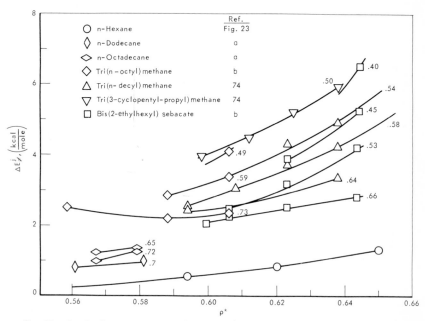

FIG. 28a. Isochoric temperature coefficient $\Delta E_\ddagger^j$ vs. packing density for liquids composed of flexible molecules. Parameter on curves is T^*. (a) API Research Project 42, at Pennsylvania State University, 1962. (b) D. Bradbury, M. Mark, and R. V. Kleinschmidt, *Trans. Am. Soc. Mech. Engrs.* **73**, 667 (1951).

High-molecular-weight poly(dimethyl siloxane), by contrast, exhibits so low a level of $\Delta E_\ddagger^j$ that it can only be associated with the movement of individual segments and, hence, with the barrier to internal rotation, which is known to be quite low, although not zero,[38] as had been supposed by many. At $\rho^* < 0.58$ one finds the unusual condition $(\partial \Delta E_\ddagger^j / \partial T^*)_v > 0$. At $\rho^* = 0.597$ one obtains over a 200°C range $(\partial \Delta E_\ddagger^j / \partial T^*)_v \approx 0$. The effects

[38] P. J. Flory, V. Crescenzi, and J. E. Mark, *J. Am. Chem. Soc.* **86**, 146 (1964).

of other functional groups on $\Delta E_{\ddagger}{}^j$ at $\rho^* < 0.58$ have not been studied. The interesting part of Fig. 28 at $\rho^* > 0.58$ will be discussed in Section III,2,c.

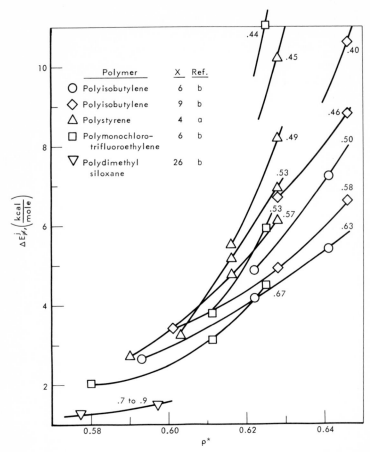

FIG. 28b. Isochoric temperature coefficient $\Delta E_{\ddagger}{}^j$ 25 packing density for various liquid polymers. Parameter on curve is T^*. (a) J. W. M. Boelhouwer, Koninklijke Shell Laboratorium, Amsterdam, private communication, 1950. (b) See Ref. (b) of Fig. 28a.

k. Pressure on the Viscosity of Flexible Molecules

Here, as before, the effect of pressure is best discussed simultaneously with that of temperature. Inspection of the experimental viscosity–pressure isotherms of different flexible compounds shows that $\ln \eta \propto p^{1/n}$. Accordingly, a trial plot of $\ln \eta_F^*$ vs. $(P^*)^{1/2}/T^*$ has been made and is shown in Fig. 29.[27] The representation on this graph is quite good except at low reduced temperature. The physical meaning of the proportionality to a low power of the pressure is obscure.

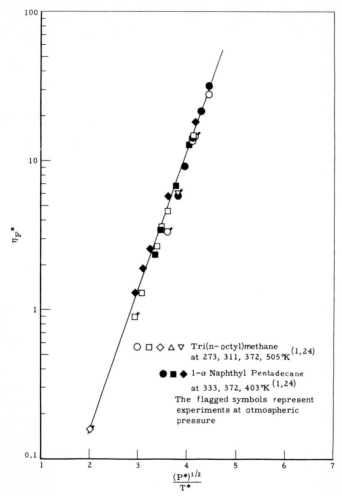

FIG. 29. Generalized viscosity-pressure-temperature relation for liquid composed of flexible hydrocarbon molecules.[27]

l. Summary of High-Temperature Characteristics

Comparison of viscosity data with the various molecular properties that determine their magnitude leads quickly to the conclusion that through its appearance in the exponential term the ratio E°/c is of overwhelming importance, the preexponential factor

$$\frac{(E^\circ M)^{1/2}}{A_w^{1/2} N_A d_w}$$

playing a more subordinate role. Hence, any increase in intermolecular forces has, through E°, a large effect on viscosity, especially in the case of rigid molecules, in which it alone determines the order of magnitude and the temperature coefficient of η. Those polar groups which raise E° without affecting c very much raise the viscosity correspondingly. However, the greater "smoothness" of many polar functional groups leads to viscosity increases somewhat lower than expected. Superimposed on the large exponential effects cited is the smaller (linear) effect of molecular weight, which is built into the viscosity-reducing parameter, such that for liquids composed of rigid molecules

$$\eta = A'\left(\frac{mE^\circ}{A_w}\right)(1/d_w)\exp(B'E^\circ/5cRT) \tag{3a}$$

and for liquids composed of flexible molecules

$$\eta = A''[(ME^\circ)^{1/2}/N_A d_w^{\,2}]\exp(B''E^\circ/5cRT) \tag{4}$$

The exponential dependence of the viscosity of a substance at a fixed temperature on the ratio E°/c accounts for the fact that the stiff cyclic compounds with $c \approx 2$ have generally a much higher viscosity than the flexible straight-chain compounds of the same molecular weight and, hence, of similar E° value, but for which $c > 2$. Naphthene rings can carry out a number of internal motions associated with cis-trans (chair-boat) isomerization and are therefore not as rigid as the aromatic, especially condensed cyclic, ring systems. The conversion of long chains to branch chains affects the magnitude of E° and c often in a compensatory way and therefore affects the viscosity only when the number of side branches is sufficiently large to rigidify the backbone structure of the molecule. Replacements of single bonds with double bonds generally increases c because of the low barrier to rotation around the $=CH-CH_2-$ bond, while E° remains unaffected, thus decreasing the viscosity, as is generally observed.

The well-known relation between the viscosity of liquids and their vapor pressure[39,40] is, of course, directly connected with present correlation, since $\log p_v$ is directly related to the variable RT/E°. The reduced temperature at which the correlation of viscosity vs. vapor pressure ceases to be valid is

[39] A. H. Nissan et al., J. Inst. Petrol. **26**, 155 (1940); Phil. Mag. [7] **32**, 441 (1941).
[40] G. W. Nederbragt and J. W. M. Boelhouwer, Physica **13**, 305 (1947).

virtually identical with the low-temperature limit of the present reduced-variable correlation. All other correlations of viscosity with equilibrium properties (Eyring, Souders, and Thomas, see Section VIII) have about the same low-temperature limit of validity. The reason for this phenomenon will be discussed in the next section.

2. THE LOW-TEMPERATURE–HIGH-PACKING DENSITY REGIME

An increase in the packing density through temperature reduction or through pressure application, such that $\rho^* \rightarrow > 0.6$ or thereabouts, reduces

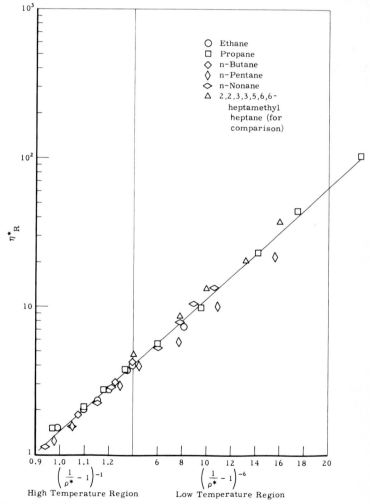

FIG. 30. Reduced viscosity of short-chain normal paraffins as functions of reduced density.

the number of degrees of freedom excited per molecule. As a result the density increases (at atmospheric pressure) faster than a rate corresponding to the generalized ρ^*-T^* curve.[41] This is particularly noticeable with high polymers but is also true of monomeric compounds. At a packing density corresponding to $V/V_0 = 1.14$, where $V_0 = V_{cryst}$ at $T = 0°K$ all external molecular motion stops and the liquid turns into a glass.[41] One generally connects the loss of free volume with the onset of the glassy state. A relation between packing density (in this case expressed as a power function of the fluctuation volume) and the viscosity of lower paraffins is presented in Fig. 30. The particular form of the volume function employed at low temperatures can be related to Bueche's theory of cooperative phenomena.[42] The striking effect of increasing the packing density on the viscosity of the easily interlocking dimethylsiloxane polymers is shown in Fig. 31. Owing to the very high compressibility of this series of liquids, one obtains the somewhat confusing picture of siloxane viscosity vs. the more conventional temperature and pressure scales.

A theory that connects rate processes, such as diffusion and flow, with the available volume has been proposed by Cohen and Turnbull[43] and is of the form

$$D = ga^*(3kT/m)^{1/2} \exp\left(\frac{-\gamma v^*}{\bar{v}_m\bar{\alpha}(T - T_0) - \bar{v}_p\beta\,\Delta P}\right) \tag{5a}$$

$$\frac{1}{\eta} = ga^{*2}(3T/km)^{1/2} \exp\left(\frac{-\gamma v^*}{\bar{v}_m\bar{\alpha}(T - T_0) - \bar{v}_p\beta\,\Delta P)}\right) \tag{5b}$$

where g is a geometric factor of the order of unity, a^* is approximately one molecule diameter, $v^* \approx 10v_f$, $\gamma v^* \approx v_L$ (the liquid volume per molecule), $\bar{v}_m$ is the mean molecular volume over the temperature range considered, such that $v_f = \bar{v}_m\alpha(T - T_0)$, $\bar{\alpha}$ is the mean coefficient of expansion, $T_0 = T$, at which $v_f \to 0$, $\bar{v}_p$ is the mean molecular volume over the pressure range ΔP, and β is the mean compressibility over the same ΔP range. The temperature $T_0 < T_g$. The assumptions of constant α and $\bar{v}_m$ are not very good.

Although one prediction of this theory, that at constant volume $\eta \sim (1/T)^{1/2}$, is not borne out by experiment, we shall use this simple relation as a starting point for the correlation of viscosity data obtained at low temperatures and high pressures.

The Cohen–Turnbull (C-T) treatment gives some physical meaning to the constants of an equation proposed by Vogel[44] to describe the viscosity-temperature function of oils. The utility of this equation has been rediscovered

[41] A. Bondi, *J. Polymer Sci.* **A2**, 3159 (1964).
[42] F. Bueche, *J. Chem. Phys.* **30**, 748 (1959).
[43] M. H. Cohen and D. Turnbull, *J. Chem. Phys.* **31**, 1164 (1959).
[44] H. Vogel, *Physik. Z.* **22**, 645 (1921).

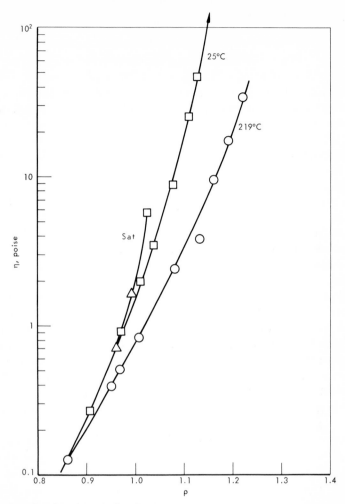

FIG. 31. Viscosity of a polydimethylsiloxane as a function of density along saturation line and at constant temperature. (From data of Ref. (b) of Fig. 28a.)

by Guttman and Simmons,[45] and employed, in a complicated garb, by Ferry,[46] as W-L-F equation to describe the viscous properties of polymers of $T_g < T < (T_g + 100°C)$. The C-T treatment also provides a rationale for Doolittle's proportionality between the logarithm of viscosity and a suitably chosen free volume function.[47]

[45] F. Guttman and L. R. Simmons, J. Appl. Phys. 23, 977 (1952).
[46] J. D. Ferry, "Viscoelastic Properties of Polymers." Wiley, New York, 1961.
[47] A. K. Doolittle, J. Appl. Phys. 22, 1031 and 1471 (1951); 23, 418 (1952).

Stated in the reduced coordinates of the present review, Vogel's equation is of the form

$$\log \eta^* = \frac{\mathscr{B}}{T^* - T_0^*} - \mathscr{A} \tag{6}$$

where the constants $\mathscr{B}$ and $\mathscr{A}$ can be expressed in terms of the constants A and B of the high temperature regime, if one postulates that the deviation from the Arrhenius curve begins at $T^* = 0.7$. Then

$$\log \eta^* = B[1 - 1.43T_0^*]^2 \left\{ \frac{1}{T^* - T_0^*} + \left[\frac{1}{0.70 - T_0^*} - \frac{1.43}{[1 - 1.43T_0^*]^2} \right] \right\} - A \tag{7}$$

so that the dominant new feature is T_0^*. Several authors[48(a)-(c)] tried, so far unsuccessfully, to establish a close coupling between T_0 and T_g. With liquids composed of comparatively flexible molecules, one often finds $T_0 \sim 0.8T_g$.

Before examining the relation of T_0^* to molecular structure, a few inherent limitations of this otherwise very significant new approach to the description of low temperature viscosity should be mentioned. A minor point is that the near constancy of the product $\bar{\alpha}\mathscr{B}$ demanded by equation (5) is practically never found in practice, even among chemically very similar compounds. A major problem is the recent discovery by several groups[49(a)-(c)] that equation (6) describes the viscosity of certain liquids only down to a certain temperature, below which all three constants take on drastically different values.

Inspection of the data shows that the incidence of this break is a characteristic of liquids composed of rather rigid molecules, such as short chain alkyl benzenes, alkyl phthalates and di- and tri-aryl-benzenes. The effect disappears when the alkyl chains are lengthened to $N \geqslant 4$.[49(a)] Hence it is understandable that the effect has not been observed with most typical vinyl polymers. In the reduced temperature scale the break, when it occurs, is observed at $T^* \approx 0.38 \pm 0.03$. Since few liquids are used at such low temperature, this limitation is less of practical interest than of intellectual concern. Far more work is obviously needed before the problem of low temperature viscosity can be considered fully clarified, even in principle.

[48] (a) J. H. Gibbs and E. A. di Marzio, *J. Chem. Phys.* **28**, 373 (1958). (b) R. S. Stearns *et al.*, *ACS Div. of Petroleum Chemistry Preprints* **11**, 5 (1966). (c) A. B. Bestul and S. S. Chang, *J. Chem. Phys.* **40**, 3731 (1964).

[49] (a) A. J. Barlow, J. Lamb, and A. J. Matheson, *Proc. Roy. Soc.* **A292**, 322 (1966). (b) D. J. Plazek and J. H. Magill, *J. Chem. Phys.* **45**, 3038 (1966). (c) T. G. Fox and G. C. Berry, *Fortschr. Hochpolym. Forschg.* **5** (1967).

With these limitations in mind, a search for a correlation of T_0^* with molecular structure was undertaken. Guided by earlier work on the reduced glass-transition temperature T_g^*,[41] the temperature T_0^* was compared with molecule size and flexibility. The correlation in Fig. 32 shows that T_0^* is primarily a function of the "flexibility density" $3c/V$. From the glaring exception of the series 2,2,3,3,5,6,6-heptamethylheptane → polyisobutylene one sees that an additional factor matters. As suggested earlier,[41,48a] this factor may be ΔE_{iso}, the energy of rotational isomerization per chain segment, which is nearly zero for this series but may be of the order 0.5 to 1 kcal/mole of repeating units for many of the nonrigid molecules of the other series. With rigid molecules, for all of which $3c \approx 6$, the temperature T_0^* is primarily a function of molecule size.

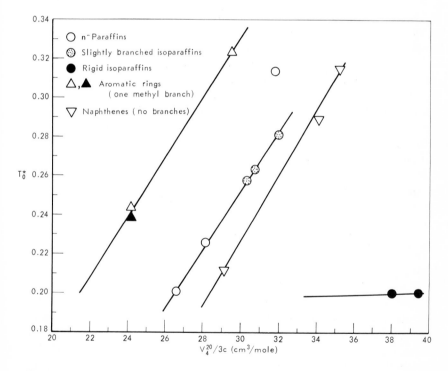

FIG. 32. Reduced equilibrium glass transition temperature T_0^* as a function of the flexibility density $V/3c$ for parent compounds alkyl derivatives fall into the appropriate regions.

The effect of pressure on the viscosity in the high packing density regime is qualitatively apparent from equation (5). The quantitative relations to molecular structure require replacement of $\bar{v}_p \beta \, \Delta P$ by a more accurate equation of state, whence

$$\log \eta = \frac{\mathscr{B}'}{\alpha(T - T_0) - [(P/K_0) + 1]^{1/K_1} - 1 + 0.05} - \mathscr{A}' \tag{8}$$

where the molecular structure dependence of the bulk modulus K_0 is effectively suppressed by the magnitude of K_1, the pressure coefficient of K_0. The coefficient K_1 is nearly a universal constant for liquids, namely, ≈ 10. The number 0.05 has been added to prevent the equation from "blowing up" at high pressures. When equation (8) is expressed in reduced units the resulting α^* is practically independent of molecular structure. Hence one sees that the pressure dependence of viscosity in the high packing density regime is dominated by the same factor (T_0) responsible for the temperature dependence of viscosity, in excellent agreement with a long established experimental fact.[4,5] If equation (8) were as reliable quantitatively as its meaning is transparent qualitatively, all points for a given liquid, shown in Fig. 33, would fall on a single curve. They obviously do not.

The lack of coincidence of curves of a given liquid in Fig. 33 shows that these, like all other free-volume equations, are good only for qualitative discussions because, as mentioned above, viscosity is not only a function of free volume. This becomes particularly noticeable at high pressures. Hence, the term $\Delta v(P)/v$ or its equivalent compressibility term should be multiplied by a temperature-dependent coefficient. This was done, in effect, by Webb and Dixon[50] in the form

$$-\ln \eta = C_1(T) + \frac{C_2(T)V(T)}{V(T) - V_w} \tag{9}$$

but there is no theoretical guideline to relate such a parameter to molecular structure.

a. Rigid Nonpolar and Slightly Polar Compounds

Since only a few data[51] are available for rigid, slightly polar compounds that melt at sufficiently low temperature to permit evaluation of the constants of the low-temperature viscosity equation (6), most of the available generalizations have been derived from the plentiful hydrocarbon data. The previously described dependence of T_0^* on $3c/V$ means that for rigid molecules (for which $3c \approx 6$) the T_0^* is primarily a function of V. If the "effective glass transition temperature" T_0^* rose without limit with increasing van der Waals' volume, the absolute glass transition temperature would make such

---

[50] W. Webb and J. A. Dixon, *Proc. Am. Petrol. Inst. Sect. III* **42**, 146 (1962).

[51] The reason is that just very few irregularly shaped low-melting polar compounds have been synthesized and their properties investigated.

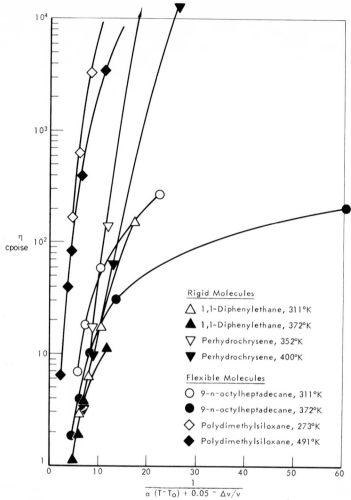

FIG. 33. Viscosity isotherms vs. free volume parameter for various liquids.

substances glassy resins at ordinary temperature and at comparatively low molecular weights. Subcooled 1,3,5-trinaphthyl benzene,[50] for which $T_g = 343°K$ and $T_g^* = 0.366$ and various abietic acid resins are typical examples. In the limit this would mean that at some high molecular weight the glass transition temperature might even exceed the atmospheric boiling (sublimation) point.

Rotation of connected aromatic rings is hindered by potential-energy barriers of $\geqslant 3.9$ kcal/mole,[51a] the barrier in diphenyl, which makes the

[51a] K. E. Howlett, *J. Chem. Soc.* p. 1655 (1960).

polyphenyls and their analogues effectively rigid over a wide temperature range. Deviations of the generalized viscosity from that of diphenyl, shown in Fig. 34, largely reflect the increased stiffness of the interring bond from diphenyl up. The difference between 1- and 2-phenylnaphthalene and that between 1,1- and 2,2-dinaphthyl are in line with the extra hindrance to rotation caused by the H-8 for 1-substituents on the naphthalene ring. The lack of coincidence of the curves for diphenyl, 2-phenylnaphthalene, and 2,2-dinaphthyl in spite of their probably very similar barrier to internal rotation shows that an additional factor influences the level of η_s^* at fixed T^*. The numerical data in the linear portion of the curve of $\ln \eta^*$ vs. $(T^*)^{-1}$ are in agreement with the supposition that the moment of inertia in the viscosity-reducing parameter (see Section III) would have achieved better coincidence of the curves for compounds with equal V°/RT.

The nearly complete coincidence of the generalized viscosity–temperature curves in the series 1,1-diparatolylethane to 1,1-diorthotolylethane,[52] in spite of the concurrent sharp increase in bond stiffness, suggests that the effects shown in Fig. 34 are indeed quite specific.

The three-dimensionally bunched molecules of the series o-terphenyl, 1,3,5-trinaphthylbenzene and, to some extent 1,1-dinaphthyl must have the high glass transition point (the measured T_g^* of the trinaphthylbenzene being 0.366) that is associated with that structure and high values of $V/3c$. The steep upward sweep of the viscosity curves is simply the consequence of a rapid approach to T_g^*, where $\eta_s^* \to 10^{15}$. Since the onset of glass transition can be described without recourse to cluster formation,[53] there seems to be no need of this additional hypothesis.

b. Flexible Nonpolar and Slightly Polar Compounds

The reduced effective glass transition temperature T_0^* of compounds with flexible backbones reaches an upper limit, because in any given polymer homologous series the concentration of excited degrees of freedom reaches an upper limit, whence it stays independent of molecular weight. This limiting value of T_0^* and the magnitude of the molecular weight at which it is reached are, of course, functions of molecular structure. Typical examples of the limiting values of T_g^* have been assembled on Table IX.

The comparatively narrow range of these reduced glass transition temperatures permits approximate prediction of T_g^* and T_0^* from molecular-structure information. The effect of polar groups on T_g^* and T_0^* seems to be largely absorbed in the reducing parameter $E^\circ/5cR$. With B' coupled to T_0^*

[52] Data by R. J. Best, *J. Chem. Eng. Data* **8**, 267 (1963).
[53] E. McLaughlin and A. R. Ubbelohde, *Trans. Faraday Soc.* **54**, 1804 (1958).

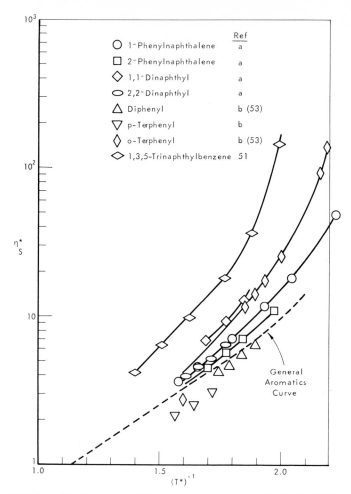

FIG. 34. Effect of hindrance to rotation between rings on viscosity. (a) E. H. Binns and K. H. Squire, *Trans. Faraday Soc.* **58**, 762 (1962). (b) M. Andrews and A. R. Ubbelohde, *Proc. Roy. Soc.* **A228**, 435 (1955).

through the previously described relation to the slope constant *B* of the high-temperature region, the temperature dependence of viscosity in the low-temperature region seems to be well correlated with molecular structure.

The absolute level of the viscosity is fixed by the requirement that the low-temperature part of the viscosity curve connect smoothly with the high-temperature part, e.g., at $T^* = 0.7$, where it is well covered by the simple relations discussed in Section V,1.

The viscosity of liquids composed of flexible long-chain molecules, which are long enough for their configuration to be treated by the statistics of

TABLE IX
LIMITING VALUES OF T_g^* AT $M \to \infty$

Polymer	T_g^*	ΔE_{iso}, kcal/mole
Polyisobutene	0.216	0
cis-1,4-Polyisoprene	0.27	
Polyethylene	0.27	0.77*
Poly(dimethylsiloxane)	0.30	0.80†
Poly(methyl methacrylate)	0.33	⩾0.8
Poly(n-butyl methacrylate)	0.33	
Polypropylene	0.35	
Poly(monochlorotrifluoroethylene)	0.36	
Poly(vinyl acetate)	0.39	≈1.0
Poly(vinyl chloride)	0.42	1.9‡
Polystyrene	0.43_5	

* Commonly accepted datum for liquid n-paraffins.
† Ref. 38.
‡ H. Germar, Kolloid Z. **193,** 25 (1963). This figure is somewhat high;
1.2 kcal/mole is more likely correct.

long-chain molecules yet shorter than the critical-entanglement length[53a] (beyond which they would fall outside the scope of this section), can be related to molecular structure in a somewhat different way from that given so far. Starting from a consideration of the flow of a chain of segments past other segment chains, Bueche[34] and others showed that the viscosity of such a polymer melt should be given by

$$\eta = \left(\frac{kT}{6v}\right)\frac{S^2}{\delta^2}\left[\frac{N^*}{\phi_0}\exp(\varepsilon^*/kT)\right] \tag{10}$$

where S^2 is the mean square distance of segments from the center of mass of the polymer coil, a number which is derived from light-scattering measurements of polymer solutions (at the so-called θ temperature) and can be estimated for simple chains from a priori arguments, δ^2 is the mean square distance moved by the center of mass of the coil per segment jump, ϕ_0 is the usual preexponential jump frequency factor ($\sim 10^{12}$ sec^{-1}), N^* is the effective number of segments per molecule, and ε^* is the critical energy per segment jump and is equal to $4\pi\varepsilon b^{*2}$, b^* being the radius of critical hole and ε the pair potential between two nonbonded segments. The term in square brackets in equation (10) can easily be related to the corresponding magnitudes of our high-temperature viscosity equations—but applied to

[53a] Given in Table VI.

the repeating unit. The quotient $S^2/v\delta^2$ is specifically related to the coiled configuration of the long-chain molecule and can be estimated from chain segment geometry, chain length, and ΔE_{iso}, the energy of rotational isomerization for the connected segments. The details of these relations should be obtained from the references, since their description would go beyond the scope of the present article. Suffice it to state that S^2/v rises proportionately to the length of the chain and is the larger the greater the chain stiffness. Chain branching (at a fixed molecular weight) increases the viscosity by a factor $g = (S_b/S_L)^2$, which is obtainable from light-scattering data, subscript b referring to the branched compound and L to the linear.

c. Isochore at High Packing Densities

While the curve of $\Delta E_{\ddagger}{}^j$ vs. ρ^* of liquids composed of rigid molecules (Fig. 23) passes smoothly from the low to the high packing-density region, a drastic change of slope is observed with most liquids composed of flexible molecules at $\rho^* > 0.58$ to 0.60, as shown in Fig. 28. Here, too, the n-paraffins with $N_c < 8$ act like rigid molecules. Particularly noteworthy is the large temperature coefficient of $\Delta E_{\ddagger}{}^j$ at high values of ρ^* and the consequent fanning out of the curves for any one compound or group of compounds toward higher packing densities. Normal hexadecane is the only compound on which the transition from the low-density to the high-density regions can be clearly observed. With $(n\text{-}C_8)_3C$ and $(c\text{-}C_5\text{—}C\text{—}C\text{—}C)_2C(n\text{-}C_8)$ the somewhat uncertain high-temperature data (not shown) suggest a flattening out at the 2.5 kcal/mole level rather than a steep dip at $\rho^* < 0.58$.

The sequence of $\Delta E_{\ddagger}{}^j$ levels (at constant ρ^* and T^*) clearly follows the order given by the (lowest) barrier to internal motion of the molecules. The highest level is found for the stiff poly(monochlorotrifluoroethylene), followed by that for polyisobutylene (with a neopentane-like barrier of ≈ 5 kcal/mole between chain segments) and the nearly superimposable curve of perhydrochrysene shown on Fig. 23. The internal motions of the latter are probably between cis and trans conformations. The low position of the diester may well be associated with the low barrier to rotation around the

$$\begin{array}{c} O \\ \parallel \\ -O-C- \end{array}$$

ester bond ($V^\circ \approx 0.5$ kcal/mole). The incompleted continuation of the poly-(dimethylsiloxane) curve puts it into the position appropriate to its low barriers to internal rotation.

Theoretical analysis of this complicated situation is still outstanding. The intuitive notion that $(\partial \Delta E_{\ddagger}{}^j/\partial T)_v$ is a measure of the number of degrees of freedom that have to participate in the translation across a potential-energy barrier $\Delta E_{\ddagger}{}^{j}$ is not immediately applicable, because the facts are not in agreement with the requirement $\Delta E_{\ddagger}{}^{j}/RT \geqslant 1 + R^{-1}(\partial \Delta E_{\ddagger}{}^{j}/\partial T)_{v}$. The practical importance of obtaining a better understanding of this problem is emphasized by the role which the same phenomena play in determining the mechanical properties of polymers, where $\Delta E_{\ddagger}{}^{j}$ is rather more difficult to measure, but where the few early experiments show exactly the same kind of behavior[54] as those noted here for liquids.

d. A More Complete Formulation for Viscosity at High Packing Densities

The magnitude of the isochoric temperature coefficient of viscosity $\Delta E_{\ddagger}{}^{j}$, the potential-energy barrier to motion in the liquid, described earlier, suggests an important reason for the failure of the free-volume treatment of viscosity to represent the data at elevated pressures. This failure had been noted by Fox et al.,[55] who supplemented the W-L-F free-volume equation with a term E/RT, where E was defined as the energy required to jump into the hole created by free-volume fluctuations. In a recent paper Fox and Allen[56] give numerical values for E obtained from polymer melt viscosity data. All of the E values parallel, but are not identical with, the known barriers to internal rotation for the segments under consideration. Macedo and Litovitz[57] suggest the form

$$\ln \eta = A + H^{*}/RT + 1/f \tag{11}$$

where f is the fractional free volume discussed in the introduction to Section V,2 and $H^{*} \approx \Delta H_{\text{vap}}/5.3$.

Both these treatments relegate the strong density and temperature dependence of the energy term to the free-volume term. This procedure provides convenient algebra but is probably responsible for the confusion regarding the meaning of the numerical values of f besides T_{0} in equations expressing f as $\sim \alpha(T - T_{0})$. Hence, it is proposed to write

$$\log \eta^{*} = A - \frac{B_{v}(\rho^{*}, T^{*})}{T^{*}} - \frac{L}{\alpha^{*}(T^{*} - T_{0}{}^{*})} \tag{12}$$

where $B_{v} = \Delta E_{\ddagger}{}^{j}/4.57\theta_{L}$ and L is a constant not far from unity. In the absence of a form for the function $B_{v}(\rho^{*}, T^{*})$ this equation is just a research program. It should be noted, however, that in the few cases for which data

[54] G. Williams, National Physical Laboratory, Teddington, England, private communication, 1964.

[55] T. G Fox, S. Gratch, and S. Loshaek, in "Rheology: Theory and Applications" (F. R. Eirich, ed.), Vol. 1, p. 484. Academic Press, New York, 1956.

[56] T. G Fox and V. R. Allen, J. Chem. Phys. 41, 344 (1964).

[57] P. B. Macedo and A. T. Litovitz, J. Chem. Phys. 42, 245 (1965).

were available the equation yielded values for T_0^* that were quite close to the measured T_g^*. Because T_g^* covers so narrow a range of values, especially for simple compounds, that it can be guessed from molecular structure,[41] and the low-temperature limit of α^* is essentially a universal constant the free-volume term is easily related to molecular structure. There is reason to hope that a way will be found to relate the welter of ΔE_+^j data in Figs. 23 and 28 to molecular properties. Then equation (12) may be considered one further step toward better understanding of the effect of molecular structure on the viscosity of liquids in the high packing-density regime.

VI. Effect of Association (Hydrogen Bonding)

A corresponding states treatment of the properties of associating compounds is excluded almost by definition because of the superposition of the highly specific dissociation equilibrium on whatever individual properties the monomer and association polymers might have. Techniques for determining association equilibria in pure liquids have been developed very recently, and not enough data have been assembled to permit a rational treatment of as complicated a property as viscosity, when even the pressure–volume–temperature properties of alcohols cannot yet be calculated. The relations of viscosity to the molecular structure of hydrogen-bonding substances given in this section are therefore rather crude.

1. STABLE DIMERS

The simplest kind of association to handle is that of comparatively stable dimers such as are formed by certain carboxylic acids. Over the temperature range over which association persists to better than 90% such dimers can be treated as but slightly polar compounds with the molecular structure of the dimer. The cohering functional group

$$R-C\overset{\displaystyle O-H-O}{\underset{\displaystyle O-H-O}{}}C-R$$

may be considered a planar six-membered ring, like a benzene ring, and all properties calculated accordingly.[3] Their viscosity properties then fall reasonably well into the expected patterns, as indicated in Table X.

Any broad application of this reasoning must always be preceded by an experimental determination of its premise, namely, that stable dimers are indeed formed in the liquid state. Abietic acid is a typical case of the absence of stable dimerization because of steric hindrance. Other acids with very bulky radicals must be treated with corresponding suspicion before their

TABLE X

EVIDENCE OF THE EXISTENCE OF FATTY ACIDS AS
DOUBLE MOLECULES BY COMPARISON OF VISCOSITY DATA
WITH THOSE OF HYDROCARBONS AND ESTERS

	M^*	$t, °C$	η, poises
Propionic acid	(148)	20	0.011
Propionic anhydride	130	20	0.011
n-Decane	142	20	0.0091
Valeric acid	(204)	20	0.022
n-Tetradecane	198	20	0.022
n-Heptylic acid	(260)	20	0.043
n-Octadecane	254	20	0.045
Oleic acid	(564)	20	0.318
Oleyl oleate	532	20	0.330

* Data in parentheses are calculated for the double molecule.

viscosity is estimated. One of the best-established methods of determining association equilibria of this type is the sound absorption in liquids.[58]

2. CHAIN ASSOCIATION

One of the major difficulties with substances in the chain-association class is the absence of reliable data on the degree of association prevailing in the pure liquid (in contrast to the rather good data on their association equilibria in dilute solutions). Moreover, there is no good method of eliminating the nonspecific intermolecular force effects, as the corresponding states principle permits one to do in the case of nonassociating substances.

A direct investigation of molecular motions by NMR and dipole-relaxation spectroscopy may serve as a guide to the problem. In the case of n-octyl alcohol Powles and Hartland[59] found that the rotational motion of the hydrocarbon tail (at 25°C) is about thirty times faster than that of the hydroxyl group.

As a first approximation it appears reasonable, however, to compare the viscosity of associating substances with those of their hydrocarbon homomorphs[59a] at the same packing density[59b] $\rho^* \equiv V_w/V$, which means in essence

---

[58] K. F. Herzfeld and T. A. Litovitz, "Absorption and Dispersion of Ultrasonic Waves." Academic Press, New York, 1959.

[59] J. G. Powles and A. Hartland, *Proc. Phys. Soc.* **75**, 617 (1960).

[59a] A hydrocarbon homomorph is a hydrocarbon of the same molecular geometry as the polar compound; e.g., toluene is the homomorph of phenol.

[59b] Here V_w is corrected for the contraction due to hydrogen bond formation, as indicated in Ref. 9.

at the same "free volume." This difference $\eta(\text{ass.}) - \eta(\text{hom.})|_{\rho*} \equiv \Delta\eta^h$ may well be considered a measure of the extra drag due to hydrogen bond formation over and beyond that caused by the general force field of the neighboring molecules.[59c] The curves in Fig. 35 show the rather interesting result of such a comparison. A downward drift of $\Delta\eta^h$ of the normal primary alcohols with increasing molecular weight renders its asymptotic limit at about C_8, as is evident from the location of $\Delta\eta^h$ for cetyl alcohol. The location of the ethylene glycol data on this graph is rather unexpected, pointing perhaps to water contamination of the glycol sample used in the measurement. If the data are genuine, two other possibilities are intramolecular hydrogen bonding,

which should be very weak and still leave one hydroxyl group for chain formation, or ring formation,

The comparison of the latter case would then be with cyclooctane, which at the same packing density as the glycol exhibits somewhat higher viscosity, as one might expect because of the greater rigidity of the cyclooctane than of the association ring. The primary amine data are based on V_w, corrected for the formation of two hydrogen bonds. Assumption of only a single bond leads to unreasonably high packing densities and thus to homomorph viscosities higher than those of the amine. As it is, the $\Delta\eta^h$ for amines is almost negligible in magnitude. Although most $\Delta\eta^h$ curves fall in the well-known order of hydrogen bond strengths, one should not lose sight of the fact that the association equilibrium responsible for $\Delta\eta^h$ depends on the absolute temperature level.[60] Hence, before truly quantitative comparisons can be made, a way will have to be found to compensate for temperature effects. Another way of examining the data is through the ratio $\eta^*(\text{ass.})/\eta^*(\text{hom.})$ at constant ρ^*. According to the definition of the reduced viscosity, this ratio might be taken as $X_a^{1/2}$, where X_a is the number of alcohol molecules in an association complex. Such a plot is shown in Fig. 36. Beyond accentuating the absence of a specific hydrogen-bonding effect

---

[59c] This treatment is somewhat similar to that of L. H. Thomas, *J. Chem. Soc.* p. 4906 (1960), who obtains the viscosity increment at constant P_v.

[60] G. C. Pimentel and A. L. McClellan, "The Hydrogen Bond." Freeman, San Francisco, 1960.

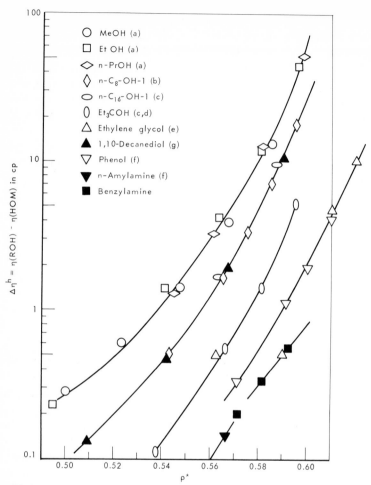

FIG. 35. Hydrogen bond contribution to viscosity, $\Delta\eta^h$ vs. packing density of hydrogen bonding substances. (a) T. Tonomura, *Sci. Rept. Tohoku Imp. Univ. First Ser.* **22**, 104 (1932). (b) E. C. Bingham and L. B. Darrall, *J. Rheol.* **1**, 174 (1930). (c) E. C. Bingham and H. L. De Turck, *J. Rheol.* **3**, 479 (1932). (d) L. Grunberg and A. H. Nissan, *World Petrol. Proc., 3rd, The Hague, 1951*, Vol. 6, p. 279, Brill, Leiden, 1951. (e) E. C. Bingham and H. J. Fornwalt, *J. Rheol.* **1**, 372 (1930). (f) E. C. Bingham and L. W. Spooner, *Physics* **4**, 387 (1933). (g) J. W. M. Boelhower, G. W. Nederbragt, and G. Verberg, *Appl. Sci. Res.* **A2**, 249 (1950).

with the glycol data, this graph does not teach anything new in comparison with Fig. 35. The thought of connecting this information with the best-established hydrogen bond characteristic, the excess energy of vaporization,[61] is discouraged somewhat by the very small association effect indicated by

[61] A. Bondi and D. J. Simkin, *A.I.Ch.E. J.* **3**, 473 (1957).

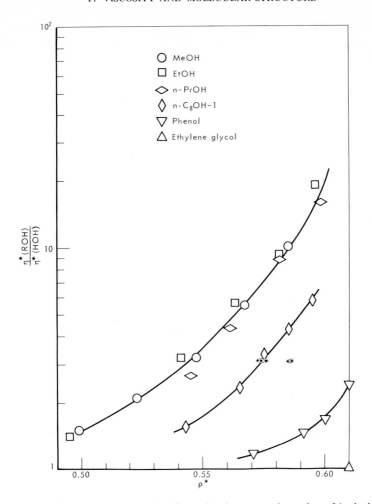

FIG. 36. Ratio of reduced viscosity of associated compounds to that of its hydrocarbon homomorph at fixed packing density ρ^*.

the viscosity graphs at $\rho^* = 0.50$, which is at or near the atmospheric boiling point. The entropy of vaporization suggests still significant association (or very strongly hindered rotation) at that point.

The specific effects of association on viscosity, other than those due to high packing density, which have been described in this section, are most notable for their comparative constancy within a given group of compounds. This should enable one to make reasonably good estimates of viscosity if one knows that of the corresponding homomorph at the appropriate packing density. The choice of the homomorph is here quite important.

Whereas for organic compounds the hydrocarbon series generally provide good homomorph data, the homomorphs of inorganic halides are preferably taken from the corresponding perhalide, known to act "normally."

VII. Viscosity of Mixtures and Solutions

1. HIGH-TEMPERATURE RANGE

a. Perfect Liquids

An easily manageable mixture rule for viscosity has not yet been developed from Stuart Rice's molecular theory of viscosity, described in an earlier section.[61a] Earlier molecular theories had been interpreted to yield mixture rules (including the more recent effort by Collins and Sedgwick[62]) which, as I mentioned in Volume I of this series, are at variance with experimental data.

In the absence of guidance from theory the problem might profitably be examined in terms of the corresponding states principle. If one assumes that the general relation

$$\ln \eta^* = B/T^* - A \tag{13}$$

where A and B are the "universal constants," is as valid for mixtures as for pure liquids, then the mixture viscosity should be related to the component properties m_i, σ_i, ε_i through the additivity rules appropriate to them, namely, $\bar{m} = (\sum_i w_i/m_i)^{-1}$, where w_i is the weight fraction of component i, $\bar{\sigma} = i^{-1} \sum_i \sigma_i$.

The additivity rule for ε (and $E°$) depends on the range of σ values involved. Restricting the discussion, for the sake of simplicity, to binary mixtures, one finds for the case $\sigma_A \approx \sigma_B$, or $d_w(A) \approx d_w(B)$ and $A_w(A) \approx A_w(B)$, that

$$\bar{\varepsilon} = X_A^2 \varepsilon_A + 2X_A X_B (\varepsilon_A \varepsilon_B)^{1/2} + X_B^2 \varepsilon_B \tag{14}$$

and the same for $\bar{E}°$, where $E°_{A,B}$ replace $\varepsilon_{A,B}$. Here X_i is a mole fraction. When $\sigma_A \neq \sigma_B$, the additivity rule for ε (and $E°$) becomes rather more complicated and requires the assumption of a potential law; here we shall assume the convenient 6:12 potential. Then, according to Prigogine,[6]

$$\bar{\varepsilon} = \frac{[X_A^2 \varepsilon_A \sigma_A^6 + 2X_A X_B (\varepsilon_A \varepsilon_B)^{1/2} \sigma_{AB}^6 + X_B^2 \varepsilon_B \sigma_B^6]^2}{X_A^2 \varepsilon_A \sigma_A^{12} + 2X_A X_B (\varepsilon_A \varepsilon_B)^{1/2} \sigma_{AB}^{12} + X_B^2 \varepsilon_B \sigma_B^{12}} \tag{15}$$

and the corresponding rule for $E°$ and d_w.

[61a] The mixture rules in the paper by S. A. Rice and A. R. Allnat, *J. Chem. Phys.* **34**, 409 (1961) require input data that are not generally available.

[62] F. C. Collins and T. O. Sedgwick, U.S. Air Force Contract AF 33(616)-3897, December 1961.

Then the mixture viscosity $\bar{\eta}$ is given by

$$\ln \bar{\eta} = \ln \frac{(\bar{\varepsilon}\bar{m})^{1/2}}{\sigma^2} + \frac{B\bar{\varepsilon}}{kT} - A \tag{16}$$

or the corresponding averaged quantities for the M, E°, d_w, A_w system. This relation would be valid only for ideal mixtures of "perfect liquids," which, in Pitzer's sense of the term, means substances that follow the corresponding states principle exactly and for which A and B would therefore be universal constants.

Nonideal mixture components interact with lesser or greater potential than $\bar{\varepsilon}$ (or $\bar{E}^\circ$), depending on the sign of the excess free energy of mixing F^E. Hence, a more general additivity rule for perfect liquids must contain the potential energy in the form

$$\bar{\varepsilon}_e = \bar{\varepsilon} - w'F^E(T) \quad\text{and}\quad \bar{E}_e^\circ = \bar{E}^\circ - wF^E(T) \tag{17}$$

where the constants w' and w can, in principle, be calculated from a relation involving the number of nearest neighbors. Empirical comparison with experiment, first proposed by Powell et al.,[63] but more accurately with recent data,[63a] suggests that $w \approx 0.9(z/2)c \approx 4.5c$. A test of this relation on the mixture data for perfect liquids[64a,b] is not possible because of insufficient supplementary data. However, the qualitative consequence that a negative excess free energy of mixing increases, and a positive excess free energy of mixing decreases, the mixture viscosity relative to that for the ideal mixture is in good agreement with data for many systems.

b. Real Liquids

There is no difference, in principle, between mixtures of real liquids to all components of which the same value of A and B applies and the mixtures of the so-called perfect liquids. A new problem arises when the mixture components belong to groups of different A and B values. Since these differ only by small amounts, it is probably safe to assume linear mole fraction additivity for both constants. Then one obtains for the mixture viscosity (rigid molecules)

$$\ln \bar{\eta} = \ln \left[\left(\frac{\bar{M}\bar{E}^\circ}{\bar{A}_w} \right)^{1/2} \frac{l}{d_w} \right] + \frac{\bar{B}\bar{E}_e^\circ}{5cRT} - \bar{A} \tag{18}$$

[63] R. E. Powell, W. E. Rosevaere, and H. Eyring, Ind. Eng. Chem. 33, 430 (1941).
[63a] R. J. Fort and W. R. Moore, Trans. Faraday Soc. 62, 112 (1966).
[64a] J. P. Boon and G. Thomaes, Physica 28, 1074 (1962); 29, 208 (1963).
[64b] G. Thomas and J. van Itterbeek, Mol. Phys. 2, 372 (1959).

A reasonable mixture law for c is $\bar{c} = x_A c_A + x_B c_B$, at least for rigid molecules that differ comparatively little in c. When F^E, which is needed for the estimate of $\bar{E}_e^\circ$ according to equation (17), is not available directly, it can be estimated for many systems from correlations with the molecular structure of the mixture components.[65a,b] With flexible molecules all the additivity rules should use the volume fraction ϕ_i instead of mole fractions as concentration unit:

$$\bar{E}^\circ = \phi_A{}^2 E_A^\circ + 2\phi_A\phi_B(E_A^\circ E_B^\circ)^{1/2} + \phi_B{}^2 E_B^\circ$$

$$\bar{c} = \phi_A c_A + \phi_B c_B, \quad \text{etc.}$$

(19)

Then for mixtures of liquids composed of flexible molecules

$$\ln \bar{\eta} = \ln \frac{(ME_e^\circ)^{1/2}}{N_A \bar{d}_w{}^2} + \frac{\bar{B}\bar{E}^\circ}{5\bar{c}RT} - \bar{A} \qquad (20)$$

Volume fraction additivity should also be expected for mixtures of rigid (especially small rigid) and flexible molecules. A serious difficulty in this case is the difference in viscosity-reducing parameters for the two types of molecules. In view of the dominating effect of the flexible molecule component on the properties of the mixture it seems appropriate to use the reducing system for flexible molecules and, hence, equation (19).

One might summarize the comparatively meager data on the viscosity of mixtures of nonassociating compounds in the high-temperature range by the statement that virtually all effects of molecular structure are described by the (equilibrium) properties of the components and by the excess free energy of mixing.

2. LOW-TEMPERATURE RANGE

The viscosity of liquids at $T^* < 0.6$ depends primarily on the property T_0, for which additivity rules with any theoretical significance are only just beginning to be developed. Most of that work is concerned with the effect of plasticizers on the glass transition temperature of polymers. The present examination of the effects of molecular structure on the low-temperature mixture viscosity of largely nonpolymeric liquids is based on the supposition that the free-volume theory of glass transitions is applicable.

The free-volume treatment of the glass transition temperatures, given by Kanig,[66] may serve as a useful starting point. An advantage of that theory is its provision of a term for an interaction energy (equal to the excess free

[65a] G. J. Pierotti, C. H. Deal, and E. L. Derr, *Ind. Eng. Chem.* **51**, 95 (1959).
[65b] O. Redlich, E. L. Derr, and G. J. Pierotti, *J. Am. Chem. Soc.* **81**, 2283 (1959).
[66] G. Kanig, *Kolloid-Z.* **190**, 1 (1963).

energy of mixing), which permits prediction of minima or maxima in the curve of T_g vs. composition. For the purposes of this discussion we make the assumption that Kanig's additivity law for T_g can be applied to T_0 with only minor modifications. Then one obtains for a binary mixture

$$\bar{T}_0 = \left(\frac{L_1 V_f(A)}{V_f(B)} - \frac{1}{\Phi_f(B)}\right)^{-1} \left[T_0(A)\left(1 - \frac{1}{\Phi_f(B)}\right) - T_0(B)\left(1 - \frac{L_1 V_f(A)}{V_f(B)}\right)\right.$$

$$\left. - \Phi_f(A)f(A_{ij})\right] \qquad (21)$$

Here $V_f = V_g - V_0$, where V_g and V_0 are the molal volumes at T_0 and at $0°K$, respectively. The packing density $\rho_0{}^* = V_w/V_0$ of most organic substances at $0°K$ covers a very narrow range (Table XII), so that V_0 can be estimated easily by using the established data for the van der Waals volume V_w.[9] By definition $V_g = V_T[1 - \alpha_L(T - T_0)]$, where $V(T)$ is the molal volume of the liquid at some temperature T not very far above T_0, α_L is the expansion coefficient of the liquid, L_1 is a constant of the order of -1.5.

$$\Phi_f(A) = \frac{\phi_A}{\phi_A + \phi_B\left|\dfrac{\alpha_L(B) - \alpha_g(B)}{\alpha_L(A) - \alpha_g(A)}\right|}$$

and $\Phi_f(B) = 1 - \Phi_f(A)$, where α_g, and α_L are the expansion coefficients in the glass and liquid states, respectively. Since α_g is rarely known for non-polymeric compounds, it can be approximated by the expansion coefficient of the crystal which, if not known experimentally, can be estimated quite easily.[66a] Finally,

$$f(A_{ij}) = \frac{L_1}{R}(A_A + A_B - 2A_{AB}) \qquad (22)$$

where the "affinities" A_A and A_B are defined as $0.664RT_0$, and the interaction term A_{AB} is a function of the excess free energy of mixing, perhaps of the form

$$A_{AB} = \phi_A{}^2 A_A + 2\phi_A\phi_B(A_A A_B)^{1/2} + \phi_B{}^2 A_B - w''F^E \qquad (23)$$

A comparison of A_i with the characteristic terms of the writer's description of glass transitions[41] leads to the approximate relation $A_i \approx E_i^°/19c$. Hence, for miscible systems for which $F^E/X_A X_B < 2RT$ one expects $w''F^E < 0.1X_A X_B RT$. Attractive interactions between the mixture components will thus increase A_{AB}, whereas repulsive interactions (poor miscibility) will decrease it. The qualitative relations between the molecular–structure–dependent properties of mixture components and their effect on T_0 are apparent from the equations. They show that the curve of T_0 vs. x can have

[66a] A. Bondi, J. Appl. Phys. **37**, 4643 (1966).

TABLE XI

PACKING DENSITY OF VARIOUS ORGANIC SUBSTANCES AT 0°K (CRYSTALLINE STATE) AND AT T_g (AMORPHOUS STATE)

Aliphatic hydrocarbons

	C_2H_4	C_2H_6	C_3H_8	$n\text{-}C_4H_{10}$	$1\text{-}C_4H_8$	$n\text{-}C_5H_{12}$	$i\text{-}C_5H_{12}$	$n\text{-}C_6H_{14}$	2-Me-pent	3-Me-pent	$n\text{-}C_7H_{16}$
ρ_0^*	0.670	0.684	0.695	0.725	—	0.711	0.702	0.722	—	—	0.725
ρ_g^*	—	—	—	—	0.673	—	0.678	—	0.688	0.678	—

	3-Me-hex	2,2-di-Me-pent	$n\text{-}C_8H_{18}$	$n\text{-}C_9H_{20}$	4-Me-non	$n\text{-}C_{62}H_{126}$	$(CH_2)_\infty$	cyclo-C_6H_{12}	PIB
ρ_0^*	—	—	0.735	0.727	—	0.762	0.762	0.726	—
ρ_g^*	0.678	0.689	—	—	0.670	—	—	—	0.680

Aromatic hydrocarbons

	$\varphi\text{-}H$	Me-φ	Et-φ	Polystyrene	n-Pr-φ	1,3,5-Me$_3$-φ	$C_{10}H_8$	$C_{14}H_{10}$	1,3,5$(C_{10}H_7)_3$-φ
ρ_0^*	0.697	0.715	0.713	0.715	0.688	0.736	0.721	0.733	0.73 (est.)
ρ_g^*	—	—	—	0.625	—	—	—	—	0.640

Alcohols

	Me-OH	Et-OH	n-Bu-OH	i-Bu-OH	$n\text{-}C_{12}OH$	Cellulose	φ-CH$_2$OH	φ-OH
ρ_0^*	0.677	0.698	0.720	—	0.736	0.720	0.716	0.690
ρ_g^*	—	0.648	—	0.664	—	—	—	—

a minimum if the components are not very compatible, and that T_0 can go through a maximum if the components associate strongly. These results are not surprising and are in the same direction as the mixture effects on viscosity in the high-temperature region.

The complicated form of equation (21) hides the practically important effect of comparatively small concentrations of substances having large free volume V_f and low T_0, such as liquids composed of flexible molecules, in depressing the T_0 of liquids of rigid molecules having small V_f and high T_0. The dominating effect of the lower T_0 in mixtures of components differing widely in T_0 is apparent from the well-known mixture rule

$$\frac{1}{T_0} = \frac{\phi_A}{T_0(A)} + \frac{\phi_B}{T_0(B)} \tag{24}$$

which is just a special case of equation (21) when the $F^E \approx 0$ and $V_f(A) \approx V_f(B)$.

Although there is a great deal of information on mixture viscosities of ill-defined systems in qualitative agreement with the conclusions arrived at above, very few data have been published on blends of pure compounds,[67] by which equation (23) might be tested experimentally.

3. MIXTURES OF ASSOCIATING SUBSTANCES

There are three kinds of mixtures involving associating substances: solutions of associating substances in nonassociating liquids, mixtures of associating substances with each other, and mixtures whose components combine into liquid (association) complexes.

The viscosity of none of these mixtures can be treated quantitatively at present. A combination of the very extensive association equilibrium measurements of the first type of solution with viscosity measurements might yield a basis for predictive calculations, but the problem is not very interesting, because only monotonical changes of viscosity with concentration and temperature have been observed (and can be expected).

The best-known example of the second type is aqueous alcoholic solution. The peculiar maximum in its curve of η vs. x is in qualitative agreement with thermodynamic data, but no convincing quantitative analysis of the viscosity data has been published.

The most interesting systems are those of the third type, whose mixture components enter into more or less strong chemical (association) compound formation. In this case F^E is unrelated to viscosity, and no attempt need be made to insert it into the mixture equation. The examples given in Fig. 37 show the not infrequent dramatic maxima in the plots of η vs. x of such

---

[67] There is, of course, a natural reluctance to contaminate expensive pure substances by mixing them with others.

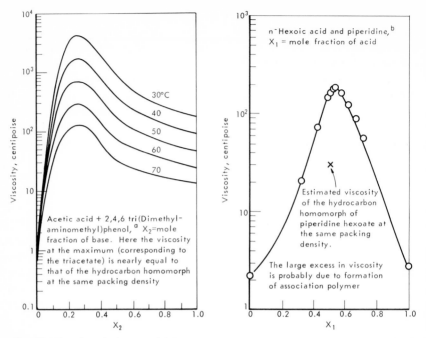

FIG. 37. Viscosity of mixtures with formation of stable compounds. (a) A. Bondi and H. L. Parry, *J. Phys. Chem.* **60**, 1906 (1956). (b) E. B. R. Prideaux and R. N. Coleman, *J. Chem. Soc.* **1936, 1946, 1937**, 4.

systems. Although exact quantitative prediction of the maximal viscosity is quite impossible, one can guess the order of magnitude of the expected maximal viscosity from a consideration of the viscosity of known pure liquids, the molecules of which have the approximate molecular structure of the association complex. In the absence of model compounds one can construct the appropriate relations of η^* vs. T^*, described in Section V, for a compound of the structure of the association complex. The viscosity increase is naturally very large when the association complex is polymeric. Here is an area worthy of further investigation.

4. COMPARISON WITH OTHER ADDITIVITY RULES

The viscosity additivity rule for perfect liquids presented as equation (15) is closely related to the well-known Arrhenius mixture rule

$$\ln \bar{\eta} = \sum x_i \ln \eta_i \qquad (25)$$

if one replaces the geometric mixing rule for E_0, with the arithmetic, and if one ignores the additive constants. A modification of the Arrhenius equation, very similar to equation (17), has been proposed by Powell et al.,[63] in which η was also expressed in terms of the energies of vaporization of the components, augmented by the excess free energy of mixing.

One of the advantages of formulating the viscosity additivity rules in the shapes of equations (18) and (20) is the opportunity to estimate mixture viscosities, even if one component in equilibrium with the solution at the prevailing temperature and pressure exists only in vapor or in solid phase. The older methods, which rely on the availability of viscosity data to estimate blending values of $\bar{\eta}$ are not useful if viscosity data are neither available nor calculable for one of the blend components.

VIII. Comparison with Previous Work

Early in this century various organic chemists developed schemes for relating the viscosity of liquids to molecular structure with the goal of estimating viscosity from group increments or, conversely, of using viscosity data measurement as a means to molecular structure determination. It turned out, however, that viscosity is too insensitive for the latter purpose in the high-temperature range, in which alone consistent schemes could be built. Hence, little use was made of them. Later on, chemical engineers were somewhat more successful because of a better choice of mathematical functions.[68,69] Yet in spite of the introduction of the density as an additional correlating parameter the latter, like the earlier methods, are only interpolation or, at best, extrapolation schemes, because they require viscosity data for some members of the particular family of chemical compounds under consideration. Among the chemical engineering schemes one finds that the Souders–Lagemann[70] relation

$$\log^2 \eta(T) = \frac{a(R) + b}{V(T)} - 2.9 \qquad (26)$$

yields a particularly simple connection with molecular structure, because the molar refractivity (R) and the molal volume V can be built up from structure increments; the homologous group constants a and b, however, must be derived from viscosity data.

The first significant step forward was made by Eyring and his co-workers,[71] who postulated that molecular motion in viscous flow is determined by the rate at which molecules pass the potential barrier formed by their nearest

[68] M. Souders, Jr., J. Am. Chem. Soc. **60**, 154 (1938).

[69] L. H. Thomas, J. Chem. Soc. p. 573 (1946).

[70] R. T. Lagemann, J. Am. Chem. Soc. **67**, 409 (1945).

[71] H. Eyring, J. Chem. Phys. **4**, 283 (1936); R. H. Ewell and H. Eyring, ibid. **5**, 720 (1937).

neighbors. The magnitude of that barrier should be proportional to the energy of vaporization, and the surprisingly simple relation

$$\eta = (h/v)\exp(\Delta E_v/2.45RT) \tag{27}$$

was found to correlate the viscosity of liquids (in the high-temperature range) that were composed of small, essentially rigid molecules to better than within an order of magnitude. This relation contains but a single constant (2.45) derived from viscosity measurements; the dependence of viscosity upon molecular structure is thus directly determined by the well-understood effect of molecular structure on the energy of vaporization and the molecular volume v.

Subsequently many chemists[4,5,72,73] contemplated the effect of molecular-structure variations on the heat $\Delta H_{\ddagger}$ and entropy $\Delta S_{\ddagger}$ of activation in the more detailed Eyring equation

$$\eta = (h/v)\exp(-\Delta S_{\ddagger}/R)\exp(\Delta H_{\ddagger}/RT) \tag{28}$$

The separation of $\Delta H_{\ddagger}$ into the very density-dependent isochore $\Delta H_{\ddagger}{}^j \equiv R(\partial \ln \eta/\partial l/T)_V$ and the remaining $\Delta H_{\ddagger}{}^{h}$ [3,4] showed that equation (27) and Nissan and Grunberg's closely related correlations (viscosity vs. vapor pressure)[5,41] could be valid only in the high-temperature range (low packing density), where $\Delta H_{\ddagger}{}^j \ll \Delta E_v$. A particularly detailed analysis of $\Delta H_{\ddagger}{}^j = f(\rho,$ molecular structure) was made by Eirich and co-workers.[74] These correlations of $\Delta H_{\ddagger}$, $\Delta H_{\ddagger}{}^j$, and $\Delta S_{\ddagger}$ with the molecular structure of many pure compounds gave chemists a feeling for the broadly generalizable viscosity relations. In some instances quantitative relations between the effect of functional groups on ΔE_v and on $\Delta H_{\ddagger}$ could be detected.[4] The present work indicates that the often-considered ratio $\Delta E_v/\Delta H_{\ddagger}$ is approximately $5c/2.3B$, giving a quantitative account of the well-known effects of molecular flexibility. However, in spite of much effort, no consistent quantitative system of viscosity prediction could be achieved. Numerous reasons for this failure of the activated-state theory are now apparent. Closer examination of the nature of transport phenomena in liquids has shown that the high-temperature phenomena could be described in terms of simple molecular properties without recourse to activated states. Those simple relations had been obscured by the earlier preoccupation with activated-state processes.

An analysis by Turnbull and Cohen[75] puts the various treatments into perspective by considering the energetics of the volume redistribution required for creation of the void into which a molecule can move. At a

[72] E. M. Griest, W. Webb, and R. W. Schiessler, *J. Chem. Phys.* **29**, 711 (1958).

[73] E. Kuss, *Z. Angew. Phys.* **7**, 372 (1955).

[74] F. R. Eirich, R. Simha, R. Ullman, *et al.*, A.P.I.-Grant-in-aid No. 6, 1954/57.

[75] D. Turnbull and M. H. Cohen, *J. Chem. Phys.* **34**, 120 (1959).

temperature T substantially above T_g a liquid has the volume

$$V_T = V_0 + V_e + V_f'$$ (29)

whereas at T_g it has

$$V_T = V_g = V_0 + V_e$$ (30)

Almost no energy is required for the redistribution of V_f' to produce the voids permitting molecular transport. However, the expansion volume V_e, available beyond the closely packed zero-point volume V_0, is so small that considerable energy is required for its redistribution into voids of the required size.

In the temperature range in which V_f' is the largest component of the empty volume the molecular transport may be represented by the Doolittle[47] equation,

$$\eta = A \exp(BV_0/V_f)$$ (31)

which, because of $V_0/V_f = f(\varepsilon/kT)$ for simple liquids, at high reduced temperatures[12a,76] can also be written

$$\eta = A \exp[Bf(\varepsilon/kT)]$$ (32)

which is the form employed for the high temperature correlations of this chapter.

The incidence of an exponential temperature dependence of viscosity along an isochore indicates the extent of energy requirement, that is, the height of a potential-energy barrier for volume redistribution or molecule movement, or both even in the temperature region dominated by $V_f'^2$. To that extent it is justifiable to discuss flow as a rate process involving an activated state. With simple substances this potential barrier $\Delta E_\ddagger^j \equiv R(\partial \ln \eta / \partial l/T)_v$ is a function of volume only.[74,77] However, with larger molecules $\Delta E_\ddagger^j$ at a given volume is also a function of temperature, indicative of the need to activate internal degrees of freedom to pass over the potential-energy barrier. The available data are insufficient to establish reliable correlations with molecular structure.

In the temperature range $0.6 > T^* > T_g^*$ in which $V_f' \to 0$, molecular transport is an activated process primarily because of the large amount of energy required for redistribution of the small amount of expansion volume V_e. The notion that in this temperature region viscosity depends only on the volume found its expression in equations (4) to (8), which have given us excellent qualitative guidance but are now known to be quantitatively

[76] J. A. Barker, "Lattice Theories of the Liquid State." Macmillan, New York, 1963.

[77] A. Jobling and A. S. C. Lawrence, *Proc. Roy. Soc.* **A206**, 257 (1951); *J. Chem. Phys.* **20**, 1296 (1952).

inadequate. Their introduction of T_g (as T_0) into the viscosity discussion has proven to be very fruitful, but it is also apparent that an additional parameter is needed, possibly of the $\exp(E/RT)$ variety as suggested by Macedo and Litovitz.[57]

The theoretical justification of their formulation, as in equation (11) and to a lesser extent equation (12), is still under debate.[78,79] But from the practical point of view it should be noted that such equations are useful only if the theory is good enough for the energy term to be supplied from extraneous, preferably equilibrium, measurements. Otherwise one has too many disposable constants for meaningful data fitting. In other words, the inquiry into the detailed nature of low temperature (high packing density) flow is again wide open.

A model differing somewhat from that of Eyring and co-workers has been developed by Panchenkov.[80] In the nomenclature of the present work his viscosity equation reads

$$\eta = 11\left(\frac{RT}{M}\right)^{1/2} \sigma\rho(\rho^*)^{1/3} \exp(\Delta S/R)\exp(\varepsilon_0/RT)[1 - \exp(-\varepsilon_0/RT)]^n \quad (33)$$

where $\Delta S/R$ is dominated by $(\Delta H_v(T) - pV')/RT$ and the pair potential $\varepsilon_0 = 2\Delta H_0^\circ/Z$, ΔH_0° is the heat of vaporization at $0°K$, and Z is the coordination number of the molecule in the liquid state. In earlier papers $n = 2$ was used, but more recently Panchenkov derived $n = 1$. The coordination number Z is treated as a temperature-dependent disposable constant.[81] The joint action of ΔS, which tends to zero at $T = T_c$, and of the term in square brackets appears to yield a reasonably good representation of the sharp downward trend of the viscosity in the region between T_b and T_c. No other simple equation has so far been presented which will do that. The sharp upward sweep of ΔS when hindered external rotation becomes important (as discussed in Section II,1,c) may well describe the viscosity behavior in the low-temperature region. In the case of long-chain n-paraffin hydrocarbons it is likely to predict too steep a viscosity–temperature curve. No published test of the ability of the Panchenkov equation to describe the viscosity in the low-temperature region could be found.

[78] S. B. Brummer, J. Chem. Phys. **42**, 4316 (1965).

[79] H. S. Chung, J. Chem. Phys. **44**, 1362 (1966).

[80] G. M. Panchenkov, Izv. Akad. Nauk SSSR, Ser. Fiz. **63**, 701 (1948); Zh. Fiz. Khim. **24**, 1390 (1950).

[81] The form of the temperature function proposed by Panchenkov and Erchenkov,[82] $z = A(T_c - T)$, is not only wrong, because $z > 0$ at T_c and is usually about 3 or 4,[83] but represents their own data less well than a function that would permit $z > 0$ at $T = T_c$.

[82] G. M. Panchenkov and V. V. Erchenkov, Russ. J. Phys. Chem. **36**, 455 (1962).

[83] R. F. Kruh, Chem. Rev. **62**, 319 (1962).

IX. Conclusions

Within the framework of a corresponding states system one finds that the viscosity of liquids can be examined most easily if one confines comparisons to specific regions. For convenience, we divide the liquid range into three regions: the near-critical region, between about the atmospheric boiling point T_b and the critical temperature, the "high-temperature" region, between about $0.6\theta_L$ and $1.0\theta_L$, and the "low-temperature" region, between the glass transition temperature and $0.6\theta_L$, where θ_L is the reducing temperature $E°/5cR$.

The near-critical region is uninteresting from the point of view of molecular-structure correlations, because in appropriately reduced coordinates there seem to be only small effects of molecular structure, suggesting that, at least for nonassociating substances, viscosity in the near-critical region is determined primarily by the magnitude of molecule dimensions and of intermolecular force constants. Even the effect of association is small.

The viscosity in the high-temperature region is also dominated by molecule dimension and intermolecular force constants, but even in reduced coordinates significant effects of molecular structure can be observed. The largest difference, between rigid and flexible molecules, can be handled quantitatively through the use of the number of external degrees of freedom excited in the liquid state, an independently determined molecular property. Smaller specific structure effects exist within these two groups but cannot yet be quantified in terms of molecule properties.

The viscosity in the low-temperature region, although still affected by molecule dimension, intermolecular force and the number of external degrees of freedom per molecule, is dominated by what may be loosely called the "equilibrium glass transition temperature," T_0. Even in the reduced form $5cRT_0/E°$ this temperature parameter covers a range and can be related only very approximately to currently known structure parameters, primarily the number of external degrees of freedom per unit volume.

Strongly associating, especially hydrogen-bonding, substances exhibit viscosity characteristics that can be related to molecular structure only in a still somewhat approximate manner. Their quantitative understanding will have to wait for better data on the hydrogen-bonding equilibrium constants in pure liquids than are available now.

The relation between the viscosity of mixtures and the molecular structure of their components appears to be broadly understood and can be treated reasonably quantitatively. That treatment involves the correction of the interaction potential by means of the excess free energy of mixing, F^E. Hence, mixture viscosity may be predictable to the extent to which the thermodynamic properties of the mixture can be predicted from molecular-structure information.

Nomenclature

a	Empirical constant	P^*	$5(P + P_i)V_w/E°$, reduced pressure
a_F	Empirical constant	r	Intermolecular distance
A	Dimensionless intercept of η^* vs. T^* relation	r_w	van der Waals' radius of an atom
A_F	Dimensionless intercept of η_F^* vs. T^* relation	R	Gas constant
		R	Intermolecular separation
A_w	Surface area per mole of molecules	ΔS_v	Entropy of vaporization
b	Empirical constant	$\Delta S_v{}^i$	Excess entropy of vaporization
b_F	Empirical constant	T	Temperature
B	Dimensionless slope constant of η^* vs. T^* relation	T^*	kT/ε for simple spherical molecules
		T^*	$5cRT/E°$ for polyatomic molecules
B_F	Dimensionless slope constant of η_F^* vs. T^* relation	T_g	Glass transition temperature
		$T_g{}^*$	$5cRT_g/E°$
B_v	Dimensionless slope constant of η^* vs. T^* at constant volume	T_0	Effective glass transition temperature
		$T_0{}^*$	$5cRT_0/E°$
c	One third of number of external degrees of freedom per molecule	T_b	Atmospheric boiling point (°K)
		T_c	Critical temperature
C_V	Heat capacity at constant volume	T_R	T/T_c
d	Empirical constant	v	All lower case v in equations (4), (5), et seq. are liquid, etc., volume per molecule
d_F	Empirical constant		
d_w	Molecule thickness		
$\overline{d_w}$	Average molecule thickness	V	Molal volume at T
$E°$	Standard energy of vaporization, $\Delta H_v - RT$, at temperature at which $\rho^* = 0.588$	$V°$	Barrier to internal rotation
		V_w	van der Waals' volume
		V_0	Molal volume at 0°K
		V_f	Free volume
ΔE_{iso}	Energy of rotational isomerization	V_g	Molal volume at $T = T_g$
$\Delta E_{\ddagger}$	$R(\partial \ln \eta/\partial T^{-1})_p$	w	Width of a molecule
$\Delta E_{\ddagger}{}^j$	$R(\partial \ln \eta/\partial T^{-1})_v$	w_i	Weight fraction of component i
f	Fractional free volume	x_i	Mole fraction of component i
F^E	Excess free energy of mixing	Z_c	Entanglement chain length
g	Empirical constant	α	Cubic thermal expansion coefficient
g_F	Empirical constant	α_j	Cubic thermal expansion coefficient at $T < T_g$
$g(R)$	Radial distribution function		
ΔH_v	Heat of vaporization	α_L	Cubic thermal expansion coefficient at $T > T_g$
I_i	Principal moment of inertia of a molecule around axis i		
		β	Compressibility
k	Boltzmann constant	ε	Pair potential between two molecules
L	Length of outstretched molecule	η	Viscosity
m	Mass per molecule	η^*	$\eta\sigma^2/(m\varepsilon)^{1/2}$
M	Molal weight	$\eta_s{}^*$	$\eta d_w(A_w/mE°)^{1/2}$
n	Number of skeletal atoms per repeating unit in polymer	$\eta_F{}^*$	$\eta N_A \overline{\alpha_w}^2/(ME°)^{1/2}$
		θ_L	$E°/5cR$
N_A	Avogadro number	ξ	$T_c^{1/6}/M^{1/2}P_c^{2/3}$ = viscosity reducing parameter[25]
N_c	Number of carbon atoms per molecule		
N_s	Number of rigid skeletal units per molecule	ρ	Density
		ρ^*	V_w/V
P	Pressure (external)	$\rho_0{}^*$	V_w/V_0
P_i	$(\partial E/\partial V)_T$, internal pressure	$\rho_g{}^*$	V_w/V_g

σ Distance between molecules at steepest descent of repulsion branch of potential well

τ_e Rotational relaxation time of an electric dipole

$\tau_i{}^*$ $\tau_e(E^\circ/I_i)^{1/2}$, where the index i refers to the common indices A, B, C of the appropriate principal moment of inertia

τ_{ABC}^* $\tau_i{}^*$ for which $I_i = (I_A \cdot I_B \cdot I_C)^{1/3}$

CHAPTER 2

THE MICRORHEOLOGY OF DISPERSIONS

H. L. Goldsmith and S. G. Mason

I. Scope

The general problem of microrheology is the prediction of the macroscopic rheological properties of a material from a detailed description of the behavior of the elements of which it is composed. In the present chapter the materials considered are suspensions of rigid and deformable spheres and spheroids, the elements are the elementary volumes of Newtonian fluids each containing a particle. The behavior of these elements as seen under the microscope, first in isolation from each other and then in interaction, in dilute and concentrated suspensions undergoing laminar shear flow will be described. Throughout the chapter it has been the aim of the authors to relate the observed motions at the microrheological level to the macroscopic rheological properties of dilute and concentrated suspensions.

With the exception of Section VII, which deals with the effects of inertia on suspension behavior, the m   erial relates to the motion of viscous, incompressible liquids at small   rticle Reynolds numbers when the term $\rho \mathbf{V} \cdot \nabla \mathbf{V}$ in the Navier–Stokes ϵ ıtion can be neglected, leaving a linear equation

$$\rho \frac{\partial \mathbf{V}}{\partial t} + \nabla p = \eta \nabla^2 \mathbf{V} \tag{I1}$$

with the continuity equation for incompressible fluids

$$\nabla \cdot \mathbf{V} = 0 \tag{I2}$$

where η is the fluid viscosity, p the viscous pressure, and $\mathbf{V}$ the velocity field. Equation (I1), however, contains the local acceleration term $\rho \, \partial \mathbf{V}/\partial t$, which also either vanishes (for steady-state motion) or is of the same order of smallness as $\rho \mathbf{V} \cdot \nabla \mathbf{V}$ (quasistatic motion),[1] leaving

$$\nabla p = \eta \nabla^2 \mathbf{V} \tag{I3}$$

Throughout Sections II to VI the validity of equations (I2) and (I3), known as the creeping-flow equations, is assumed. Furthermore, in all systems studied the suspended particles were large enough for Brownian motion to be negligible.

Wherever possible, the experimental results are discussed in the light of available theory. In some cases, as in those of rigid spheres, spheroids, and

[1] H. Brenner, *Chem. Eng. Sci.* **18**, 2 (1963).

deformable drops, when interactions are neglected, particle behavior has been worked out in detail by hydrodynamic theory. Thus, the motion of a small ellipsoidal particle in a uniformly sheared viscous fluid at low Reynolds number was studied by Jeffery,[2] who derived equations for the torque acting on the particle suspended in a liquid undergoing flow with arbitrary components of rotation and dilatation. In this way equations were obtained for the time dependence of the orientation of ellipsoids of revolution (prolate and oblate spheroids) when no external torque acts. These equations have provided the starting point of a number of investigations of particle orientation, of which those in Couette and hyperbolic flow are described in Section II.

Jeffery's theory can be extended to the case in which the spheroid is subject to an external torque. Equations for the motion of ellipsoids in a combined shear and electric field have been derived[3-5] and were tested by applying an external electric field to a particle rotating in Couette flow.[4,5]

A theoretical treatment of the system of velocity gradients established by internal circulation inside a fluid drop suspended in a liquid undergoing shear flow has been given by Taylor.[6] Calculations were also made[7] of the deformation of the drop associated with the fluid stresses generated by the flow. Although the theory is applicable only to very small deformations, the investigations of deformable drops in hyperbolic and Couette flow described in Section III included observations at higher shear rates, at which bursting of drops occurred.

The analogous situation of drop deformation in an electric field, due to the existence of electrical stresses in the suspending fluid, has also been treated theoretically[8-10] and compared with experiment, as described in Section III.

The above-mentioned theories, which apply to a uniform shear field as in Couette and plane hyperbolic flow, have also been tested in Poiseuille flow,[11] in which the components of the fluid rate of shear are not constant in space (Section IV). It was found that, providing the particle size is small compared to that of the tube, the rotation of rigid spheres and spheroids and the deformation of drops is the same as that found in Couette flow. However, deformable liquid drops migrated across the planes of shear toward the tube axis,

[2] G. B. Jeffery, *Proc. Roy. Soc.* **A102,** 161 (1922).
[3] S. T. Demetriades, *J. Chem. Phys.* **29,** 1054 (1958).
[4] R. S. Allan and S. G. Mason, *Proc. Roy. Soc.* **A267,** 62 (1962).
[5] C. E. Chaffey and S. G. Mason, *J. Colloid Sci.* **19,** 525 (1964).
[6] G. I. Taylor, *Proc. Roy. Soc.* **A138,** 41 (1932).
[7] G. I. Taylor, *Proc. Roy. Soc.* **A146,** 501 (1934).
[8] C. G. Garton and Z. Krasucki, *Proc. Roy. Soc.* **A280,** 211 (1963).
[9] C. T. O'Konski and H. C. Thacher, *J. Phys. Chem.* **57,** 955 (1953).
[10] G. I. Taylor, *Proc. Roy. Soc.* **A291,** 159 (1965).
[11] H. L. Goldsmith and S. G. Mason, *J. Colloid Sci.* **17,** 448 (1962).

whereas at these negligibly small particle Reynolds numbers rigid particles translated along paths parallel to the tube axis. This effect can be explained by taking into account both the drop deformation and the presence of the wall. Drop migration away from the wall of a Couette apparatus was also observed.

The interactions between the particles in dilute and concentrated suspensions of rigid spheres and spheroids in Couette and Poiseuille flow are described in Section V. Here there exists no theory based on the creeping-flow equations that can predict the translational and rotational motions of the components of the disperse phase while they interact and collide with each other. By using the experimental results, however, it was possible to construct a semiempirical and geometrical theory of flow behavior. Some of the observations made in dilute and concentrated suspensions are of interest in connection with measurements of suspension viscosity; this is separately discussed in Section VI.

In most of the experiments described in Sections II to V reversibility in particle rotational and translational displacements was found; that is, if the flow was reversed, the motions were the reverse in time of those which were observed in the forward flow, as if a cinefilm of the sequence were run backwards. Such time-reversed flows occurred not only in the simple case of the periodic motions of single prolate and oblate spheroids, considered by Bretherton,[12] but also in the apparently very complex case of the continually changing configurations of particles in flowing concentrated suspensions. A formalized theoretical basis of time-reversed flows has been given by Slattery[13] and is based on the linearized Stokes–Navier equation, equation (I3), which requires that flow reversal and time reversal should have the same effect.

Section VII deals with the flow of suspensions in the transition region, where the particle Reynolds numbers, although still small, are not negligible, and where inertial effects are present. Of particular interest is the radial migration of rigid spheres and spheroids observed in suspensions undergoing steady[14] and oscillatory[15] flow in a circular tube.

Finally, the application of some of the described phenomena to the rheology of blood, with special reference to the flow behavior of the mammalian red blood cell, is briefly considered.

[12] F. P. Bretherton, *J. Fluid Mech.* **14**, 284 (1962).

[13] J. C. Slattery, *J. Fluid Mech.* **19**, 625 (1964).

[14] G. Segré and A. Silberberg, *J. Fluid Mech.* **14**, 136 (1962).

[15] B. Shizgal, H. L. Goldsmith, and S. G. Mason, *Can. J. Chem. Eng.* **43**, 97 (1965).

II. Rotation of Rigid Spheroids in Shear Flow

1. GENERAL

Since Jeffery's theoretical investigation into the motion of ellipsoids in shear flow the behavior of rigid particles suspended in a viscous fluid undergoing slow shear flow has been studied from a more general point of view by several authors. Giesekus[16] has classified the various types of shear flow, discussing the conditions for the existence of steady orientations of particles with ellipsoidal symmetry, and has also developed an apparatus[17] for producing two-dimensional shear flows with rotational and dilatational components, which may be continuously varied relative to each other and which has been used to confirm the general theory. Brenner[18] has presented a method of calculating linear and angular velocities of an arbitrary particle suspended in shear flow in the creeping-flow regime from a knowledge of the force and torque acting on it. This method was used[19] in calculating the angular velocities of spheroids in special shear flows, for which the vorticity vector is parallel to one of the principal axes of dilatation. With increasing rotational parts such flows range from irrotational rectangular and three-dimensional hyperbolic flow, through hyperbolic flow, Couette flow and elliptic flow, to pure rotation.

In the following we shall first discuss the behavior of spheroids in liquids undergoing Couette and hyperbolic shear flow in the absence of an external torque. The calculations are then extended to the case in which an external torque, supplied by an electric field arbitrarily oriented with respect to Couette flow, acts on the particles.

2. EQUATIONS OF MOTION

Jeffery[2] related the hydrodynamic torque resulting from fluid stresses on the surface of a rigid ellipsoid to the spins about its axes. He assumed creeping motion of the suspending fluid (which is Newtonian, having viscosity η_0, and is incompressible), no slip at the particle–liquid interface, and that the ellipsoid is nonsedimenting and has its center at a point in the fluid where its translational velocity is zero, Brownian motion being absent. The three principal semiaxes of the ellipsoid, b_1, b_2, b_3, were chosen to be at any time coincident with the Cartesian coordinate system $X_1{}^0$, $X_2{}^0$, $X_3{}^0$, so that $(X_1{}^0)^2/b_1{}^2 + (X_2{}^0)^2/b_2{}^2 + (X_3{}^0)^2/b_3{}^2 = 1$.

The first component of the hydrodynamic torque Γ derived by Jeffery

[16] H. Giesekus, *Rheol. Acta* **2**, 101 (1962).

[17] H. Giesekus, *Rheol. Acta* **2**, 112 (1962).

[18] H. Brenner, *Chem. Eng. Sci.* **19**, 631 (1964).

[19] C. E. Chaffey, M. Takano, and S. G. Mason, *Can. J. Phys.* **43**, 1269 (1965).

may be written as[20]

$$\Gamma_1 = \frac{16\pi\eta_0(b_2{}^2 + b_3{}^2)}{3(b_2{}^2\alpha_2 + b_3{}^2\alpha_3)}(B_1 D_{23}^0 + Z_1{}^0 - \omega_1{}^0) \tag{II1}$$

where D_{23}^0 and $Z_1{}^0$ are the respective components of dilatation (or rate of shear) and rotation of the fluid referred to the $X_1{}^0$, $X_2{}^0$, $X_3{}^0$ coordinate system given by

$$D_{11}^0 = \frac{\partial V_1}{\partial X_1{}^0}, \quad D_{23}^0 = \frac{1}{2}\left(\frac{\partial V_3}{\partial X_2{}^0} + \frac{\partial V_2}{\partial X_3{}^0}\right), \quad Z_1{}^0 = \frac{1}{2}\left(\frac{\partial V_3}{\partial X_2{}^0} - \frac{\partial V_2}{\partial X_3{}^0}\right), \quad \text{etc.}$$

where V_1, V_2, and V_3 are the components of fluid velocity; $\omega_1{}^0$ is the spin of the ellipsoid about the $X_1{}^0$ axis, and the constants α_1, B_1, etc., characteristic of the ellipsoid, are defined by

$$\alpha_1 = \int_0^\infty (b_1{}^2 + \zeta)^{-3/2}(b_2{}^2 + \zeta)^{-1/2}(b_3{}^2 + \zeta)^{-1/2}\, d\zeta \tag{II2}$$

$$B_1 = \frac{(b_2{}^2 - b_3{}^2)}{(b_2{}^2 + b_3{}^2)} \tag{II3}$$

If the ellipsoid is free from external forces, the resultant torque must vanish, and equation (II1) reduces to

$$\omega_1{}^0 = Z_1{}^0 + B_1 D_{23}^0, \quad \text{etc.} \tag{II4}$$

These general equations of motion of the ellipsoid have been given by Giesekus[16] and Bretherton[12] in tensor form, in which B_1, B_2, B_3 are components of a third-rank tensor **B**, such that

$$\boldsymbol{\omega} = \mathbf{Z} + \tfrac{1}{2}\mathbf{B}:\mathbf{D}$$

The interpretation of **B** has been discussed by Bretherton[12] and by Brenner,[18] who derived the analogue of equation (II1) for an arbitrary particle. Solutions to equation (II4) for various shear flows have been obtained for the case of an ellipsoid of revolution having principal semiaxes b_1 and $b_2 = b_3$, when the ellipsoid integrals reduce to

$$\alpha_1 = \frac{2}{b_1 b_2{}^2} - 2\alpha_2, \quad \alpha_2 = \alpha_3 \tag{II5}$$

[20] Vector formulas are written out by giving the first component in full. The second and third components are obtained by cyclic permutation of indices.

and

$$B_1 = 0,$$

$$B_2 = -B_3 = B = \frac{b_1^2 - b_2^2}{b_1^2 + b_2^2} = \frac{r_p^2 - 1}{r_p^2 + 1} \tag{II6}$$

where $r_p = b_1/b_2$ is the particle axis ratio, which is > 1 for a prolate spheroid and < 1 for an oblate spheroid.

In order that the quantities in equation (II4) may be expressed in terms of the fixed coordinate system X_1, X_2, X_3, in which the components of fluid rotation and dilatation are Z_1 and D_{23}, respectively, the spherical polar coordinates ϕ_1 and θ_1 having X_1 as the polar axis are used. These are, respectively, the angles between the $X_1 X_2$ and $X_1 X_1^0$ planes and the X_1^0 and X_1 axes, as illustrated in Fig. 1.[21] It can then be shown[2,19] that

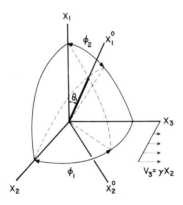

FIG. 1. The relative orientations of the fixed coordinate system X_1, X_2, X_3 of the external flow field, in this case Couette flow, and the particle coordinate system X_1^0, X_2^0, X_3^0 so chosen that the X_1^0 axis coincides with the spheroid principal semiaxis b_1 (heavy line), and the X_2^0 axis lies in the $X_1^0 X_1$ plane. The coordinates ϕ_1, θ_1 are the polar coordinates with respect to X_1 as the polar axis, and ϕ_2 is the coordinate having X_2 as the polar axis.

$$\dot\phi_1 = -\omega_2^0 \operatorname{cosec} \theta_1, \qquad \dot\theta_1 = \omega_3^0 \tag{II7}$$

and, with the aid of equations (II4) and (II7) and by transformation of coordinates, that the general equations for ϕ_1 and θ_1 are[19]

[21] The numerical subscripts 1, 2, 3, are used (i) for components of vectors and tensors, (ii) for quantities related to the three different axes, such as ϕ_1, ϕ_2, and ϕ_3, for the polar coordinates referred to the respective polar axes X_1, X_2, and X_3, and (iii) for the principal axes of the ellipsoid b_1, b_2, and b_3.

$$\dot{\phi}_1 = Z_1 - Z_2 \cos \phi_1 \cot \theta_1 - Z_3 \sin \phi_1 \cot \theta_1$$

$$+ B[-D_{12} \sin \phi_1 \cot \theta_1 + D_{23} \cos 2\phi_1 + D_{31} \cos \phi_1 \cot \theta_1$$

$$-\tfrac{1}{2}(D_{22} - D_{33}) \sin 2\phi_1] \tag{II8}$$

$$\dot{\theta}_1 = -Z_2 \sin \phi_1 + Z_3 \cos \phi_1$$

$$+ B[D_{12} \cos \phi_1 \cos 2\theta_1 + \tfrac{1}{2}D_{23} \sin 2\phi_1 \sin 2\theta_1 + D_{31} \sin \phi_1 \cos 2\theta_1$$

$$+\tfrac{1}{4}(D_{22} - D_{33}) \cos 2\phi_1 \sin 2\theta_1 + \tfrac{3}{4}(D_{22} + D_{33}) \sin 2\theta_1] \tag{II9}$$

The differential equation of the particle orbit, obtained by eliminating dt from equations (II8) and (II9), cannot be integrated readily, in general. The time dependence of the orientations ϕ_1 and θ_1 of bodies of revolution may, however, be obtained in the special case of shear flows in which the polar or X_1 axis is parallel to the fluid vorticity and one principal axis of dilatation, $Z_2, Z_3 = 0$.[16,19] Such flows include a number of cases of practical interest that have been investigated, three of which are considered below.

3. COUETTE FLOW

In the case of the two-dimensional flow between parallel plates moving in opposite directions, shown in Fig. 2(a), the field of flow is defined by

$$V_1, V_2 = 0, \qquad V_3 = \gamma X_2 \tag{II10}$$

where V_1, V_2, and V_3 are the respective components of fluid motion along the X_1, X_2, and X_3 axes and γ is the rate of shear (or velocity gradient), which is constant. The flow is rotational, and the nonzero components of rotation and dilatation are

$$Z_1 = \tfrac{1}{2}\gamma, \qquad D_{23} = D_{32} = \tfrac{1}{2}\gamma$$

Fields of laminar Couette flow have been realized experimentally by means of a parallel-band apparatus[6] or a double coaxial cylinder device,[22] such as that sketched in Fig. 2(b), in which observations are made in a direction parallel to the planes of shear, that is, along the X_1 axis of the field of flow defined in equation (II10). The suspension of particles in a viscous medium is placed in the annular gap between the two concentric cylinders, which are rotated in opposite directions and whose speeds may be continuously varied. It is thus possible to bring the stationary layer into coincidence with the center of a particle selected for viewing in the field of the microscope, thereby arresting its translational motion. Similar devices, in which the

---

[22] S. G. Mason and W. Bartok, in "Rheology of Disperse Systems" (C. C. Mill, ed.), Chap. 2. Macmillan (Pergamon), New York, 1959.

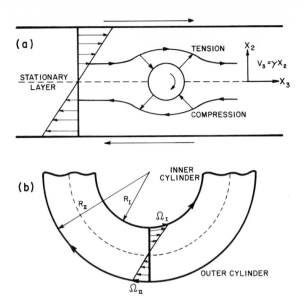

FIG. 2. (a) Diagram of laminar Couette flow of a Newtonian fluid between two parallel plates moving in opposite directions. A rigid sphere rotating in the stationary layer and the principal axes of the fluid rate of deformation are shown, the latter are oriented at $\pm 45°$ to the X_2 axis, producing tension and compression in alternate quadrants. This is further considered in Section III in connection with the deformation in flow of fluid drops. (b) Schematic representation of a counterrotating cylinder device used for observing particle motion in Couette flow. The suspension is placed in the annulus between the outer and inner cylinders, and the motions are observed in a microscope, which may be aligned along either the X_1 or the X_2 axis.

particle motions are viewed from a direction normal to the planes of shear (i.e., along the X_2 axis) have also been described.[23–26]

The dimensions in these devices are such that the variation in the rate of shear across the annulus cannot be ignored. The velocity profile $V_3(R)$ $= \Omega(R)R$, where $\Omega(R)$ is the angular velocity of an elementary annulus of fluid at a radial distance R from the center of rotation of the apparatus and is given by

$$\frac{\Omega(R) - \Omega_{\mathrm{I}}}{\Omega_{\mathrm{I}} - \Omega_{\mathrm{II}}} = \frac{R_{\mathrm{II}}^2}{R_{\mathrm{II}}^2 - R_{\mathrm{I}}^2}\left(1 - \frac{R_{\mathrm{I}}^2}{R^2}\right) \tag{II11}$$

[23] E. Anczurowski and S. G. Mason, *J. Colloid Interface Sci.* **23**, 522 (1967).
[24] E. Anczurowski, Ph.D. thesis, McGill University, Montreal, Canada, 1966.
[25] C. L. Darabaner, J. K. Raasch, and S. G. Mason, *Can. J. Chem. Eng.* **45**, 3 (1967).
[26] B. J. Trevelyan and S. G. Mason, *J. Colloid Sci.* **6**, 354 (1951).

Ω_I, Ω_{II}, R_I, and R_{II} being the respective angular velocities and radii of the inner and outer cylinder. Hence the rate of shear is

$$\gamma(R) = R\frac{d\Omega(R)}{dR}$$

$$= \frac{2}{R^2}\frac{(\Omega_I + \Omega_{II})R_I{}^2 R_{II}^2}{(R_{II}^2 - R_I{}^2)}$$

Thus, $\gamma(R)$ decreases with increasing R, and at the stationary layer $R = R^*$ has the value

$$\gamma(R^*) = \frac{2(R_I{}^2\Omega_I - R_{II}^2\Omega_{II})}{(R_{II}^2 - R_I{}^2)} \tag{II12}$$

By inserting $Z_1 = \gamma/2$ and $D_{23} = \gamma/2$ into (II8) and (II9) one obtains the equations for the angular rotations of the axis of revolution or axis of symmetry (the b_1 axis) of the spheroid

$$\dot{\phi}_1 = \frac{\gamma}{r_p{}^2 + 1}(r_p{}^2\cos^2\phi_1 + \sin^2\phi_1) \tag{II13}$$

$$\dot{\theta}_1 = \frac{\gamma(r_p{}^2 - 1)}{4(r_p{}^2 + 1)}\sin 2\phi_1 \sin 2\theta_1 \tag{II14}$$

which is the result first obtained by Jeffery.[2] At the same time the spheroid undergoes spin about its axis of symmetry which, from equation (II4), is

$$\omega_1{}^0 = \tfrac{1}{2}\gamma\cos\theta_1 \tag{II15}$$

a. Spheres

For the simple case of a sphere, $r_p = 1$, equations (II13) and (II14) reduce to

$$\dot{\theta}_1 = 0, \qquad \dot{\phi}_1 = \omega_1{}^0 = \gamma/2 \tag{II16}$$

That is, the particle spins with constant angular velocity $\omega_1{}^0$, and its period of rotation T is

$$T = 4\pi/\gamma \tag{II17}$$

Equation (II17) has been verified experimentally with glass and plastic[26-28] spheres 50 to 300 microns in diameter by visually timing the rotation of reference marks on the particles. The results agree with theory within the limit of error of 1 : 2000 of measuring γ.

[27] W. Bartok and S. G. Mason, *J. Colloid Sci.* **12**, 243 (1957).
[28] R. St. J. Manley and S. G. Mason, *J. Colloid Sci.* **7**, 354 (1952).

b. *Prolate and Oblate Spheroids*

Integration of equation (II13) shows the rotation of the axis of revolution of the spheroid to be periodic,

$$\tan \phi_1 = r_p \tan \frac{\gamma t}{r_p + 1/r_p} \qquad (II18a)$$

with a period of rotation

$$T = (2\pi/\gamma)(r_p + 1/r_p) \qquad (II19)$$

whence

$$\tan \phi_1 = r_p \tan (2\pi t/T) \qquad (II18b)$$

According to equation (II13), $\dot{\phi}_1$ for a spheroid has maxima at $\phi_1 = 0, \pi$, or $\phi_1 = \pi/2, 3\pi/2$, according as the spheroid is prolate ($r_p > 1$) or oblate ($r_p < 1$). The corresponding minima of $\dot{\phi}_1$, which are not zero, occur at $\phi_1 = \pi/2, 3\pi/2$, or $\phi_1 = 0, \pi$. Thus, the angular velocities of two spheroids, one prolate and the other oblate, whose r_p are reciprocals are identical when their ϕ_1 orientations are $\pi/2$ apart (cf. Fig. 5); the periods of rotation given by equation (II19) are identical. Dividing equations (II13) and (II14) and integrating yields

$$\tan \theta_1 = \frac{Cr_p}{(r_p^2 \cos^2 \phi_1 + \sin^2 \phi_1)^{1/2}} \qquad (II20)$$

showing that the ends of the axis of revolution describe a pair of symmetrical spherical ellipses whose eccentricity is defined by the integration constant C and whose principal axes for any spheroid are $\tan^{-1} Cr_p$ at $\phi_1 = \pi/2$, $3\pi/2$ and are $\tan^{-1} C$ at $\phi_1 = 0, \pi$.

The constant of integration defined as the orbit constant may take on values between the limits 0 ($\theta_1 = 0$ at all ϕ_1), corresponding to steady spin of the axis of revolution about the X_1 axis at a rate $\omega_1{}^0 = \gamma/2$, and ∞ ($\theta_1 = \pi/2$ at all ϕ_1), corresponding to variable $\dot{\phi}_1$ without axial spin.

At intermediate C it can be shown from equations (II15), (II18b), and (II20) that the number of complete axial spins $\mathfrak{n}$ executed by the spheroid during one complete rotation is[23,29]

$$\mathfrak{n} = \frac{r_p^2 + 1}{\pi r_p (C^2 + 1)^{1/2}} K \left[\left(\frac{C^2(1 - r_p^2)}{C^2 + 1} \right)^{1/2} \right], \qquad \text{for} \quad r_p < 1$$

$$\mathfrak{n} = \frac{r_p^2 + 1}{\pi r_p (C^2 r_p^2 + 1)^{1/2}} K \left[\left(\frac{C^2(r_p^2 - 1)}{C^2 r_p^2 + 1} \right)^{1/2} \right], \qquad \text{for} \quad r_p > 1 \qquad (II21)$$

[29] O. L. Forgacs and S. G. Mason, *J. Colloid Sci.* **14**, 457 (1959).

where

$$K(k) = \int_0^{\pi/2} d\zeta/(1 - k^2 \sin^2 \zeta)^{1/2}$$

is the complete elliptic integral of the first kind.

Finally, it may be shown from equations (II18) and (II20) and the transformation equations

$$\tan \phi_2 = \tan \theta_1 \sin \phi_1, \qquad \tan \phi_3 = \tan \theta_1 \cos \phi_1$$

that the equations of angular motion of the axis of revolution with the X_2 and X_3 axes as the respective polar axes are

$$\tan \phi_2 = Cr_p \sin(2\pi t/T) \tag{II22}$$

$$\tan \phi_3 = C \cos(2\pi t/T) \tag{II23}$$

As will become evident, these equations are useful in describing particle orientations.

For experimental purposes it has proven difficult to make small regular ellipsoids of revolution; instead, rods prepared by microtoming continuous filaments of glass, Dacron, nylon, and Orlon,[26,27,29,30] and discs made by compressing plastic spheres between the heated platens of a hydraulic press[31] have been used as models of prolate and oblate spheroids, respectively.

It has, however, been shown[24] that a prolate spheroid $r_p = 1.8$ made by polymerizing a liquid droplet deformed in an electric field[8,24] has the period of rotation predicted by equation (II19) and that the variation of ϕ_1 with time is given by equation (II18).

c. Rods and Discs

In observing a suspension of rods or discs in Couette flow the particle motions, although clearly periodic, at first sight appear rather complex. Without the aid of Jeffery's theory,[2] which shows that the movement of the axis of revolution is of a general character, somewhat like the motion of a precessing top, it would have been difficult to give a rational description of the orbits of rods and discs.

A series of investigations have shown that the axial spin[29,31] and rotation[5,26–29,31] of rigid rods and discs of varying axis ratios at very low Reynolds numbers are in excellent agreement with the predictions of equations (II18) to (II22), *provided* that an "equivalent ellipsoidal" axis

[30] S. G. Mason and R. St. J. Manley, *Proc. Roy. Soc.* **A238,** 117 (1956).
[31] H. L. Goldsmith and S. G. Mason, *J. Fluid Mech.* **12,** 88 (1962).

ratio r_e, calculated from the measured period of rotation T and equation (II19), is used instead of the particle axis ratio r_p. Thus, all aspects of Jeffery's theory for Couette flow have been confirmed experimentally.

(1) *Rotation observed in the X_2X_3 plane.* When viewed along the X_1 axis, the axis of symmetry of a rod, which is coincident with the particle major axis, describes a complete rotation (Figs. 3 and 4a) with $\dot\phi_1$ a maximum at $\phi_1 = 0$ and π and a minimum at $\phi_1 = \pi/2$ and $3\pi/2$. The X_2X_3 projection of the ends of the rod describes an ellipse, the major axis of which lies on the X_3 axis [Fig. 3(b)] and has an axis ratio of $(1 + C^2)^{1/2}/(1 + C^2r_e^2)^{1/2} > 1$. The projection of unit length of the axis of revolution of the rod, l_{23}, is

$$l_{23} = \frac{b_1{}'(\phi_1)}{b_1} = \sin\theta_1 \qquad (II24)$$

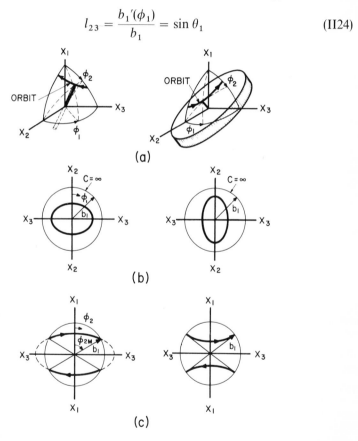

(a)

(b)

(c)

FIG. 3. Spherical elliptical orbits of rods (left) and discs (right). (a) Schematic representation of the orbit of the particle axis of revolution. (b) X_2X_3 projection of the orbit shown in (a); a plane ellipse with major axis along X_3 axis ($r_p > 1$) and X_2 axis ($r_p < 1$). Also shown is the orbit projection for $C = \infty$ ($\theta_1 = \pi/2$), which is a circle of radius b_1. (c) X_1X_3 projection of the orbits giving arcs of a plane ellipse ($r_p > 1$) and hyperbola ($r_p < 1$).

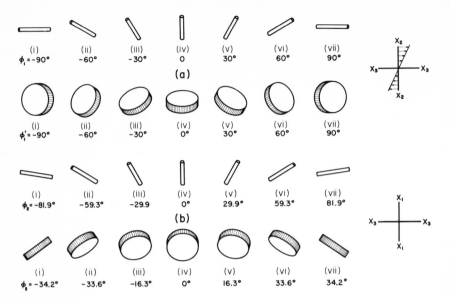

FIG. 4. Particle orientations, calculated from Jeffery's theory,[2] during one half-rotation from $\phi_1 = -\pi/2$ to $+\pi/2$, as they are seen in the X_2X_3 plane (upper half, a) and X_1X_3 plane (lower half, b). Rod: $r_p = 10$ and $r_e = 7$ in an orbit having $C = 1$. Disc: $r_p = 0.2$ and $r_e = 0.34$ in an orbit having $C = 2$. The angular velocities $\dot{\phi}_1$ are greatest at (iv) for the rod and at (i) for the disc.

where $2b_1'(\phi_1)$ is the projected length of the particle at ϕ_1 which, from equation (II20), is a maximum when the particle is aligned with the flow ($\phi_1 = \pi/2, 3\pi/2$) and a minimum when it is aligned normal to the flow ($\phi_1 = 0, \pi$), see Fig. 4.

For a disc the axis of revolution is normal to the particle equatorial plane [Fig. 3(a)] and the X_2X_3 projection of its ends describes an ellipse of axis ratio < 1 whose major axis lies on the X_2 axis; see Fig. 3(b). The X_2X_3 projection of the equatorial plane seen as the upper face of the disc, of radius b_2, is an ellipse whose projected area relative to that of the face of the disc is given by

$$A_{23} = \frac{b_2'(\phi_1)}{b_2} = \cos\theta_1 \qquad (II25)$$

where $b_2'(\phi_1)$ is the projected length of the equatorial radius at ϕ_1. As is evident from Fig. 4b, A_{23} is a minimum at $\phi_1 = 0$, where the projected diameter is aligned with the flow and the angular velocity $\dot{\phi}_1$ is least; at $\phi_1 = \pi/2, 3\pi/2$, when the diameter is at right angles to the flow, A_{23} and $\dot{\phi}_1$ are maxima.

Two examples of the results obtained in an experimental test of these parts of the theory are illustrated in Figs. 5 and 6, giving respectively the variation of ϕ_1 with time, equation (II13) and of θ_1 with respect to ϕ_1, equation (II20).

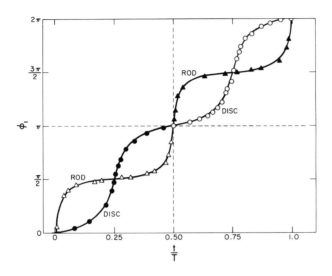

FIG. 5. Measured ϕ_1 versus t/T for a rod, $r_e = 8.2$ (triangles), and a disc, $r_e = 0.25$ (circles) The points are experimental, and the lines drawn are calculated from theory. The graph also illustrates the fact that the angular velocities $\dot{\phi}_1$ for rods and discs are $\pi/2$ out of phase. (After Karnis et al.[31a]).

(2) *Rotation observed in the $X_1 X_3$ plane.* The same motions have quite a different appearance when viewed along the X_2 axis: the particles are seen to rock to and fro between the angles $\pm\phi_{2M} = \tan^{-1} Cr_e$ [see Figs. 3(c) and 4(b)], as predicted by equation (II22), with maximum velocity $\dot{\phi}_2$ at $\phi_2 = 0$. As shown in Fig. 3(c), the projection of the ends of the rod describe an arc of an ellipse, and the ends of the disc (two points on the axis of revolution equidistant from the center) describe an arc of a hyperbola. Here the respective $X_1 X_3$ projections of unit length of the axis of revolution l_{31} and unit area of the equatorial plane A_{31} are

$$l_{31} = \frac{b_1{}'(\phi_2)}{b_1} = \sin\theta_2 \qquad (\text{II}26a)$$

$$= 2b_1(1 - \sin^2\theta_1 \cos^2\phi_1)^{1/2} \qquad (\text{II}26b)$$

[31a] A. Karnis, H. L. Goldsmith, and S. G. Mason, *Can. J. Chem. Eng.* **44**, 181 (1966).

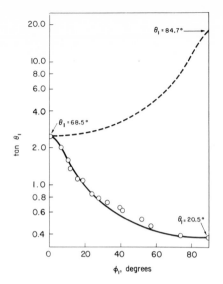

FIG. 6. Variation of the measured θ_1 with ϕ_1 illustrated for a disc, $r_e = 0.15$, in an orbit of $C = 2.6$ (solid curve). The dotted line is calculated for a rod having $r_e = 1/0.15$. (After Goldsmith and Mason.[31])

$$A_{31} = \frac{b_2'(\phi_2)}{b_2} = \cos\theta_2 \qquad\qquad (II27a)$$

$$= \sin\theta_1 \cos\theta_1 \qquad\qquad (II27b)$$

where θ_2 is the angle between the axis of revolution (X_1^0 axis) and the X_2 axis, and $2b_1'(\phi_2)$ and $b_2'(\phi_2)$ are the respective projected length and equatorial radius of the particle.

It follows from equations (II26) and (II20) that $\theta_2 = \pi/2$ at ϕ_{2M} and, hence, that at this point in its orbit the rod lies in the X_1X_3 plane and the disc is seen edge on (Fig. 4b). At $\phi_2 = \phi_1 = 0$ the projected length l_{31} of the rod is a minimum, $= 2b_1/(1 + C^2)^{1/2}$, and A_{31} of the disc passes through a maximum.

The motions viewed along the X_3 axis are quite similar: the axis of revolution rocks to and fro between the angles $\pm\phi_{3M} = \tan^{-1} C$, and for a rod its X_1X_2 projection forms part of a hyperbola;[32] for a disc it is part of an ellipse.

d. Fused Spherical Doublets

Among the original stock of glass or plastic spheres there are many examples of two spheres that have become fused together (during manufac-

[32] F. P. Bretherton and Lord Rothschild, *Proc. Roy. Soc.* **B153**, 490 (1961).

ture) at the surface without distorting the shape of either sphere. Such rigid dumbbells rotate as prolate spheroids of equivalent axis ratio $r_e = 2$.[22,27] This is not surprising since, according to Eirich et al.,[33] the liquid contained in cavities and slots of a particle is immobilized, and the hydrodynamic shape of such a particle is given by its outer contour. On this basis a doublet of equal-size spheres would correspond to an ellipsoid of $r_e = 2$.

e. Variation of Orbit Constant with Time

According to Jeffery's theory, all values of C between 0 and ∞ are possible, but there is no indication of their relative probability. Jeffery considered this indeterminacy a limitation of the theory of the slow viscous motion of a liquid, in which terms involving squares and products of velocities are neglected, but pointed out[2] that "a more complete investigation would reveal the fact that the particles do tend to adopt special orientations with respect to the motion of the surrounding fluid. The suggestion is that *the particles will tend to adopt that motion which, of all the motions possible under the approximated equations, corresponds to the least dissipation of energy.*" The motion that gives minimum overall energy dissipation was shown to correspond to $C = 0$ for a prolate spheroid, $C = \infty$ for an oblate spheroid, and vice versa for maximum average energy dissipation. In the present discussion it is assumed that for long rods and thin discs the orbit constants corresponding to maximum and minimum energy dissipation are the same as those for prolate and oblate spheroids. No rigorous proof for this assumption has been given, although Burgers[34] has given an approximate calculation of the influence of long rods on the suspension viscosity and finds that it is the same as for a spheroid with $r_e/r_p = 0.74$.

Experiments with Newtonian liquids[35] have demonstrated that in the absence of particle sedimentation and interaction and at very low Reynolds numbers *a single rod or disc will continue to rotate indefinitely without any drift in orbit constant toward either of the limiting orbits.* Where changes in C, as measured by changes in ϕ_{2M} or in A_{23}, were observed to occur, they were random and arose as a result of particle interactions.[28,31]

Saffman[36] in a discussion of possible reasons for orbital drift concluded that *non-Newtonian* viscosity of the suspending medium can account for the drift of orbits and that the rate of change in C should be independent of the size of the particle. Experiment confirmed this prediction and showed

[33] F. Eirich, H. Mark, and T. Huber, *Papier Fabr.* **35**, 251 (1937).

[34] J. M. Burgers, *in* "Second Report on Viscosity and Plasticity," Chapt. 3. North-Holland Publ., Amsterdam, 1938.

[35] E. Anczurowski, R. G. Cox, and S. G. Mason, *J. Colloid Interface Sci.* **23**, 547 (1967).

[36] P. G. Saffman, *J. Fluid Mech.* **1**, 540 (1956).

that prolate and oblate spheroids assumed orbits of minimum energy dissipation (as had previously been reported by Taylor,[37] who may also have been using a non-Newtonian medium).

Single rods and discs have also been observed to drift into orbits of minimum energy dissipation in elasticoviscous liquids undergoing Couette and Poiseuille flow[38,39] and into orbits of maximum energy dissipation in Newtonian liquids undergoing shear at higher Reynolds numbers, at which inertial effects become important,[38] as will be described in Section VII.

The orbits assumed by the particles in very dilute Newtonian suspensions undergoing Couette flow appear to be determined by the conditions of initial release. It is found that rods (and discs) of a given particle axis ratio have constant values[31] of $T\gamma/2\pi$ and, hence, r_e [see equation (II19)] is independent of C. The distributions of orbit constants in assemblies of discs and rods, when interactions can occur, are considered in Section V.

f. Equivalent Ellipsoidal and Particle Axis Ratios

Bretherton has shown on theoretical grounds[12] "that the motion in a uniform simple shear of a rigid body of revolution is mathematically identical, at least insofar as rotation is concerned, with that of some ellipsoid of revolution." Only one constant, B, is needed to describe its motion and, provided $|B| < 1$, it may be shown[12] that

$$B = \frac{r_e^2 - 1}{r_e^2 + 1} \tag{II28}$$

Although no theoretical calculations of B for right circular cylinders have been made, an indication of its behavior has been obtained in the observed variation of r_e with r_p for rods[26,28,29,31] and discs.[19,31] It was found that for discs $r_e > r_p$, whereas for rods $r_e < r_p$; in both cases r_e/r_p showed a significant increase with decreasing r_p, as shown in Fig. 7, which gives a plot of r_e/r_p versus $\ln r_p$. A smooth curve may be drawn through the points obtained in different investigations; the scatter is probably due to imperfections or irregularities in the particles.

Some ambiguity arises in distinguishing a "rod" from a "disc"; the simplest way to resolve the problem is to define a disc as a cylinder having $r_e < 1$ and a rod as one having $r_e > 1$. Experiments have shown that the transition from disc to rod occurs at a value of $r_p = 0.86$.[24]

Bretherton has also shown that when $|B| < 1$, as is the case with spheroids, the ends of the axis of symmetry rotate periodically. However, a body of

[37] G. I. Taylor, *Proc. Roy. Soc.* **A103**, 58 (1923).

[38] A. Karnis, H. L. Goldsmith, and S. G. Mason, *Nature* **200**, 159 (1963).

[39] A. Karnis and S. G. Mason, *Trans. Soc. Rheology* **10**, 571 (1966).

revolution lacking a plane of symmetry perpendicular to its axis, such as a hemisphere, does not rotate about a fixed point but translates back and forth across the streamlines.

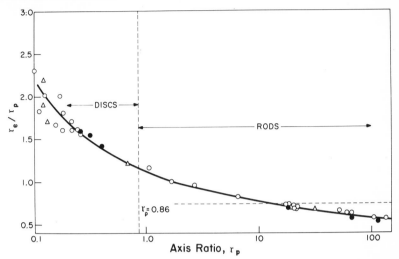

FIG. 7. The ratio r_e/r_p as a function of r_p for rods ($r_e > 1$) and discs ($r_e < 1$). Measurements in Couette flow (circles) and Poiseuille flow (triangles) yielded the points, through which was drawn by eye the best curve, apparently not passing through the point (1, 1) corresponding to a sphere but asymptotically approaching $r_e/r_p = 0.5$ for very long rods. The horizontal dashed line corresponds to the value $r_e/r_p = 0.74$ calculated from Burger's approximate theory[34] for thin rigid rods.

4. RECTANGULAR HYPERBOLIC FLOW

The field of motion is defined by

$$V_1' = 0, \qquad V_2' = -\tfrac{1}{2}\gamma X_2', \qquad V_3' = \tfrac{1}{2}\gamma X_3' \qquad (II29)$$

where now V_1', V_2', and V_3' are the respective components of fluid velocity along the X_1', X_2', and X_3' axes. The streamlines are given by $X_2'X_3' = $ constant, being rectangular hyperbolas, as shown in Fig. 8, with X_2' the inflow axis and X_3' the outflow axis.

This two-dimensional flow was first realized experimentally by Taylor,[7] who used a four-roller apparatus consisting of four identical cylinders mounted vertically at the corners of a square and immersed in the suspending fluid (Fig. 8) contained in a rectangular box. The rollers were driven at identical speeds with adjacent pairs in opposite directions, as shown in the

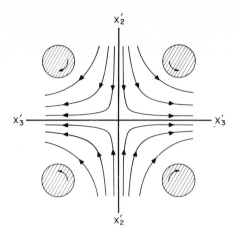

FIG. 8. Principle of the four-roller apparatus for producing a field of rectangular hyperbolic flow by having adjacent rollers (shown shaded) rotating in opposite directions.

figure.[40] The experimental γ may be obtained by timing the passage of small tracer particles along various rectangular hyperbolic paths and using the integrated form of equation (II29).[7,41]

The only nonzero components of fluid dilatation in plane hyperbolic flow are $D_{33} = -D_{22} = \gamma/2$ and, when substituted into equations (II8) and (II9), they yield the following expressions for the angular velocities of the axis of revolution of a spheroid at the center of the flow field:[19]

$$\dot{\phi}_1' = \frac{\gamma B}{2}\sin 2\phi_1' \qquad (\text{II}30)$$

$$\dot{\theta}_1' = -\frac{\gamma B}{4}\sin 2\theta_1'\cos 2\phi_1' \qquad (\text{II}31)$$

where ϕ_1' and θ_1' are the spherical polar coordinates in plane hyperbolic flow with X_1' as polar axis, and B is defined by equation (II28). Equations (II30) and (II31) may be derived directly from equations (II13) and (II14) by making use of the transformation of coordinates from Couette to plane hyperbolic flow. As pointed out by Taylor,[6] if the coordinate axes X_2' and X_3' of the field of rectangular hyperbolic flow are rotated with constant

[40] As mentioned in Section I, Giesekus[17] has demonstrated a whole variety of two-dimensional flows in a four-roller apparatus when the diagonally opposite cylinders are mechanically coupled. When the speed of one cylinder pair is varied between the limits $-\Omega$ to Ω while the speed of the other pair is maintained at Ω, the flows range from pure deformation to pure rotation. Illustrations of the streamline patterns in the various flows with the aid of tracer particles have been given.[17]

[41] F. D. Rumscheidt and S. G. Mason, *J. Colloid Sci.* **16**, 210 (1961).

angular velocity $\gamma/2$, and if the coordinate axes X_2 and X_3 of the uniform laminar shear field defined by equation (II10) lie instantaneously at 45° to the axes X_2' and X_3', then the two fields of flow become identical (Fig. 9). It follows that the coordinates of the two fields are related by

$$X_1 = X_1', \qquad X_2 = (1/\sqrt{2})(X_2' + X_3'), \qquad X_3 = (1/\sqrt{2})(X_3' - X_2') \quad \text{(II32a)}$$

$$\phi_1 = \phi_1' - \pi/4, \qquad \theta_1 = \theta_1' \quad \text{(II32b)}$$

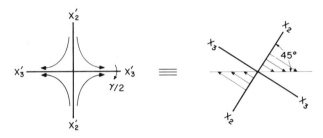

FIG. 9. Transformation of hyperbolic flow (left) to shear flow (right) by rotation of the coordinate axes. (After Bartok and Mason.[42])

and the components of fluid velocity by[42]

$$V_2 = (1/\sqrt{2})[V_2' + V_3' + (\gamma/2)(X_2' + X_3')]$$
$$V_3 = (1/\sqrt{2})[V_3' - V_2' + (\gamma/2)(X_2' + X_3')] \quad \text{(II33)}$$

Integrating equations (II30) and (II31) after division by equation (II30) yields, respectively,

$$\ln \tan \phi_1' = \frac{\gamma(r_e^2 - 1)}{r_e^2 + 1}(t - t_0) \quad \text{(II34)}$$

$$\ln \tan \theta_1' = -\tfrac{1}{2} \ln \sin 2\phi_1' + \tfrac{1}{2} \ln C \quad \text{(II35)}$$

where t_0 and C are integration constants and r_e is interpreted as the axis ratio of the spheroid having the same B.

According to equation (II34), particles with $r_e > 1$ move such that ϕ_1' asymptotically approaches $\pi/2$, the axis of revolution rotating toward the $X_1'X_3'$ plane, in which fluid flows radially outward from the X_1' axis; see Fig. 10. The axis of an oblate spheroid or disc ($r_e < 1$) approaches the

[42] W. Bartok and S. G. Mason, *J. Colloid Sci.* **13**, 293 (1958).

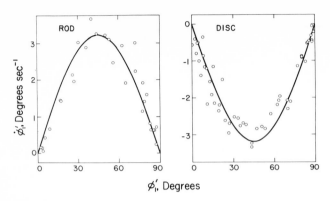

FIG. 10. The rotation of rods and discs in plane hyperbolic flow. Variation of $\dot{\phi}_1'$ with ϕ_1'. The lines are calculated from equation (II30), according to which $\dot{\phi}_1'$ is a minimum at $\phi_1' = \pi/4$. (After Chaffey et al.[19])

$X_1'X_2'$ plane, in which fluid flows inward ($\phi_1' = 0$). Rods at $\phi_1' = 0$ and discs at $\phi_1' = \pi/2$ are in unstable equilibrium orientations, corresponding to $t = -\infty$ in equation (II34). For the limiting orientations $\phi_1' = 0$ and $\pi/2$, equation (II35) is indeterminate; in this case θ_1' is found by integrating equation (II31):

$$\ln \tan \theta_1' = \frac{\gamma(r_e^2 - 1)}{2(r_e^2 + 1)}(t - t_0) \qquad \text{for} \quad \phi_1' = \pi/2 \qquad (II36a)$$

$$\ln \tan \theta_1' = -\frac{\gamma(r_e^2 - 1)}{2(r_e^2 + 1)}(t - t_0) \quad \text{for} \quad \phi_1' = 0. \qquad (II36b)$$

The stable equilibrium position of θ_1' is thus $\pi/2$ for both rods and discs. Equation (II30) shows that the effectiveness of a dilatation flow in rotating a particle is greater, the larger the magnitude of B or the more eccentric the spheroid: a sphere ($r_e = 1, B = 0$) does not rotate.

a. Observations with Rods and Discs

Quantitative agreement with the theory described above has been observed with rods, discs, and spheres.[19]

When viewed along the X_1' axis, rods whose axes of revolution are initially aligned close to the X_2' axis near $\phi_1' = 0$, rotate away from that position at first slowly, then more rapidly, and finally more slowly again, as ϕ_1' approaches $\pi/2$ (Fig. 10) and they become aligned with the X_3' axis;

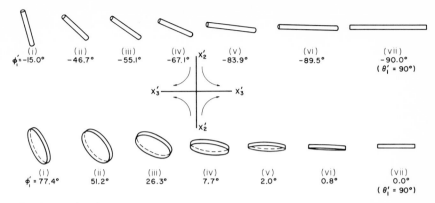

FIG. 11. Tracings from photomicrographs of the successive orientations of a rigid rod (upper) and disc (lower) in plane hyperbolic flow. The particles align themselves in the plane of fluid outflow, the $X_1'X_3'$ plane; θ_1' at first decreases, as shown by the foreshortening of the rod and the increased area of the upper face of the disc at (ii), and then it finally tends to $\pi/2$.

see Fig. 11. Discs behave similarly, but in this case the axis of revolution rotates from $\phi_1' = \pi/2$ to 0, so that the equatorial diameter of the particle becomes aligned with the X_2' axis.

During rotation the inclination of the axis of revolution to the X_1' axis (θ_1') changes until it reaches $\pi/2$; that is, the long axes of the rods lie in the $X_2'X_3'$ plane and the discs are seen edge on (Fig. 11). As required by equation (II35), θ_1' decreases at first, increases when $\phi_1' > \pi/4$, and finally approaches $\pi/2$. After ϕ_1' has attained its equilibrium value, $\ln \tan \theta_1'$ becomes a linear function of time; see equation (II36a).

b. Equivalent Ellipsoidal Axis Ratio

The values of r_e calculated in hyperbolic flow from B by means of an experimental plot of equation (II34) are comparable to those obtained for the same particle in Couette flow from the period of rotation and equation (II19).[19]

5. HYPERBOLIC RADIAL FLOW

The three-dimensional analogue of plane hyperbolic flow, also irrotational, is a form of hyperbolic radial flow, which occurs in threads of fluid that are being uniformly extended and is of practical importance in the spinning of fibers from molten polymers. The theory of the behavior of rigid ellipsoids suspended in such a flow has been developed by Takserman-Krozer and

Ziabicki.[43] With the X_3 axis, the direction of flow, as the polar axis the velocity field is described by

$$V_1 = -\frac{\gamma}{2}X_1, \qquad V_2 = \frac{\gamma}{2}X_2, \qquad V_3 = \gamma X_3$$

and $D_{11} = D_{22} = -\frac{1}{2}\gamma$ and $D_{33} = \gamma$. The equations of motion are[43]

$$\dot\phi_3 = 0, \qquad \dot\theta_3 = -\frac{3}{2}\gamma B \sin 2\theta_3 \tag{II37}$$

where, on integration,

$$\dot\phi_3 = \text{constant}, \qquad \tan\theta_3 = \exp[-\tfrac{3}{2}\gamma B(t - t_0)] \tag{II38}$$

Thus, only the orientation of the axis of revolution with respect to the outflow axis changes with time. The equilibrium orientations assumed by the particles are such that θ_3 is equal to 0 for rods and to $\pi/2$ for discs; that is, rods are completely oriented, their long axes coincident with the X_3 axis, whereas for the discs, in contrast to plane hyperbolic flow, there remains a freedom with respect to the angle ϕ_3, and the axis of revolution lies in the $X_1 X_3$ plane.

6. ROTATION OF RIGID PARTICLES IN SHEAR AND ELECTRIC FIELDS

The angular rotation and orientation of the axis of revolution of an ellipsoid placed in a combined electric and shear field may be calculated from the electrostatic and hydrodynamic torques since, on the assumption of negligible inertia, the latter, given by equation (II1), is exactly balanced by the external torque, which can be calculated from electrostatic theory. Applying this result, Demetriades[3] found the angular velocities of an ellipsoid in Couette flow with an electric field normal to the velocity and vorticity (i.e., acting along the X_2 axis). The theory was confirmed in the case of rigid conducting rods[4] and, as shown below, has since been extended to the more general case of an electric field arbitrarily oriented with respect to Couette flow.[5]

a. Equations of Motion

The components of the electrostatic torque $\mathbf{\Gamma}^{(e)}$ acting on an isotropically dielectric spheroid of dielectric constant $q\varepsilon$ and having no permanent dipole moment, suspended in a medium of dielectric constant ε, is given by Stratton[44]:

$$\Gamma_1^{(e)} = \tfrac{4}{3}\pi b_1 b_2{}^2(q - 1)\varepsilon(e_2{}^0\mathscr{E}_3{}^0 - e_3{}^0\mathscr{E}_2{}^0) \tag{II39}$$

---

[43] R. Takserman-Krozer and A. Ziabicki, *J. Polymer Sci.* **1A**, 491 (1963).
[44] J. A. Stratton, "Electromagnetic Theory," 1st ed. McGraw-Hill, New York, 1941.

where $\mathscr{E}_2{}^0$ and $\mathscr{E}_3{}^0$ are the components of the external electrical field $\mathscr{E}^0$ referred to the $X_1{}^0$, $X_2{}^0$, $X_3{}^0$ coordinate system and considered to be parallel and uniform, which produces the field $\mathbf{e}^0$ inside the spheroid, whose components are

$$e_1{}^0 = [1 + \tfrac{1}{2}b_1 b_2{}^2 \alpha_1 (q - 1)]^{-1} \mathscr{E}_1{}^0$$

$$e_2{}^0 = [1 + \tfrac{1}{2}b_1 b_2{}^2 \alpha_2 (q - 1)]^{-1} \mathscr{E}_2{}^0 \qquad \text{(II40)}$$

$$e_3{}^0 = [1 + \tfrac{1}{2}b_1 b_2{}^2 \alpha_2 (q - 1)]^{-1} \mathscr{E}_3{}^0$$

Since the hydrodynamic and electrostatic torques are equal and opposite, the components of $\Gamma^{(e)}$ may be substituted into equation (II1) and, on the assumption that $\mathbf{r}_p$ may be replaced with $\mathbf{r}_e$, to be determined experimentally, the resulting equations giving the particle spins are[5]

$$\omega_1{}^0 = Z_1{}^0$$

$$\omega_2{}^0 = Z_2{}^0 - BD^0_{31} + 2\mathfrak{P}(q, \mathfrak{r}_e)\varepsilon \eta_0^{-1}\mathscr{E}_3{}^0\mathscr{E}_1{}^0 \qquad \text{(II41)}$$

$$\omega_3{}^0 = Z_3{}^0 + BD^0_{12} - 2\mathfrak{P}(q, \mathfrak{r}_e)\varepsilon \eta_0^{-1}\mathscr{E}_1{}^0\mathscr{E}_2{}^0$$

Here B is as defined by equation (II28) and $\mathfrak{P}(q, \mathfrak{r}_e)$ is given by

$$\mathfrak{P}(q, \mathfrak{r}_e) = \frac{[2\mathfrak{r}_e{}^2 + (1 - 2\mathfrak{r}_e{}^2)\mathfrak{A}]\mathfrak{f}(q, \mathfrak{r}_e)}{32\pi(\mathfrak{r}_e{}^2 + 1)} \qquad \text{(II42)}$$

where

$$\mathfrak{f}(q, \mathfrak{r}_e) = (q - 1)[1 + \tfrac{1}{2}b_1 b_2{}^2(\alpha_2 - \alpha_1)(q - 1)]^{-1}$$

$$= \frac{(3\mathfrak{A} - 2)(q - 1)^2}{[(q - 1)\mathfrak{A} - q][2 + (q - 1)\mathfrak{A}]} \qquad \text{(II43}a\text{)}$$

$$\mathfrak{A} = b_1 b_2{}^2 \alpha_2$$

$$= \frac{\mathfrak{r}_e{}^2}{\mathfrak{r}_e{}^2 - 1} - \frac{\mathfrak{r}_e \cosh^{-1} \mathfrak{r}_e}{(\mathfrak{r}_e{}^2 - 1)^{3/2}} \qquad \text{(II43}b\text{)}$$

Here $\mathfrak{P}(q, \mathfrak{r}_e)$ is >0 when $\mathfrak{r}_e < 1$ and is <0 when $\mathfrak{r}_e > 1$.[5] If $q = \infty$, for conducting particles in a dielectric liquid $\mathfrak{P}(q, \mathfrak{r}_e)$ reduces to a limiting form $\mathfrak{P}(\mathfrak{r}_e)$, and $\mathfrak{f}(\mathfrak{r}_e)$ is given by

$$\mathfrak{f}(\mathfrak{r}_e) = \frac{3\mathfrak{A} - 2}{\mathfrak{A}(\mathfrak{A} - 1)} \qquad \text{(II44)}$$

As in the case of shear flow, by transformation of coordinates and use of equation (II8) one obtains from equation (II41) the angular velocities in a combined Couette and electric field, whose components $\mathscr{E}_1$, $\mathscr{E}_2$, and $\mathscr{E}_3$ are referred to the fixed coordinate system X_1, X_2, X_3:

$$\dot{\phi}_1 = \dot{\phi}_{1(\text{shear})} + \dot{\phi}_{1(\text{electric})}$$

$$= \frac{\gamma}{(r_e^2 + 1)} (r_e^2 \cos^2 \phi_1 + \sin^2 \phi_1) + \mathfrak{P}(q, r_e)\varepsilon \eta_0^{-1}[(\mathscr{E}_3^2 - \mathscr{E}_2^2) \sin 2\phi_1$$

$$+ \mathscr{E}_2 \mathscr{E}_3 \cos 2\phi_1 + (\mathscr{E}_3 \mathscr{E}_1 \cos \phi_1 - \mathscr{E}_1 \mathscr{E}_2 \sin \phi_1) \cot \theta_1] \tag{II45}$$

$$\dot{\theta}_1 = \dot{\theta}_{1(\text{shear})} + \dot{\theta}_{1(\text{electric})}$$

$$= \frac{\gamma(r_e^2 - 1)}{4(r_e^2 + 1)}(\sin 2\phi_1 \sin 2\theta_1) - \mathfrak{P}(q, r_e)\varepsilon \eta_0^{-1}[(\mathscr{E}_2^2 \cos^2 \phi_1 + \mathscr{E}_3^2 \sin^2 \phi_1$$

$$- \mathscr{E}_1^2 + \mathscr{E}_2 \mathscr{E}_3 \sin 2\phi_1) \sin 2\theta_1 + (\mathscr{E}_3 \mathscr{E}_1 \sin \phi_1 + \mathscr{E}_1 \mathscr{E}_2 \cos \phi_1) \cos 2\theta_1] \tag{II46}$$

The axial spin of the spheroid is unaffected by the electric field, because no torque may be exerted by it about the axis of revolution, and it is given by equation (II15). Equations (II45) and (II46) have been integrated for the following two special cases.

b. Electric Field Directed Along X_1 Axis

In this case $\mathscr{E}_2 = \mathscr{E}_3 = 0$, and $\dot{\phi}_1$ is not affected by the electric field, so that integrating equation (II45) yields equation (II18), the period of rotation T being given by equation (II19). Integration of equation (II46) leads to

$$\tan \theta_1 = \frac{C_0 r_e \exp(-2\mathscr{F}t)}{(r_e^2 \cos^2 \phi_1 + \sin^2 \phi_1)}$$

$$= \frac{C r_e}{(r_e^2 \cos^2 \phi_1 + \sin^2 \phi_1)} \tag{II47}$$

where $\mathscr{F} = -\mathfrak{P}(q, r_e)\mathscr{E}_0^2/\eta_0$, $\mathscr{E}_0$ being the magnitude of the electric field, and C_0 is the orbit constant defined by equation (II18), whereas $C = C_0 \exp(-2\mathscr{F}t)$ is not a constant but tends to zero for a rod and to infinity for a disc as t approaches ∞.

c. Electric Field Perpendicular to the X_1 Axis

(1) Angular velocity. When $\mathscr{E}_1 = 0$, then $\mathscr{E}_0^2 = \mathscr{E}_2^2 + \mathscr{E}_3^2$ and, with the substitutions $\tan \delta = \mathscr{E}_3/\mathscr{E}_2$ and $H = \gamma/(r_e^2 + 1)$, equation (II45) reduces to
$$\dot{\phi}_1 = H(r_e^2 \cos^2 \phi_1 + \sin^2 \phi_1) - \mathscr{F} \sin 2\delta(\cos^2 \phi_1 - \sin^2 \phi_1) - \mathscr{F} \cos \delta \sin 2\phi_1$$

which may be integrated to give

$$\tan \phi_1 = \frac{\chi r_e \tan[\chi r_e H(t - t_0)] + (\mathscr{F}/H) \cos 2\delta}{1 + (\mathscr{F}/H) \sin 2\delta} \tag{II48}$$

with a period of rotation T now given by

$$T = \frac{2\pi}{\chi\gamma}(r_e + 1/r_e), \qquad \chi \leqslant 1 \tag{II49}$$

Thus T is increased by a factor $1/\chi$, where χ is defined by

$$\chi = \frac{1}{r_e}\left[r_e{}^2 - \left(\frac{\mathscr{F}}{H}\right)^2 + \frac{\mathscr{F}}{H}(r_e{}^2 - 1)\sin 2\delta\right]^{1/2} \tag{II50}$$

Experimentally, the behavior of conducting rods[4] and discs[5] was observed in an electric field directed along the X_2 axis of the motion, that is, across the annulus of the Couette apparatus [Fig. 2(b)], so that $\mathscr{E}_3 = 0$ and equations (II45), (II48), and (II50) reduce to

$$\dot{\phi}_1 = H(r_e{}^2 \cos^2 \phi_1 + \sin^2 \phi_1) - \mathscr{F} \sin 2\phi_1 \tag{II51}$$

$$\tan \phi_1 = \chi r_e \tan[\chi r_e H(t - t_0)] + \frac{\mathscr{F}}{H} \tag{II52}$$

$$\chi = \left[1 - \left(\frac{\mathscr{F}}{Hr_e}\right)\right]^{1/2} \tag{II53}$$

The retardation of the rotation is not uniform: the time spent by the particle in the first and third quadrants calculated from equation (II52) is $(\chi r_e H)^{-1}[\pi - \tan^{-1}(\chi r_e H/\mathscr{F})]$. Subtracting this from $T/2$ gives the time spent in the second and fourth quadrants. Thus, as found experimentally[4] and illustrated in Fig. 12(a), the presence of the electric field decreases the time of rotation of the rod ($\mathscr{F} > 0$) in the second and fourth quadrants and increases it in the first and third, where the electrostatic and hydrodynamic torques are in opposite directions. The reverse is true for a disc[5] [$\mathscr{F} < 0$, Fig. 12(b)].

(2) *Particle orbits.* The particle orbit is found by integrating equation (II46), with $\mathscr{E}_1, \mathscr{E}_3 = 0$, and this yields

$$\tan \theta_1 = \frac{C_0 r_e \exp(\mathscr{F} t)}{r_e{}^2 \cos^2 \phi_1 + \sin^2 \phi_1 - (\mathscr{F}/H)\sin 2\phi_1} \tag{II54}$$

The ultimate motion is thus determined by the exponential factor: the axis of revolution of the rod finally lies in the $X_2 X_3$ plane ($C = \infty$), whereas the

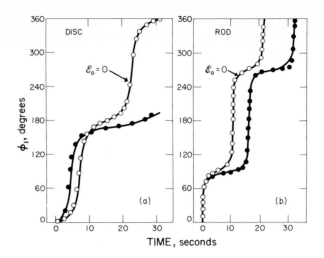

Fig. 12. Measured rotation of rods and discs in a combined shear and electric field, showing the variation of ϕ_1 with time. For the rod the time spent in the first and third quadrants is greater than that spent in the absence of the field; the reverse holds true for the disc. The lines drawn have been calculated with equation (II52). (After Allan and Mason[4] and Chaffey and Mason.[5])

disc, as shown in Fig. 13, finally spins about its axis of revolution, which coincides with the X_1 axis ($C = 0$).

(3) *Critical field for impeded rotation.* As the ratio $\mathscr{F}/H$ is increased, χ decreases and T progressively increases, until at a critical value of $\mathscr{F}/H$, when $\mathscr{E}_0 = \mathscr{E}_{0,\text{crit.}}$, χ vanishes and T becomes infinite. The motion is then impeded, and particle no longer executes a complete rotation but moves until it attains a constant angle $\phi_1 = \phi_{\text{crit.}}$ which, from equations (II52) and (II53) for $\chi = 0$, is

$$\phi_{\text{crit.}} = \tan^{-1}(\mathscr{F}/H)$$

$$= \tan^{-1} \mathfrak{r}_e \tag{II55}$$

whence the magnitude of the critical electrical field is given by

$$\mathscr{E}_{0,\text{crit.}}^2 = -\frac{\mathfrak{r}_e \gamma \eta_0}{(\mathfrak{r}_e^2 + 1)\mathfrak{P}(q, \mathfrak{r}_e)\varepsilon}$$

Upon further increase of the electric field, χ and T become imaginary quantities. The particles move to fixed orientations at a rate that may be obtained from equation (II52) by substituting $\chi' = \chi/i$:

$$\tan \phi_1 = \chi' \mathfrak{r}_e \tanh \chi' \mathfrak{r}_e H(t - t_0) + \mathscr{F}/H \tag{II56}$$

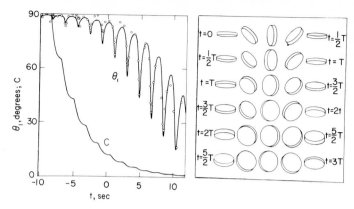

FIG. 13. Orientations of a rigid disc in successive orbits in a combined shear and electric field. With increase in time more and more of the upper face of the particle is visible during each orbit, corresponding to a progressively lower θ_1 and C; see equation (II54). (After Chaffey and Mason.[5])

and the steady-state angle ϕ_i is found by setting $t = \infty$ in equation (II56):

$$\phi_i = \tan^{-1}(\chi' r_e + \mathcal{F}/H) \tag{II57}$$

An unstable orientation ϕ_i'' corresponding to $t = -\infty$ also exists:

$$\phi_i'' = \tan^{-1}(\mathcal{F}/H - \chi' r_e) \tag{II58}$$

As predicted by the theory, rods align themselves at an angle ϕ_i, which decreases as $\mathcal{F}/H$ increases, the converse being true of discs. However, there is a discrepancy between the values of $\mathcal{F}$ and, hence, of $\mathfrak{P}(q, r_e)$, obtained from the experimentally measured $\phi_{\text{crit.}}$, ϕ_i, and $\mathcal{E}_{0,\text{crit.}}$, and those calculated from theory, equation (II42). Just as r_e is chosen to fit the measured T in shear only, making H an experimental quantity, so in shear and electric fields, when an experimental $\mathcal{F}$, calculated from equation (II53) with χ or χ' given by the measured period of rotation [equation (II49)] or ϕ_i [equation (57)] is chosen, good agreement with equations (II52), (II54), and (II56) results.

d. Rotation in Electric Field Only

As might be expected, in the absence of the shear field the spheroids tend to align themselves in the direction of the electric field. When the field is parallel to the X_2 axis, equations (II51) and (II52) may be integrated to give

$$\ln \tan \phi_1 = -2\mathcal{F}(t - t_0) \tag{II59}$$

$$\ln \tan \theta_1 = -\tfrac{1}{2}\ln \sin 2\phi_1 - \tfrac{1}{2}\ln C \tag{II60}$$

which are of the same form as equations (II34) and (II35), describing particle behavior in rectangular hyperbolic shear flow, but in this case, when $t = \infty$, then $\phi_1 = 0$ for a rod ($\mathscr{F} > 1$) and $\phi_1 = \pi/2$ for a disc ($\mathscr{F} < 1$).

As in hyperbolic flow, θ_1 for the limiting orientations is given by equation (II36), so that $\theta_1 = \pi/2$ when $t = \infty$. However, this behavior is shown only by rods;[4] discs rotating to $\phi_1 = \pi/2$ assume a variety of different orientations with respect to θ_1, depending on the initial condition of release.[5] The values of $\mathfrak{P}(q, \mathfrak{r}_e)$ and $\mathfrak{P}(\mathfrak{r}_e)$ obtained from experimental plots of equation (II59) differ from the theoretical ones.[5] The discrepancy cannot be attributed solely to a shape effect (as with $\mathfrak{r}_e$ and rotation in shear only), because it seems that the measured $\mathfrak{P}(q, \mathfrak{r}_e)$ depends not only on q and $\mathfrak{r}_e$ but also on the suspending fluid. Theoretical considerations show[5] that a more reasonable quantity to compare with measurements is $\mathfrak{P}(\mathfrak{r}_{e,\mathfrak{P}})$ based on $\mathfrak{r}_{e,\mathfrak{P}}$, the equivalent ellipsoidal axis ratio chosen to give the correct $\mathfrak{P}(\mathfrak{r}_e)$ when substituted into equations (II42) and (II44). The particle would then be characterized by two equivalent ellipsoidal axis ratios.

III. Stresses on Particles in Laminar Shear

1. GENERAL

In the previous section we considered the rotations of rigid spheroids and their relation with torques due to the action of fluid stresses on the particle surface when the suspending liquid is subjected to laminar flow. We shall now be concerned with the calculation of these stresses in the case of suspended liquid drops and spheroidal particles and with the deformation associated with them. As before, the theory is limited to plane hyperbolic and Couette flows, both of which have been studied experimentally.

When a particle is placed at the origin of either of these two flow fields, the disturbance to the fluid motion in its neighborhood generates a stress system, which may be resolved into tangential and normal components acting at the surface. In the case of a liquid drop whose interface is uncontaminated by impurity or surfactant the tangential stresses are continuous across the interface, so that a system of velocity gradients is established inside the drop by internal circulation. The normal stresses, on the other hand, are discontinuous at the interface and generate a pressure difference across it given by the Laplace equation

$$\Delta p = \sigma(b_{(1)}^{-1} + b_{(2)}^{-1}) \tag{III1}$$

where $b_{(1)}$ and $b_{(2)}$ are the principal radii of curvature and σ is the interfacial tension. The drop is deformed in such a way that the stresses generated by the flow are balanced by the interfacial tension; Taylor[7] showed that stresses can be balanced when the surface of the drop assumes a form that is approximately ellipsoidal at very small deformations. An analogous situation, in which the drop is deformed in an electric field owing to electrical stresses in the suspending fluid or is deformed in a combined shear and electric field, is also considered.

Lastly, we shall investigate the stresses acting on a rigid rod in Couette flow and show that there exists a critical value of $\gamma \eta_0$, at which particles of a given axis ratio r_p and tensile and bending moduli will buckle. As illustrated in Fig. 2(a), the principal axes of fluid rate of strain in Couette flow are oriented such that the compressive stresses are a maximum at $\phi_1 = -\pi/4$; it is at this orientation that the rods are first observed to buckle while rotating.

2. INTERNAL CIRCULATION IN FLUID DROPS

a. Irrotational Shear Flow

Taylor[6] extended Einstein's theoretical treatment[45,46] of the viscosity of a suspension of rigid particles to the case of small immiscible fluid spheres. To render the problem tractable it was assumed that the drop radius and velocity of distortion are so small that the interfacial tension keeps the drops nearly spherical, that there is no slip at the interface between drop and suspending medium, and that the tangential stress parallel to the surface is continuous across the surface of the drop. With these boundary conditions Taylor was able to calculate the velocity components V_1', V_2', V_3' outside, and v_1', v_2', v_3' inside, a drop of radius b situated at the center of a field of irrotational flow given by equation (II29).[6] To simplify the mathematical analysis, the following treatment is restricted to the equatorial or $x_2'x_3'$ plane of the drop defined by $(x_2'^2 + x_3'^2) = r^2$ and $x_1' = 0$, where r is the radius vector measured from the drop center.

In plane polar coordinates the external velocity field was calculated to be[6,42]

$$V_1' = 0$$

$$V_2' = \frac{\gamma r \cos \phi_1'}{4(\lambda + 1)} \left[\frac{5b^3}{r^3} \left(\lambda + \frac{2}{5} - \frac{\lambda b^2}{r^2} \right) \cos 2\phi_1' + 2\lambda \left(\frac{b^5}{r^5} - 1 \right) - 2 \right]$$

$$V_3' = \frac{\gamma r \sin \phi_1'}{4(\lambda + 1)} \left[\frac{5b^3}{r^3} \left(\lambda + \frac{2}{5} - \frac{\lambda b^2}{r^2} \right) \cos 2\phi_1' - 2\lambda \left(\frac{b^5}{r^5} - 1 \right) + 2 \right] \tag{III2}$$

where λ is the viscosity ratio of suspended phase to suspending phase, and the components of the fluid radial stress vector $\mathbf{S}_r'$ outside the drop was

[45] A. Einstein, *Ann. Physik* [4] **19**, 289 (1906).
[46] A. Einstein, *Ann. Physik* [4] **34**, 591 (1911).

calculated to be

$$
S'_{r2} = -\frac{5\eta_0 \gamma r \cos \phi_1'}{2b(\lambda + 1)} \left(\lambda + \frac{16r^2}{5b^2} \cos 2\phi_1'\right)
$$

$$
S'_{r3} = \frac{5\eta_0 \gamma r \sin \phi_1'}{2b(\lambda + 1)} \left(\lambda - \frac{16r^2}{5b^2} \cos 2\phi_1'\right)
$$

(III3)

The velocity field inside the drop is given by

$$
v_1' = 0
$$

$$
v_2' = \frac{\gamma r \cos \phi_1'}{4(\lambda + 1)} \left(\frac{r^2}{b^2}(2 \cos 2\phi_1' - 5) + 3\right)
$$

$$
v_3' = \frac{\gamma r \sin \phi_1'}{4(\lambda + 1)} \left(\frac{r^2}{b^2}(2 \cos 2\phi_1' + 5) - 3\right)
$$

(III4)

By using the following transformation equations giving the angular velocity on a streamline,

$$
\dot\phi_1' = \omega_1 = \frac{-v_2'}{r \sin \phi_1' - \cos \phi_1' \, dr/d\phi_1'} = \frac{v_3'}{r \cos \phi_1' + \sin \phi_1' \, dr/d\phi_1'}
$$

(III5)

we obtain from equation (III4) the differential equation for the streamlines,

$$
\frac{(10r^2/b^2) - 6}{r(1 - r^2/b^2)} \, dr = 6 \cot 2\phi_1' \, d\phi_1'
$$

which upon integration yields[42]

$$
(r^3/b^3)(1 - r^2/b^2) = k_s'(\sin 2\phi_1')^{-3/2}
$$

(III6)

The streamlines calculated from equation (III6) for various k_s', the internal streamline constant, are shown in Fig. 14. They give rise to a characteristic quadrant pattern: the circulation path of a fluid element is a closed curve in each quadrant, and the streamlines are symmetrical about the line $\phi_1' = \pi/4$. Close to the drop center, where equation (III6) reduces to

$$
\frac{r}{b} = (k_s')^{1/3}(\sin 2\phi_1')^{-1/2}
$$

they approximate to rectangular hyperbolas. At a given ϕ_1' the constant k_s' is a maximum at $r/b = (3/5)^{1/2}$, and at this radius each streamline is radial and four stagnation points exist at $\phi_1' = \pm 3\pi/4$ in addition to that at the drop center. Unlike the internal streamlines in Couette flow given below, those in plane hyperbolic flow are independent of the viscosity ratio λ.

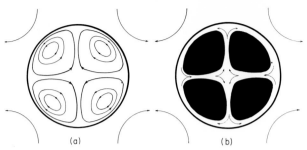

FIG. 14. Circulation inside a fluid drop situated at the origin of a field of plane hyperbolic flow showing the characteristic quadrant pattern. The streamlines calculated from equation (III6) are shown on the left (a). The pattern observed with emulsified water droplets in a castor oil drop is shown on the right (b). The light portions indicate regions containing no emulsion droplets. (After Rumscheidt and Mason.[41])

b. Couette Flow

Taylor has pointed out[7] that effects which depend on the instantaneous distribution of velocity, such as internal circulation (the internal velocity vector) in undeformed fluid drops are unaffected by rotation of the whole system and will thus be identical in plane hyperbolic and Couette flow. However, effects that depend both on the instantaneous distribution and on the sequence of such distributions (such as rotation of a spheroid and appreciable deformation of fluid drops) will be different.

With the aid of the transformation equations (II32) and (II33) we may derive the velocity components inside and outside a drop situated at the center of a field of Couette flow, equation (II10).

(1) *External streamlines.* The velocity components outside the drop in the equatorial plane are[42]

$$V_1 = 0$$

$$V_2 = \frac{\gamma r \sin \phi_1}{2(\lambda + 1)}\left[5\frac{b^3}{r^3}\left(\frac{\lambda b^2}{r^2} - \lambda - \frac{2}{5}\right)\cos^2 \phi_1 - \frac{\lambda b^5}{r^5}\right] \tag{III7}$$

$$V_3 = \frac{\gamma r \cos \phi_1}{2(\lambda + 1)}\left[5\frac{b^3}{r^3}\left(\frac{\lambda b^2}{r^2} - \lambda - \frac{2}{5}\right)\sin^2 \phi_1 - \frac{\lambda b^5}{r^5} + 2(\lambda + 1)\right]$$

the stress components being given by

$$S_{r2} = \frac{5\eta_0 \gamma r \sin \phi_1}{2b(\lambda + 1)}\left(\lambda + \frac{16}{5}\cdot\frac{r^2}{b^2}\cos^2 \phi_1\right)$$

$$S_{r3} = \frac{5\eta_0 \gamma r \cos \phi_1}{2b(\lambda + 1)}\left(\lambda + \frac{16}{5}\cdot\frac{r^2}{b^2}\sin^2 \phi_1\right)$$

$$\tag{III8}$$

As before, we make use of the equation for Couette flow corresponding to equation (III5) and obtain the following differential equation for the external streamlines:

$$\frac{d\phi_1}{dr} = -\frac{\dfrac{\lambda b^5}{2(\lambda + 1)}\cos 2\phi_1 + r^5 \cos^2 \phi_1}{\left(\dfrac{3\lambda b^5}{2(\lambda + 1)}r - \dfrac{(5\lambda + 2)b^3}{2(\lambda + 1)}r^3 + r^6\right) \sin \phi_1 \cos \phi_1}$$

This may be integrated by neglecting terms higher than $(b/r)^5$ to give

$$\frac{r}{b} = \left\{\frac{K_s^{\,3}}{\cos^3 \phi_1} + \frac{5\lambda + 2}{2(\lambda + 1)}\right\}^{1/3} \tag{III9}$$

where K_s is the external streamline constant x_2/b, in which x_2 is the displacement along the X_2 axis from the drop center. Equation (III9) is an exact solution in the case of $\lambda = 0$; for $\lambda \neq 0$ numerical solutions have been worked out.[41] As is evident from equation (III9) and borne out by experiment,[41] the disturbance to a given streamline caused by the sphere is least when $\lambda = 0$ and greatest when $\lambda = \infty$; see Fig. 15. Hence, as calculated by Taylor and demonstrated experimentally,[47] the increase in viscosity of a fluid due to the presence of spheres will be least for $\lambda = 0$ and greatest for $\lambda = \infty$. The viscosity η of a suspension of volume fraction c small enough for particle interactions to be negligible was calculated from equation (III2) and (III4)

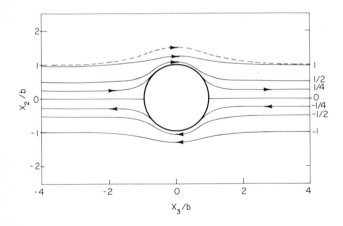

FIG. 15. External streamlines around a fluid sphere at the origin of a field of Couette flow. The lines are calculated from equation (III9) for $\lambda = 0$ and may be compared with the streamline $X_2/b = 1$ for a rigid sphere, shown dashed. (After Bartok and Mason.[42])

[47] M. A. Nawab and S. G. Mason, *Trans. Faraday Soc.* **54**, 1712 (1958).

by Taylor[6] and is

$$\eta = \eta_0 \left(1 + \frac{(5\lambda + 2)}{2(\lambda + 1)}c\right) \qquad \text{(III10)}$$

and thus the intrinsic viscosity, given by

$$[\eta] = \frac{5\lambda + 2}{2(\lambda + 1)} \qquad \text{(III11)}$$

varies between the limits 5/2 for $\lambda = \infty$, as calculated by Einstein,[45,46] and 1 for $\lambda = 0$.

The time of passage along an external streamline may be calculated from equation (III5) by means of equation (III7). Rather surprisingly, in the case of an inviscid drop ($\lambda = 0$) the time of passage along all streamlines, including the drop surface, is found to be unaffected by the presence of the drop; that is, it has the value, obtained from equation (II10), of $X_3/V_3 = \tan \phi_1/\gamma$.

(2) *Internal streamlines.* The velocity components in the equatorial plane within the drop are [41,42]

$$v_1 = 0$$

$$v_2 = \frac{\gamma r \sin \phi_1}{4(\lambda + 1)} \left|\frac{r^2}{b^2}(5 - 4\cos^2 \phi_1) - (2\lambda + 5)\right| \qquad \text{(III12)}$$

$$v_3 = \frac{\gamma r \cos \phi_1}{4(\lambda + 1)} \left|\frac{r^2}{b^2}(5 - 4\sin^2 \phi_1) + (2\lambda - 1)\right|$$

and the differential equation for the streamlines has been shown to be[42]

$$\frac{d\phi_1}{dr} = \frac{(5r^2/b^2)\cos 2\phi_1 + 6\sin^2 \phi_1 + (2\lambda - 1)}{3r[(r^2/b^2) - 1]\sin 2\phi_1} \qquad \text{(III13)}$$

which on integration yields

$$[(r^2/b^2)\cos 2\phi_1 + \lambda + 1]^3[(r^2/b^2) - 1]^2 = k_s \qquad \text{(III14)}$$

Here k_s is the internal streamline constant in Couette flow, varying continuously from 0 at the periphery to $(\lambda + 1)^3$ at the center. Equation (III14) shows that the streamlines are roughly elliptical in shape, the major axis along the X_2 axis, as illustrated in Fig. 16(a), and the eccentricity decreasing with displacement from the drop center. Furthermore, when $\lambda < \frac{1}{2}$, two pockets of fluid circulate around two stagnation points close to the center of the drop; see Fig. 16(b). At a given k_s the streamlines become more and more eccentric as λ decreases.

(3) *Periods of circulation.* The angular velocity ω_1 along a streamline close to the surface inside the drop may be calculated from the equation

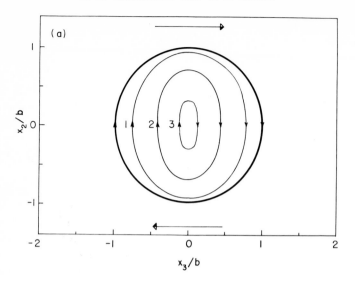

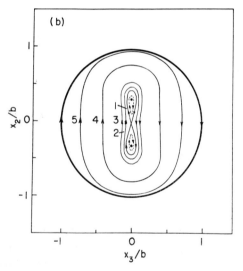

FIG. 16. Circulation inside fluid spheres in Couette flow: (a) for $\lambda = 1$, showing the elliptical streamlines calculated from equation (III14), where k_s increases from streamline 1 to 3, and (b) for $\lambda = \frac{1}{4}$ with increasing k_s, showing two pockets of circulation near the drop center. (After Bartok and Mason.[42])

corresponding to equation (III5) for Couette flow and is a maximum at $\phi_1 = 0$ and a minimum at $\phi_1 = \pi/2$; the reverse is true closer to the drop center and at an intermediate $k_s = 2.0$; ω_1 remains approximately constant.

When $\phi_1 = \pm\pi/4$, then $\omega_1 = \gamma/2$ for all λ along all streamlines. This fact is a consequence of the equivalence of plane hyperbolic and Couette flow when the field described by equation (II29) is rotated to yield that given by equation (II10).

The times of passage along an internal streamline from 0 to ϕ_1 and the period of circulation T around a complete streamline, defined by

$$T = \int_0^{2\pi} \frac{d\phi_1}{\omega_1}$$

can be evaluated analytically for the limiting cases of the interface and close to the center; at intermediate k_s they have been obtained by numerical methods.[41] In terms of the circulation number $\mathfrak{m} = T\gamma/4\pi$ one obtains:

When $k_s = 0,$ $\mathfrak{m} = \dfrac{\lambda - 1}{[\lambda(\lambda + 2)]^{1/2}}$

(III15)

when $k_s = (\lambda + 1)^3,$ $\mathfrak{m} = \dfrac{2(\lambda + 1)}{[(2\lambda + 5)(2\lambda - 1)]^{1/2}}$

The circulation number decreases from the periphery, until a minimum is reached in the vicinity of the streamline $k_s = (\lambda + 1)^{3/4}$, and then increases again to a value greater than that for $k_s = 0$. When $\lambda = \infty$, then $\mathfrak{m} = 1$ everywhere, and the particle is a rigid sphere, for which ω_1 and T are given by equations (II16) and (II17), respectively.

At the interface the condition of continuity of velocity is satisfied by equations (III7) and (III13), since at $r = b$ both reduce to

$$V_1 = v_1 = 0$$

$$V_2 = v_2 = \frac{\gamma b \sin \phi_1}{(\lambda + 1)} \left(-\cos^2 \phi_1 - \frac{\lambda}{2} \right)$$

(III16)

$$V_3 = v_3 = \frac{\gamma b \cos \phi_1}{(\lambda + 1)} \left(-\sin^2 \phi_1 + \frac{\lambda}{2} + 1 \right)$$

and the resulting angular velocity at the surface is

$$\omega_1 = \frac{\gamma(\lambda + 2 - 2 \sin^2 \phi_1)}{2(\lambda + 1)}$$

(III17)

That is, the fluid circulates with variable and positive ω_1, never zero, with a period of circulation given by equation (III15). It is also of interest to note that, when $\lambda = 0$, the angular velocity at the surface is zero at $\phi_1 = \pi/2$.

(4) *Observations with fluid drops.* The characteristic quadrant pattern in hyperbolic flow and the elliptical pattern ($\lambda > \frac{1}{2}$) in Couette flow, besides

the variation in angular velocity along given streamlines, have been observed inside drops (Figs. 14 and 16).[41,48] Quantitative agreement with equation (III14) was obtained in one system in which $\lambda = 1$, but generally measurements proved difficult to make, because internal circulation at low deformation was progressively inhibited with drop age and often ceased altogether. That this was due to an accumulation of surface-active impurities at the interface was shown by the addition of emulsifiers, which markedly inhibited internal circulation.[41,48] Such inhibition may be explained by considering the shear stresses acting at the drop surface following the procedure used by Linton and Sutherland[49] for sedimenting fluid drops.

In plane polar coordinates the tangential shear stress, or ϕ_1 component, of the internal radial stress vector s_r at the surface is

$$s_{r\phi} = \eta_0 b (\partial \omega_1 / \partial r)_{r=b}$$

where $\partial \omega_1 / \partial r$ may be obtained by eliminating $dr/d\phi_1$ from equation (III5):

$$\omega_1 = \frac{1}{r}(V_3 \cos \phi_1 - V_2 \sin \phi_1)$$

From this, using equation (III16), one obtains

$$s_{r\phi} = \frac{5\eta_0\gamma}{2(\lambda + 1)}\cos 2\phi_1 \tag{III18}$$

According to equation (III18), there are four points of zero tangential stress: at ϕ_1 equal to $\pm \pi/4$ and $\pm 3\pi/4$, which define four quadrants, in each of which the direction of the stress is constant and of maximum magnitude at its midpoint, at ϕ_1 equal to 0, $\pm \pi/2$, and π; see Fig. 17. The total shear force P_ϕ acting per unit width in one of these quadrants is, therefore,

$$P_\phi = \int_{-\pi/4}^{\pi/4} s_{r\phi} b \, d\phi_1$$

$$= \frac{5\eta_0\gamma b}{2(\lambda + 1)} \tag{III19}$$

The direction of the shear force reverses between adjacent quadrants, so that the net moment on the drop is zero, as required for equilibrium. When internal circulation is unrestricted, as assumed by the theory,[6,42] the shear forces are resisted by the establishment of the appropriate set of velocity gradients inside the drop.

[48] W. Bartok and S. G. Mason, *J. Colloid Sci.* **14,** 13 (1959).

[49] M. Linton and K. L. Sutherland, *Proc. 2nd Intern. Congr. Surface Activity, London, 1957,* p. 494 *et seq.* Academic Press, New York, 1957.

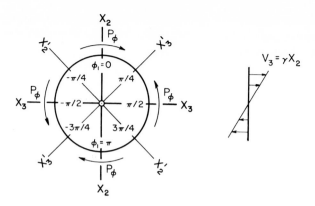

FIG. 17. System of tangential stresses acting at the surface of a drop at the origin of a field of Couette flow. The shear stress $s_{r\phi}$ is a maximum at the points $\phi_1 = 0$, $\pm\pi/2$, and π, and zero at the points $\pm\pi/4$ and $\pm3\pi/4$. When the $X_2'X_3'$ coordinate system describing rectangular hyperbolic flow is rotated in the clockwise direction at the rate $\gamma/2$, the flow becomes identical with that in the Couette field. (After Rumscheidt and Mason.[41])

When internal circulation is *completely stopped*, however, the drop behaves as though $\lambda = \infty$, so that

$$P_\phi = \tfrac{5}{2}\eta_0\gamma b \tag{III20}$$

In the presence of ionic and nonionic surfactants such behavior has been indicated by measurements of emulsion viscosity.[47] A progressive increase in intrinsic viscosity with emulsifier concentration from the value predicted by Taylor's theory, equation (III11), to that for rigid spheres has been reported.[47,50] It has also been observed that emulsion droplets containing tracer particles in systems obeying equation (III11) when settling under gravity exhibit internal circulation patterns rather like smoke rings and similar to the patterns described by several authors[51–54] for sedimenting drops.

When internal circulation is suppressed, the shear force P_ϕ must be balanced by a difference in interfacial tension $\Delta\sigma$ across each quadrant: $-\Delta\sigma = P_\phi$. This is possible if a surface-active agent is present to render the interface viscoelastic, thereby establishing a corresponding difference in surface pressure, $-\Delta p_0 = \Delta\sigma$.

[50] M. Van der Waarden, *J. Colloid Sci.* **5**, 448 (1950).
[51] F. H. Garner and A. H. P. Skelland, *Chem. Eng. Sci.* **4**, 149 (1955).
[52] F. H. Garner and P. J. Haycock, *Proc. Roy. Soc.* **A252**, 457 (1959).
[53] P. Savic and G. T. Boult, *Nat. Res. Council Can. Rept.* **MT-26** (1955).
[54] R. H. Magarvey and J. Kalejs, *Nature* **148**, 377 (1963).

If c_0 is the concentration of surfactant at the interface and ω_1 the peripheral angular velocity, and one may neglect diffusion in and out of the interface,[55] and the equation of continuity becomes

$$\omega_1 c_0(\phi_1) = \text{constant} \tag{III21}$$

Then, if η_s is a combination of the two-dimensional shear viscosity and area viscosity of the surface film,[56,57] it can be shown [41] that

$$\Delta p_0 = \frac{1}{\kappa} \Delta \ln c_0 + \eta_s \, \Delta \left| \frac{d\omega_1}{d\phi_1} \right| \tag{III22}$$

where κ is the surface compressibility.[57] For a purely elastic film $\eta_s = 0$; the internal circulation is inhibited when a compression of the film occurs over a quadrant, such that

$$P_\phi = \Delta p_0 = \frac{1}{\kappa} \Delta \ln c_0 \tag{III23}$$

A calculation based on the experimental conditions[41] shows that, as the element of interface rotates, for example, from $\phi_1 = -\pi/4$ to $\phi_1 = \pi/4$, a linear compression of only 1.25%, followed by the same expansion between $\phi_1 = \pi/4$ and $3\pi/4$, is sufficient to prevent internal circulation. The corresponding change in ω_1, which satisfies equation (III21), would result in a 1.25% higher velocity at $\phi_1 = -\pi/4$ and $3\pi/4$ than at $\phi_1 = \pi/4$ and $-3\pi/4$, and this would be difficult to detect. The amount of surfactant required to cover the interface and give rise to these effects is estimated to be of the order of 6×10^{-8}g for a molecular weight of 300 and a surface concentration of $10 \, \text{Å}^2$ per molecule. It is thus clear why mere traces can have such a profound effect.

3. DEFORMATION AND BURST OF FLUID DROPS

a. The Deformation Equations

The respective normal stresses in the suspending liquid and the drop acting at the surface $r = b$ in plane hyperbolic flow are

$$S'_{rr} = -\frac{5}{2}\gamma\eta_0 \left(\frac{\gamma + 8/5}{\lambda + 1} \right) \cos 2\phi_1' \tag{III24}$$

$$s'_{rr} = \frac{9}{4}\gamma\eta_0 \left(\frac{\lambda}{\lambda + 1} \right) \cos 2\phi_1' \tag{III25}$$

[55] J. L. Moilliet and B. Collie, "Surface Activity," p. 86 et seq. Spons, London, 1951.

[56] W. D. Harkins, "The Physical Chemistry of Surface Films," p. 135. Reinhold, New York, 1952.

[57] J. D. Oldroyd, Proc. Roy. Soc. A232, 567 (1955).

with a resulting pressure difference across the interface, which is given by

$$\Delta p_{rr} = s'_{rr} - S'_{rr}$$

$$= -4\gamma\eta_0 \frac{(19\lambda + 16)}{(16\lambda + 16)} \cos 2\phi_1' \qquad (III26)$$

It is evident from equation (III26) that $\Delta p_{rr} < 0$ when $-\pi/4 < \phi_1' < \pi/4$ and the drop is subject to compressive stresses tending to contract it along the X_2' axis (Fig. 18), and that $\Delta p_{rr} > 0$ when $\pi/4 < \phi_1' < 3\pi/4$ and the drop is subject to tensile stresses tending to extend it along the X_3' axis. The problem of relating these finite differences in normal stress across the interface to the deformation of the drop was solved by Taylor,[7] who assumed that the stresses generated by the flow are balanced by the interfacial tension. Thus, the drop will undergo a change in curvature to satisfy the Laplace equation, equation (III1), in the form

$$\sigma(b_{(1)}^{-1} + b_{(2)}^{-2}) = \Delta p_{rr} + \text{constant} \qquad (III27)$$

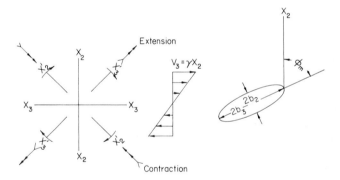

FIG. 18. Principal axes of deformation in plane hyperbolic and Couette flow, shown by the double arrows. The parameters of a deformed fluid drop are shown on the right. (After Rumscheidt and Mason.[58])

It was shown that for small deformations, when equation (III27) is satisfied, that is, when the variation in curvature is proportional to $\cos 2\phi_1'$, the cross section of the drop assumes an elliptical form, given by the polar equation

$$r = b(1 - \mathfrak{D}_1 \cos 2\phi_1') \qquad (III28)$$

[58] F. D. Rumscheidt and S. G. Mason, J. Colloid Sci. 16, 238 (1961).

where r is the radius vector. Here $\mathfrak{D}_1$ is the deformation in the equatorial $X_2'X_3'$ plane, defined as

$$\mathfrak{D}_1 = \frac{b_3 - b_2}{b_3 + b_2} \qquad (III29)$$

where b_2 and b_3 are the principal semiaxes of the deformed drop aligned with the X_2' and X_3' axes, respectively. The third principal axis of the drop, b_1, aligned with the X_1' axis, equals the radius b of the original sphere.[59] The deformation in the $X_1'X_3'$ plane is

$$\mathfrak{D}_2 = \frac{b_3 - b}{b_3 + b} \qquad (III30)$$

For small deformations

$$\mathfrak{D}_1 = \frac{\Delta p_{rr} b}{4\sigma} = \mathfrak{D} \qquad (III31)$$

where $\mathfrak{D}$ is the ratio of viscous forces to surface-tension forces, given by

$$\mathfrak{D} = \frac{\gamma b \eta_0}{\sigma} \cdot \frac{19\lambda + 16}{16\lambda + 16} \qquad (III32)$$

If $\mathfrak{D}$ is small, equations (III24) to (III32) apply to Couette flow when $\sin 2\phi_1$ is substituted for $-\cos 2\phi_1'$, provided that the rotation of the particle is slow enough to allow it sufficient time to accommodate its shape to the changing stresses. The b_3 axis of the drop is then at $\phi_1 = \pi/4$, that is, along the principal axis of elongation, as in hyperbolic flow. Unlike hyperbolic flow, however, where the line of fluid elements extending at the greatest rate (those parallel to the X_3' axis) remain in the direction of the maximal rate of extension, in Couette flow the line of fluid elements being extended most rapidly is at $\phi_1 = \pi/4$ (Fig. 17) and is being continuously rotated away toward $\phi_1 = \pi/2$, where there is neither extension nor contraction. Owing to the finite rate of accommodation of the particle shape to the deforming stresses, the orientation $\phi_1 = \phi_m$ of the principal axis of the drop increases beyond $\pi/4$ as γ increases. This effect has been investigated in the case of an elastic sphere by Cerf,[60] who pointed out that alignment at $\phi_m = \pi/4$ arises from taking only first-order terms in γ in the stress tensor; when second-order terms are taken into account,

$$\phi_m = \pi/4 + (1 + 2\lambda/5)\mathfrak{D} \qquad (III33a)$$

$$= \pi/4 + (1 + 2\lambda/5)\mathfrak{D}_1 \qquad (III33b)$$

[59] C. E. Chaffey, H. Brenner, and S. G. Mason, *Rheol. Acta* **4**, 56 (1965).
[60] R. Cerf, *J. Chim. Phys.* **48**, 59 (1951).

if equation (III31) applies. In irrotational hyperbolic flow $\phi_m' = \pi/2$ for all values of $\mathfrak{D}$.

b. Deformation at Burst

Taylor suggested that, when the maximum value of Δp_{rr} tending to disrupt the drop exceeds the force, due to surface tension, which tends to hold it together, the drop will burst. Assuming equation (III26) to hold and neglecting deviation from spherical shape, this occurs when

$$4\gamma\eta_0 \frac{19\lambda + 16}{16\lambda + 16} > \frac{2\sigma}{b}$$

and, if equation (III31) still holds, the drop will burst in both Couette and hyperbolic flow when

$$\mathfrak{D}_{1,B} = \mathfrak{D}_B = \tfrac{1}{2} \tag{III34}$$

c. Deformation at High λ

An approximate analysis of a special case, in which λ and $\gamma b\eta_0$ are so high that the surface tension forces opposing deformation are negligible compared with those due to viscosity, and there is continuity of normal stress across the interface, was given by Taylor.[7] The velocity components at the drop surface are now

$$v_2' = \frac{5\gamma b}{2(2\lambda + 3)} \sin \phi_1'$$

$$v_3' = -\frac{5\gamma b}{2(2\lambda + 3)} \cos \phi_1'$$

whence in hyperbolic flow the drop "bursts" at a rate of radial extension given by

$$\left(\frac{\partial r}{\partial t}\right)_{\phi_1'} = -\frac{5\gamma b}{2(2\lambda + 3)} \cos 2\phi_1' \tag{III35}$$

In Couette flow, as found experimentally,[7,58] the particle rotates and assumes a steady shape in space, such that at a fixed ϕ_1 the radial component of velocity due to rotation is equal to that due to deformation; see equation (III35). This condition is satisfied by the surface

$$r = b\left(1 - \frac{5}{2(2\lambda + 3)} \cos 2\phi_1\right) \tag{III36}$$

corresponding to a limiting deformation, which is independent of γ and is given by

$$\mathfrak{D}' = \frac{5}{2(2\lambda + 3)} \qquad (III37)$$

and an orientation along the direction of flow ($\phi_m = \pi/2$).

d. Observed Deformations

At low velocity gradients the behavior of the drops in both hyperbolic and Couette flow has been found to be in good accord with equations (III31) to (III33).[7,58] Thus, the values of σ computed from the observed $\mathfrak{D}_1$ at low velocity gradients in nineteen systems were in remarkable agreement with values measured by means of the ring or the pendant drop method.[58] In Couette flow the drop is first deformed into an ellipsoid, its longest axis initially aligned at $\phi_m = \pi/4$ and both $\mathfrak{D}_1$ and ϕ_m increasing with γ. In some systems, however, in which internal circulation is known to be inhibited, the measured $\mathfrak{D}_1$ is found to be higher than that given by equation (III31). This can be explained if the drop behaves as though $\lambda = \infty$, when the normal and tangential stresses at the interface are then

$$S_{rr} = \tfrac{5}{2}\gamma\eta_0 \sin 2\phi_1$$
$$S_{r\phi} = \tfrac{5}{2}\gamma\eta_0 \cos 2\phi_1 \qquad (III38)$$

The radial stress thus has magnitude $S_r = \tfrac{5}{2}\gamma\eta_0$, its direction making an angle $\tfrac{1}{2}\pi - 2\phi_1$ with the normal.

If $S_{r\phi}$ is borne by the interface, the total deforming stress S_r is thus greater than that assumed by the theory, except at $\phi_1 = \pm\pi/4$, where it is the same (zero). Hence, the measured value of $\mathfrak{D}_1$ should be greater than that given by equation (III31), and σ calculated from $\mathfrak{D}_1$ should be too small, as is found.

As the deformation increases beyond the range of applicability of equation (III31), three distinct modes of behavior are observed; these depend mainly[7,58] on λ and are illustrated in Figs. 19 and 20.

(1) *Class A.* At low values of λ (< 0.2), as the velocity gradient exceeds the value γ_B corresponding to burst, the drops in hyperbolic flow develop pointed ends [Fig. 19(a)], from which fragments of disperse phase are released. In Couette flow the drops assume a sigmoid shape with pointed ends [Fig. 20(a)].

(2) *Class B.* Here $\lambda > 0.2$. In one type of breakup, class B-1, observed only in Couette flow,[58] the central portion of the drop suddenly begins to extend into a cylinder, which then progressively necks off in the middle, until two nearly identical daughter drops are formed, which are separated by three

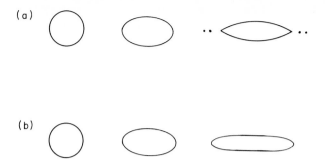

FIG. 19. Deformation and burst of drops in plane hyperbolic flow, as seen in the equatorial plane. (a) Water drop in a viscous oil, $\lambda \approx 10^{-4}$, which forms pointed ends at breakup, small droplets being ejected. (b) Oil in oil system, $\lambda = 1$, in which the drop after breakup becomes extended into a long cylindrical thread, which eventually breaks up. (After Rumscheidt and Mason.[58])

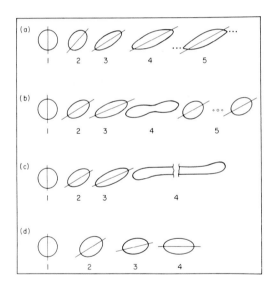

FIG. 20. Changes in drop shape seen during deformation and breakup in Couette flow. (a) Class A deformation corresponding to (a) in Fig. 19. (b) Class B-1 deformation (not observed in hyperbolic flow), in which necking-off results in satellite drops. (c) Class B-2 deformation corresponding to (b) in Fig. 19. (d) Class C deformation, in which no burst is seen in Couette flow. (After Rumscheidt and Mason.[58])

satellite droplets; see Fig. 20(b). In the other type, class B-2, common to both flows, the rate of growth of the disturbance is too slow for necking to occur, and the drop extends into a thread, which progressively increases in length until, at a sufficiently small diameter, it breaks up into tiny droplets; see Figs. 19(b) and 20(c). If flow is arrested before this occurs, the thread after a short time becomes varicose and finally breaks up into a series of regularly spaced drops. The breakup of such stationary threads has been treated theoretically[61] and the theory confirmed by experiment.[62]

In Couette flow class B-2 breakup occurs in systems having $\lambda < 2$, whereas in hyperbolic flow, as predicted by equation (III35), drops in systems of high λ are also drawn out into threads.

For drops in twenty systems showing class A and B breakup mean values of $\mathfrak{D}_{1,B} = 0.59 \pm 0.06$ and $\mathfrak{D}_B = 0.52 \pm 0.10$ (based on the measured γ_B and σ) have been found. Considering the fact that equation (III31) really only applies at small deformation, the agreement with equation (III34) is good.

(3) *Class C.* When $\lambda > 2$, in Couette flow, the behavior of the drops is as predicted by the theory, equation (III36): as γ is increased, the ellipsoidal drop reaches an upper limit of deformation at which $\phi_m = \pi/2$; see Fig. 20(d).

4. Deformation and Burst in Shear and Electric Fields

a. Electric Field Only

Several theoretical and experimental investigations into the deformation, under the action of a uniform electric field, of suspended incompressible fluid drops in a nonconducting medium have been given. Results were obtained for both conducting drops,[8,9,63,64] where it is agreed that deformation into a prolate spheroid occurs, and for nonconducting drops,[8,9,63,65,66] where it has been stated[8,9] that in all cases the spheroid is prolate. In an extensive series of experiments,[63] however, it was shown that oblate spheroids resulted when q, the dielectric constant ratio of drop phase to suspending phase, was less than unity. Neither the calculations of drop deformation, based on energy equations,[9,65] nor those based on force equations,[8,63,64] one of which is presented below,[63] which assume a uniform electric field inside the drop, could account for this behavior. It was suggested by O'Konski and Harris[67] that such an effect could be brought about by the modification of the electric fields inside and outside the drops by electrical conduction. However, the necessary conditions were not present in the systems in which oblate spheroids were observed.[63] Moreover, as Taylor pointed out,[10] in

[61] S. Tomotika, *Proc. Roy. Soc.* **A153**, 302 (1936).

[62] F. D. Rumscheidt and S. G. Mason, *J. Colloid Sci.* **17**, 260 (1962).

[63] R. S. Allan and S. G. Mason, *Proc. Roy. Soc.* **A267**, 45 (1962).

[64] G. I. Taylor, *Proc. Roy. Soc.* **A280**, 383 (1964).

[65] N. K. Nayyar and G. S. Murty, *Proc. Natl. Inst. Sci. India Pt. A* **25**, 373 (1959).

[66] E. S. Rajagopal, *J. Indian Inst. Sci.* **40**, 152 (1958).

[67] C. T. O'Konski and F. E. Harris, *J. Phys. Chem.* **61**, 1172 (1957).

common with other theories employing energy equations, that of O'Konski and Harris,[67] dealing with conducting fluids, neglects the existence of a surface charge on the drop. One effect of such a charge is to produce circulatory currents inside and outside the drop. For equilibrium to be maintained the interfacial electric stresses acting on the deformed drop can only be balanced by a variable pressure difference between the drop and the surrounding medium. Therefore, energy principles should not be applied when irreversible processes of viscous flow and electrical conduction are constantly occurring. Taylor[10] calculated the equilibrium conditions for a spherical and for a slightly deformed drop and found that the function that discriminated between prolate and oblate forms was independent of the magnitude of the electric field and was given by

$$\mathfrak{f}(q, \lambda, \mathscr{R}) = \frac{1}{q}(\mathscr{R}^2 + 1) - 2 + 3\left(\frac{\mathscr{R}}{q} - 1\right)\left(\frac{3\lambda + 5}{5\lambda + 5}\right) \qquad (\text{III}39)$$

where $\mathscr{R}$ is the resistivity ratio of suspending phase to suspended phase. When $\mathfrak{f}(q, \lambda, \mathscr{R})$ is negative, the drop will be oblate; when positive, it will be prolate. Taylor showed[10] that the theory correctly predicts the drop deformation observed in the investigation[63] described below, in which circulation currents in both phases had been seen.[10] Before discussing these experimental results an approximate theory developed from force equations, which predicts the observed deformation in many systems, is given.

The normal electric stress acting on the interface of a drop of dielectric constant $q\varepsilon$ in a medium of dielectric constant ε subjected to a uniform electric field of strength $\mathscr{E}_0$ directed along the X_2 axis of the rectangular coordinate system X_1, X_2, X_3 is given by Smythe[68]:

$$S_{rr}^{(e)} = \frac{q-1}{8\pi} \cdot \frac{\mathscr{D}_\phi{}^2}{\varepsilon} + \frac{\mathscr{D}_r{}^2}{q\varepsilon} \qquad (\text{III}40)$$

Here $\mathscr{D}_\phi$ and $\mathscr{D}_r$ are the respective tangential and normal electric displacements $\mathscr{D}_\phi = \varepsilon\mathscr{E}_\phi$ and $\mathscr{D}_r = \varepsilon\mathscr{E}_r$, and $\mathscr{E}_\phi$ and $\mathscr{E}_r$ are the tangential and normal components of the local field strength at the sphere surface, which may be calculated from the potential V_e outside the sphere by means of the relations

$$\mathscr{E}_\phi = \frac{1}{b}\left(\frac{\partial V_e}{\partial \phi_1}\right)_{r=b} = \frac{3}{q+2}\mathscr{E}_0 \sin \phi_1$$

$$\mathscr{E}_r = \frac{1}{b}\left(\frac{\partial V_e}{\partial r}\right)_{r=b} = \frac{3q}{q+2}\mathscr{E}_0 \cos \phi_1 \qquad (\text{III}41)$$

[68] W. R. Smythe, "Static and Dynamic Electricity," 2nd. ed. McGraw-Hill, New York, 1953.

where V_e is given by the well-known electrostatic equation

$$V_e = \left(1 - \frac{b^3}{r^3} \frac{(q-1)}{(q+2)}\right) \mathscr{E}_0 r \cos \phi_1$$

On combining equations (III40) and (III41) and simplifying one obtains

$$S_{rr}^{(e)} = \Psi^{(e)}\left(\frac{q+1}{q-1} + \cos 2\phi_1\right) \tag{III42}$$

where

$$\Psi^{(e)} = \frac{9\varepsilon\mathscr{E}_0{}^2(q-1)^2}{16\pi(q+2)^2}$$

As in shear flow, we require the drop curvature to change, so that the Laplace equation, now in the form

$$\sigma(b_{(1)}^{-1} + b_{(2)}^{-1}) = \Psi^{(e)} \cos 2\phi_1 + \text{constant} \tag{III43}$$

is satisfied. The deformation in a stress system in which

$$\sigma(b_{(1)}^{-1} + b_{(2)}^{-1}) = \Psi^{(s)} \sin 2\phi_1 + \text{constant}$$

and in which $\Psi^{(s)} = 4\gamma\eta_0(19\lambda + 16)/(16\lambda + 16)$ has already been calculated, so that in this case the equation of the drop is again given by equation (III28) with the electrical deformation $\mathfrak{D}_1^{(e)}$

$$\mathfrak{D}_1^{(e)} = \frac{b\Psi^{(e)}}{4\sigma} = \frac{9b\varepsilon\mathscr{E}_0{}^2(q-1)^2}{64\pi\sigma(q+2)^2} \tag{III44}$$

The theory thus predicts deformation into a prolate ellipsoid whose major and minor axes in the equatorial plane are given by equation (III29) provided the deformation is small, and with $\phi_m = 0$. This result is identical with that obtained by O'Konski and Thacher[9] in an earlier theory, in which they used energy instead of force equations.

As the field strength is increased, the stresses on the drop increase and when these exceed the forces due to surface tension, the drop bursts. This will occur when

$$\Psi_B^{(e)}\frac{q+1}{q-1} = \frac{2\sigma}{b}$$

whence from equation (III44)

$$\mathfrak{D}_B^{(e)} = \frac{1}{2} \cdot \frac{q-1}{q+1} \tag{III45}$$

and the electric field strength at burst is

$$\mathscr{E}_{0,B} = \frac{32\pi\sigma(q + 2)^2}{9b\varepsilon(q - 1)(q + 1)} \tag{III46}$$

When $q = \infty$, the deformation at burst is $\frac{1}{2}$ that in shear flow. This compares with the values for $\mathfrak{D}_B^{(e)}$ of 0.30, given by Garton and Krasucki,[8] and of 0.31, calculated by Taylor.[64] However, equations (III45) and (III46) are only rough approximations, because they are based on a spherical model which also neglects the internal pressure, whereas the drops are ellipsoidal.

b. Combined Shear and Electric Fields

When the normal stresses due to shear flow and electric field are added, we may write, taking into account drop rotation in shear to $\phi_m > \pi/4$,

$$\sigma(b_{(1)}^{-1} + b_{(2)}^{-1}) = \Psi^{(e)} \cos 2\phi_1 + \Psi^{(s)} \sin 2(\phi_1 + \psi) + \text{constant}$$

$$= \Psi^{(n)} \cos 2(\phi_1 - \phi_n) + \text{constant} \tag{III47}$$

where

$$\psi = (1 + \tfrac{2}{5}\lambda)\mathfrak{D}$$

$$\Psi^{(n)} = (\Psi^{(e)2} + 2\Psi^{(e)}\Psi^{(s)} \sin 2\psi + \Psi^{(s)2})$$

$$\phi_n = \frac{1}{2}\tan^{-1}\left(\frac{\Psi^{(s)} \cos 2\psi}{\Psi^{(e)} + \Psi^{(s)} \sin 2\psi}\right)$$

By analogy with deformation in shear flow the drop assumes an ellipsoidal shape with a net deformation

$$\mathfrak{D}^{(n)} = (\mathfrak{D}_1^{(e)2} - 2\mathfrak{D}_1\mathfrak{D}_1^{(e)} \sin \psi + \mathfrak{D}_1{}^2)^{1/2} \tag{III48}$$

and is oriented at an angle

$$\phi_n = \frac{1}{2}\tan^{-1}\left(\frac{\mathfrak{D}_1 \cos 2\psi}{\mathfrak{D}_1^{(e)} - \mathfrak{D}_1 \sin 2\psi}\right) \tag{III49}$$

When the deformation is very small and $\psi \approx 0$, these equations reduce to

$$\mathfrak{D}^{(n)} = (\mathfrak{D}_1^{(e)2} + \mathfrak{D}_1{}^2)^{1/2} \quad \text{and} \quad \phi_n = \tan^{-1}\left(\frac{\mathfrak{D}_1}{\mathfrak{D}_1^{(e)}}\right)$$

c. Observed Drop Deformation

Liquid drops having dielectric constants greater than the suspending phase are observed to deform in an electric field in a manner predicted by the theory given above, and for conducting drops ($q = \infty$) the interfacial tensions calculated from electrical deformation agree with those obtained from deformation in shear, equation (III31). The effect of the electric field, when superimposed upon the shear field in rotating the drops toward lower orientation angles can be demonstrated, and the measured net deformation $\mathfrak{D}^{(n)}$ and orientations ϕ_n agree with those given by equations (III49) and (III50). However, as mentioned above, drops having a lower dielectric constant than the suspending phase, $q < 1$, are deformed into *oblate spheroids*,[63] where $\mathfrak{D}_1 < 1$; see Fig. 21(c).

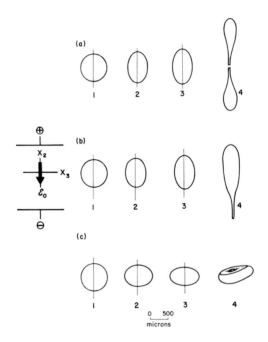

FIG. 21. Deformation and breakup of drops in an electric field applied along the X_2 axis: (a) $q = \infty$; the drop separates into halves with many satellite drops between; (b) $1 < q < \infty$; the drop is drawn out into a thread beginning at the end nearest the negative electrode (followed by extension of the other end to the positive electrode); (c) $q < 1$; the drops flatten into a sheet initially in the $X_2 X_3$ plane, which then folds over until no longer coplanar and eventually breaks up unevenly. (After Allan and Mason.[63])

Three distinct modes of behavior have been observed as $\mathscr{E}_0$ is increased and breakup finally occurs:

(1) $q = \infty$. The drop remains prolate-spheroidal as $\mathscr{E}_0$ increases, until suddenly it separates into two approximately equal parts with a larger number of fine droplets in between; see Fig. 21(a). The two principal daughter drops move apart rapidly as a result of the separation of charge.

(2) $1 < q < \infty$. The deformation is as in (1), but at breakup the end of the drop near the negative electrode is pulled out into a thin thread, which moves directly toward the negative electrode [Fig. 21(b)], indicating a negative charge leakage from the other end of the drop. At higher $\mathscr{E}_0$ the other end of the drop also extends into a thread, which moves toward the positive electrode. The threads break up only after discharge of the field.

(3) $q < 1$. At low λ the drop deforms into a sheet at breakup which is initially aligned in the $X_2 X_3$ plane; see Fig. 21(c). The flattened drop then folds over and twists, until it is no longer coplanar, and eventually breaks up unevenly. At high λ the drops flatten and fold over, but without bursting, up to high $\mathscr{E}_0$.

Drop breakup in combined shear and electric fields is found to be similar to that in the electric field, with some modifications due to the velocity gradient, which leads to a greater variety of modes of breakup than is possible in an electric field alone.

In all cases of drops' exhibiting positive deformations the values of $\mathfrak{D}_1^{(e)}$ and $\mathscr{E}_0$ at burst are appreciably lower than those predicted by equations (III45) and (III46). Considering that the theory assumes the drop to be spherical and that there is an increase in $S_{rr}^{(e)}$ due to field enhancement as the drop extends toward the electrodes, this result is to be expected.

Mention should be made of the investigations reported on the electrical dispersion of liquid threads emerging from a capillary,[69–71] the bursting of soap bubbles in electric fields,[72] and the disintegration of water drops in an electric field.[64]

5. Forces Acting on Rigid Spheroids in Couette Flow

a. Stresses at the Surface

Jeffery[2] has given the equations for the components of the stress S acting on the surface of an ellipsoid at the center of a field of fluid motion with arbitrary components of dilatation and rotation. Referred to the X_1^0, X_2^0, X_3^0 coordinate system, in the case of a spheroid in Couette flow these reduce to

$$S_{11} = \frac{4\eta_0 J}{b_1 b_2{}^2} \{2j_{11} - b_1 b_2{}^2 [j_{11}(\alpha_1 - \alpha_2)]\} \qquad (\text{III}50a)$$

$$S_{12} = S_{21} = \frac{8\eta_0 J}{b_1 b_2{}^2} j_{12} \qquad (\text{III}50b)$$

[69] M. A. Nawab and S. G. Mason, J. Colloid Sci. 13, 179 (1958).
[70] B. Vonnegut and R. L. Neubauer, J. Colloid Sci. 7, 616 (1952).
[71] J. Zeleny, Phys. Rev. 10, No. 2, 1 (1917).
[72] C. T. R. Wilson and G. I. Taylor, Proc. Cambridge Phil. Soc. 22, 728 (1925).

Here J, which is the perpendicular distance of the tangent plane at the point (X_1^0, X_2^0, X_3^0) from the spheroid center, is given by

$$J = [b_1^4/X_1^{0^2} + b_2^4/(X_2^{0^2} + X_3^{0^2})]^{1/2}$$

and j_{11}, j_{12}, etc., are functions of the spheroid integrals α_1, α_2, etc., and of the components of dilatation or rate of shear:

$$j_{11} = \frac{D_{11}^0}{6\alpha_2''},$$

$$j_{22} = \frac{a_2''(2D_{22}^0 - D_{33}^0) - \alpha_1'' D_{11}}{6\alpha_2''(\alpha_2'' + 2\alpha_1'')}, \qquad \text{(III51a)}$$

$$j_{33} = \frac{\alpha_2''(2D_{33}^0 - D_{22}^0) - \alpha_1'' D_{11}^0}{6\alpha_2''(\alpha_2'' + 2\alpha_1'')}$$

$$j_{12} = \frac{D_{12}^0}{2\alpha_2'(b_1^2 + b_2^2)}, \qquad j_{13} = \frac{D_{13}^0}{2\alpha_2'(b_1^2 + b_2^2)}, \qquad j_{23} = \frac{D_{23}^0}{4\alpha_1' b_2^2} \qquad \text{(III51b)}$$

The components of the rate of shear expressed in the polar coordinates ϕ_1 and θ_1 of the coordinate system shown in Fig. 1 are

$$D_{11}^0 = \frac{\gamma}{2} \sin 2\phi_1 \sin^2 \theta_1, \qquad D_{12}^0 = \frac{\gamma}{4} \sin 2\phi_1 \sin 2\theta_1$$

$$D_{22}^0 = \frac{\gamma}{2} \sin 2\phi_1 \cos^2 \theta_1, \qquad D_{13}^0 = \frac{\gamma}{2} \cos 2\phi_1 \sin \theta_1 \qquad \text{(III52)}$$

$$D_{33}^0 = \frac{\gamma}{2} \sin 2\phi_1, \qquad D_{23}^0 = \frac{\gamma}{2} \cos 2\phi_1 \cos \theta_1$$

Finally, the spheroid integrals α_1, α_2, etc., are defined by equations (II2) and (II5), and in terms of the parameter $\mathfrak{A}$ of equation (II43) and the particle axis ratio r_p they can be expressed as[73]

$$b_1 b_2^2 \alpha_1 = 2 - 2\mathfrak{A}, \qquad\qquad b_1 b_2^2 \alpha_2 = \mathfrak{A}$$

$$b_1 b_2^4 \alpha_1' = \frac{\frac{1}{2} r_p - \frac{3}{4} \mathfrak{A}}{r_p^2 - 1}, \qquad\qquad b_1 b_2^4 \alpha_2' = \frac{3\mathfrak{A} - 2}{r_p^2 - 1} \qquad \text{(III53)}$$

$$b_1 b_2^2 \alpha_1'' = \frac{(r_p^2 - \frac{1}{4})\mathfrak{A} - \frac{1}{2} r_p^2}{r_p^2 - 1}, \qquad b_1 b_2^2 \alpha_2'' = \frac{2r_p^2 - (2r_p^2 + 1)\mathfrak{A}}{r_p^2 - 1}$$

The components of the force $\mathbf{p}$ per unit area acting on a point on the surface of the spheroid may then be calculated from the stress components by means

[73] C. E. Chaffey and S. G. Mason, *J. Colloid Sci.* **20**, 330 (1965).

of the following equations[2]:

$$p_1 = \frac{S_{11}X_1{}^0}{b_1{}^2} + \frac{1}{b_2{}^2}(S_{12}X_2{}^0 + S_{13}X_3{}^0)$$

$$p_2 = \frac{S_{12}X_1{}^0}{b_1{}^2} + \frac{1}{b_2{}^2}(S_{22}X_2{}^0 + S_{23}X_3{}^0) \qquad \text{(III54)}$$

$$p_3 = \frac{S_{13}X_1{}^0}{b_1{}^2} + \frac{1}{b_2{}^2}(S_{23}X_2{}^0 + S_{33}X_3{}^0)$$

Since apart from J, which is even in $X_1{}^0$, $X_2{}^0$, $X_3{}^0$, only first powers of the spheroid coordinates appear in equation (III54), they will separately vanish on integration over the whole particle surface. Hence no *resultant* force acts on the spheroid.

b. Forces on Portions of the Surface

In the discussion of the behavior of liquid drops in laminar flow it was shown that the particle in Couette flow is subjected to a system of normal and tangential stresses, resulting in tensile forces in the first and third quadrants, and corresponding compressive forces in the second and fourth quadrants (Fig. 18).

In the case of a rotating rigid spheroid it is possible that during a part of its orbit compressive stresses acting along one of its principal axes will lead to bending of the particle. The necessary conditions for such shear-induced bending, or buckling, are considered below. First, however, it is necessary to calculate the resultant force **P** acting on a portion of the surface, in particular the component P_1 acting in the direction of the axis of revolution on the central cross section, $X_1{}^0 = 0$.

The net $X_1{}^0$ component of force acting on a portion of the surface is

$$P_1 = \int p_1 \, dA \qquad \text{(III55)}$$

where the element of surface area, dA, in terms of the *spheroid* polar coordinates θ and ϕ $(X_1{}^0 = b_1 \cos\theta,\ X_2{}^0 = b_2 \sin\theta \sin\phi,\ \text{and}\ X_3{}^0 = b_2 \sin\theta \cos\phi)$ is

$$dA = \sin\theta\,(b_1{}^2 b_2{}^2 \sin^2\theta + b_2{}^4 \cos^2\theta)^{1/2}\,d\theta\,d\phi$$

which, it can be readily shown, may be written as

$$dA = \frac{b_1 b_2{}^2}{J} \sin\theta\,d\theta\,d\phi$$

Since area is symmetric with respect to $X_2{}^0$ and $X_3{}^0$, terms odd in these do not contribute and, hence, in substituting for p_1 and dA in equation (III55)

138   H. L. GOLDSMITH AND S. G. MASON

one obtains, for the force acting on a portion $X_1^0 < b_1$, of the surface,

$$P_1(X_1^0) = \frac{b_2^2}{J} S_{11} \int_0^{2\pi} d\phi \int_0^{\cos^{-1}(X_1^0/b_1)} \sin\theta\cos\theta \, d\theta$$

$$= \frac{\pi b_2^2}{J} S_{11} \left(1 - \frac{X_1^{02}}{b_1^2}\right) \tag{III56}$$

Upon substitution of S_{11} from equation (III50a) and simplification this yields, for the force acting on half the ellipsoid surface,

$$P_1 = \pi\eta_0\gamma b_2^2 \sin 2\phi_1 \sin^2\theta_1 \frac{(r_p^2 - 1)\mathfrak{A}}{2r_p^2 - \mathfrak{A}(2r_p^2 + 1)} \tag{III57}$$

Similarly, it can be shown that the X_2^0 and X_3^0 components of the force **P** acting on half the ellipsoid surface are

$$P_2 = \pi\eta_0\gamma b_2^2 \sin 2\phi_1 \sin 2\theta_1 \frac{r_p^2 - 1}{(r_p^2 + 1)(3\mathfrak{A} - 2)} \tag{III58}$$

$$P_3 = 2\pi\eta_0\gamma b_2^2 \cos 2\phi_1 \sin\theta_1 \frac{r_p^2 - 1}{(r_p^2 + 1)(3\mathfrak{A} - 2)} \tag{III59}$$

Equation (III57) shows that the axial force P_1 depends on the dimensions of the particle, the product $\gamma\eta_0$, and its orientation with respect to ϕ_1 and θ_1. When $r_p \gg 1$, equation (III57) may be simplified by making use of the formula

$$\cosh^{-1} r_p = \ln[r_p + (r_p^2 - 1)^{1/2}]$$

and by keeping only terms larger than $r_p^{-2} \ln r_p$ in $\mathfrak{A}$, equation (II43), one obtains

$$P_1' = \frac{\pi\eta_0\gamma b_1^2 \sin 2\phi_1 \sin^2\theta_1}{2(\ln 2r_p - 1.5)} \tag{III60}$$

This may be compared with an expression

$$P_1' = \frac{\pi\eta_0\gamma b_1^2 \sin 2\phi_1 \sin^2\theta_1}{2(\ln 2r_p - 1.75)} \tag{III61}$$

previously deduced[29] from Burgers' approximate theory[34] for thin rigid rods, in which the thickness of the particle is neglected. Equation (III61) reduces to equation (III60), when one substitutes for r_p an equivalent ellipsoidal axis ratio such that $\ln(2r_e) - 1.5 = \ln(2r_p) - 1.75$, giving $r_e/r_p = 0.78$. This quotient is different from that characteristic of cylindrical rods' rotational orbits in shear flow (Fig. 7), where $r_e/r_p = \frac{1}{2}$, roughly, when $r_p \gg 1$.

The effect of orbit constant on the magnitude of P_1' may be obtained by substituting $\sin^2 \theta_1 = \tan^2 \theta_1/(\tan^2 \theta_1 + 1)$ in equation (III60) and using equation (II20) to get

$$P_1' = \frac{\pi \eta_0 \gamma b_1^2}{2(\ln 2r_p - 1.5)} \cdot \frac{\sin 2\phi_1 C^2 r_e^2}{C^2 r_e^2 + r_e^2 \cos^2 \phi_1 + \sin^2 \phi_1}$$

$$= \frac{\pi \eta_0 \gamma b_1^2}{\ln 2r_p - 1.5} \cdot \frac{C^2 \sin 2\phi_1}{2(C^2 + 1 - \sin^2 \phi_1)}, \quad \text{for} \quad r_p^2 \gg 1$$

Thus at a given C the quantity P_1' is a maximum at $\phi_1 = \pi/4$ (tension) and a minimum at $\phi_1 = -\pi/4$ (compression); at a given ϕ_1 it decreases with decreasing C, the greatest stresses being exerted on a particle that rotates wholly in the $X_2 X_3$ plane ($C = \infty$); when at $\phi_1 = \pm \pi/4$ P_1' has the value

$$P'_{1M} = \pm \pi \eta_0 \gamma b_1^2/2(\ln 2r_p - 1.5) \tag{III62}$$

c. Shear-Induced Buckling of Rods

It is now possible to calculate the least force required to buckle a rod in the orbit $C = \infty$ by assuming that r_p in equations (III60) and (III62) may be replaced with r_e. Consider the rod to be hinged at the ends and subject to a system of axial forces P_1 acting on an area element dA, as shown in Fig. 22. The total compressive force at a point X_1^0 acting on a cross section of the rod is, from equation (III56), with the approximations introduced for $r_p \gg 1$,

$$P_1'(X_1^0) = \frac{\pi \eta_0 \gamma (b_1^2 - X_1^{02})}{2(\ln 2r_e - 1.5)} \tag{III63}$$

The equation for the shape of the rod suffering small deformations under compressive forces is given by Euler's classical equation,[74]

$$E_b I \frac{d^2 X_2^0}{dX_1^{02}} + P_1' X_2^0 = 0 \tag{III64}$$

where E_b is the bending modulus of the rod, I the moment of inertia of the smallest cross section, and X_2^0 the displacement at X_1^0 (Fig. 22). By a change of variable equation (III64) can be transformed into a differential equation

[74] J. Prescott, "Applied Elasticity," Chapt. 4. Dover, New York, 1946.

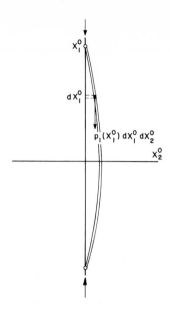

FIG. 22. A rod hinged at both ends, used as a model in the theory of shear-induced buckling. Euler's equation, equation (III64), is derived by assuming that the rod is slightly bent under axial compression. (After Forgacs and Mason.[29])

having solutions in terms of Hermite polynomials.[29] The solution corresponding to the first mode of buckling of the rod is[29]

$$(P_1'/E_bI)(b_1{}^2 - X_1^{0^2})b_1{}^4 = 1$$

and, since $I = \pi b_2{}^4/4$, the minimum value $(\gamma\eta_0)_{\text{crit.}}$ at which rodlike particles of axis ratio r_p will buckle at the position of maximum compression $(\phi_1 = -\pi/4)$ is

$$(\gamma\eta_0)_{\text{crit.}} = \frac{E_b(\ln 2r_e - 1.50)}{2r_p{}^4} \tag{III65}$$

It follows from equation (III65) that for a given value of $\gamma\eta_0$ and E_b there exists also a critical value of $r_p = r_{\text{crit.}}$, at which shear-induced buckling will set in.

Experiments performed with Dacron and nylon filaments having r_p in the range[29] 150 to 300 have shown quite good agreement with equation (III65). The calculated values of the bending moduli, although of the same order as the corresponding values of the tensile moduli, are about 2 to 4 times greater.[29] In each sample E_b is found to decrease slightly with increasing r_p, perhaps owing to the greater incidence of slight permanent deformations in the long samples.

An interesting illustration of shear-induced buckling of rodlike particles is provided by the rotation, in Couette flow, of chains of rigid spheres.[75] As will be described in Section V, doublets and higher-order multiplets of spheres can be made by bringing together metal-coated spheres in a straight chain under the action of an electric field directed across the annulus of a Couette apparatus (along the X_2 axis). When the field is turned off, it is found that at low shear rates these "n-lets" ($n = 2$ to 10) rotate as rigid rods in Couette flow and follow Jeffery's equation for ϕ_1, equation (III3).

On the basis of the creeping motion equations and lubrication theory it has been demonstrated[75] that a straight chain of spheres in which the particles are in contact should rotate as a rigid body, without relative rotation of particles, in a spherical elliptical orbit similar to that predicted by Jeffery[2] for a prolate spheroid. However, chains having finite but small gap widths between spheres will exhibit periodic stretching, with the chain length a *minimum* at $\phi_1 = 0$ and *maximum* at $\phi_1 = \pi/2$, and bending which is generally progressive.

The chains of rigid spheres are indeed found to rotate as rigid rods, and the chain length varies periodically[75,75a] as predicted by the theory; however, as with flexible fibers[85] there is a drift of C towards zero. At high shear rates the chains buckle under compression and subsequently break up under tension with the separation of part of the chain. This is illustrated in Fig. 23a for an 8-let rotating in the X_2X_3 plane.

Several investigations into the degradation of long-chain molecules in solutions subjected to shear flow have been reported.[76–79] It has been shown that in the case of a rodlike macromolecule such as desoxyribonucleic acid there exists a critical shear rate, above which the molecule breaks into nearly perfect half-molecules. This is expected, because the distribution of stress in a straight rod in shear flow is parabolic with the peak at the center; see equation (III63). In most cases, however, the situation is complicated by the coiling of the chain, and the distribution of stress in the molecule is difficult to predict. It has been shown that here the degradation may be correlated with an average force acting on a long-chain solute molecule.[80]

[75] I. Y. Zia, R. G. Cox, and S. G. Mason, *Science* **153**, 1405 (1966).
[75a] I. Y. Zia, R. G. Cox, and S. G. Mason, *Proc. Roy. Soc.* In press.
[76] L. F. Cavalieri and B. H. Rosenberg, *J. Am. Chem. Soc.* **81**, 5136 (1959).
[77] A. D. Hershey and E. Burgi, *J. Mol. Biol.* **2**, 143 (1960).
[78] W. R. Johnson and C. C. Price, *J. Polymer Sci.* **45**, 217 (1960).
[79] D. Levinthal and P. F. Davison, *J. Mol. Biol.* **3**, 674 (1961).
[80] R. E. Harrington and B. H. Zimm, *J. Phys. Chem.* **69**, 161 (1965).

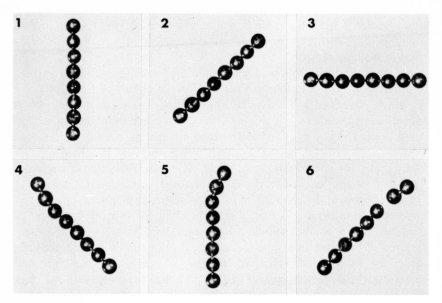

FIG. 23a. Shear-induced buckling and subsequent breakup of an 8-let of rigid spheres in Couette flow. With a probe the eight metal-coated plastic spheres were first roughly lined up with the X_2 axis of the Couette field, along which an electric field was then applied. This brought the particles together in a straight chain, which remained intact after the field was turned off and the 8-let rotated in shear flow as a rigid rod in the $X_2 X_3$ plane. The angle ϕ_1 is $+2°$ at 1, 45° at 2, and 90° at 3. However, under compression in the second quadrant buckling occurred ($\phi_1 = -45°$ at 4), which in turn led to separation into a 6-let and doublet under tension in the third quadrant. Angle ϕ_1 is 0° and 45° at 5 and 6, respectively.

d. Orbits of Flexible Threadlike Particles

When the critical conditions for bending are only slightly exceeded, rods undergoing rotation in Couette flow are observed to bend in the second and fourth quadrants; such orbits have been called "springy"[81] [they are illustrated in Fig. 24(a)]. The loci of the ends of the particles about the X_2 axis are therefore asymmetrical, as shown in Fig. 25; at $\gamma\eta_0$, close to the critical value, bending appears to set in at about $\phi_1 = -\pi/4$, and with increasing $\gamma\eta_0$ deformation is more pronounced and the onset of bending is close to $\phi_1 = -\pi/2$.

As the flexibility of the rod or filament is increased further by increasing either $\gamma\eta_0$ or r_p, a stage is reached ($r_p/r_{crit.} > 1.5$) at which the two ends appear capable of independent movement, the particle still straightening out at $\phi_1 = \pm\pi/2$. Particles rotating in these so-called "snake orbits" [Fig. 24(b)] tend to drift to orbits lying wholly in the $X_2 X_3$ plane.

[81] O. L. Forgacs and S. G. Mason, J. Colloid Sci. 14, 473 (1959).

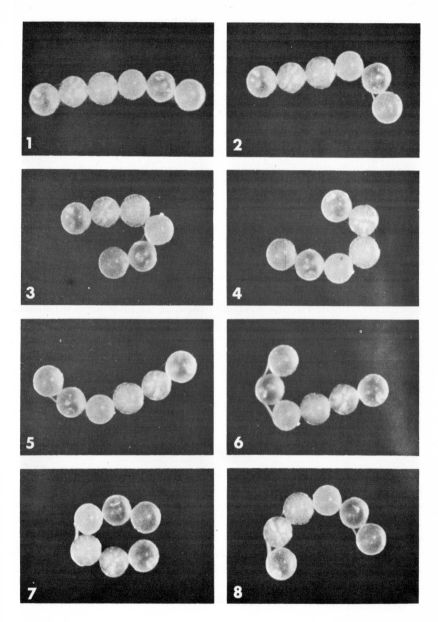

FIG. 23b. Rotation of a flexible chain of 6 plastic spheres suspended in a silicone oil and bridged by aqueous menisci which can be seen in the photograph. The independent movement of the two ends is characteristic of the "snake turn" with flexible filaments shown in Fig. 24(c).

Such snake orbits have been observed with chains of rigid spheres in which a liquid immiscible with the suspending medium is introduced to provide a meniscus which binds the particles.[75,75a] The large bending, illustrated in Fig. 23b for a chain of 6 spheres, probably results from the low viscosity of the fluid introduced, the possible increase in gap width and the interfacial tension which caused the chain, once bent, to bend even more. As described in Section VIII linear aggregates of red blood cells (rouleaux) also exhibit snake turns in shear flow.

At values of $r_p/r_{crit.} > 3$ the filaments no longer straighten out between half-rotations but assume coiled configurations [Fig. 24(c)] and, if sufficiently long and flexible, are observed to rotate as a helix about the X_1 axis.

Finally, in the case of very long elastomer filaments it is found that at the onset of flow an initial alignment with the X_3 axis takes place, after which complicated entanglements, formed during coiling at each end, prevent the formation of the spiral configuration [Fig. 24(d)]. As the coils grow, they rotate bodily and move toward each other, winding up the yet extended part of the filament, and eventually entangle with one another.

The various orbits described above have been used as a basis of a method of measuring the flexibility spectrum of a suspension of wood pulp fibers.[82,83] Here the classification of orbits is complicated by the presence of permanent distortions, weak or damaged areas, and the presence of both elastic and plastic deformations. For instance, fibers are observed to bend segmentally, indicating that weak points are distributed along the fiber length. These may be due to anatomical features of the fiber or to damage resulting from mechanical treatment, or both. Forgacs[84] showed that in softwood fibers weakness may be related to the points at which the tracheids were crossed by ray cells in the wood.

e. Effect of Deformation on the Period of Rotation

Unlike rigid rotating rods the product $T\gamma$ is not constant for a given particle, equation (II19), but in general is found to increase with increasing $\gamma\eta_0$[75a,81,85] much as the period of circulation of the interface of a fluid drop increases with increasing deformation.[48,58] It is also found that $T\gamma$ is very sensitive to small deformations: as the rods become permanently bent during rotations, $T\gamma$ falls off appreciably, and its value corresponds to the axis ratio r_e of the figure formed by rotating the X_2X_3 projection of the filaments about the chord joining their ends. As illustrated in Fig. 26, the intrinsic

[82] O. L. Forgacs and S. G. Mason, *Tappi* **41**, 695 (1958).

[83] A. A. Robertson, E. Meindersma, and S. G. Mason, *Pulp Paper Mag. Can.* **62**, No. 1, T3 (1961).

[84] O. L. Forgacs, *Tappi* **44**, 112 (1961).

[85] A. P. Arlov, O. L. Forgacs, and S. G. Mason, *Svensk Papperstid.* **61**, No. 3, 61 (1958).

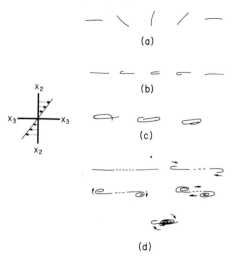

Fig. 24. Tracings of photomicrographs of the orbits of flexible filaments viewed along the X_1 axis. (a) Springy: Dacron rod, $r_p = 180$, rotating in the $X_2 X_3$ plane and showing compression in the first quadrant, $-\pi/2 < \phi_1 < 0$. (b) Snake: elastomer filament, $r_p = 250$, undergoing a snake turn through $\phi_1 = \pi$. (c) Helix: elastomer filament, $r_p = 330$, no longer straightening out but assuming a coiled configuration; viewed along X_2 axis, it is seen to rotate as a helix. (d) Coil formation: a very long elastomer filament, $r_p \approx 800$, is shown at the various stages of coiling. (After Forgacs and Mason.[81])

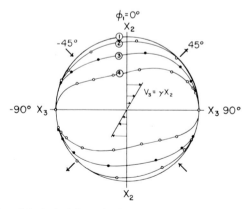

Fig. 25. Polar plot of the loci of the ends of a nylon filament ($r_p = 170$) during rotation in a "springy" orbit [Fig. 24(a)] in the $X_2 X_3$ plane. Curve 1 (a circle) represents rigid rotation at $\gamma < \gamma_{crit.}$. Curves 2, 3, and 4 were obtained at successively higher γ, all above the critical value. The theory predicts that deviation from curve 1 will commence at $\phi_1 = -45°$ when γ approaches the critical value. (After Forgacs and Mason.[29])

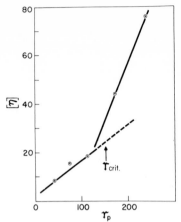

FIG. 26. Intrinsic viscosity of rayon filaments in castor oil suspension at various axis ratios[86] below and above $r_p = r_{crit.}$. (After Forgacs and Mason.[81])

viscosity of suspensions of filaments has been found to increase linearly with increasing r_p up to $r_{crit.}$;[86] after this it increases much more rapidly, presumably because of the permanent deformation from the action of the gradient, which results in the sweeping out of a greater volume of liquid during each orbit.

Provided the filaments are not permanently deformed, $T\gamma$ also increases with r_p over the whole range of orbits, its value deviating increasingly from the predicted value for rigid rods as r_p increases.

IV. Particle Motions in Nonuniform Shear Fields

1. GENERAL

To complete the discussion of single particles it remains to consider the motion under viscous forces of rigid and deformable spheres and spheroids in nonuniform shear fields, that is, fields in which the components of the fluid rate of shear are not constant in space. The most important example of such a field occurs in the Poiseuille flow of a liquid, when the velocity distribution and variation in the rate of shear are functions of the radial distance R from the tube axis, as illustrated in Fig. 27. Another example, which is briefly considered below, occurs when a liquid is subjected to laminar flow between two counter-rotating discs (see Fig. 31), γ increasing linearly with the radial distance from the center of rotation of the discs.

With the frequent use of capillary instruments in the measurement of viscosity considerable interest has been shown in the mechanism of the flow

[86] M. A. Nawab and S. G. Mason, J. Phys. Chem. 62, 1248 (1958).

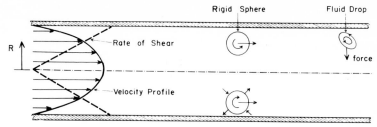

FIG. 27. Diagram of Poiseuille flow in the median plane of the tube. The velocity distribution is parabolic, and the rate of shear varies linearly with the radial distance from the tube axis. Rigid spheres (left) are shown rotating in opposite directions on either side of the tube axis. A fluid drop (right) deforms into an ellipsoid during flow and, unlike the rigid sphere, at low particle Reynolds number it migrates toward the tube axis.

of suspensions and emulsions through tubes, including that of blood and other biological suspensions. It is ironical that the law enunciated by Poiseuille from his work on simple Newtonian liquids should fail to apply to suspensions of red blood cells,[87,88] since the object of his research was to arrive at an understanding of flow in the living vessels. Since the time of Bingham,[89] however, it has become apparent that there are many suspensions, of large and small particles, that do not obey the Poiseuille–Hagen equation[90] relating the pressure gradient along the tube (in the X_3 direction; Fig. 27), the volume flow rate Q, the viscosity η, and the tube radius R_0:

$$\frac{\Delta p}{\Delta X_3} = \frac{8Q\eta}{\pi R_0^{\,4}} \qquad\qquad \text{(IV1)}$$

As in the case of blood, the apparent viscosities of such suspensions, calculated from equation (IV1), vary with tube radius, length, and flow rate.[86,91–98] To account for such anomalies it has been postulated that there

[87] L. E. Bayliss, in "Deformation and Flow in Biological Systems" (A. Frey Wyssling, ed.), p. 362. North-Holland Publ., Amsterdam, 1952.

[88] J. M. L. Poiseuille, Compt. Rend. 15, 1167 (1842).

[89] E. C. Bingham and H. Green, Proc. Am. Soc. Test Mater. [2] 19, 640 (1919).

[90] H. Lamb, "Hydrodynamics," 6th ed. Dover, New York, 1945.

[91] E. C. Bingham, "Fluidity and Plasticity." McGraw-Hill, New York, 1922.

[92] E. Hatschek, Kolloid-Z. 13, 88 (1913).

[93] G. H. Higginbotham, D. R. Oliver, and S. G. Ward, Brit. J. Appl. Phys. 9, 372 (1958).

[94] A. Müller, "Abhandlungen zur Mechanik der Flüssigkeiten," Vol. I. Univ. of Fribourg, Fribourg, Switzerland, 1936.

[95] A. D. Maude, Brit. J. Appl. Phys. 10, 371 (1959).

[96] R. K. Schofield and G. W. Scott Blair, J. Phys. Chem. 34, 248 and 1505 (1930); 39, 973 (1935).

[97] G. Segré and A. Silberberg, J. Colloid Sci. 18, 312 (1963).

[97a] F. Eirich and O. Goldschmid, Kolloid. Z. 81, 7 (1937).

[98] V. Vand, J. Phys. Colloid Chem. 52, 300 (1948).

exists at the wall of the vessel a particle-depleted layer, which is formed
either by particle migration across the planes of shear away from the tube
wall or by a mechanical entrance effect.[99,100]

In some systems, as in suspension of rigid spheres[14,31a,38,39,97a,101–103]
and spheroids,[31a,38,94] and of deformable drops[11,38] and wood pulp
fibers,[104] radial migration has been observed visually.

To avoid confusion, a distinction must be made between particle behavior,
and especially radial migration in the tube, in the creeping or quasistatic
flow regime and that which occurs at higher Reynolds numbers from inertial
effects; the latter being separately considered in a later section. As will
become evident, however, the presence of the vessel walls is a cardinal
feature of the radial particle migration at all Reynolds numbers. Thus,
particle migration across the planes of shear near the wall of a Couette
apparatus, where the flow is rendered nonuniform by the particles, has
also been observed. The effect of a redistribution of the disperse phase
on the measurement of viscosity in capillary and Couette viscometers is
described in Section VI.

2. Rigid Spheres, Rods, and Discs in Poiseuille Flow

a. Rotation and Translation

Particle motions in flow through tubes have been followed with a traveling
microscope, whose speed can be continuously varied and, when matched
with that of a particle, enables continuous viewing for a distance of about
50 cm.[11,105] Steady flow rates are achieved by means of an infusion-with-
drawal pump. To prevent optical distortion, the precision-bore glass tubes
used pass through a square cell containing a liquid having the same refractive
index as glass and, when possible, as the liquid in the tube.

The rate of shear in Poiseuille flow increases linearly with radial distance
R from the tube axis according to the relation

$$\gamma(R) = \frac{dU_3}{dR} = \mathscr{K} R \qquad (IV2)$$

[99] A. D. Maude and R. L. Whitmore, *Brit. J. Appl. Phys.* **7,** 98 (1956).

[100] R. L. Whitmore, in "Rheology of Disperse Systems" (C. C. Mill, ed.), Chapt. 3, Macmillan
(Pergamon), New York, 1959.

[101] D. R. Oliver, *Nature* **194,** 1269 (1962).

[102] G. Segré and A. Silberberg, *Nature* **190,** 1095 (1961).

[103] G. Vejlens, *Acta Pathol. Micrbiol. Scand.* Suppl. **XXXIII,** 159–190 (1938).

[104] O. L. Forgacs, A. A. Robertson, and S. G. Mason, *Pulp Paper Mag. Can.* **59.** No. 5, 117
(1958).

[105] H. L. Goldsmith and S. G. Mason, 2nd European Conf. Microcirculation, Pavia, 1962.
Bibliotheca Anat. Fasc. **4,** 462 (1964).

where $\mathcal{K} = 4Q/\pi R_0^4$ and U_3, the component of fluid velocity in the X_3 direction (Fig. 27), is, by integration of equation (IV2),

$$U_3 = \frac{\mathcal{K}}{2}(R_0^2 - R^2)$$

$$= U_{3M}(1 - R^2/R_0^2) \tag{IV3}$$

U_{3M} being the center line, maximum fluid velocity.

Provided the particles are small compared to the tube radius (for example, $b/R_0 < 0.05$ for spheres), the rotational motions are identical with those seen in Couette flow. Quantitative agreement with Jeffery's theory has been obtained when the value of γ from equation (IV2) is inserted into equations (II16) to (II20) and the equivalent ellipsoidal axis ratio is used for the rods and discs.[11] Thus, as illustrated in Fig. 28, the angular velocity of single small rigid spheres, $= \gamma/2$, decreases linearly with decreasing radial distance and is of opposite sign in the two halves of the median plane of the tube (Fig. 27).

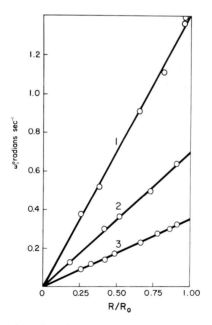

FIG. 28. The measured angular velocities of neutrally buoyant small rigid spheres in a suspension undergoing Poiseuille flow, as a function of the radial distance from the tube axis. The lines drawn are calculated from equation (II16) with $\gamma = 4QR/\pi R_0^4$, where Q, the volume flow rate, is successively doubled from curve 3 to curve 1. (After Goldsmith and Mason.[11])

Over a range of tube Reynolds numbers from 10^{-5} to 10^{-3} the particles travel with a velocity u_3, which is equal to the local undisturbed fluid velocity at their center of rotation, U_3, and over distances of as much as 10^4 particle diameters there is no measurable migration across the planes of shear.

b. Wall Effects

When the particle size becomes appreciable compared to the distance of its center of rotation from the tube wall [$b/(R_0 - R) > 0.1$ for spheres], the translational and angular velocities are smaller than those calculated from the theory for unbounded flow. For rigid spheres, both the "translational slip" velocity, defined as $U_3 - u_3$, and the difference in angular velocities between the undisturbed fluid, given by $\gamma/2$, and sphere, $\omega_1{}^0$, increase with increasing $b/(R_0 - R)$.[11] Similarly, an increase in the period of rotation, T, is observed with rods and discs rotating close to the tube wall. In the case of rods the decrease in $\dot{\phi}_1$ occurs only during that part of the orbit close to $\phi_1 = \pm\pi/2$, when the particle is aligned with the tube wall.[11]

These decreases in angular and translational velocities are brought about by an interaction between the particle and the wall; similar effects are observed close to the walls of a counterrotating-cylinder Couette apparatus.[25] In the latter case the additional velocity field due to the introduction of a sphere at the center of a field of unbounded Couette flow has been shown to be[34,106]

$$V_1 = \frac{5\gamma}{2}\left[\frac{b^3}{r^5}\left(\frac{b^2}{r^2} - 1\right)X_1 X_2 X_3\right]$$

$$V_2 = \frac{\gamma X_3}{2}\left[\frac{5b^3}{r^5}\left(\frac{b^2}{r^2} - 1\right)X_2{}^2 - \frac{b^5}{r^5}\right] \tag{IV4}$$

$$V_3 = \frac{\gamma X_2}{2}\left[\frac{5b^3}{r^5}\left(\frac{b^2}{r^2} - 1\right)X_3{}^2 - \frac{b^5}{r^5}\right]$$

where $r^2 = (X_1{}^2 + X_2{}^2 + X_3{}^3)$. The values for V_2 and V_3 given by equation (III7) reduce to those in equation (IV4) when $\lambda = \infty$. Owing to the perturbed field, the boundary condition at the wall, $X_2 = -l$, in the absence of a sphere, viz. $V_3 = -\gamma l$, is no longer satisfied, and a second-order field which, when added to the original field, restores the above condition, can be calculated. We shall define the effect of such a field as a wall effect.[99] Using the approximate

[106] V. Vand, J. Phys. Colloid Chem. **52**, 277 (1948).

method of "reflections" first used by Lorentz,[107] several authors[98,108–112] have calculated the secondary field to varying degrees of approximation. These solutions are found with the linearized Navier–Stokes equations and hence are valid only at very low Reynolds numbers. The method assumes that the velocity and pressure fields, $\mathbf{V}$ and p, respectively, can be decomposed into a sum of fields, each term of which separately satisfies the equations of motion.[111] The individual fields are determined successively by the following boundary conditions:

$$\mathbf{V}^{(1)} = \mathbf{v} \text{ on the particle surface}$$

$$\mathbf{V}^{(2)} = -\mathbf{V}^{(1)} \text{ on the vessel wall}$$

$$\mathbf{V}^{(3)} = -\mathbf{V}^{(2)} \text{ on the particle surface}$$

$$\mathbf{V}^{(4)} = -\mathbf{V}^{(3)} \text{ on the vessel wall, etc.}$$

(IV5)

where

$$\mathbf{V} = \mathbf{V}^{(1)} + \mathbf{V}^{(2)} + \mathbf{V}^{(3)} + \mathbf{V}^{(4)} + \cdots + \qquad \text{(IV6)}$$

and $\mathbf{V}^{(1)}$ is the perturbed velocity field due to the particle in unbounded flow, for example, equation (IV4) in Couette flow, $\mathbf{v}$ being the particle velocity. Furthermore, each of the individual fields $\mathbf{V}^{(n)}$ must vanish as $r \to \infty$.

The odd-numbered fields involving boundary conditions on the particle introduce the characteristic particle dimension, for example, through such terms as $(b/r)^n$ for spheres; the even-numbered fields introduce the surface characteristic dimensions, such as the tube radius in Poiseuille flow. Each successive pair of reflections increases the overall accuracy of the solution by contributing terms in the ratio particle dimension to vessel dimension, whose dominant powers are of higher order than those arising from preceding reflections. Thus, equation (IV6) amounts to a series expansion in powers of the above-given ratio, which converges to the solution of the original boundary-value problem.

Lorentz[107] first obtained equations for the force acting on a rigid sphere approaching a plane wall and for the fall of a sphere parallel to a plane wall by employing the first reflection. Maude[110] has used the same treatment for a sphere placed in an arbitrary flow in front of a plane. When the fluid undergoes Couette flow and the sphere is at the origin of the field, the

[107] H. A. Lorentz, *Verslagen Kgl. Akad. Wetenschap. Amsterdam* **5,** 168 (1896); "Collected Papers," Vol. 4, pp. 7–14. Martinus Nijhoff, The Hague, Holland, 1937.

[108] H. Brenner, *Advan. Chem. Eng.* **6,** 287 (1965).

[109] H. Brenner and J. Happel, *J. Fluid Mech.* **4,** 195 (1958).

[110] A. D. Maude, *Brit. J. Appl. Phys.* **14,** 894 (1963).

[111] H. Brenner, *J. Fluid Mech.* **12,** 35 (1962).

[112] S. Wakiya, C. L. Darabaner, and S. G. Mason, *Rheol. Acta* **6** (1967).

equations for the respective angular and translational velocities are

$$\omega_1{}^0 = \frac{\gamma}{2}\left[1 - \frac{5}{16}\left(\frac{b}{l}\right)^3 + 0\left(\frac{b}{l}\right)^3\right] \tag{IV7}$$

$$v_3 = -\frac{5}{4}\gamma b\left(\frac{b}{l}\right)^2 + 0\left(\frac{b}{l}\right)^2 \tag{IV8}$$

where γ is the undisturbed velocity gradient. The change in angular velocity from that given by $\gamma/2$, due to the presence of a wall, was regarded by Vand[106] in an earlier treatment as a change in the effective velocity gradient. A formula similar to equation (IV7) was given, although in Vand's calculation only the normal component, V_2 in equation (IV4) of the additional field $V^{(1)}$ was reflected on the vessel wall, the tangential component not being considered. Vand[106] proceeded to show that the enhancement of the effective velocity gradient in the fluid surrounding the sphere by the additional field is equivalent to having a layer of thickness $h = 1.301b$ at each wall, which does not contribute to the energy dissipation due to the presence of the sphere, that is, which can be regarded as having the viscosity of the suspending phase. Hence, "the measured uncorrected values of viscosity of suspensions will tend to be smaller in smaller instruments than in larger ones."[106]

Furthermore, Vand suggested that "in the region of high concentrations, considerable slip at the wall might develop due to the layers of low viscosity along the walls which might finally completely overshadow the effects of shear inside of the suspension, making the measurements useless." Recent measurements of the velocity profiles of rigid spheres in both Poiseuille and Couette flow, which will be described in Section V, have shown this view to be substantially correct.

Good experimental agreement with equation (IV7) has been obtained[112] for values of $b/l < 0.5$; up to $b/l = 0.8$ agreement is obtained with equations derived by Wakiya et al.,[112] which are similar to equations (IV7) and (IV8) but correct to $(b/l)^8$. An exact solution to the problem of the rotation and translation of a neutrally buoyant sphere in shear flow near a plane wall has been obtained by Goldman et al.[113] by linearly superimposing three independent solutions for (1) the rotation of a sphere about an axis parallel to a nearby plane in an infinite quiescent fluid; (2) the translation of a sphere in a direction parallel to a nearby plane, in an infinite quiescent fluid; (3) the force and torque exerted by a Couette field on a stationary sphere near a plane. When the sphere is very close but not exactly touching the wall

$$\lim_{l \to b} (\omega_1{}^0 b/v_3) = 0.568$$

[113] A. J. Goldman, R. G. Cox, and H. Brenner, *Chem. Eng. Sci.* **22**, 637, 653 (1967).

i.e., the translational velocity is about double that corresponding to rolling motion along the wall. This result is in accord with experiment.[127] Using a lubrication theory approximation[113] both $\omega_1{}^0$ and v_3 are shown to become zero when the sphere is in contact with the wall, i.e., the theory predicts that the particle cannot move because an infinite force and torque would be required to sustain the motion.

Raasch also obtained a complete solution in the two-dimensional case of the infinitely long cylinder, whose axis lies in the median plane of a region of Couette flow near a plane, by using a system of bipolar coordinates.[114] The respective angular and translational velocities of the cylinder, radius b, are

$$\omega_1{}^0 = \frac{\gamma}{2}\left[1 - \left(\frac{b}{X_2}\right)^2\right]^{1/2} \tag{IV9}$$

$$v_3 = \gamma X_2\left[1 - \left(\frac{b}{X_2}\right)^2\right]^{1/2} \tag{IV10}$$

where X_2 is now the distance of the cylinder center from the origin of the field at the wall. As has been shown experimentally,[25] equation (IV9) holds for all values of b/X_2. It should also be noted that equation (IV9) and (IV10) indicate that $\lim_{l\to b}(\omega_1{}^0 b/v_3) = \frac{1}{2}$. The theory also predicts that when in contact with the wall the cylinder will stick without rotating.

Brenner and Happel,[109] using the method of reflections, have derived approximate equations for the force and torque on a sphere sedimenting without rotation in a tube parallel to its axis; the fluid far from the sphere may be at rest or be in uniform Poiseuille flow. The force on the nonrotating sphere is parallel to the tube axis and of magnitude

$$6\pi\eta_0(U_3 - u_3)[1 + \Xi(b/R_0) + \cdots +]$$

where $(b/R_0)^2$ is neglected, compared to 1. The dependence of Ξ on R/R_0 has been given by Brenner[111]:

$$\Xi = 2.1044 - 0.6977(R/R_0)^2 + \cdots +, \qquad R/R_0 \ll 1$$

$\Xi = 2.079$ when $R/R_0 = 0.2$, increasing to 2.165 when $R/R_0 = 0.6$ and, for $R/R_0 \to 1$,[115]

$$\lim_{R/R_0 \to 1}\left|\frac{(R_0 - R)\Xi}{R_0}\right| = \frac{9}{16}$$

[114] J. Raasch, Z. Angew. Math. Mech. **41**, T147 (1961); Dissertation, Karlsruhe Technical University, 1961.
[115] J. Famularo, Eng. Sci. D. dissertation, New York University, New York, 1962.

In common with the calculations in Couette flow those of Brenner and Happel in tube flow[109] reveal that there is no force normal to the wall. For a neutrally buoyant sphere freely suspended in Poiseuille flow the translational slip velocity, $U_3 - u_3$, is found by setting the force equal to zero; thus, $U_3 - u_3$ differs from zero by a quantity of the order of $(b/R_0)^2$ at most. In the special case of a sphere on the tube axis the value of this term has been obtained from higher-order calculations of the wall effect by Bohlin[116] and Haberman and Sayre.[117] Thus, at the axis

$$U_{3M} - u_3 = \frac{2}{3}\left(\frac{b}{R_0}\right)^2 U_{3M} \qquad (IV11)$$

This formula was previously given by Simha,[118] who considered a sphere suspended in an *unbounded* fluid having a Poiseuille velocity profile.

The effect of the calculated additional field is, then, to impart a drag and a torque to the sphere, which give rise to a positive translational slip and a decrease in the angular velocity. Since this result has been obtained by neglect of fluid inertia, it is valid only as the relative particle Reynolds number, based on translational slip, tends to zero,

$$\mathrm{Re}_p = 2\frac{\rho}{\eta_0}b(|U_3 - u_3|) \qquad (IV12a)$$

$$= \frac{4}{3}\cdot\frac{\rho}{\eta_0}b\left(\frac{b}{R_0}\right)^2 U_{3M} \qquad (IV12b)$$

if the slip velocity given by equation (IV11) is used.[102] In fact, the measured slip velocity in Poiseuille flow at $\mathrm{Re}_p < 10^{-5}$ is found to increase[11] with increasing R from the value predicted by equation (IV11) at the tube axis; see Fig. 29.

The absence of any tendency of the sphere to migrate toward the tube axis can be explained quite simply from the properties of the creeping flow equation, equation (I3), by an argument valid for any spherical particle, whether a rigid sphere or a fluid drop; if there were migration, for example, away from the wall, then, when flow is reversed, the migration should still be away from the wall. On the other hand, if the direction of time were reversed, the apparent migration should be toward the wall. But equation (I3) requires that flow reversal and time reversal should have the same effect; this will only be so if a spherical drop does not migrate. With a deformed

[116] T. Bohlin, *Kgl. Tek. Hogskol. Handl.* No. 155 (1960).

[117] W. L. Haberman and R. M. Sayre, David Taylor Model Basin Rept. No. 1143, U.S. Navy Dept., Washington, D.C., 1958.

[118] R. Simha, *Kolloid-Z.* **76**, 16 (1936).

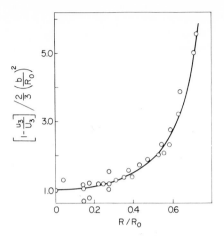

FIG. 29. Wall effect for neutrally buoyant rigid spheres in Poiseuille flow. The relative translational slip plotted against radial distance; $b/R_0 > 0.1$. The intercept at the tube axis, as required by theory,[111, 118] is 1.0. (After Goldsmith and Mason.[11])

drop, having no principal axis perpendicular to the wall, as will be seen below, the additional perturbation of the field by the deformation can be shown to give rise to migration away from the wall. Further, at finite but still low values of the relative particle Reynolds number ($Re_p = 10^{-4}$ to 1) radial migration of *rigid* spheres has been observed,[14,31a,101,119,120] the rate of migration increasing approximately linearly with Re_p.

3. DEFORMABLE SPHERES AND RODS IN POISEUILLE FLOW

a. Deformation of Fluid Drops

The deformation, orientation, and breakup of single small drops suspended in liquids undergoing Poiseuille flow corresponds exactly to that described above for Couette flow.[11] Thus, when viewed in the median plane, the drop initially aligns itself at an angle $\phi_1 = \pi/4$ to the R axis (Fig. 27), the deformation and ϕ_1 increasing with increasing flow rate.

b. Axial Migration

In contrast to the flow of rigid spheres in a tube, migration of the deformed drops across the planes of shear toward the axis occurs. The migration rate increases with increasing drop radius, flow rate, and radial distance (Fig. 30)

[119] R. V. Repetti and E. F. Leonard, *Nature* **203**, 1346 (1964).
[120] R. C. Jeffrey, Ph.D. dissertation, Cambridge University, England, 1964.

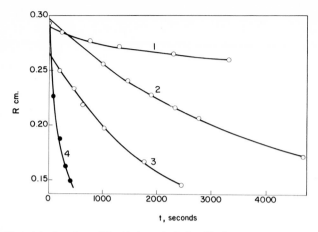

FIG. 30. Axial migration of liquid drops in Poiseuille flow, as shown by a plot of the radial distance of the center of deformed water drops in silicone oil versus time. The migration rate at a given flow rate increases markedly with drop size, as illustrated by curves 1 to 3, in which b is 175, 300, and 390 microns, respectively; curve 4 is for a drop, $b = 350$ microns, at four times the flow rate. (After Goldsmith and Mason.[11])

and, it must be emphasized, occurs at limiting, almost zero, values of Re_p. Thus, for instance, a neutrally buoyant rigid sphere ($b/R_0 = 0.14$, $R/R_0 = 0.86$) at $Re_p \approx 10^{-5}$, after traveling 48 hours for a distance of 3500 cm, did not measurably migrate (i.e., $\Delta R/R_0 < 0.25\%$), whereas a water drop in the same tube ($b/R_0 = 0.10$, $R/R_0 = 0.65$) at $Re_p \approx 10^{-6}$ migrated radially 1300 microns to $R/R_0 = 0.33$, half way to the tube axis, in only 20 minutes, or after only 150 cm of travel.[11,121] Equilibrium is attained only when the particle has reached $R = 0$.

Similarly, flexible fibers rotating in the springy and snake orbits described in the previous section and also flexible filaments having a very high r_p migrate toward the tube axis.

Ideally, a theoretical treatment of drop migration in Poiseuille flow involves the determination of a velocity field satisfying certain conditions on the deformed drop surface and vanishing on the tube wall. Calculations by the method of reflections, so as to satisfy boundary conditions on a cylindrical surface, are very elaborate, as the investigation of Brenner and Happel[109] has shown, and no such analysis of the drop migration has been undertaken. However, a simpler discussion of the problem has been put forward,[11] in which it is pointed out that the fluid stresses at the drop surface must give rise to a net force tending to impel the drop toward the tube axis (Fig. 27). In applying this principle both drop deformation and the presence of the

[121] H. L. Goldsmith and S. G. Mason, *Biorheology* 3, 33 (1965).

tube wall should be taken into account. Deformation of an elastic sphere has been considered by Cerf;[60] it is very similar to that of a drop. From the assumptions that the deformation is given by Cerf's result and that the effect of the tube wall is a variation in velocity gradient across the drop a formula has been developed,[11] which describes many of the features of the observed migration. Because this formula predicts that, as the viscosity becomes infinite, the axial force becomes infinite, it was recognized that a more fundamental analysis of the problem was in order. This has actually only been done, to a first-order approximation, for a drop suspended near a plane wall bounding a fluid in Couette flow—a simple situation, to which the Poiseuille-flow situation reduces for a small drop near the wall of a large tube. The basic idea is that, although there is no migration in the simple case of creeping flow around a sphere, nevertheless the effect of the drop deformation is a perturbation in the flow that introduces additional terms to the flow field, and the reflections of the largest of these terms off the wall yields a finite migration velocity, and this contribution to the migration is larger than that arising from other terms.

The velocity fields to first order in the deformation parameter $\mathfrak{D}$, equation (III32), inside and outside the drop at the center of an unbounded field of Couette flow are calculated, as was the zero-order field for the spherical drop, by Taylor,[6] who used the general solution to equation (I3) given by Lamb.[90] The result for the X_2 component of the first-order field outside the drop is[122]

$$V_1^{(2)} = \frac{\gamma b^3 X_2}{35 r^5 (\lambda + 1)^2} [\tfrac{1}{2}(25\lambda^2 + 41\lambda + 4)X_1^{\,2}$$
$$+ (54\lambda^2 + 102\lambda + 54)X_2^{\,2} - (79\lambda^2 + 143\lambda + 58)X_3^{\,2}] \qquad (IV13)$$

This component is reflected off the wall, $X_2 = -l$, to restore the boundary condition $V_3 = -\gamma l$, by using general equations for the first reflected field $\mathbf{V}^{(2)}$ given by Maude[110]:

$$V_2^{(2)} = \frac{Y^2}{\eta_0}\frac{\partial P}{\partial Y} - 2Y\frac{\partial V_2^{(1)}}{\partial Y} + V_2^{(1)} \qquad (IV14)$$

where $Y = X_2 + l$. The resulting reflected component evaluated at the drop center is

$$V_2^{(3)} = \gamma \frac{b^3}{l^3} \cdot \frac{99(9\lambda^2 + 17\lambda + 9)}{140(\lambda + 1)^2}$$

If the drop is freely suspended, the migration velocity v_2 is simply equal to

[122] C. E. Chaffey, H. Brenner, and S. G. Mason, *Rheol. Acta* **4**, 64 (1965); **6**, 100 (1967).

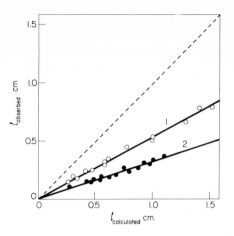

FIG. 31. Observed radial migration of liquid drops away from the wall in Couette flow, compared with the theoretical predictions, equation (IV15). The dashed line is the 45° line of perfect correlation. The experimental points are (1) $\lambda = 2 \times 10^{-4}$ for a water-in-oil system and (2) $\lambda = 1.4$ for an oil-in-oil system. (After Karnis.[122a])

$\mathfrak{D}V_2^{(3)}$ and by substituting for $\mathfrak{D}$ from equation (III32) one finally obtains

$$v_2 = -\dot{l} = \frac{\gamma b^4 \eta_0}{\sigma} \cdot \frac{1}{l^2} \cdot \frac{99(9\lambda^2 + 17\lambda + 9)(19\lambda + 16)}{2240(\lambda + 1)^3} \qquad \text{(IV15)}$$

Thus the drop, deformed because of the velocity gradient, always migrates perpendicularly away from a plane wall bounding the flow. Experiments on drop migration in Couette flow have confirmed this result, and the measured velocities are in fair agreement with equation (IV15),[122a] as shown in Fig. 31.

This result may also be applied to Poiseuille flow, since although the unbounded field $\mathbf{V}^{(1)}$ contains additional terms from the quadratic terms in equation (IV3), these will be of an order of magnitude in b/R smaller than the terms from the linear part of the flow, and, hence, need not be considered. When $b/R_0 \ll 1$, and as $R/R_0 \to 1$, the expression for the migration velocity in the tube can then be expected to reduce to equation (IV15) which, when $\gamma = \mathscr{K}R$ and $l = R_0 - R$, are substituted in, yields[122]

$$-\dot{R} = \frac{\mathscr{K}^2 b^4 \eta_0}{\sigma} \cdot \frac{R^2}{(R_0 - R)^2} \cdot \frac{99(9\lambda^2 + 17\lambda + 9)(19\lambda + 16)}{2240(\lambda + 1)^3} \qquad \text{(IV16)}$$

[122a] A. Karnis, Ph.D. thesis, McGill University, Montreal, Canada, 1966.

and upon integrating one obtains

$$\left(\frac{R}{R_0}\right)^{-1} + 2\ln\left(\frac{R}{R_0}\right) - \frac{R}{R_0}$$

$$= \frac{\mathscr{K}^2 b^4 \eta_0}{\sigma R_0} \cdot \frac{99(9\lambda^2 + 17\lambda + 9)(19\lambda + 16)}{2240(\lambda + 1)^3} (t - t_0) \qquad (IV17)$$

where t_0 is an integration constant.

The available experimental results in Poiseuille flow[11] show that the measured $\dot{R}$ are greater than those predicted from equation (IV16).[122] However, in the experiment R/R_0 was not close to unity, as is required by the theory given above.

It is interesting to note that these theories predict migration of rigid rods toward the wall in the quadrant $0 < \phi_1 < \pi/2$ and away from the wall in the quadrant $-\pi/2 < \phi_1 < 0$ (and vice versa for discs). Since, however, the orbits of the particles are closed, and since the parts of the orbit previous to, and subsequent to, the orientation $\phi_1 = 0$ are mirror images in the $X_1 X_2$ plane, as illustrated in Fig. 4, there is no more time spent pointing upstream than downstream, and drift across the streamlines in one quadrant would be accurately canceled out by a reverse drift in the succeeding quadrant. One would, therefore, expect the center of rotation of rods and discs to describe a sinusoidal path. The fact that such behavior has not been observed experimentally in Poiseuille flow[11] indicates that the amplitude of these displacements is very small.

The foregoing argument concerning the symmetry of the orbits of rods and discs is a particular application of a mirror-symmetry time-reversal theorem given by Bretherton,[12] which states that "when moving in a steady unidirectional shear flow at small Reynolds number under the action of viscous forces alone, to every orbit of a given finite rigid body there corresponds one of the body of opposite mirror-symmetry. The corresponding orbits are 'mirror images' obtained by reflection in a plane perpendicular to the streamlines, but are traversed in opposite senses."

4. MIGRATION IN SHEAR FLOW BETWEEN PARALLEL DISCS

Although drop migration in the nonuniform shear field in viscous flow through a tube cannot be calculated without explicitly considering the effect of the wall, migration of a drop suspended in a liquid contained between counterrotating discs can be shown to arise solely from the nonuniform shear field.[59] This system, moreover, can be realized experimentally.

We consider a liquid undergoing laminar viscous flow between two infinite parallel discs rotating with counterclockwise angular velocities Ω_U

and Ω_L about a common perpendicular axis, as illustrated in Fig. 32(a). With reference to the cylindrical polar coordinates X_2, R, ϕ_2 [Fig. 32(b)], the undisturbed fluid velocity at a radial distance R from the center of rotation and X_2 from the stationary plane is

$$V_2(R, X_2) = \gamma(R)X_2 \tag{IV18}$$

where the velocity gradient $\gamma(R)$ is given by

$$\gamma(R) = \mathcal{K}R$$

$$= \left(\frac{\Omega_U + \Omega_L}{\Delta X_2}\right)R \tag{IV19}$$

and ΔX_2 is the distance between the respective disc surfaces.

The migration velocity of a drop is calculated as follows.[59] The velocity field is separated into two parts, one being the uniform field of Couette flow at a given R, the other being the field which at a given X_2 varies with the radial distance. The drop is deformed during flow according to equation

(a)

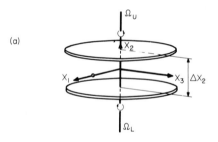

(b)

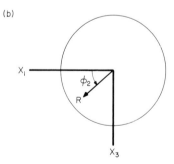

FIG. 32. Coordinate system for shear flow between two parallel counterrotating discs. (a) Cartesian coordinates X_1, X_2, X_3. (b) Cylindrical polar coordinates X_2, R, ϕ_2. (After Chaffey et al.[122])

(III32), and the velocity fields to first order in the deformation $\mathfrak{D}$ inside and outside the drop are found, as was done in the case of the drop near a wall, by means of Lamb's solution in spherical harmonics.[90] The uniform part of the field is not considered in the calculation since, because of the symmetry of the configuration, no force can act in the X_1, X_2, or X_3 direction.

The harmonics required in the solution are found first by solving the problem of an infinitely viscous drop, which is the same as a rigid body; the solution for drops of any viscosity is then determined by using the same harmonics but with different coefficients. Finally, with the use of a theory of Brenner[123] for the force and torque acting on a slightly deformed sphere a force is found to act on a fixed drop along the radial coordinate. The velocity at which a free drop migrates in the radial direction is then found by using the solution of Hadamard and Rybczinski[90] for a sedimenting liquid sphere, giving the following result:

$$\dot{R} = -\frac{\mathscr{K}^2 b^3 R \eta_0}{\sigma} \cdot \frac{(3\lambda^2 + 14\lambda - 2)(19\lambda + 16)}{19(3\lambda + 2)(\lambda + 1)(\lambda + 4)} \qquad \text{(IV20)}$$

Upon integration this yields

$$\ln\frac{R}{R'} = \frac{\mathscr{K}^2 b^3 \eta_0}{\sigma} \cdot \frac{(3\lambda^2 + 14\lambda - 2)(19\lambda + 16)}{19(3\lambda + 2)(\lambda + 1)(\lambda + 4)} t \qquad \text{(IV21)}$$

where R' is the initial radial distance at time $t = 0$. This result is remarkable, because it indicates that for large λ the direction of migration is toward the axis of rotation of the discs, the rate decreasing until at $\lambda = 0.139$ the function of λ on the right-hand side of equation (IV20) vanishes, and the drop should not migrate. For $0 < \lambda < 0.139$ outward migration is predicted.

The experiments, conducted in a counterrotating disc Couette, have so far yielded inconclusive results on drop migration.

V. The Kinetics of Flowing Dispersions

1. GENERAL

The preceding three sections have described the behavior of single rigid and deformable particles in suspensions undergoing laminar viscous flow. It was shown that in many cases, for example with rigid spheroids and slightly deformed drops, the particle motions could be predicted from hydrodynamic theory on the basis of the linearized Stokes–Navier equation, equation (I3). We consider now the interactions between the particles of the disperse phase, as they are carried into close proximity to each other by the flow. Here, with

[123] H. Brenner, *Chem. Eng. Sci.* **19**, 519 (1964).

one exception (the two-dimensional case of infinitely long cylinders, their long axes parallel to the X_1 axis in Couette flow, which has been treated by Raasch[114]), there exists no theory based on the creeping-flow equations that can predict the translational and rotational motions of the interacting particles when in close proximity. From observations, however, of the behavior of spheres in dilute suspensions it has been possible to construct a kinetic theory of particle interactions in laminar flow with the calculation of such quantities as the two-body collision frequency, the mean free path and distribution of doublet life times. Two-body collisions between rods or discs are more complex and difficult to describe than those between spheres. Nevertheless, it has been possible to relate the observed particle interactions qualitatively to the distribution of orientations measured in dilute suspensions in the steady state.

As in the case of the periodic motions of single rigid spheres and spheroids in closed orbits under the action of viscous forces alone, which were shown to be reversible, it is found that collisions between rigid spheres, rods, and discs are reversible. Even in concentrated suspensions, where at first sight the phenomena appear to be very complex, it has been shown that a surprising degree of order exists, strikingly manifested by the reversibility of particle orientations and paths. As pointed out by Slattery[13] and discussed below, such reversibility is a consequence of the reversibility properties of the creeping-motion equations.

2. Collisions of Rigid Spheres in Couette Flow

Two-body collisions between rigid spheres in dilute suspensions subject to shear are well defined since, as described below, there is a marked change in the translational and rotational motion of each sphere, and the particles are in apparent contact for a certain time. Detailed studies[27,28,48,124,125] have shown that there are two classes of two-body collision doublets, which will be designated standard and nonseparating doublets.

a. Standard Collision Doublets of Uniform Spheres

When viewed along the X_1 axis of the motion (Fig. 33a), the $X_2 X_3$ projection of the paths of approach of the spheres is curvilinear when the axial separation ΔX_3 is within two particle diameters, the separation of particle centers along the X_2 axis gradually increasing. Upon apparent contact at an orientation $\phi_1 = \phi_1{}^0$ and $\theta_1 = \theta_1{}^0$ there is no discontinuity in the motion, and the spheres rotate as a rigid dumbbell, i.e., without sliding or rolling

[124] R. S. Allan and S. G. Mason, *J. Colloid Sci.* **17,** 383 (1962).
[125] R. St. J. Manley and S. G. Mason, *Can. J. Chem.* **33,** 763 (1955).

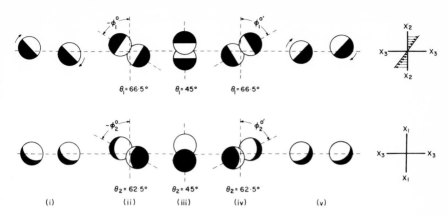

FIG. 33. Two-body standard collisions between equal-sized rigid spheres, as seen in the X_2X_3 plane (a) and the X_1X_3 plane (b). The stages shown are (i) approach, (ii) apparent contact at $\phi_1{}^0 = -60°$ and $\phi_2{}^0 = -59°$, (iii) mid-point of doublet rotation as rigid spheroid $r_e = 2$ (in an orbit here having $C = 2.0$), (iv) point of separation at $\phi_1^{0'} = 60°$ and $\phi_2^{0'} = 59°$, and (v) recession. The collision process is symmetrical and reversible. For convenience the origin is placed at the mid-point between the sphere centers.

motion and with variable angular velocities $\dot{\phi}_1$ and $\dot{\theta}_1$ (Fig. 34), as would a spheroid of axis ratio $r_e = 2.0$; that is,

$$\tan \phi_1 = 2 \tan(\tfrac{2}{5}\gamma t) \tag{V1}$$

$$\tan \theta_1 = 2C(4 \cos^2 \phi_1 + \sin^2 \phi_1)^{1/2} \tag{V2}$$

Thus, the rotation is symmetrical about the X_1X_2 plane, θ_1 decreasing to a minimum at $\phi_1 = 0$ and increasing again, until at an angle $\phi_1 = \phi_1^{0'}$, which is the reflection of the apparent collision angle, $\theta_1 = \theta_1{}^0$, and the particles separate. After separation the paths of recession are curvilinear, and mirror images of the paths of approach, until the axial separation exceeds $4b$, when the X_2 positions and translational velocities v_3 existing before collision are regained. Because of the curvilinear paths before and after doublet formation the "rectilinear" collision angles ϕ_r calculated from the measured ΔX_2 at $\Delta X_3 > 4b$,

$$\phi_r = \cos^{-1}(\Delta X_2/2b) \tag{V3}$$

are up to 20° greater than $\phi_1{}^0$.

If the flow is reversed, the particles are found to recollide in the exact reverse of the first collision. Interrupting the motion after a doublet has formed and then restoring flow has no effect on the paths of the particle centers. These findings make it unlikely that there is true physical contact

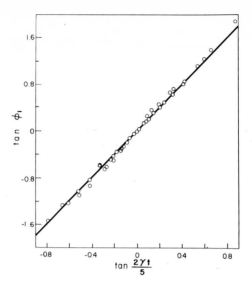

FIG. 34. Rotation of a standard doublet of rigid spheres as a spheroid of equivalent axis ratio $r_e = 2$ according to equation (V1). The points are experimental, and the line is calculated from equation (V1) by means of the measured γ. (After Bartok and Mason.[27])

between the interacting spheres during collision. The symmetry and reversibility of the collision process can be explained qualitatively in terms of the symmetry properties of the creeping-flow equations; this is discussed below in connection with other reversibility phenomena. The following argument shows that the film thickness separating the sphere surfaces varies symmetrically with ϕ_1.

Assuming that the equation for the axial force P_1 generated by the suspending liquid along the major axis of a prolate spheroid given by equation (III57) can be applied to a collision doublet of spheres of radius b, $r_e = 2$, we may write

$$P_1 = \pi\eta_0\gamma b^2 \sin 2\phi_1 \sin^2 \theta_1 \left(\frac{3\mathfrak{A}}{8 - 9\mathfrak{A}}\right) \tag{V4}$$

which from equation (II43) is

$$P_1 = 4.41\pi\eta_0\gamma b^2 \sin 2\phi_1 \sin^2 \theta_1 \tag{V5}$$

Compressive forces acting along the line joining the centers thus act to push the spheres together when $\phi_1 < 0$, and tensile forces act to pull them apart when $\phi_1 > 0$, as shown in Fig. 35 for the equatorial plane. It has been

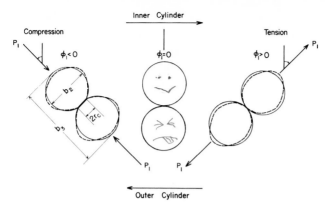

FIG. 35. Forces acting on colliding spheres in Couette flow. In the case of fluid spheres, as shown, the result is a flattening in a circle of radius r_c when the drops are pressed together ($\phi_1 < 0$). The observed deformation into oblate spheroids ($\phi_1 < 0$) and prolate spheroids ($\phi_1 > 0$) is shown by the dashed contour. (After Allan and Mason.[124])

shown[126] that this occurs at a rate

$$\dot{h} = \frac{2P_1 h}{3\pi\eta_0 b^2} \tag{V6}$$

where h is the minimum distance of separation of the surfaces of the two spheres. Substituting for P_1 in equation (V6), combining with equation (II13) for $\dot{\phi}_1$, and simplifying then yields, for the rate of approach in the equatorial plane,

$$\frac{dh}{d\phi_1} = \frac{29(4h \sin 2\phi_1)}{3 \cos 2\phi_1 + 5} \tag{V7}$$

indicating that the variation in h is independent of b, η_0, and γ. Integration of equation (V7) yields

$$h = \frac{122(3h_0)}{(\cos 2\phi_1 + \frac{5}{3})^{4.90}} \tag{V8}$$

showing that h_0 is the closest distance of approach and that h varies symmetrically about $\phi_1 = 0$. Thus, the paths of approach and recession of colliding rigid spheres are mirror images of each other. It should be noted, however, that equation (V8) is only an approximation and may be expected to hold best at small values of ϕ_1. The symmetry and reversibility of such

[126] G. D. M. MacKay and S. G. Mason, *J. Colloid Sci.* **16**, 632 (1961).

collisions is illustrated in Fig. 36, which gives a dimensionless plot of the path of the sphere centers about the mid-point of the doublet axis.

Using the method of reflections, a more rigorous treatment of the rotations and paths of two interacting spheres has been given by Wakyia *et al.*,[112] but is valid only for particle center separations greater than 1.4 sphere diameters. The paths of approach and recession are shown to be curvilinear and the theory reveals the existence of permanent interactions resulting in closed orbits. The properties of such nonseparating doublets are considered below.

The collision process, as seen along the X_2 axis of the motion,[28] is illustrated in Fig. 33(b). The X_1X_3 projections of the path of approach and recession of particle centers are slightly curvilinear, ΔX_1 is a maximum at

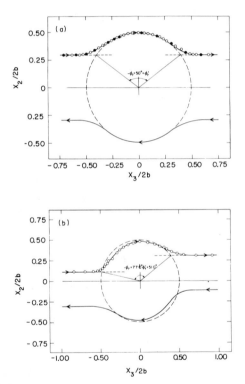

FIG. 36. Dimensionless plot of the observed curvilinear paths of approach and recession of the centers of colliding rigid spheres (a) and fluid spheres (b) in Couette flow. The collision between rigid spheres (open circles) is symmetrical, and during the period of apparent contact the paths lie on a circle of radius $\frac{1}{2}$ shown in the diagram. Further, on reversal of flow (closed circles) the particles recollide in the exact reverse of the first collision. In the case of fluid spheres $-\phi_1{}^0 > \phi_1^{0\prime}$, and owing to deformation of the drops into oblate spheroids when $\phi_1 < 0$ (Fig. 35), the path lies inside the circle.

$\phi_1 = \phi_2 = 0$, and the variation of ϕ_2 as given by

$$\tan \phi_2 = 2C \sin (\tfrac{2}{5}\gamma t) \tag{V9}$$

with the separation angle $\phi_2^{0'}$, the reflection of the apparent collision angle $-\phi_2^0$.

The lifetime τ of the doublet is, from equation (V1),

$$\tau = (5/\gamma) \tan^{-1}(\tfrac{1}{2} \tan \phi_1^0) \tag{V10}$$

Below $\phi_1^0 = 73°$ this yields a τ less than, and above it a τ greater than, the time of rotation of a single sphere, $r_e = 1$. As will be shown later, the mean lifetime $\langle \tau \rangle = 5\pi/6\gamma$ is five sixths that of an isolated sphere. Despite this the mean periods of rotation of a given sphere in dilute suspensions are found to be unaffected by collisions,[27,28] suggesting that the mean velocity of rotation over the entire interaction is, within limits of the experimental error, the same as that for a free sphere.

b. Nonseparating Collision Doublets

Observations made in dilute suspensions[27,125,127] have revealed the existence of a number of collision doublets in which the particles continue to rotate about each other with variable angular velocities and without separating, so that twice during each orbit, at $\phi_1 = -\pi/2$ and $\phi_1 = \pi/2$, the particles are in a "head-on" position. Observations of the same doublet in the $X_2 X_3$ and $X_1 X_3$ planes[127,128] show that the spheres are visibly separated throughout much of the orbit, the doublet axis being a maximum at a ϕ_1 of $\pi/2$ and $\phi_2 = \phi_{2M}$ and a minimum, the spheres being sometimes in apparent contact, at ϕ_1 and $\phi_2 = 0$. When separated, the spheres do not rotate as a rigid dumbbell. Despite the variation in length of the doublet axis it appears to describe the spherical elliptical orbit of a prolate spheroid given by equation (II20), with C approximately constant throughout the orbit but a value of $r_e > 2$, where r_e is larger in those doublets showing a higher particle-center separation. Thus, in most cases there is a variation in θ_1 during each orbit, the $X_2 X_3$ projection of the particle centers about the mid-point of the doublet axis (Fig. 37) being an ellipse; cf. Fig. 3(b). Careful analysis has shown, however,[127] that the agreement with Jeffery's theory for doublets having large separation of sphere centers is satisfactory only when $\phi_1 < 60°$. As demonstrated by Zia et al.,[75a] the use of Jeffery's equations is strictly justified only when the spheres are in contact and do not rotate with respect to each other.

[127] C. L. Darabaner and S. G. Mason, Rheol. Acta 6 (1967).
[128] H. L. Goldsmith and S. G. Mason, Proc. Roy. Soc. A282, 569 (1964).

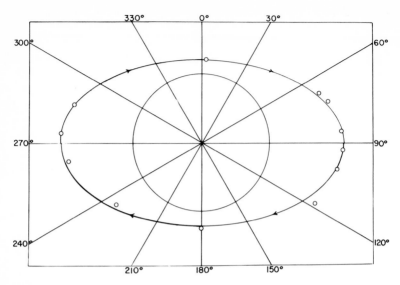

FIG. 37. Rotation of a nonseparating doublet in Couette flow. Plot of the X_2X_3 projection of the path of particle centers around the mid-point of the doublet axis. The circle drawn has diameter $2b$; thus, at $\phi_1 = \pm\pi/2$ there is separation of the spheres. (After Mason and Bartok.[22])

In the absence of buoyancy effects there is no tendency for r_e to increase with time,[127] and nonseparating doublets continue to rotate indefinitely with periodically varying $\dot{\phi}_1$ and $\dot{\phi}_2$, unless a collision with a third sphere occurs. Such three-body interactions may result in a change of orbit constant or in the separation of the spheres and are observed to be reversible; thus, it has been shown that upon reversal of flow and recollision a nonseparating doublet is re-formed by the same sphere that brought about its destruction.[127,128]

A limiting case of a nonseparating doublet of rigid spheres is provided by a doublet of conducting spheres brought together by the application of an electric field along the X_2 axis of the Couette flow field.[4,75] At sufficiently low shear rates these doublets rotate, as do the fused doublets described in Section II; that is, $h = 0$. Upon increasing γ a point is reached at which the spheres separate. The values of P_{1M} at the γ corresponding to separation, calculated from equation (V3), giving an order of magnitude of the forces holding the spheres together, ranged from 0.01 to 0.2 dyne in these experiments.[4]

As mentioned in Section III, it is possible to form n-lets of conducting spheres, $n \leqslant 10$, by applying an electric field, and these also rotate as rigid rods at low γ.[75,75a]

Further insight into the possible mode of formation of nonseparating doublets is provided by considering the analogous situation in two-dimensional flow. This has been treated by Raasch,[114] who has calculated the curvilinear paths of approach and recession in the encounter of neutrally buoyant, infinitely long cylinders whose long axes are aligned with the X_1 axis of the Couette field of flow (Fig. 38). It is found that there exist closed orbits, in which the doublet axis has finite angular velocity at $\phi_1 \pm \pi/2$ and the cylinders do not separate. The variation of the distance y between the particle centers (length of the doublet axis) is given by

$$\dot{y} = \gamma \frac{\sinh^2 \ell}{\cosh^2 \ell} \tanh \ell \sin 2\phi_1 \qquad \text{(V11)}$$

and the respective angular velocities of the cylinder, $\omega_1{}^0$, and doublet axis ω_D are

$$\omega_1{}^0 = \frac{\gamma}{2}[1 + (2\mathscr{S} - 1)\cos 2\phi_1] \qquad \text{(V12)}$$

$$\omega_D = \frac{\gamma}{2}[1 + (2\mathscr{S} - 1 + \tanh \ell)\cos 2\phi_1] \qquad \text{(V13)}$$

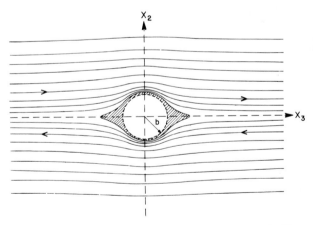

FIG. 38. Two-dimensional analogue of standard and nonseparating doublets: two-body collisions between very long cylinders whose long axes are aligned with the X_1 axis of the Couette field. The paths drawn are calculated according to the theory of Raasch,[114] which predicts the existence of closed orbits (some of which are contained within the shaded portion) corresponding to nonseparating cylinder doublets, which have been observed. (After Goldsmith and Mason.[128])

where

$$\ell = \cosh^{-1}\left(\frac{y}{2b}\right)$$

$$\mathscr{S} = 2\sinh^2\ell \sum_1^\infty \frac{n}{\sinh 2n\ell + n\sinh 2\ell}$$

It follows from equations (V12) and (V13) that

$$\omega_1{}^0 - \omega_D = \frac{\gamma}{2}\tanh\ell\,\cos 2\phi_1$$

and, hence, that there is rolling of the cylinders of the doublet except when $\ell = 0$ and $y/2b = 1$, that is, when they are touching. In this case $\omega_1{}^0 - \omega_D = 0$, and there is dumbbell rotation. Moreover, when $\ell = 0$, equation (V12) reduces to Jeffery's equation, equation (III13), in the form

$$\dot{\phi}_1 = \frac{\gamma}{2}(1 + B\cos 2\phi_1)$$

where $B = (r_e{}^2 - 1)/(r_e{}^2 + 1) = 0.537$ and, hence, $r_e = 1.83$.

Furthermore, equation (V11) predicts that the paths are symmetrical about $\phi_1 = 0$, and equation (V13) predicts that orbits at low values of ℓ are closed. In general, the condition for closed orbits, that is, $\omega_D > 0$ for all ϕ_1, is that $(\mathscr{S} + \frac{1}{2}\tanh\ell) < 1$. The existence of closed orbits and also the validity of equations (V11) to (V13), has been experimentally demonstrated.[25]

Approximate expressions similar to equations (V12) and (V13) and valid for $y/2b > 1.4$ have been derived for interacting rigid spheres[112] and found to give better agreement with experiment than Jeffery's equations in the case of nonseparating doublets having high r_e.[127]

c. Collision Frequency

The experimentally observed collision process provides a basis for the theory of the kinetics of interaction in flowing suspensions of rigid spheres. This theory, however, is purely geometrical and simplifies the calculations by assuming a rectilinear approach of the spheres up to the instant of apparent contact.

One considers a suspension of uniformly dispersed and equal-sized spheres of radius b containing n particles per unit volume. After the rectilinear approach of two sphere centers to within a distance $2b$ a standard collision doublet is formed at $-\phi_1{}^0, \theta_1{}^0$. The doublet rotates with angular velocity $\dot{\phi}_1$, given by equation (V1), and separates at $\phi_1{}^0, \theta_1{}^0$, the reflection of the angles of contact.

Thus, a reference sphere placed at the origin of the Couette-flow field collides with all particles whose centers pass through a "collision disc" of radius $2b$ and origin coincident with that of the reference sphere [Fig. 39(a)] lying in the $X_1 X_2$ plane normal to the direction of flow.

Taking cylindrical polar coordinates X_3, r, θ_2 with origin at the reference-sphere center, the number of particles passing through an area element of the disc in unit time is $df(r, \theta_2) = n|V_3(r, \theta_2)|r\, dr\, d\theta_2$. The absolute value of V_3 is taken, since all collisions are counted positive, whether they occur in a region of positive flux f^+ ($V_3 > 0$) or negative flux f^- ($V_3 < 0$). It

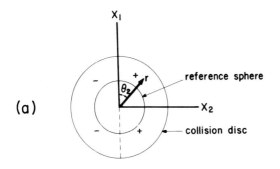

(a)

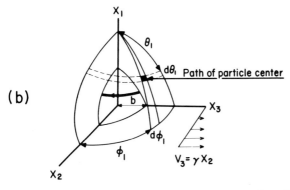

(b)

FIG. 39. Coordinate system for describing collisions in Couette flow. (a) Cylindrical polar coordinates X_3, r, θ_2 showing a "collision disc" drawn about the reference sphere in the $X_1 X_2$ plane normal to the direction of flow (the X_3 axis is directed into the page). All spheres whose centers lie on a path having $r \leqslant b$ are assumed to collide with the reference sphere. The number of such collisions is equal on both sides, $(+)$ and $(-)$, of the $X_1 X_3$ plane. (b) Spherical polar coordinates ϕ_1, θ_1 with X_1 as the polar axis. The number of particles passing through the collision sphere surface in unit time in the interval $d\theta_1, d\phi_1$ shown by the black area is given by equation (V21). The orbit of the doublet axis is indicated by the heavy line. (After Manley and Mason.[28])

follows that the two-body collision frequency f is

$$f = f^+ - f^- = 2f^+ - f' \tag{V14}$$

where

$$f' = f^+ + f^-$$

Since $V_3(r, \theta_2) = \gamma r \sin \theta_2$, one obtains

$$f^+ = n\gamma \int_0^{2b} \int_0^{\pi} r^2 \sin \theta_2 \, dr \, d\theta_2$$

$$= \tfrac{16}{3} n\gamma b^3 \tag{V15}$$

and similarly it is found that

$$f^- = -\tfrac{16}{3} n\gamma b^3 \tag{V16}$$

Thus, the same *number* of collisions occur on each side of the $X_1 X_3$ plane in unit time ($f' = 0$), and f is given by

$$f = \tfrac{32}{3} n\gamma b^3 \tag{V17a}$$

$$= 8c\gamma/\pi \tag{V17b}$$

where c is the volume concentration of singlet spheres. The two-body collision frequency per unit volume is given by $fn/2 = 16n^2\gamma b^3/3$. As will be shown below, in a nonuniform velocity gradient, as in Poiseuille flow, $f^+ \neq f^-$. Equations (V17) were first derived by von Smoluchowski.[129] If the suspension contains spheres of two sizes having radii b_A and b_B, then the number of collisions suffered by a type A sphere with type B sphere is[125]

$$f_{AB} = \tfrac{2}{3} n_B{}^2 \gamma (b_A + b_B)^3 \tag{V18a}$$

$$= 3c_B{}^2 \gamma (b_A + b_B)^3 / 8\pi b_B{}^3 \tag{V18b}$$

where n_B and c_B are the respective number and volume concentration of type B spheres.

Despite the simplifying assumptions of the treatment, the measured collision frequencies in suspensions of volume concentration $<3\%$ are found to be in excellent agreement with equations (V17) and (V18).[28]

The mean free path $\langle l \rangle$, that is, the average distance ΔX_3 traveled by the reference sphere between collisions, is

$$\langle l \rangle = V_3/f$$

$$= 3X_2/32nb^3 \tag{V19}$$

[129] M. von Smoluchowski, *Physik.-Z.* **17**, 557, 583 (1916); *Z. Physik. Chem.* **92**, 129 (1917).

Equation (V19) may be compared to the expression for the mean free path of a molecule in a gaseous system containing only one species:

$$\langle l \rangle = \sqrt{2}\pi n b^2 / 4 \tag{V20}$$

where n is now the number of molecules per unit volume having radius b. The difference in the power of b in equations (V19) and (V20), together with the presence of the coordinate X_2 in equation (V19), are, of course, related to the nature of the shear flow and the two different collision mechanisms involved.

d. Distribution of Doublet Lives

With respect to the spherical polar coordinates ϕ_1 and θ_1 [Fig. 39(b)], the number of collisions with the reference sphere in four of the eight octants in which collisions are possible in the interval $d\theta_1{}^0$, $d\phi_1{}^0$, $df(\theta_1{}^0, \phi_1{}^0)$ is given by[28]

$$df(\theta_1{}^0, \phi_1{}^0) = 16 n \gamma b^3 \sin^3 \theta_1{}^0 \sin 2\phi_1{}^0 \, d\theta_1{}^0 \, d\phi_1{}^0 \tag{V21}$$

and, hence, the fraction of collisions in the interval $d\theta_1{}^0$, $d\phi_1{}^0$, when the limits of $\theta_1{}^0$ and $\phi_1{}^0$ are assumed to be between 0 and $\pi/2$, is, from equations (V21) and (V17)

$$g(\theta_1{}^0, \phi_1{}^0) = \frac{df(\theta_1{}^0, \phi_1{}^0)}{f} = \frac{3}{2}\sin^3 \theta_1{}^0 \sin 2\phi_1{}^0$$

The fraction having initial values less than $\theta_1{}^0$, $\phi_1{}^0$ is

$$G(\theta_1{}^0, \phi_1{}^0) = \int_0^{\theta_1{}^0} \int_0^{\phi_1{}^0} g(\theta_1{}^0, \phi_1{}^0) \, d\theta_1{}^0 \, d\phi_1{}^0$$

$$= \tfrac{1}{2}[2 - \cos \theta_1{}^0 (\sin^2 \theta_1{}^0 + 2)] \sin^2 \phi_1{}^0 \tag{V22}$$

Since it is known that the doublet life τ is a function of $\phi_1{}^0$ only [equation (V10)], one can calculate the integral distribution function $G(\tau)$ from $G(\phi_1{}^0)$ which, after insertion of the limits of $\theta_1{}^0$ into equation (V22), is

$$G(\phi_1{}^0) = \sin^2 \phi_1{}^0$$

whence

$$G(\tau) = \sin^2 \left[\tan^{-1} \left(2 \tan \frac{\tau \gamma}{5} \right) \right] \tag{V23}$$

The mean doublet life $\langle \tau \rangle$ is given by

$$\langle \tau \rangle = \int_0^{\pi/2} \tau(\phi_1{}^0) \, g(\phi_1{}^0) \, d\phi_1{}^0$$

which, on substituting in $g(\phi_1{}^0)\,d\phi_1{}^0 = \sin 2\phi_1{}^0\,d\phi_1{}^0$ and $\tau(\phi_1{}^0)$ from equation (V10) yields

$$\langle\tau\rangle = 5\pi/6\gamma \qquad (V24)$$

and the maximum lifetime $\tau_M = 5\pi/2\gamma$.

These equations are, of course, only approximate, because the distribution with respect to $\phi_1{}^0$ is based on rectilinear approach, and for a given doublet this would yield a value of $\phi_1{}^0$ greater than is actually observed.

e. Steady-State Concentration of Doublets

The existence of spherical doublets for a finite time means that a steady-state concentration of doublets is built up as the suspension flows. At statistical equilibrium the number of doublets n' per unit volume is

$$n' = \langle\tau\rangle fn/2$$

$$= 5c^2/2\pi b^3 \qquad (V25)$$

when $\langle\tau\rangle$ and f are substituted from equations (V24) and (V17), respectively.

It is easily shown,[125,128] then, that if c' is the steady-state concentration of doublets,

$$c' = \tfrac{20}{3}c^2 \qquad (V26)$$

Equation (V26) can be valid only at low concentrations, because the depletion of singlets in the system and the formation of triplets and higher-order multiplets have been ignored. However, as is evident from Table I, even at low concentrations an appreciable fraction of single spheres exists in doublet form.

f. Distribution of Orbit Constants

During the period of apparent contact the collision doublet is known to rotate in the spherical elliptical orbit given by equation (V2). Assuming rectilinear approach, the fraction of collisions having orbit constants less than C can then be calculated from the differential distribution in $\theta_1{}^0, \phi_1{}^0$:

$$G(C) = \int_0^{\pi/2} \int_0^{\theta_1{}^0(\phi_1{}^0)} g(\theta_1{}^0, \phi_1{}^0)\,d\theta_1{}^0\,d\phi_1{}^0$$

where $\theta_1{}^0(\phi_1{}^0)$ is given by equation (V2). Substituting for $g(\theta_1{}^0, \phi_1{}^0)$ from equation (V22) and integrating gives[27,130]

$$G(C) = 1 + \tfrac{1}{3}(1 + 4C^2)^{-1/2} - \tfrac{4}{3}(1 + C^2)^{-1/2} \qquad (V27)$$

[130] C. E. Chaffey, private communication, 1964.

TABLE I

STEADY-STATE CONCENTRATION OF DOUBLETS IN A
SUSPENSION OF UNIFORM SPHERES

c, vol.-%	c', vol.-%	Fraction of spheres existing as doublets
2	0.3	15
4	1.1	28
8	4.3	54
12	9.6	80

The distribution $G(C)$ is of importance, in calculations of the contribution of collision doublets, to the viscosity of a suspension of spheres, which is discussed in the succeeding section.

g. Sphere Paths

From the collision frequency and collision course one can compute the mean lateral displacements of the spheres in suspensions undergoing Couette flow. For simplicity it is assumed that upon collision the doublet rotates as $r_e = 1$ with $\dot{\phi}_1 = \gamma/2$, $\dot{\theta}_1 = 0$ and, hence, that the path of the approaching sphere center about the reference sphere is an arc of a circle of radius $2b \sin \theta_1{}^0$.

At any time during a collision the displacement along the X_2 axis is

$$\Delta X_2 = \pm b \sin \theta_1{}^0 (\cos \phi_1 - \cos \phi_1{}^0) \qquad (V28)$$

the positive sign applying to one sphere and the negative to the other. Considering only magnitudes of displacement, without regard to sign, the time-average displacement $\langle |\Delta X_2| \rangle$ is given by

$$\langle |\Delta X_2| \rangle = \frac{1}{\tau} \int_0^\tau |\Delta X_2|\, dt = \frac{2}{\tau\gamma} \int_{-\phi_1{}^0}^{\phi_1{}^0} |\Delta X_2|\, d\phi_1$$

since, for $r_e = 1$, $2\, d\phi_1/\gamma = dt$. Upon substitution of $|\Delta X_2|$ from equation (V28) and integration one obtains

$$\langle |\Delta X_2| \rangle = \frac{4b}{\tau\gamma} \sin \theta_1{}^0 (\sin \phi_1{}^0 - \phi_1{}^0 \cos \phi_1{}^0) \qquad (V29)$$

In deriving the time-average mean displacement over all collisions, $\langle\!\langle |\Delta X_2| \rangle\!\rangle$, one eliminates τ from the integral as a result of the consideration that we must sum over all collisions in unit time:

$$\langle\!\langle|\Delta X_2|\rangle\!\rangle = \sum_{n=1}^{n=f} \langle|\Delta X_2|\rangle_n \tau_n$$

$$= \int_0^{\pi/2} \int_0^{\pi/2} \frac{4b}{\gamma} \sin \theta_1{}^0 (\sin \phi_1{}^0 - \phi_1{}^0 \cos \phi_1{}^0) \, df(\theta_1{}^0, \phi_1{}^0)$$

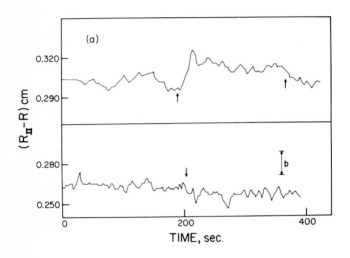

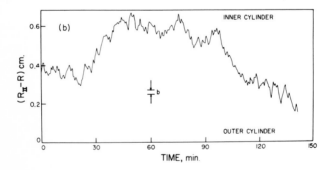

FIG. 40. (a) Variation of the position of the center of a rigid sphere along the X_2 coordinate (R coordinate in cylinder Couette device) due to interactions in a dilute suspension; $c = 3.5\%$, undergoing Couette flow ($\gamma = 0.4 \, \text{sec}^{-1}$, $\Delta R = 0.683 \, \text{cm}$). The arrows indicate the occurrence of a three-body or higher-order collision. (b) The same plot, but in a concentrated suspension, $c = 19\%$, where interactions leading to the formation of triplets and higher-order multiplets result in large ΔX_2. (After Karnis et al.[131])

which, upon substitution of $df(\theta_1{}^0, \phi_1{}^0)$ from equation (V21), yields[131]

$$\langle\!\langle|\Delta X_2|\rangle\!\rangle = \tfrac{8}{3}\pi b^4 n \qquad\qquad (V30a)$$

$$= 2bc \qquad\qquad (V30b)$$

By following the paths of tracer spheres in dilute suspensions ($c < 3\%$) equation (V30) has been tested experimentally[131] and the measured $\langle\!\langle|\Delta X_2|\rangle\!\rangle$ found to be greater than predicted. This is due both to neglect of the curvilinear paths of approach and to neglect of displacements, which occur through interactions of two spheres which, according to the rectilinear approach theory, do not collide. Figure 40a shows the recorded variations of the positions of sphere centers along the X_2 coordinate.

3. Collisions of Fluid Spheres in Couette Flow

a. Standard Doublets

Unlike collisions between rigid spheres, those between fluid drops are unsymmetrical about the $X_1 X_2$ plane with the angle of separation $\phi_1^{0\prime}$ numerically lower than the apparent contact angle $-\phi_1{}^0$, resulting in a displacement of the particles in equal and opposite directions from their initial position along the X_2 axis [Fig. 36(b)].[48,124] The asymmetry is most pronounced in equatorial collisions ($\theta_1 = \pi/2$) and also increases with increasing deformation of the drops.[48,124] Moreover, each time the drops are brought back into collision by reversal of flow, they separate at successively increasing values of ΔX_2, until eventually there is visible separation at $\phi_1 = 0$.

During doublet rotation, as shown in Fig. 35, when $\phi_1 < 0$, the drop surface is flattened at the point of apparent contact, and the particles bulge slightly into oblate spheroids; when $\phi_1 > 0$, they are deformed into prolate spheroids, the point of contact being no longer flattened, demonstrating the existence of the respective compressive and tensile forces used above to explain the symmetry in the variation of h with ϕ_1 during the collisions between rigid spheres. Here, however, because of the flattening of the drops when $\phi_1 < 0$, due to the axial force P_1, an asymmetry is introduced into the collision process. To take into account the flattening of the drops we assume the equation for the radius of contact r_c of the drops,[132]

$$r_c = \left(\frac{bP_1}{2\pi\sigma}\right)^{1/2}, \qquad \text{for} \quad P_1 < 0 \qquad\qquad (V31)$$

[131] A. Karnis, H. L. Goldsmith, and S. G. Mason, *J. Colloid Interface Sci.* **22**, 531 (1966).
[132] G. E. Charles and S. G. Mason, *J. Colloid Sci.* **15**, 235 (1960).

to apply to the present case and that the velocity of approach is given by

$$\dot{h} = \frac{2P_1 h^3}{3\pi\eta_0 r_c^4} + \frac{2P_1 h}{3\pi\eta_0 b^2} \qquad (V32)$$

The first term on the right corresponds to the parallel-disc approach[132] and the second to the sphere–sphere approach.[126] Thus, in the absence of deformation, $r_c = 0$, as when $\phi_1 > 0$ and equation (V32) reduces to the case of the rigid sphere given by equation (V6) but, when $r_c > 0$ ($\phi_1 < 0$), the rate of thinning is slowed down, and the variation in h is no longer symmetric about $\phi_1 = 0$.

Proceeding as before, one finds that

$$\frac{dh}{d\phi_1} = \frac{5h^3 \sin 2\phi_1}{(3\cos 2\phi_1 + 5)(0.93\eta_0\gamma^2 b^4 \sin^2 2\phi_1/\sigma^2 + 0.17h^2)} \qquad (V33)$$

Equation (V33) indicates a paradox, namely that, as γ is increased, the drops tend to remain farther apart. This arises from the increase in drop deformation. When $\gamma = 0$, equation (V33) reduces to equation (V7). By integrating equation (V33) numerically and assuming an arbitrary initial value of $h(-\phi_1{}^0)$ the paths of approach and recession have been calculated, and these approximate the observed paths quite closely.[124]

b. Nonseparating Doublets

The behavior of nonseparating doublets of fluid drops is similar to that described above for rigid spheres, with the exception that r_e is sometimes observed to decrease with time and to lead to coalescence of the droplets.

c. Coalescence in Shear and Electric Fields

In freshly prepared and rigorously cleaned systems coalescence of droplets during doublet rotation can be observed at very low shear rates.[48,124] After some time, however, presumably because of the accumulation of surface-active impurities at the interface, such coalescence becomes rare.[48,124] Coalescence is promoted by the application of an electric field along the X_2 axis, as shown by the sequence of photomicrographs for colliding drops in Fig. 41.

In electric and shear fields the total force is $P_1 + P_1^{(e)}$, where $P_1^{(e)}$ is the electrostatic force of attraction directed along the line joining particle centers.[124,133] If the deformation r_c is not affected by the electric field (as

[133] M. H. Davis, Rand Corp. Publ. RM-2607 (1961).

found experimentally[124]), it may be shown from equations (V4) and (V32) that

$$\dot{h} = \frac{(P_1 + P_1^{(e)})h^3}{8.82\pi\eta_0 b^2(0.93\eta_0\gamma^2 b^4 \sin^2 2\phi_1/\sigma^2 + 0.17h^2)} \tag{V34}$$

Since $P_1^{(e)}$ is always negative, the electrostatic force reinforces P_1 when

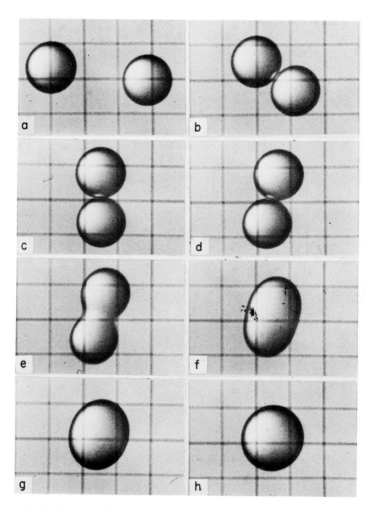

FIG. 41. Coalescence of colliding water drops in a silicone oil under the influence of an electric field directed along the X_2 axis. The stages shown are (a) approach, (b) doublet formation, (c) rotation to $\phi_1 = 0$, and (d) to (h) coalescence in a time interval of 0.4 sec. (After Allan and Mason.[124])

$\phi_1 < 0$, and opposes it when $\phi_1 > 0$; at all events, it acts to reinforce the rate of thinning of the suspending phase film separating the drops.

For equatorial collisions the angle ϕ_1 at which coalescence occurs at a given $\mathscr{E}_0$ and γ shows a distribution about a mean value, which decreases from positive to negative ϕ_1 as $\mathscr{E}_0$ is increased and at a given $\mathscr{E}_0$ increases as γ is increased. This indicates that the drop separation at which coalescence occurs has a distribution of values just as has been reported in gravity and electric fields and in the combination of the two.[132,134] At sufficiently high $\mathscr{E}_0$ the drops coalesce immediately on contact. When $\theta_1{}^0 < \pi/2$, the absolute values of both P_1 and $P_1^{(e)}$ decrease, and there is a consequent sharp decrease in coalescence angle.

The effect of a third diffusing component on the coalescence of colliding drops in Couette flow has also been investigated.[124,135] It had earlier been shown in gravity fields[136] that the diffusion of a mutually soluble component which lowers the interfacial tension between the drop and the suspending phase promotes the coalescence of drops when it diffuses out of the drop and hinders coalescence when it diffuses into the drop. This occurs as a result of the Marangoni effect, whereby an interfacial flow is established by gradients in interfacial tension caused in turn by concentration changes in the diffusing component around the drop. The results in Couette flow,[135] however, appear to be more complex but, in general, are opposite to those in gravity fields, that is, diffusion out of the drops inhibits coalescence, and vice versa.

4. Collisions in Poiseuille Flow

Providing b/R_0 is small, standard and nonseparating two-body collisions between small, uniform, rigid, and deformable spheres in dilute suspensions undergoing Poiseuille flow are the same as those described above in Couette flow.[128] In the case of rigid spheres the only exceptional behavior is observed when one of the colliding particles is almost touching the wall of the tube. In that case, as shown in Fig. 42, the doublets separate at a value of $\phi_1{}^{0'}$ which is numerically lower than the apparent collision angle $-\phi_1{}^0$, resulting in a displacement of the sphere having the higher u_3 to a position of lower R, that is, away from the wall. A reversal of flow in the tube does not bring about a reversal of the first collision. Since the creeping-motion equations predict reversibility, this behavior is exceptional.

In the case of fluid drops the mid-point of the doublet axis during the paths of approach and recession continuously migrates to positions of lower

[134] R. S. Allan and S. G. Mason, *Trans. Faraday Soc.* **57**, 2027 (1961).
[135] G. D. M. MacKay and S. G. Mason, *Kolloid-Z.* **195**, 138 (1964).
[136] G. D. M. MacKay and S. G. Mason, *J. Colloid Sci.* **18**, 674 (1963).

R, resulting in unequal radial displacements of the drops after collision [Fig. 42(c)]. This effect is attributed to the axial migration of the individual drops,[11,128] described in Section IV.

a. Collision Frequency

With the proviso that b/R_0 is small, the assumptions made in the calculation of f in Couette flow are valid in the tube. One considers a reference sphere in a median plane of the tube at a radial distance R^* [Fig. 43(a)], which is rendered stationary by moving the tube with a velocity $-U_3(R^*)$, so that the new velocity distribution becomes

$$U_3'(R) = U_3(R) - U_3(R^*) = \tfrac{1}{2}\mathcal{K}(R^{*2} - R^2) \tag{V35}$$

As in Couette flow, cylindrical polar coordinates X_3, r, θ_2 with origin at the sphere center are taken, but now because of the curvature of the field the regions of positive and negative flux on the collision disc are separated by the arc of a circle, radius R^* [Fig. 43(b)]. Making use of the transformation equation

$$U_3'(\theta_2) = \tfrac{1}{2}\mathcal{K}(2rR^* \sin \theta_2 - r^2)$$

one arrives at the following expressions[128] for f^+ and f'

$$f^+ = \tfrac{16}{3}n\mathcal{K}R^*b^3(1 - m^2)^{1/2} - 4n\mathcal{K}b^4 \cos^{-1} m + \tfrac{1}{3}n\mathcal{K}R^{*4}$$

$$\times [\tfrac{3}{2}\sin^{-1} m - (1 - m^2)^{1/2}(m^3 + \tfrac{3}{2}m)] \tag{V36}$$

$$f' = -4n\mathcal{K}\pi b^4 \tag{V37}$$

where $m = b/R^*$. Substituting these into equation (V14) yields the total collision frequency f:

$$f = \tfrac{32}{3}n\mathcal{K}R^*b^3(1 - m^2)^{1/2} - 8n\mathcal{K}b^4 \cos^{-1} m + \tfrac{2}{3}n\mathcal{K}R^{*4}$$

$$\times [\tfrac{3}{2}\sin^{-1} m - (1 - m^2)^{1/2}(\tfrac{3}{2}m + m^3)] + 4n\mathcal{K}\pi b^4, \quad 0 < m \leqslant 1 \tag{V38}$$

The solution for f^+ does not apply when $m > 1$, that is, when the center of the reference sphere lies within one particle radius of the tube axis. In that case[128]

$$f^+ = \tfrac{1}{4}n\mathcal{K}\pi R^{*4} \tag{V39}$$

and results in a collision frequency given by

$$f = n\mathcal{K}\pi(\tfrac{1}{2}R^{*4} + 4b^4) \tag{V40}$$

equation (V38) reducing to equation (V40) when $m = 1$.

X_3 cm.

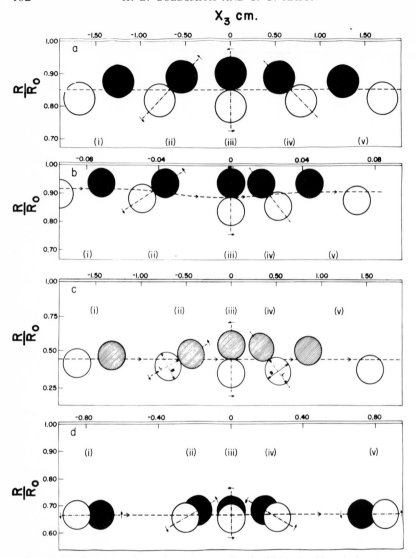

FIG. 42. Tracings taken from photomicrographs of collision doublets during approach (i), collision (ii) to (iv), and recession (v) in the median $(X_2 X_3)$ plane of the tube. The spheres and their separation distances are drawn to scale with reference to the R axis; the X_3 axis, drawn coincident with the tube wall (at rest) has been expanded (a, c, d) or compressed (b) and indicates only the successive positions of the mid-point of the axis joining the sphere centers, the path of which is shown by the dashed line. (a) Symmetrical collision between rigid polystyrene spheres. (b) Unsymmetrical collision close to the tube wall. (c) Unsymmetrical collision between water drops in an oil showing deformation into oblate spheroids ($b_3/b_2 < 1$) at (ii) and into prolate spheroids ($b_3/b_2 > 1$) at (iv). (d) Half-orbit of nonseparating collision doublet of rigid polystyrene spheres. (After Goldsmith and Mason.[128])

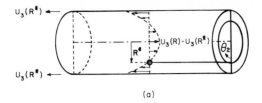

(a)

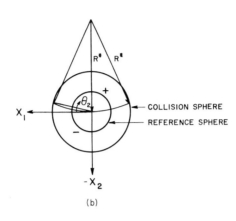

(b)

FIG. 43. Cylindrical polar coordinates X_3, r, θ_2 for describing collisions in Poiseuille flow, having origin at the reference sphere center, which is situated in the median plane of the tube. (a) The sphere at a radial distance R^* is rendered stationary by moving the tube with a velocity $-U_3(R^*)$. The new velocity profile $U_3{}'(R)$ is shown. (b) Detail of collision disc as in Fig. 39; here, because of the curvature of the velocity field the areas of positive ($+$) and negative ($-$) flux are not equal but are separated by the arc of a circle, radius R^*. (After Goldsmith and Mason.[128])

If $m = b/R^* \ll 1$, equation (V38) is closely approximated by equation (V17), the solution when curvature is neglected. As may be seen in Fig. 44, measurements of the collision frequency in dilute suspensions flowing through tubes, $b/R^* < 0.06$, have shown excellent agreement with equation (V17). The quantity $-f' = 4n\mathcal{K}\pi b^4$ gives the excess of collisions taking place on the tube wall side of the reference sphere per unit time. It is independent of radial position, but strongly dependent on particle size; the ratio $-f'/f$, calculated from equation (V17), is $3\pi b/8R^*$, so that the proportion of the total collisions that occur on the wall side of a given reference sphere increases with decreasing R^*. When the collisions are symmetrical, as with rigid spheres, the radial distribution of particles in the tube remains unaffected. However, since collisions between fluid drops are unsymmetrical,[48,124,128] the excess of collisions on the wall side of a given drop leads to a more frequent displacement toward the tube axis than toward the

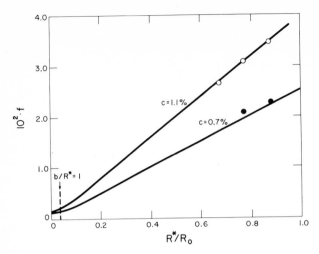

F<small>IG.</small> 44. Frequency of two-body collisions in Poiseuille flow as a function of the radial position of the reference sphere at two concentrations. The points are experimental, and the lines drawn are calculated from equation (V38) and (V40).

tube wall. This effect may be expected to enhance the net radial migration of the suspension.

The mean free path of a given sphere in the tube is given by

$$\langle l \rangle = 3(R_0{}^2 - R^2)/64Rnb^3 \tag{V41}$$

and the mean free path for all the particles in the tube is

$$\langle\!\langle l \rangle\!\rangle = \int_0^{R_0} \langle l \rangle 2\pi nR \, dR \Big/ \int_0^{R_0} 2\pi nR \, dR$$

$$= R_0/16b^3n \tag{V42}$$

b. Steady-State Concentration of Doublets

Equation (V26) will apply at low concentrations in Poiseuille flow; surprisingly, however, measurements have shown that at a given instant there are about twice as many doublets per unit length of the tube than predicted by the theory.[128] The most plausible explanation of the discrepancy lies in the existence of nonseparating doublets ignored by the theory given above. By analogy with the two-dimensional case of the rigid cylinders[114] one would expect a number of doublets to be nonseparating. Moreover, Wakiya has shown[112] that two spheres whose centers lie in the stationary layer of Couette flow have a finite X_2 component of velocity, proportional to b^5/y^5, where y is the distance between sphere centers.

c. Sphere Paths

As in Couette flow, the time-average displacement of spheres normal to the wall, $\langle\!\langle |\Delta R| \rangle\!\rangle = 2bc$, equation (V30), has been measured and shown to be larger than the theoretical.[131]

5. INTERACTIONS OF RODS AND DISCS

When rods or discs are carried into proximity by the velocity gradient, they become associated for a time and then separate with different orbit constants.[30,137] Typical results of such interactions between rigid rods and discs is shown in Fig. 45, in which $\phi_{2M} = \tan^{-1} Cr_e$ has been plotted against the number of particle rotations; moreover, the variation of ϕ_1 and ϕ_2 with time may be observed to change during the interaction period (Fig. 46), deviating from that given by equations (II18) and (II22). It should be noted that the particles spend more time in orbits having higher C; this can be related to the observed steady-state distribution in C, discussed below.

One of the most interesting aspects of two- and even three-body inter-actions of rods and discs is that, like those between rigid spheres, the original orientations of the particles are regained upon reversal of flow; that is,

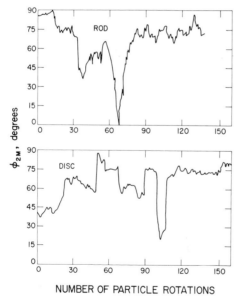

NUMBER OF PARTICLE ROTATIONS

FIG. 45. Interactions between rigid rotating rods and discs. Variations in the orbit of a rotating rod (upper) and disc (lower), as shown by plotting ϕ_{2M} against the number of particle rotations. (After Anczurowski and Mason.[137])

[137] E. Anczurowski and S. G. Mason, *J. Colloid Interface Sci.* **23**, 522 (1967).

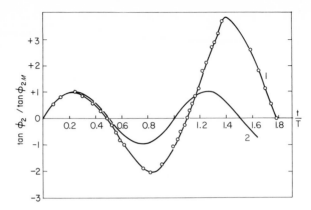

FIG. 46. Disturbance of the angular velocity $\dot{\phi}_2$ during the collision of two rods (curve 1). For comparison a plot for a noninteracting particle obeying a Jeffery's equation is shown (curve 2). (After Mason and Manley.[30])

they return to rotate in the orbits and move with the translational velocities existing prior to collision. An example of a reversible three-body collision between discs is shown in Fig. 47.

It is evident that in a dilute suspension a given particle undergoes many such interactions, and experiment has shown[30,35,137] that after a time a steady state distribution of orbit constants exists. The nature of the distribution function in C and the rate at which a steady state is reached are discussed below.

6. CONCENTRATED SUSPENSIONS

a. Spheres

As the volume concentration of rigid spheres in suspension is increased beyond 2 %, there is a marked increase in triple and higher-order collisions. In both Couette [Fig. 40(b)] and Poiseuille flow (Fig. 48) n-body interactions ($n \geq 3$), unlike two-body collisions, are unsymmetrical and result in displacements both along the X_2 (or R) and X_1 axes of some of the interacting spheres. If the particle motions are followed at increasingly greater concentrations (by matching the refractive indices of disperse and liquid phases and then introducing visible tracer spheres[105,131]), a point is reached at which the velocity distribution changes, such that the velocity gradient near the annulus or tube wall is greater than, and that in the center smaller than, the theoretical Couette or Poiseuille profiles. The following is a summary of the main experimental findings.

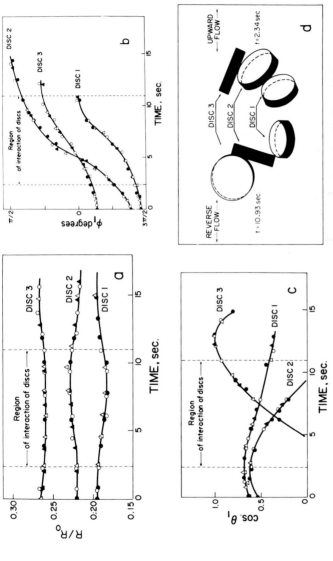

FIG. 47. Reversibility of collisions between three interacting discs in Poiseuille flow. (a) to (c) Respective variations of the radial position R, the angle ϕ_1 and $\cos\theta_1$ with time for the first collision and when the particles are three times recollided by reversal of flow. Experimental points are (circles) collisions with flow upward of the fluid. (triangles) collisions with flow downward in the tube. (d) The particle orientations as seen in the $X_2 X_3$ plane immediately before and after the first collision. (After Karnis et al.[131])

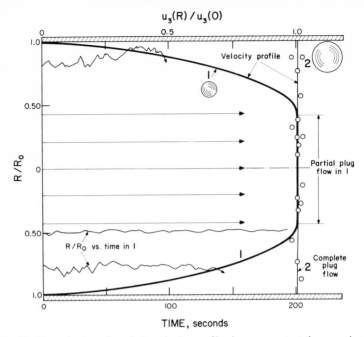

FIG. 48. Dimensionless plot of the velocity profiles in a concentrated suspension of rigid spheres ($c = 0.33$) flowing through a circular tube. As the mean-sphere to tube-size ratio increases from 0.056 in curve 1 to 0.112 in curve 2, the degree of blunting increases from partial to complete plug flow. The lines drawn are the best fit of the experimental points, shown only for curve 2. The figure also shows the variation in radial distance R with time for a tracer sphere in the suspension giving the profile 1. The radial displacements decrease in magnitude with decreasing R and disappear in the region of plug flow. (After Karnis et al.[131])

(1) *Tube flow.*[131]　At sufficiently low values of c and b/R_0 the particle and suspending fluid velocities are identical and parabolic [equation (IV3)], for example, for $b/R_0 = 0.02$ up to $c = 0.35$ and for $b/R_0 = 0.04$ up to $c = 0.20$. As c and b/R_0 or both are increased, u_3 and U_3 continue to be equal, but a pronounced blunting of the velocity profile develops in the center of the tube with a core of radius R_c, in which u_3 is constant (Fig. 48). This may be considered a "partial plug flow," although it should be emphasized that this does not mean the profile is mathematically flat when $R < R_c$ with a discontinuous drop of the velocity gradient to zero at R_c; rather, it is a region in which there is no *measurable* gradient.

When $R > R_c$, the spheres rotate although not always with steady angular velocity, because they are subject to frequent interactions, and the average measured $\omega_1{}^0$ correspond to those that may be calculated from equation (II16) using values of γ obtained from the slopes of the experimental velocity profiles.

As may be seen from Fig. 48, the spheres exhibit erratic radial displacements, the magnitude and frequency of which decrease with decreasing R, until at $R < R_c$ these disappear, and the particles move with identical velocities and zero rotation.

At a given b/R_0 the radius of the core increases with increasing c and similarly for increasing b/R_0 at constant c, until an apparent complete plug flow is present in the tube (Fig. 48). By contrast, the relative velocity profiles are *independent* of flow rate and viscosity of the suspension (provided $Re_p < 10^{-5}$, at which radial migration due to inertial effects becomes appreciable).

The observed partial and complete plug flow is not explained by a dilution of the peripheral suspension due to inward migration of spheres near the wall, since the measured concentration profiles in the tube are uniform[131] (except, of course, for the geometrical requirement that particle centers be displaced at least b from the wall), and many spheres are seen to roll along the wall.[131] Nor can it be explained by non-Newtonian properties of the suspensions, since the measured apparent viscosities in a Couette viscometer are independent of shear rate;[131] moreover, in the tube, at a given R_0, a linear relation between pressure drop and flow rate is obtained. It is concluded that one is dealing with a wall effect of the kind envisaged by Vand,[98,106] that is, that there is effectively a region of high viscosity consisting of suspension, bounded by a region of low viscosity at the wall. However, the shape of the observed velocity profiles cannot be satisfactorily accounted for by a theory of "pseudo slip" at the wall or by a two-phase flow model such as that used for flowing, neutrally buoyant bubbles in liquids undergoing Poiseuille flow.[138]

(2) *Couette flow.*[131] The results obtained in a counterrotating-cylinder Couette having a gap width ΔR provide additional evidence of the existence of a wall effect. As in the tube, the deviation from the undisturbed velocity distribution given by equation (II11) increases with increasing c and $b/\Delta R$, as shown in Fig. 49. Particle and fluid velocity profiles are identical over a range of values of Ω_I and Ω_{II}. The theoretical and experimental curves intercept each other near the center of the annulus, indicating a similar deviation at each wall. By contrast, the velocity profiles obtained from tracer particle velocities in an elasticoviscous liquid are found to deviate markedly from equation (II11) and are unsymmetrical and dependent on cylinder velocities, the profile asymmetry increasing as the stationary layer is moved away from one of the cylinders.[39]

The normal fluctuations of a sphere in concentrated suspensions in the Couette apparatus over a long period of time, shown in Fig. 40(b), allow the

[138] H. L. Goldsmith and S. G. Mason, *J. Colloid Sci.* **18**, 237 (1963).

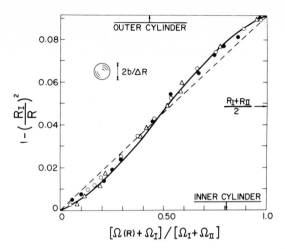

FIG. 49. Dimensionless plot of velocity profiles in concentrated suspensions of rigid spheres undergoing Couette flow, showing deviation from the theoretical curve calculated from equation (III1), shown dotted, at $2b/\Delta R = 0.29$. The profile, however, remains symmetrical and independent of the cylinder speeds of rotation, the closed circles represent points obtained when the outer cylinder speed was 3 times that for the open circles, and the open triangles represent points obtained when the inner cylinder speed was 5.5 times that for the open circles. (After Karnis et al.[131])

particle to travel almost from one wall to the other, and the time-average mean displacement is thus much larger than that given by equation (V30).

b. Rods and Discs

Concentrated suspensions of rods and discs show a similar behavior, namely, that above a certain value of c partial plug flow develops in the tube and, as shown in Fig. 50(b), with velocity profiles of the same shape as those for spheres.[131] Moreover, the velocity distribution is independent of time, indicating that the particles quickly attain the equilibrium distribution of orientations.

In the region of plug flow there is no rotation of rods or discs; the axes of revolution of most of the discs are nearly normal to the direction of flow (oriented with their faces parallel to planes passing through the tube axis, $\theta_1 = 90°$), and those of the rods are nearly parallel to the direction of flow ($\theta_1 = 90°$). At the periphery of the tube, where velocity gradients are fully established, the particle rotations, although considerably inhibited, mostly take place in orbits having high values of C. These observed orientations may be due to the effect of a convergent entry from the reservoir to the tube, as in hyperbolic radial flow,[43] or to particle–particle interactions which, as described below, shift the distribution of orbit constants toward $C = \infty$.[137]

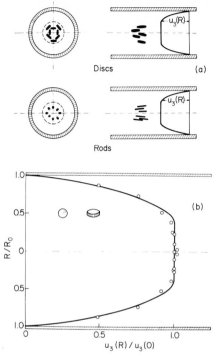

FIG. 50. (a) Steady-state orientations of rods and discs in concentrated suspensions flowing through tubes, as viewed along tube axis (left) and in the median $(X_2 X_3)$ plane (right). (b) Similarity of velocity profiles in a suspension of spheres and discs. The solid line is for the spheres, and the experimental points are for a suspension of rigid discs at the same concentration $(c = 0.30)$. The relative sizes of spheres and discs are also shown. In order to make these measurements (and those illustrated in Figs. 48 and 49) the refractive indices of the particles and suspending medium were matched, and then 0.1% by volume of particles whose surfaces had been coated with aluminum were introduced; these were easily visible and their motions could be followed. (After Karnis et al.[131])

7. REVERSIBILITY

The reversibility in particle rotational and translational displacements, which is observed in two- and three-body collisions between rigid spheres, discs, and rods at low Reynolds numbers, can also be demonstrated in concentrated suspensions, where interactions are much more complex. In the case of spheres it is clear from the analysis of doublet behavior given above that at concentrations above 12% by volume no singlet spheres can exist; above 20% it is probable that one is dealing with a system which at any given instant contains only triplets and higher-order multiplets. Such particle aggregates, it was also shown, behave unsymmetrically with respect to the displacements of the individual spheres along the axes normal to the direction of flow [cf. Figs. 40(b) and 48]. Yet, as illustrated in Fig. 51, the behavior of the individual spheres in concentrated suspensions is reversible. Both the translational (R) and rotational (ϕ_1) coordinates of a single sphere and, hence, of the complex configurations of the dynamically interacting assembly of spheres are conserved.[131] Similar behavior may be observed with a tracer disc in a concentrated suspension of discs.[131]

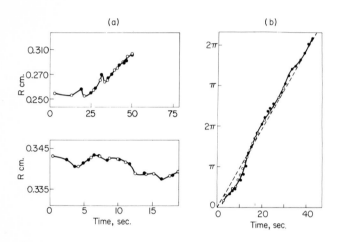

FIG. 51. Reversibility of collisions in concentrated suspensions of rigid spheres flowing through a tube, $b/R_0 = 0.04$. The open circles are experimental points obtained during flow in the upward direction; the closed circles, when the flow was reversed. (a) Variation of R with time at $c = 0.17$ (upper) and $c = 0.34$ (lower). (b) Variation of ϕ_1 with time at $c = 0.17$. The dashed line is calculated from equation (II16) for a single sphere assuming a parabolic velocity distribution. (After Karnis *et al.*[131])

Reversible motions can also be demonstrated in the domain of the suspending liquid. If a drop of suspending fluid colored with dye is injected into the suspension, at rest in the tube, in the region $R > R_c$, and flow is started, the drop is distorted due to convective dispersion by the particle motions and interactions. If, then, after a given time the flow is reversed for the same time and at the same velocity, the original drop shape is regained. Injection of the colored drop into the region of the tube, $R < R_c$, does not lead to distortion during the flow, demonstrating convincingly the absence of velocity gradients in the central core. Similar results have been obtained in a fixed particle bed.[139]

[139] J. W. Hiby, *Proc. Symp. Interaction Fluids Particles, London, 1962* p. 312. Inst. of Chemical Engineers, London, 1963.

It must be emphasized, however, that the experimental requirements in such demonstrations of reversibility are very stringent. It is important to have isothermal flow (to avoid irreversible convection currents), to match particle and liquid densities (to avoid irreversible sedimentation), and to have low flow rates (to avoid irreversible inertial effects). A formalized theoretical basis of such time-reversed flows has been given by Slattery,[13] based on the linearized form of the Stokes–Navier equation, equation (I3), and suggested by the mirror-symmetry time-reversal theorem of Bretherton.[12] Slattery[13] considered the unsteady flow developed from a given initial particle distribution and orientation in a suspension. He proved that on reversal of flow the "time-reversed" system can be thought of as a cinefilm of the "real system" run backwards. The reason for this lies in the fact that with the neglect of inertia the Navier–Stokes equations become linear and are still satisfied under time reversal. The only difference between the forward and reverse systems is that the pressure gradient in the former is the negative of that in the latter. Put another way: the velocity distribution as a function of time in the reverse motion is the reverse-chronological velocity distribution in the forward motion.

In addition to the demonstrations of reversibility described above the following two experiments may be cited.

1. An "unmixing" demonstration, first described by Heller[140] and illustrated in Fig. 52, in which several letters are drawn with a dye in a clear viscous fluid contained between two transparent concentric cylinders: after slowly rotating the outer cylinder for four rotations, until the dye appeared mixed with the suspending fluid, the motion was stopped and then reversed. The letters again appear (slightly blurred by diffusion), when the cylinder has been rotated back to the initial position.

2. Rods in a dilute suspension in a concentric-cylinder Couette device are aligned with the X_2 axis with $\theta_1 = \pi/2$ by the action of an electric field[4]: the suspension undergoes Couette flow for a short time, until there is a distribution of orientations of the rods, brought about by particle interactions, and the flow is then reversed; the rods are brought back to their initial alignment.[35]

8. STEADY-STATE ORIENTATIONS OF RODS AND DISCS

The contribution to the viscosity of a suspension made by spheroidal particles depends, not only on the particle shape and interactions, but also on their orientation with respect to the axes of the external flow field. If the concentrations are low enough to neglect particle interaction, in the absence of Brownian motion, it is possible to calculate the instantaneous

[140] J. P. Heller, *Am. J. Phys.* **28**, 348 (1960).

FIG. 52. Time-reversed flows: an "unmixing" demonstration. The letters were drawn with the aid of a syringe and needle containing a dye dissolved in the same clear viscous liquid as that in the annulus between the two transparent concentric cylinders. The outer cylinder was slowly rotated ($\frac{1}{8}$ turn at position 2, $\frac{1}{4}$ turn at position 3, and $\frac{3}{8}$ turns at position 4), until at position 5, after four revolutions, the dye is mixed with the suspending fluid. The motion was then reversed, and the positions 6, 7, and 8 correspond to 4, 3, and 2 respectively. Finally, at position 9, when the cylinder has been rotated back to its original position, the letters appear, slightly blurred by diffusion.

distribution of the particle axis of revolution with respect to the angle ϕ_1 since, as was shown in Section II, its variation with time is a function only of the velocity gradient and the equivalent ellipsoidal axis ratio. For the distributions with respect to ϕ_2, ϕ_3, and θ_1, θ_2, and θ_3, however, one also requires a knowledge of the distribution of orbit constants.

It has been shown that in laminar flow at low Reynolds numbers single rigid spheroids both in uniform- and nonuniform-shear fields undergo continuous rotation in a fixed orbit. One might therefore postulate, as Eisenschitz[141] did, that the particles are distributed isotropically before the onset of motion and that subsequently each particle moves in a fixed orbit. In actual fact experiment has shown[30,137] that the equilibrium distribution in C for both rods and discs in dilute suspensions is intermediate between that calculated from the assumptions of Eisenschitz and the minimum energy dissipation hypothesis of Jeffery.[2] The reason for this, as will be shown below, lies in the interactions between the particles, which lead to the adoption of preferred orbits. Hence, in calculating certain of the distributions of orientations it becomes necessary to use the measured distribution functions in C.

In more concentrated suspensions still another effect, that of particle crowding, which prevents the free rotation of the spheroids, must be considered, besides the possibility of entanglements, such as those observed in flowing suspensions of wood pulp fibers.[104]

a. Steady-State Distributions of ϕ_1, ϕ_2, and ϕ_3

The instantaneous distribution of the spheroid axis of revolution with respect to the angle ϕ_1 having the X_1 axis as the polar axis is given by the continuity equation

$$\frac{\partial g(\phi_1, t)}{\partial t} = \frac{-\partial g(\phi_1, t)}{\partial \phi_1} \dot{\phi}_1 \tag{V43}$$

where $g(\phi_1, t)$ is the fraction of particles having orientations in the interval $d\phi_1$ at time t. At the steady state $\partial g(\phi_1, t)/\partial \phi_1 = 0$, and

$$g(\phi_1)\dot{\phi}_1 = \text{constant} \tag{V44}$$

The equations for ϕ_2 and ϕ_3 are the same. For the angle ϕ_1 the differential and integral distribution functions are easily obtained by substituting for $\dot{\phi}_1$ from equation (II13) and noting that

$$\int_0^{2\pi} g(\phi_1)\, d\phi_1 = 1$$

$$g(\phi_1) = \frac{r_e}{2\pi(r_e{}^2 \cos^2 \phi_1 + \sin^2 \phi_1)} \tag{V45}$$

$$G(\phi_1) = \frac{1}{2\pi} \tan^{-1}\left(\frac{\tan \phi_1}{r_e}\right) \tag{V46}$$

[141] R. Eisenschitz, Z. Physik. Chem. **A158**, 85 (1932).

If the differential distribution $[g(\phi_1)]^{1/2}$ is plotted against ϕ_1 in plane polar coordinates, as shown in Fig. 53, the result is an ellipse having semiaxes $(r_e/2\pi)^{1/2}$ ($\phi_1 = \pm\pi/2$) and $(1/2\pi r_e)^{1/2}$ ($\phi_1 = 0$) with preferred orientations of particle axes of revolution in the X_3 direction for $r_e > 1$ and the X_2 direction for $r_e < 1$. For instance, the instantaneous distribution of ϕ_1 in two suspensions of rods having $r_e = 10$ and 25 ($r_p \approx 15$ and 40, respectively) is such that 45 and 73 %, respectively, of the particles would lie in orientations corresponding to $\phi_1 = 90° \pm 5°$, that is, the long axes of the rods almost aligned with the flow.

Measurements of $G(\phi_1)$ for rods in suspensions of volume concentration c from 8×10^{-6} to 6×10^{-4}, and for discs in suspensions of $c = 2.5 \times 10^{-3}$,

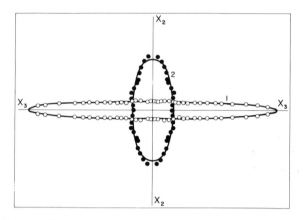

FIG. 53. Calculated differential distribution of the orientation of the axis of revolution of (1) a rod $r_e = 14.2$ and (2) disc $r_e = 0.41$, with respect to the angle ϕ_1. The points are experimental, and the lines are calculated from equation (V45). (After Anczurowski and Mason.[137])

have shown [11,30,35] (Fig. 54) that an equilibrium distribution close to the predicted value given by equation (V46) is established in a rather short time (3 rotations per particle) and is unaffected by concentration. This is in marked contrast to the equilibrium distributions in ϕ_2 and C which are established only after prolonged shearing.[30,35]

The distributions of orientations with respect to ϕ_2 and ϕ_3 are more complex and cannot be calculated a priori without knowledge of the distribution of orbit constants. The distributions of ϕ_2 and ϕ_3 corresponding to a *fixed* C, however, may be computed from equation (V44) by using equations (II22) and (II23) with the condition

$$\int_{-\phi_{2M}}^{+\phi_{2M}} g(\phi_2)\, d\phi_2 = 1, \qquad \int_{-\phi_{3M}}^{+\phi_{3M}} g(\phi_3)\, d\phi_3 = 1$$

and they are

$$g_C(\phi_2) = \frac{2\sec^2\phi_2}{\pi(C^2 r_e^2 - \tan^2\phi_2)^{1/2}}, \qquad g_C(\phi_3) = \frac{2\sec^2\phi_3}{\pi(C^2 - \tan^2\phi_3)^{1/2}} \qquad (V47)$$

$$G_C(\phi_2) = \frac{2}{\pi}\sin^{-1}\left(\frac{\tan\phi_2}{C r_e}\right), \qquad G_C(\phi_3) = \frac{2}{\pi}\sin^{-1}\left(\frac{\tan\phi_3}{C}\right) \qquad (V48)$$

the subscript C denoting that the distribution functions are evaluated at a given C. If, now, $g(C)$ and $G(C)$ are the experimentally determined differential

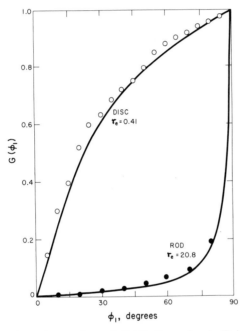

FIG. 54. Measured steady-state integral distributions of rods (Poiseuille flow) and discs (Couette flow) with respect to ϕ_1. The lines drawn are calculated from equation (V46); $r_e = 20.8$ for the rods and $r_e = 0.41$ for the discs. (After Goldsmith and Mason[11] and Anczurowski and Mason.[137])

and integral distributions of orbit constants, the integral distribution functions for ϕ_2 and ϕ_3 over all C may be written[23] as, for example, for ϕ_2,

$$G(\phi_2) = G'\left(\frac{1}{r_e}\tan\phi_2\right) + \int_{(1/r_e)\tan\phi_2}^{\infty} g(C)G_C(\phi_2)\,dC \qquad (V49)$$

The first term on the right-hand side of equation (V49) is the fraction of the assembly whose orientations cannot exceed ϕ_2, the upper bound fixed by the orbit constant corresponding to $\phi_{2M} = \phi_2$, equation (II22). The second term represents the contributions made to the orientations $\leqslant \phi_2$ by particles having orbit constants corresponding to $\phi_{2M} \geqslant \phi_2$. Differentiating equation (V49) yields the differential distribution

$$g(\phi_2) = \int_{(1/r_e)\tan\phi_2}^{\infty} g_c(\phi_2) g(C) \, dC$$

$$= \frac{2}{\pi} \int_{\phi_2}^{\pi/2} \frac{\sec^2 \phi_2}{(C^2 r_e^2 - \tan^2 \phi_2)^{1/2}} g(\phi_{2M}) \, d\phi_{2M} \qquad (V50)$$

The functions $g(\phi_1)$, $G(\phi_1)$, etc., however, do not completely describe the anisotropy of the suspension, because no allowance is made for the projected dimensions of the particles on an axis or a plane of the external flow field, which would be required, for example, in calculating the optical properties of the suspension. Useful measures of anisotropy are provided by the projections l_1 and l_{23} of unit length of the axis of revolution of a rod on the polar axis X_1 and the $X_2 X_3$ plane and the projection of unit area A_{23} of the face of a disc on the $X_2 X_3$ plane. From equations (II24) to (II27) these are

$$l_1 = \cos \theta_1, \qquad l_{23} = \sin \theta_1, \qquad A_{23} = \cos \theta_1$$

It is possible to calculate the time-average values of these quantities for particles in a given orbit; thus, for l_1 we have[23]

$$\langle (\cos \theta_1)_C \rangle = \frac{1}{T} \int_0^T (\cos \theta_1)_C \, dt = \frac{2\gamma}{\pi(r_e + 1/r_e)} \int_0^{\pi/2} (\cos \theta_1)_C \frac{dt}{d\phi_1} \, d\phi_1$$

and by substituting in for $\dot{\phi}_1$ and $\theta_1(C)$ from equations (II13) and (II20), respectively, and integrating we obtain

$$\langle (\cos \theta_1)_C \rangle = \frac{2}{\pi} (r_e^2 C^2 + 1)^{-1/2} K(k) \qquad (V51)$$

where $K(k)$ is the elliptic integral defined in equation (II21). The remaining time-average values, $(\cos \theta_2)_C$, $(\sin \theta_2)_C$, etc., are given in Table II. As in the case of the distributions of ϕ_2 and ϕ_3, the average l_1 and A_{23} over all orbit constants require a knowledge of $g(C)$; thus, for instance,

$$\langle \cos \theta_1 \rangle = \int_0^{\infty} \langle (\cos \theta_1)_C \rangle g(C) \, dC \qquad (V52)$$

TABLE II

TIME-AVERAGE VALUES OF $(\cos\theta)_C$ AND $(\sin\theta)_C$

Function		X_1 axis	X_2 axis	X_3 axis
$(\cos\theta)_C$	$r_e > 1$	$\dfrac{2}{\pi}(C^2 r_e^2 + 1)^{-1/2}K(k)^*$ $k = [C^2(r_e^2 - 1)/(r_e^2 C^2 + 1)]^{1/2}$	$\dfrac{1}{\pi}(r_e^2 - 1)^{-1/2}\ln\dfrac{(C^2 r_e^2 + 1)^{1/2} + C(r_e^2 - 1)^{1/2}}{(C^2 r_e^2 + 1)^{1/2} - C(r_e^2 - 1)^{1/2}}$	$\dfrac{2}{\pi}r_e(r_e^2 - 1)^{-1/2}\tan^{-1}[C(r_e^2 - 1)^{1/2}(C^2 + 1)^{-1/2}]$
	$r_e < 1$	$\dfrac{2}{\pi}(1 + C^2)^{-1/2}K(k)^*$ $k = [C^2(1 - r_e^2)/(C^2 + 1)]^{1/2}$	$\dfrac{2}{\pi}(1 - r_e^2)^{-1/2}\tan^{-1}[C(1 - r_e^2)^{1/2}(C^2 r_e^2 + 1)^{-1/2}]$	$\dfrac{1}{\pi}r_e(1 - r_e^2)^{-1/2}\ln\dfrac{(C^2 + 1)^{1/2} + C(1 - r_e^2)^{1/2}}{(C^2 + 1)^{1/2} - C(1 - r_e^2)^{1/2}}$

Function		$X_2 X_3$ plane	$X_1 X_3$ plane	$X_1 X_2$ plane
$(\sin\theta)_C$	$r_e > 1$	$\Lambda_0(\Phi, k)†$ $\Phi = \sin^{-1}[C^2 r_e^2/(1 + C^2 r_e^2)]^{1/2}$ $k = [(r_e^2 - 1)/r_e^2 C^2 + 1)]^{1/2}$	$r_e(r_e^2 - 1)^{-1/2}\Lambda_0(\Phi, k)†$ $\Phi = \sin^{-1}[(r_e^2 - 1)/r_e^2]^{1/2}$ $k = [C^2/(C^2 + 1)]^{1/2}$	$\dfrac{2}{\pi}\{[(r_e^2 C^2 + 1)(C^2 + 1)]^{-1/2}K(k)^* + (r_e^2 - 1)^{-1/2}KZ(\beta, k)\}‡$ $\beta = \sin^{-1}[(C^2 r_e^2 + 1)/r_e^2(C^2 + 1)]^{1/2}$ $k = [C^2 r_e^2/(C^2 r_e^2 + 1)]^{1/2}$
	$r_e < 1$	$\Lambda_0(\Phi, k)†$ $\Phi = \sin^{-1}[C^2/(C^2 + 1)]^{1/2}$ $k = [(1 - r_e^2)/(C^2 r_e^2 + 1)]^{1/2}$	$\dfrac{2}{\pi}\{[(C^2 r_e^2 + 1)(C^2 + 1)]^{-1/2}K(k)^* + r_e(1 - r_e^2)^{-1/2}KZ(\beta, k)\}‡$ $\beta = \sin^{-1}[r_e^2(C^2 + 1)/(C^2 r_e^2 + 1)]^{1/2}$ $k = [C^2/(C^2 + 1)]^{1/2}$	$(1 - r_e^2)^{-1/2}\Lambda_0(\Phi, k)†$ $\Phi = \sin^{-1}(1 - r_e^2)^{1/2}$ $k = [C^2 r_e^2/(C^2 r_e^2 + 1)]^{1/2}$

$^*\,K(k)$ = elliptic integral of first kind. $†\,\Lambda_0(\Phi, k)$ = Heuman lambda function.[142] $‡\,Z(\beta, k)$ = Jacobian zeta function.[142]

and similarly for $\langle \sin \theta_1 \rangle$, etc. Hence, before the results of measurements of orientations and projections in suspensions of rods and discs are considered the distributions of orbit constants will be discussed.

b. Steady-State Distribution of Orbit Constants

According to the Eisenschitz assumption, the integral distribution of orbit constants, $G(C)$, is given by the area on the sphere between the X_1 (polar) axis and the orbit corresponding to C (see Fig. 3), divided by the corresponding area for C's between 0 and ∞. Thus,

$$G(C) = \frac{2}{\pi} \int_0^{\pi/2} \int_0^{\theta_1(\phi_1)} \sin \theta_1 \, d\theta_1 \, d\phi_1$$

$$= \frac{2}{\pi} \int_0^{\pi/2} [1 - \cos \theta_1(\phi_1)] \, d\phi_1$$

which upon substitution of $\theta_1(\phi_1)$ from equation (II20) can be shown[23] to have the general solution

$$G(C) = 1 - \Lambda_0(\Phi, k) \tag{V53}$$

where $\Lambda_0(\Phi, k)$ is the Heuman lambda function[142] having parameter Φ and modulus k, defined by

$$r_e > 1, \qquad \Phi = \sin^{-1}(C^2 + 1)^{-1/2}, \qquad k = [C^2(r_e^2 - 1)/(C^2 r_e^2 + 1)]^{1/2}$$

$$r_e < 1, \qquad \Phi = \sin^{-1}(C^2 r_e^2 + 1)^{-1/2}, \qquad k = [C^2(1 - r_e^2)/(C^2 + 1)]^{1/2}$$

Experimentally, the distribution of C is most easily measured by measuring $\phi_{2M} = \tan^{-1} C r_e$; typical results for rods and discs are shown in Fig. 55, where $G(\phi_{2M})$ has been plotted against ϕ_{2M}, the curve arising from the Eisenschitz assumption and calculated with equation (V53) also being shown. It is evident that the measured distribution lies between the two hypotheses of instantaneously isotropic distribution and minimum energy dissipation, the latter giving $G(0) = 1$ for rods and requiring the discs to rotate in orbits of $C = \infty$, where $\phi_{2M} = \pi/2$. The results have also demonstrated that in the suspensions the frequency with which orbits of lower C appear increases as the axis ratio r_p is lowered.[23]

[142] P. F. Byrd and M. D. Friedman, "Handbook of Elliptic Integrals for Engineers and Physicists." Springer, Berlin, 1954.

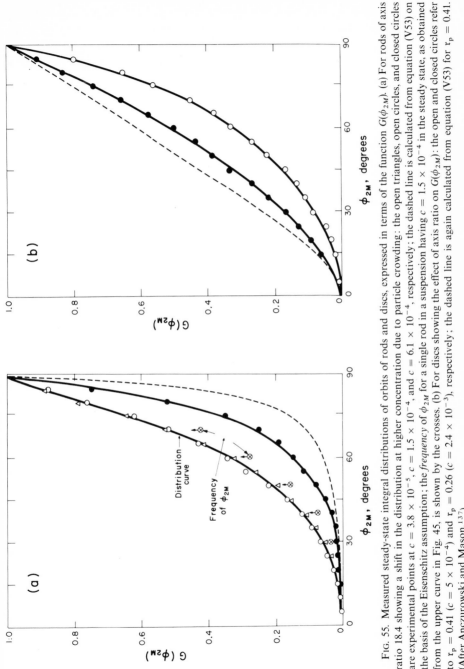

FIG. 55. Measured steady-state integral distributions of orbits of rods and discs, expressed in terms of the function $G(\phi_{2M})$. (a) For rods of axis ratio 18.4 showing a shift in the distribution at higher concentration due to particle crowding: the open triangles, open circles, and closed circles are experimental points at $c = 3.8 \times 10^{-5}$, $c = 1.5 \times 10^{-4}$, and $c = 6.1 \times 10^{-4}$, respectively; the dashed line is calculated from equation (V53) on the basis of the Eisenschitz assumption; the *frequency* of ϕ_{2M} for a single rod in a suspension having $c = 1.5 \times 10^{-4}$ in the steady state, as obtained from the upper curve in Fig. 45, is shown by the crosses. (b) For discs showing the effect of axis ratio on $G(\phi_{2M})$: the open and closed circles refer to $r_p = 0.41$ ($c = 5 \times 10^{-4}$) and $r_p = 0.26$ ($c = 2.4 \times 10^{-3}$), respectively; the dashed line is again calculated from equation (V53) for $r_p = 0.41$. (After Anczurowski and Mason.[137])

In the case of rigid rods having $r_p = 18.4$, the distributions $G(\phi_{2M})$ and $G(C)$ are independent of particle concentrations at $c = 1.5 \times 10^{-4}$. At $c = 6.1 \times 10^{-4}$, as shown in Fig. 55, the distribution curve is displaced toward higher ϕ_{2M}, that is, higher C. This is presumably an effect of particle-crowding and will be considered below.

An interesting correlation may be made between the equilibrium distribution of orbits in dilute suspension and the collisions suffered by the particles. As was pointed out above (Fig. 45), during a period of about 100 to 150 rotations, owing to collisions, ϕ_{2M} for rods and discs may vary over the entire range of values (0 to 90°); however, both rods and discs appear to spend more time in orbits having higher C. If the distribution in ϕ_{2M} calculated from the *frequency* of ϕ_{2M} during the observation of a *single* particle over 150 rotations is compared with the instantaneous distribution measured for 600 different particles in the same suspension, as illustrated in Fig. 55, the agreement obtained between the results is fairly good. It would appear then, that the observed distribution of orbits is due to the particle interactions.

Experiment has shown[137] that the measured distribution of orientations with respect to ϕ_2 and ϕ_3 and also the particle projections $\langle l_1 \rangle$, $\langle l_{23} \rangle$, and $\langle A_{23} \rangle$ are in good agreement with those that one calculates from equation (V49) and from equation (V52) using the average values listed in Table II and the measured $g(C)$. This is illustrated in Fig. 56 for the distributions with respect to ϕ_2 and ϕ_3 and in Table III for the particle projections.

Table III shows that for rods of large axis ratio there is considerable anisotropy: $(\langle l_3 \rangle - \langle l_2 \rangle)$ and $(\langle l_1 \rangle - \langle l_2 \rangle)$ are positive numbers reflecting

TABLE III

Projections of Rods ($r_p = 18.4$) and Discs ($r_p = 0.41$)
in the Steady State (Couette Flow)

Parameter	Measured	Calculated*
$\langle l_1 \rangle$	0.57	0.56
$\langle l_2 \rangle$	0.09	0.08
$\langle l_3 \rangle$	0.72	0.71
$\langle l_{23} \rangle$	0.74	0.72
$\langle l_{31} \rangle$	0.99	0.98
$\langle l_{13} \rangle$	0.59	0.58
$\langle A_{23} \rangle$	0.40	0.39
$\langle A_{31} \rangle$	0.71	0.69
$\langle A_{12} \rangle$	0.35	0.36

* Using the measured differential distributions of orbit constants $g(C)$ and equation (V52).

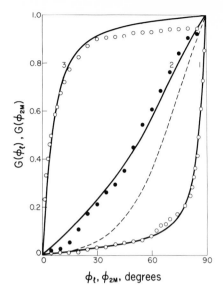

ϕ_ℓ, ϕ_{2M}, degrees

FIG. 56. Plot of the integral distributions with respect to ϕ_1 to ϕ_3 in a suspension of rods, $r_e = 14.2$, undergoing Couette flow. The higher degree of orientation of the particle axis of revolution (the long axis), shown by the $G(\phi_1)$ and $G(\phi_3)$ curves, reflects the fact that the particle spends more time aligned with the X_3 axis with θ_1 close to $\pi/2$. The lines drawn for ϕ_2 and ϕ_3 are calculated from equation (V49) using the measured distribution of orbit constants. The dashed line gives $G(\phi_{2M})$ calculated graphically from projections of rods. (After Anczurowski and Mason.[137])

the fact that the particle spends much more of its time aligned with the X_3 axis than with the X_2 axis, and during this time θ_1 is close to $\pi/2$. Thus, $\langle l_{31} \rangle > \langle l_{23} \rangle > \langle l_{12} \rangle$, and the value of $\langle l_{31} \rangle$ is close to unity, so that the rod spends most of its time near the $X_1 X_3$ plane; hence, there would be an appreciable difference between light-scattering or absorption in a suspension measured in the $X_1 X_3$ plane and in the $X_1 X_2$ plane. For the discs, too,

$$\langle A_{31} \rangle > \langle A_{23} \rangle > \langle A_{12} \rangle,$$

because their faces spend more time aligned with the flow than normal to it, and during this time θ_1 is again large. Thus, for instance, in a suspension of small discs undergoing Poiseuille flow there would be less reflection (or absorption) if the light beam were directed normal to a median plane of the tube than when directed along the tube axis.

c. Particle Crowding

During rotation, the motion of the particle axis of revolution of a spheroid resembles that of a gyrating top and thus the volume of liquid swept out by the particle in one orbit, the effective volume $\mathscr{V}_E(C)$, will be greater than the actual particle volume, $\mathscr{V}_p$. Hence, quite apart from the effect of interactions in a suspension of spheroids there exists a critical concentration, above which particles cannot undergo unimpeded rotation, because there is insufficient space to move.[143] Above this concentration they will be constrained to rotate in orbits having less volume.

Because the volume of liquid swept out by a spheroid, or the effective volume, is a function of C, its mean value in a dilute suspension will be given by[30]

$$\langle \mathscr{V}_E \rangle = \int_0^\infty \mathscr{V}_E(C) g(C)\, dC \qquad (V54)$$

For a rigid rod the orbit volume is closely approximated by a cylinder having the elliptical cross section of the $X_2 X_3$ projection of the orbit of the particle ends and height of the X_1 projection of the particle at $\phi_1 = 0$; see Fig. 57(a).[30] If the thickness or b_2 axis of the rod is taken into account, the ellipse has major and minor axes given by

$$2L_2 = 2b_2 \cos\phi_{2M} + 2b_1 \sin\phi_{2M}$$

$$2L_3 = 2b_2 \cos\phi_{3M} + 2b_1 \sin\phi_{3M} \qquad (V55)$$

where, from equations (II22) and (II23),

$$\tan\phi_{2M} = Cr_e, \qquad \tan\phi_{3M} = C$$

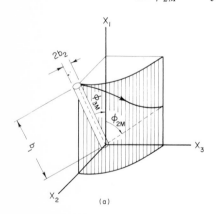 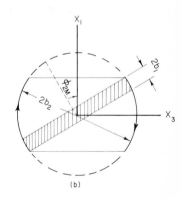

FIG. 57. Geometrical models of the volumes swept out by the particle ends during rotation in a spherical elliptical orbit. (a) Rod: cylinder having elliptical cross section of the $X_2 X_3$ projection of the orbit of the particle ends and height of the X_1 projection of the particle at $\phi_1 = 0$. (b) Discs: spherical segment of diameter $2b_2$ and height given by the X_1 projection of the diameter of the disc (parallel to the $X_1 X_3$ plane) at $\phi_2 = \phi_{2M}$.

[143] S. G. Mason, *Tappi* **37**, 494 (1954).

The height of the cylinder is $(2b_1 \cos \phi_{3M} + 2b_2 \sin \phi_{3M})$, whence the orbit volume $\mathscr{V}_E(C)$ is

$$\mathscr{V}_E(C) = 2\pi L_2 L_3 (b_1 \cos \phi_{3M} + b_2 \sin \phi_{3M})$$

$$= \frac{2\pi(b_2 + b_1 C)(b_1 + b_2 C)(b_2 + b_1 C r_e)}{(1 + C^2)(1 + C^2 r_e{}^2)^{1/2}} \qquad (V56)$$

This expression reduces to the particle volume $\mathscr{V}_p = 2\pi b_1 b_2{}^2$ when $C = 0$, the orbit volume $\mathscr{V}_E(C)$ increasing with increasing C, reaching a maximum, and then falling to a value given by the volume of a disc of thickness $2b_2$, that is $\mathscr{V}_E(\infty) = 2\pi b_1{}^2 b_2$, when the rod rotates in the $X_2 X_3$ plane. For large axis ratios equation (V56) is approximated by[30]

$$\mathscr{V}_E(C) = \frac{2\pi b_1{}^3 C^2 r_e}{(1 + C^2)(1 + C^2 r_e{}^2)^{1/2}} \qquad (V57)$$

The volume swept out by a disc approximates that of a section of a sphere of radius b_2 and thickness given by the X_3 projection of the face of the disc at ϕ_{2M} [Fig. 57(b)].

$$\mathscr{V}_E(C) = 2\pi \int X_3{}^2 \, dX_1$$

$$= 2\pi \int_0^{(b_1 \sin \phi_{2M} + b_2 \cos \phi_{2M})} (b_2{}^2 - X_1{}^2) \, dX_1$$

which upon integration yields

$$\mathscr{V}_E(C) = \frac{2\pi(b_1 + b_2 C r_e)}{(1 + C^2 r_e{}^2)^{1/2}} \left(b_2{}^2 - \frac{(b_1 + b_2 C r_e)^2}{3(1 + C^2 r_e{}^2)} \right) \qquad (V58)$$

$$= \frac{2\pi b_2{}^2 (b_2 C r_e + b_1)(2C^2 r_e{}^2 + 3)}{3(1 + C^2 r_e{}^2)^{3/2}}, \qquad r_p \ll 1 \qquad (V59)$$

Thus $\mathscr{V}_E(C)$ for a disc increases with increasing C from $\mathscr{V}_p$ at $C = 0$ to the value corresponding to a sphere of radius b_2 at $C = \infty$.

The critical concentration c^* in a suspension is reached when the orbit or effective volume fraction $\mathscr{V}_E(C)c/\mathscr{V}_p$ reaches unity;[30,143] hence, from equations (V57) and (V59) we have

$$c^* = \frac{\mathscr{V}_p}{\mathscr{V}_E(C)} = \frac{(1 + C^2)(1 + C^2 r_e{}^2)^{1/2}}{r_p{}^2 C^2 r_e} \qquad \text{for a rod} \qquad (V60)$$

$$c^* = \frac{r_p}{C r_e + r_p} \cdot \frac{3(1 + C^2 r_e{}^2)^{3/2}}{(2C^2 r_e{}^2 + 3)} \qquad \text{for a disc} \qquad (V61)$$

In a suspension in which there is a distribution in C, the critical concentration may be calculated using $\langle \mathscr{V}_E \rangle$ obtained from equation (V54). This is shown in Table IV which lists values of $\langle \mathscr{V}_E \rangle$ and c^* for dilute suspensions of uniformly sized rods and discs of various axis ratios for which $g(C)$ had been measured. It is interesting to note that in the case of the rods of $r_p = 18.4$, where $G(C)$ is concentration-dependent at $n = 8000$ particles/cm³, the volume available to each rod is still 20 times as high as the mean effective volume $\langle \mathscr{V}_E \rangle$ required for rotation. Thus, the calculated critical concentrations for rods, which vary roughly as $1/r_p{}^2$, may be considerably above the concentrations at which crowding effects set in. Table IV also includes effective volumes based on the volume of the sphere having radius given by half the largest particle dimension: $(b_1{}^2 + b_2{}^2)^{1/2}$. For discs of axis ratios between $\frac{1}{2}$ and $\frac{1}{4}$ this spherical effective volume is quite close to the mean effective volume given by equation (V54); for rods, however, it is greater and appears to provide a more realistic basis for calculating critical concentrations.

TABLE IV

EFFECTIVE VOLUMES AND CRITICAL CONCENTRATIONS FOR SUSPENSIONS OF RODS AND DISCS[137]

r_p	n, particles/ cm³	Volume available to each particle $1/n$, $10^3 \times cm^3$	$\langle \mathscr{V}_E \rangle$,* $10^3 \times cm^3$	$\langle \mathscr{V}_E \rangle / \mathscr{V}_p$	c^*	Effective volume† based on sphere, $10^3 \times cm^3$
0.26	98	10	0.063	2.52	0.40	0.068
0.32	103	9.7	0.034	2.22	0.45	0.037
0.41	52	19	0.017	1.81	0.55	0.020
18.4	500	2.0				
18.4	2000	0.50	0.006	72.4	1.38×10^{-2}	0.017
18.4	8000	0.13				
68.4	600	1.7	0.013	1300	0.77×10^{-3}	0.032
115.5	180	5.5	0.069	4000	2.50×10^{-4}	0.153

* Calculated from equation (V54) with the measured distribution $g(C)$ and $\mathscr{V}_E(C)$ given by equation (V57).

† $\mathscr{V}_E = \frac{4}{3}\pi(b_1{}^2 + b_2{}^2)^{3/2}$, that is, sphere having diameter of the largest particle dimension.

9. TRANSIENT ORIENTATIONS OF RODS

Measurements carried out in dilute suspensions of uniformly sized rigid rods have shown[35] that the process of establishing steady-state orientations in ϕ_1 to ϕ_3 and θ_1 to θ_3 from an initial isotropic state takes place in two consecutive steps:

(1) When the number of rotations $t/T < 3$, $G(\phi_1)$ changes very rapidly to its equilibrium value causing large changes in all the other orientation distributions.

(2) When $t/T > 3$ the transient integral distributions $G(\phi_2)$, $G(\phi_3)$, $G(\theta_1)$, and $G(\theta_3)$ change gradually until a final equilibrium is reached after about 500 rotations per particle at a concentration $c = 4 \times 10^{-5}$. When the concentration is doubled the orientation changes are accelerated, final equilibrium being reached after about 200 rotations per particle.

During these two stages, the distribution in orbit constants drifts progressively from the isotropic state where 80% of the particles have orbits of $\phi_{2M} > 80°$ to the equilibrium state where only 54% of particles have orbits of $\phi_{2M} > 70°$.

The two periods are reflected in changes in the anisotropy of the suspension. Thus, in the random state initially produced by thorough stirring, $\langle l_1 \rangle$, $\langle l_2 \rangle$, and $\langle l_3 \rangle$ are found to be close to the predicted value of $\frac{1}{2}$, and $\langle l_{23} \rangle$, $\langle l_{31} \rangle$, and $\langle l_{12} \rangle$ close to $\pi/4$. During 3-particle rotations large changes, due to the changes in $G(\phi_1)$, take place: at $t/T = 3 \langle l_2 \rangle = 0.11$ and $\langle l_{31} \rangle = 0.98$ (cf. Table III), that is, almost all the rods then lie in the $X_3 X_1$ plane, in which plane all the subsequent changes in orientation occur. After 3-particle rotations $\langle l_1 \rangle$ and $\langle l_{12} \rangle$ gradually increase and $\langle l_3 \rangle$ and $\langle l_{23} \rangle$ gradually decrease demonstrating that particles adopt orbits having lower C in which they become more aligned with the X_1 axis.

To explain these results it is necessary to consider the effect of collisions on the distribution of orientations. In the hypothetical case of a collisionless suspension initially isotropic, in which each particle rotates in a fixed orbit,[31,35] the orientation distributions are given by solutions of the continuity equation (V43). It has been shown[35] that this yields a $g(\phi_1)$ which is periodic in time with a period $T/2$, being random $(=\frac{1}{2}\pi)$ at $t = 0$, $T/2$, T, $3T/2$, etc., while for $t = T/4$, $3T/4$, $5T/4$, it has the value

$$g(\phi_1) = [2\pi(r_e^2 \cos^2\phi_1 + r_e^{-2} \sin^2\phi_1)]^{-1}$$

corresponding to a situation in which the particles nearly all lie in the $X_1 X_3$ plane, since $g(\phi_1)$ takes the values $\frac{1}{2}\pi r_e^2$ and $r_e^2/2\pi$ for $\phi_1 = 0$ and $\pi/2$, respectively (cf. Fig. 53). Since, however, C is constant for any particle it follows that $g(C)$ is constant for all time.

As illustrated in Figs. 45 and 46, changes in both the orbit constant and rotational phase occur as a result of collisions. The frequency of such two-body encounters between rods has been measured[35] and found to be 0.2 collisions per particle rotation at $c = 4 \times 10^{-5}$ and to increase linearly with concentration. This is too low to account for the rapid change in $G(\phi_1)$. However, one should distinguish between "close" collisions in which two rods pass with centers within about one particle diameter, and "distant" collisions when they pass within about one particle length. The collisions observed experimentally, a criterion of which is a visible change in ϕ_{2M}, orientation, translational velocity or rotational phase, correspond to both types of collisions. One may calculate the frequency of distant encounters assuming the two-body collision equations for spheres of diameter equal to the rod length [equation (V17)] and this gives about 15 times the number of collisions observed experimentally.[35] On the one hand, an appreciable damping out of the oscillating transient $g(\phi_1)$ could be produced by such distant encounters, most of which would occur with the rods in a ϕ_1-orientation close to $\pi/2$ (where $\dot{\phi}_1$ is very low) and in which a small change in ϕ_1 would make a large difference in the time at which the particle reaches $\phi_1 = 0$ where $\dot{\phi}_1$ is a maximum (cf. Fig. 5). On the other hand, changes in orbit constant due to the frequent distant encounters are small and hence the change in $G(C)$ or $G(\phi_{2M})$ will be relatively much slower and continuous until the final steady state is reached.

VI. Applications to Suspension Viscosity

1. Viscosity of Suspensions of Spheres

Since the Einstein treatment[45,46] of the viscosity of dilute suspensions of spheres several theories have been proposed that have extended the problem to higher concentrations, and very many measurements in Couette and capillary viscometers have been reported. For further details the reader is referred to two excellent reviews of the subject by Frisch and Simha[144] and Rutgers.[145] The present discussion will be confined to comments on the implications of the observed particle motions and interactions that were described in the preceding section.

a. Concentration Dependence of Viscosity

By taking into account the hydrodynamic interaction between the spheres and neglecting Brownian motion Vand[106] arrived at the following result for the relative viscosity:

$$\eta_r = 1 + 2.5c + \alpha c^2 + 0(c^3) \qquad (VI1)$$

where α is the interaction coefficient, calculated to be 7.35. However, in computing the contribution of doublets to the interaction coefficient, Vand

---

[144] H. L. Frisch and R. Simha, in "Rheology: Theory and Applications" (F. R. Eirich, ed.), Vol. 1, Chapt. 14. Academic Press, New York, 1956.
[145] R. Rutgers, Rheol. Acta **2**, 305 (1962).

assumed that in a two-body collision the participating spheres roll over each other in an arc of a great circle on each sphere and separate when the doublet axis is perpendicular to the direction of flow ($\phi_1^{0\prime} = 0$ for Couette flow). This model yields the correct collision frequency but gives values for the doublet mean life and steady-state concentrations that are $\frac{3}{5}$ of those corresponding to the observed collision mechanism. It should also be noted that Vand's collision mechanism is incompatible with the reversibility of creeping motion. When the collision time constant in Vand's theory is then multiplied by $\frac{5}{3}$, one obtains $\alpha = 9.15$. Vand's own experimental value of 7.2 was obtained with comparatively large glass spheres ($b = 65$ microns) and only after making large corrections for wall effects (and concentration changes due to particle-crowding in the capillary viscometer), and they show considerable scatter. Since then more reliable experiments, yielding values of 11.7[47] and 12.7,[146] have been obtained with small glass spheres ($b \approx 5$ microns), of 14.1 for polystyrene spheres[147] ($b = 0.26$ microns), of 6.3 to 7.6[148] and 11.7[149,150] for monodisperse latexes. The question of the value for the coefficient α has not yet been resolved.

b. Wall Effects

First discovered by Bingham and Green,[89] the "sigma" phenomenon, whereby the apparent viscosities of suspensions of clays, soils, and mineral pastes,[96] when measured in capillary viscometers, were found to decrease with decreasing tube radius, has been shown to be of more general occurrence in suspension rheology. To account for the anomalous results obtained in capillary viscometry various theories have been proposed, which may be classified under one of four headings, as follows.

(1) *Hydrodynamic wall effect*.[98,106] As outlined in Section IV, Vand showed that in the presence of a rigid boundary a flowing suspension of spherical particles behaves as if there were a layer of thickness $h = 1.301b$ at the wall having viscosity equal to that of the suspending liquid. This theory predicts that the measured viscosity decreases with decreasing vessel size and should be observable both in Couette and capillary instruments. It is readily shown[98] that the true and apparent relative viscosities η_r and η_r' are related by

$$\frac{1}{\eta_r} - 1 = \left(\frac{1}{\eta_r'} - 1 \right)\left(1 - \frac{h}{R_0} \right)^{-4} \tag{VI2}$$

[146] R. St. J. Manley and S. G. Mason, *Can. J. Chem.* **32**, 763 (1954).
[147] P. Y. Cheng and H. K. Schachman, *J. Polymer Sci.* **16**, 19 (1955).
[148] F. L. Saunders, *J. Colloid Sci.* **16**, 13 (1961).
[149] S. H. Maron and S. M. Fok, *J. Colloid Sci.* **10**, 482 (1955).
[150] S. H. Maron and R. J. Belner, *J. Colloid Sci.* **10**, 523 (1955).

for the capillary tube and by

$$\frac{1}{\eta_r} - 1 = \left(\frac{1}{\eta_r'} - 1\right)\frac{(1 - h/R_\mathrm{I})^2(1 - h/R_\mathrm{II})^2}{1 - 2h/(R_\mathrm{II} - R_\mathrm{I})} \qquad \text{(VI3)}$$

for a concentric-cylinder Couette apparatus.

(2) *Mechanical wall effect.*[99,100]   As a result of the constraint of the tube walls certain particles on entering the capillary are forced to travel on faster-moving streamlines than they would otherwise have done; their average concentration in the tube is therefore reduced relative to that in the suspension reservoir. Measurements of the concentration of rigid spheres in the tube show that this is indeed so, and the wall effect has been treated, as above, as if there were a layer of pure liquid at the wall,[99] a mean value of $h/R_0 = 0.7b$ being obtained.[93] However, a reevaluation of this hypothesis[95] has shown that the experimental results cannot wholly be accounted for by a mechanical wall effect, but that the hydrodynamic interaction of particle with wall, which leads to translational slip, also has to be taken into account.

(3) *Summation, or "sigma," hypothesis.*[151]   The summation hypothesis is based on the presumed existence of unsheared laminae within the suspension, producing a stepwise velocity distribution, the integral in the derivation of Poiseuille's equation being replaced with a summation. This approach is an empirical one and has been applied to clay pastes[151] and to flowing blood,[152,153] but it has not been possible to relate it to the problem of dilute suspensions of spheres.

(4) *Axial migration of particles.*   This hypothesis assumes that there actually exists a particle-depleted layer at the wall, brought about by radial migration of spheres during flow in the presence of normal forces. At low Re_p, $< 10^{-5}$, such migration would occur only in the case of deformable spheres and, hence, may be important in measurements of emulsion viscosity. At higher Re_p, $> 10^{-3}$, where it is known that radial migration of rigid spheres due to inertial effects takes place,[14,31a,101,119,120] the decrease in apparent viscosity with increasing flow rate in a capillary instrument has been correlated with the degree of migration of the spheres.[97]

It has been argued[99] that in a uniform shear field, as in a concentric-cylinder Couette apparatus, in which $\Delta R \ll R_\mathrm{I}$, the apparent viscosity of a suspension of spheres should be independent of the particle distribution in the direction normal to the planes of shear. The available results in

[151] F. Johnson Dix and G. W. Scott Blair, *J. Appl. Phys.* **11**, 574 (1939).

[152] R. H. Haynes, *Am. J. Physiol.* **198**, 1193 (1960).

[153] G. W. Scott Blair, *Rheol. Acta* **1**, 123 (1958).

Couette instruments[154-156] appear to support this hypothesis, because they fail to show a dependence of apparent viscosity on gap width. As described in the previous section, however, at high c and $b/\Delta R$ a wall effect has been shown to exist in Couette flow, which manifests itself by a velocity redistribution, which at a given c depends on $b/\Delta R$. Although it appears that the effect in Couette flow is smaller than in the tube,[131] nevertheless there must then be a dependence of the measured viscosity on the ratio $b/\Delta R$. One might expect that in addition the variation of suspension viscosity with c, which closely obeys the equation derived by Mooney,[157] viz.

$$\ln \eta_r = 2.5c/(1 - \Upsilon c) \qquad (VI4)$$

the constant 2.5 being approximate and Υ a crowding factor approximating a value close to the inverse of the maximum random packing fraction of spheres, would change at the point where the velocity profile deviates from the theoretical. Some evidence of such a change has been obtained.[93,131]

2. VISCOSITY OF SUSPENSIONS OF SPHEROIDS

Jeffery's equations for the motion of the axis of revolution of an ellipsoid in slow shear flow have been used in several theoretical treatments of the intrinsic viscosity of suspensions of ellipsoidal[34,158,159] and rodlike[34,141] particles. For reviews of this and other work see Frisch and Simha[144] and Sadron.[160] Jeffery[2] calculated the increase in energy dissipation in the fluid, due to the presence of the ellipsoid, and from this obtained an expression for the effective viscosity (the viscosity of a homogeneous fluid, in which the energy dissipation is the same) of a dilute suspension of spheroids undergoing Couette flow as a function of the initial particle orientations.

a. Particles Free from External Forces

Jeffery's result for the increased energy dissipation $\dot{W}$ due to the ellipsoid is

$$\dot{W} = \frac{16}{3}\pi\eta_0 \left(\frac{\alpha_1'' D_{11}^{02}}{2b_2{}^2\alpha_1'\alpha_2''} + \frac{D_{22}^{02} + 2D_{23}^{02} + D_{33}^{02}}{2b_2{}^2\alpha_1'} + \frac{2(D_{31}^{02} + D_{12}^{02})}{\alpha_2'(b_1{}^2 + b_2{}^2)} \right)$$

[154] G. F. Eveson, R. L. Whitmore, and S. G. Ward, *Nature* **166**, 1074 (1950).

[155] G. F. Eveson, S. G. Ward, and R. L. Whitmore, *Discussions Faraday Soc.* **11**, 11 (1951).

[156] K. H. Sweeney and R. D. Geckler, *J. Appl. Phys.* **25**, 1135 (1954).

[157] M. Mooney, *J. Colloid Sci.* **6**, 162 (1951).

[158] W. Kuhn and H. Kuhn, *Helv. Chim. Acta* **28**, 97 and 1533 (1945).

[159] R. Simha, *J. Phys. Chem.* **44**, 25 (1940).

[160] C. Sadron, *in* "Flow Properties of Disperse Systems" (J. J. Hermans, ed.), p. 131. North-Holland Publ., Amsterdam, 1953.

where the components of the rate of shear D^0_{11}, etc., and the spheroid integrals α_1', α_2', etc., are defined by equations (III52) and (III53), respectively. When $r_p \gg 1$, and the spheroid integrals are simplified in a manner similar to that used in arriving at equation (III58), the expression given above can be shown[2,34] to lead to the following equation for the intrinsic viscosity as a function of θ_1 and ϕ_1:

$$[\eta] = \frac{\dot{W}}{\mathscr{V}_p \eta_0 \gamma^2}$$

$$= \frac{r_p^2 (\sin^4 \theta_1 \sin^2 2\phi_1)}{6(\ln 2r_p - 1.5)} \tag{VI5}$$

This expression is almost identical with that calculated for the intrinsic viscosity of a suspension of rods by Burgers,[34] using Jeffery's equations, but neglecting the particle thickness:

$$[\eta] = \frac{r_p^2 (\sin^4 \theta_1 \sin^2 2\phi_1)}{6(\ln 2r_p - 1.8)} \tag{VI6}$$

Thus, as in the expressions for the axial force P_1 acting on a rod in Couette flow, these two equations become identical when r_p for the cylinder is replaced with the axis ratio r_e of the equivalent ellipsoid, r_e/r_p here having the value of 0.74.

The orientation factor $\sin^4 \theta_1 \sin^2 2\phi_1$ reaches a maximum when $\theta_1 = 90°$ (when $C = \infty$) and $\phi_1 = 45°$. Its time-average value, which is a function of the orbit constant, may be obtained by substituting from equations (II13) and (II20)[2]:

$$\langle(\sin^4 \theta_1 \sin^2 2\phi_1)_C\rangle = \frac{2\gamma}{r_e + 1/r_e} \int_0^{\pi/2} (\sin^4 \theta_1 \sin^2 2\phi_1)_C \frac{dt}{d\phi_1} d\phi_1$$

$$= \frac{2r_e^2}{(r_e^2 - 1)^2} \left(\frac{C^2(r_e^2 + 1) + 2}{[(C^2 r_e^2 + 1)(C^2 + 1)]^{1/2}} - 2 \right) \tag{VI7}$$

and, when $r_e \gg 1$, it reduces to[34]

$$\langle(\sin^4 \theta_1 \sin^2 2\phi_1)_C\rangle = \frac{2}{r_e} \cdot \frac{C}{(C^2 + 1)^{1/2}} \tag{VI8}$$

Thus, the time-average orientation factor for rods increases from 0 at $C = 0$ to a value of $2/r_e$ at $C = \infty$. For oblate spheroids the position of maxima and minima are reversed, as shown by Jeffery;[2] he also gave values of the minimum and maximum $[\eta]$ for spheroids of varying axis ratios, which are shown in Fig. 58.

For a calculation of the average contribution to the viscosity of all particles in the suspension the time-average orientation factor must be further averaged over all values of C; hence, the distribution of orbit constants must be known. The intrinsic viscosity for a suspension of rods may then be written

$$[\eta] = \frac{r_p{}^2}{6(\ln 2r_p - 1.8)} \int_0^\infty \langle (\sin^4 \theta_1 \sin^2 2\phi_1)_C \rangle g(C)\, dC \qquad (VI9)$$

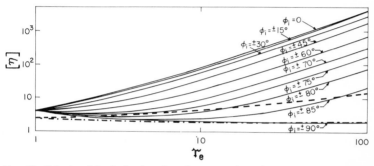

FIG. 58. Calculated intrinsic viscosity of a suspension of rods in Couette flow when the particles are held at a fixed orientation, $\phi_1 = \phi_i$ and $\theta_1 = \pi/2$, by a torque, for example, by an electric field perpendicular to the X_1 axis of the shear field, shown as a function of r_e and ϕ_i. The dashed lines are the maximum and minimum values of $[\eta]$ calculated from Jeffery's theory.[2] (After Chaffey and Mason.[73])

Burgers[34] showed that, if the particles are isotropically distributed, as postulated by Eisenschitz, the integral has the value $4/\pi r_p$ (actually, $4/\pi r_e$ should be used[86]). As described in Section V, a suspension of rods, initially in the isotropic state, becomes progressively anisotropic upon shearing with the particles adopting orbits having a lower C. Calculations of the reduced viscosity using the experimentally determined distributions in ϕ_1, θ_1, and C show that the viscosity decreases monotonically from the isotropic to the equilibrium state.[35]

The earliest measurements of the viscosity of suspensions[161] of rods were carried out with glass rods and silk and rayon fibers of diameters 50 to 90 microns and lengths of 10^2 to 10^3 microns and were probably not free from wall effects and aggregation. Later work[86] in a Couette viscometer with rayon fibers of 3.5 microns diameter and axis ratios of 45 to 350 showed that the measured $[\eta]$, under such conditions that no shear-induced buckling occurs[81] ($r_p \approx 45$), is closer to that predicted from equations (VI8) and

[161] F. Eirich, H. Margaretha, and M. Bunzl, *Kolloid-Z.* **75**, 20 (1936).

(VI9) for $C = \infty$ than to that given by the Eisenschitz orbit distribution. However, when $[\eta]$ is calculated from the measured distribution of orbits in suspensions of rods in the absence of crowding effects, the resulting value is only half that which corresponds to the Eisenschitz assumption.[30] This discrepancy is most probably due to particle-crowding effects in the rayon fiber suspensions, where the concentrations ($> 3 \times 10^6$ particles/cm^3) were well in excess of the critical value based on the spherical effective volume, in which case the distribution of C is known to be shifted toward $C = \infty$.[137]

Another result that emerged from the work on rayon fiber suspensions[86] is that the ratio of measured to calculated $[\eta]$ increases with increasing r_p. This can be explained by the permanent deformation during rotation of the fibers of large r_p,[81] described in Section III, which leads to a rapid increase in r_e over the value for the unbent fiber used in calculating $[\eta]$. Furthermore, the suspensions of flexible fibers of $r_p > 150$ are found to exhibit the Weissenberg effect, when on applying shear the liquid climbs up the inner rotating cylinder of the viscometer. The viscosity in these suspensions, in contrast to those of fibers of $r_p < 150$, is shear-dependent. Whether the effect is due to recoverable deformation of isolated particles as suggested (indirectly) by Taylor[162] or is due to the formation of a particle network is not known.

b. Particles Constrained by External Torque

If an external torque is applied to the spheroids in the suspension, by subjecting them to an electric field, for example, the orbits are modified, as described in Section II. An electric field parallel to the X_1 axis in Couette flow ($\mathscr{E}_2, \mathscr{E}_3 = 0$) causes either rods or discs to move in orbits of minimum energy dissipation, and a field perpendicular to the X_1 axis ($\mathscr{E}_1 = 0$) causes discs to spin about their axes, giving maximum energy dissipation.[4,5] Jeffery's results for these limiting orbits are therefore immediately applicable. The viscosity of a suspension of rods with an electric field normal to the X_1 axis is not so simply obtained. A solution has, however, been given[73] for the case in which the electric-field strength exceeds the critical value and the rods all lie in the $X_2 X_3$ plane at a fixed orientation $\phi_1 = \phi_i$, equation (II57), while the fluid undergoes Couette flow.

In this case the nonzero components of the fluid rotation and dilatation are[73]

$$Z_2{}^0 = \tfrac{1}{2}\gamma, \qquad D_{11}^0 = -D_{33}^0 = \tfrac{1}{2}\gamma \sin 2\phi_i, \qquad D_{31}^0 = D_{13}^0 = \tfrac{1}{2}\gamma \cos 2\phi_i$$

and, since they are not time-dependent, the rate of energy dissipation $\dot{W}$

[162] G. I. Taylor, *Proc. 2nd Intern. Congr. Rheol., Oxford, 1953* p. 1. Academic Press, New York, 1954.

and $[\eta]$ do not have to be averaged. The result for the intrinsic viscosity is[73]

$$[\eta] = \frac{(\frac{3}{2}r_e^2 - r_e^2\mathfrak{A} - \frac{5}{4}\mathfrak{A})(r_e^2 - 1)}{(r_e^2 - \frac{3}{2}\mathfrak{A})(2r_e^2 - 2r_e^2\mathfrak{A} - \mathfrak{A})} \sin^2 2\phi_i$$

$$+ \frac{(2 - \mathfrak{A})(r_e^2 - 1)}{(3\mathfrak{A} - 2)(2r_e^2 - 2r_e^2\mathfrak{A} + \mathfrak{A})} \cos^2 2\phi_i + \frac{2(r_e^2 - 1)}{2r_e^2 - 2r_e^2\mathfrak{A} + \mathfrak{A}} \cos 2\phi_i$$

$$+ \frac{r_e^2 + 1}{2r_e^2 - 2r_e^2\mathfrak{A} + \mathfrak{A}} \tag{VI10}$$

As shown in Fig. 58, at a given r_e the intrinsic viscosity $[\eta]$ increases with decreasing ϕ_i: the intrinsic viscosity of a suspension of rods held perpendicular to the streamlines would be very great. However, if an electric field is directed along the X_2 axis, rods can only be held in steady orientations such that $\tan^{-1} r_e < \phi_i < \pi/2$, and ϕ_i depends on the shear and electric fields as well as r_e. The maximum and minimum values of $[\eta]$ for freely moving rods[2] are also shown in Fig. 58; the intrinsic viscosity at a given ϕ_i always exceeds the minimum value, except at $\phi_i = \pi/2$. As r_e approaches unity, both the maximum and minimum $[\eta]$ approach 2.5, the Einstein coefficient for freely rotating spheres, and $[\eta]$ then approaches 4 at all ϕ_i, the value for non-rotating spheres, which may also be calculated from Brenner's equation for the energy dissipation around an arbitrarily translating and rotating sphere in shear flow.[163] Measurements of the viscosity of dilute suspensions of rods in an electric field have shown qualitative agreement with the theory.[164]

VII. Inertial Effects

1. GENERAL

Sections II to VI have dealt with particle behavior in the creeping (or Stokes) flow regime, where the velocity fields, forces, and torques are independent of the Reynolds number. We now consider certain effects that arise from inertia during flow at higher, but still small, Reynolds numbers. Among these is the lateral migration of a freely rotating sphere in a fluid undergoing Poiseuille flow, first observed for neutral buoyancy by Eirich[97a] and studied quantitatively by Segré and Silberberg.[14,102] It was shown that the spheres reached a stable equilibrium radial position $R^\dagger = 0.6R_0$ independent of the initial positions of release. Thus, particles introduced near the wall migrated inward, and those near the axis migrated outward, during flow, leading to an accumulation of spheres in an intermediate annulus. The

[163] H. Brenner, *Phys. Fluids* **1**, 338 (1958).
[164] Y. El-Tantawy and S. G. Mason, Unpublished Work, 1965.

phenomenon was termed the "tubular pinch effect." A number of subsequent investigations confirmed the result[31a,38,101,105,120,165] and showed that it applied also to rigid rods and discs[31a,38,165] but not to deformable spheres[31a, 38,105,121] and flexible fibers,[31a] which always migrated to the tube axis, as in the creeping-flow region. The experiments have also been performed in rectangular ducts[119] and have revealed phenomena analogous to those in circular tubes. The studies were extended to buoyant spheres,[101,119,120, 166,167] in which the particles migrated to the wall or the tube axis, depending on whether they were lighter or denser than the fluid and whether Poiseuille flow was up or down the tube. Finally, two-way migration of neutrally buoyant rigid spheres, rods, and discs has been observed in oscillatory[15,165] and pulsatile flow.[168]

Although Segré and Silberberg were the first to report two-way migration of rigid particles in the tube, it should be noted that Müller,[94] working with suspensions of small rubber discs, Vejlens,[103] with single rigid spheres, and Starkey,[169,170] with carbon black suspensions, had observed inward migration near the wall in tube flow in the same range of Reynolds numbers. The explanations advanced by these workers, however, invoking the principle of least action[170] or that of minimum energy dissipation in the flow[94] to account for migration, are known to be incorrect, because it has been shown on theoretical grounds[12,36,109] that no sidewise or lift forces can arise with rigid particles under comparable conditions in the creeping-flow regime. The above-described effects are therefore necessarily inertial in origin. A complete discussion of the experimental and theoretical work on particle behavior in the presence of inertial forces has been given by Brenner.[108] The following is a summary of the main findings.

2. RADIAL MIGRATION IN POISEUILLE FLOW

a. Neutrally Buoyant Spheres

Segré and Silberberg,[14,102] working in the range $Re_p = 10^{-3} - 5 \times 10^{-2}$, found the mean stable equilibrium position to be at $R^\dagger/R_0 = 0.63$ and to be relatively insensitive to tube Reynolds number $Re_t (= 2R_0 U_{3M} \rho/\eta_0)$ and b/R_0. The measured radial migration velocities $\dot{R} = u_2$ were correlated by

[165] H. L. Goldsmith and S. G. Mason, 3rd European Conf. Microcirculation, Jerusalem, 1964. *Bibliotheca Anat.* Fasc. **7**, p. 353 (1965).

[166] C. D. Denson, Ph.D. thesis, University of Utah, Salt Lake City, 1965.

[167] R. Eichhorn and S. Small, *J. Fluid Mech.* **20**, 513 (1964).

[168] M. Takano and S. G. Mason, Pulp and Paper Res. Inst. Canada Tech. Rept. 487 (1966).

[169] T. V. Starkey, *Brit. J. Appl. Phys.* **7**, 52 (1956).

[170] T. V. Starkey, V. A. Hewlett, J. H. A. Roberts, and R. E. James, *Brit. J. Appl. Phys.* **12**, 545 (1961).

the following empirical relation:

$$\frac{u_2}{U_{3M}} = 0.34 \mathrm{Re}_t \left(\frac{b}{R_0}\right)^{2.84} \frac{R}{R_0} \left(1 - \frac{R}{R^\dagger}\right) \tag{VII1}$$

Thus, u_2 increases with increasing flow rate and b/R_0; see Fig. 59. However, it appears that, when $b/R_0 < 0.4$, the migration velocity does not increase monotonically with distance from the equilibrium position, as required by equation (VII1), but shows a maximum at some intermediate R/R_0,[31a] as illustrated by an inflection in the plot of R/R_0 vs. time, Fig. 59.

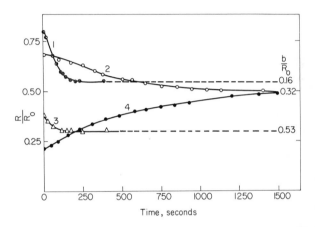

FIG. 59. Two-way radial migration of rigid spheres in Poiseuille flow at $\mathrm{Re}_p = 7.4 \times 10^{-3}$ (curves 2 and 4), $\mathrm{Re}_p = 3.0 \times 10^{-3}$ (curve 1), and $\mathrm{Re}_p = 5.5 \times 10^{-2}$ (curve 3). The points of inflection in curves 1 and 2 indicate that u_2 is maximum at an intermediate R/R_0. The effect of increasing particle size in decreasing the equilibrium radial position $R^\dagger/R_0$ is also evident.

Subsequent work has shown that $R^\dagger/R_0$ decreases with increasing b/R_0 from the value of about 0.7 at $b/R_0 = 0.05$[120] to about 0.5 at $b/R_0 = 0.3$ and falls to 0.15 when $b/R_0 = 0.78$.[31a]

Oliver[101] conducted experiments in a vertical tube with spheres whose centers of mass were eccentrically located, to prevent rotation. Again, both inward and outward migration were observed, although the final equilibrium position was much nearer the tube axis. The presence of lift forces in the case of a nonrotating sphere was confirmed by Theodore,[171] who measured the force directly by connecting the sphere in a horizontal tube via a thin vertical wire to an analytical balance. The force was found to be always inwardly directed.

[171] L. Theodore, Eng. Sci. D. dissertation, New York University, New York, 1964.

The tubular-pinch effect is not exhibited by neutrally buoyant deformable liquid drops in systems having a viscosity ratio $\lambda \leqslant 10$.[31a,38,165] The drops always migrate to the tube axis, as had been previously found at $Re_p < 10^{-6}$ and was described in Section IV. Moreover, as shown in Fig. 60, at a given Re_p and R/R_0 the value of u_2 for the drops is much higher than the inward migration rates observed with rigid spheres. At high values of $\lambda (\approx 50)$, however, at which deformation is negligible, the drops behave as rigid spheres.

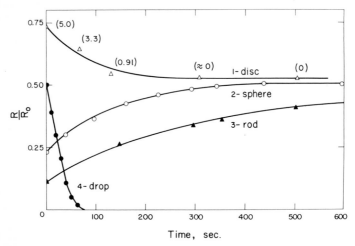

Fig. 60. The tubular-pinch effect for spheres, rods, and discs. Radial migration inward (curve 1) and outward (curves 2 and 3) to the equilibrium position $R^\dagger \approx 0.5R$ at $Re_p > 10^{-3}$. By contrast, a drop of glycerol ten times as viscous as the suspending oil migrated rapidly and steadily inward (curve 4). The values of the orbit constant C of the disc are shown above curve 1; they decrease to $C = 0$ during migration. (After Karnis et al.[38])

b. Neutrally Buoyant Rods and Discs[31a,38]

Two-way migration to an equilibrium position at R/R_0 of about 0.5 was found with rigid discs having $b_1/R_0 = 0.03$ and b_2/R_0 ranging from 0.17 to 0.23 and with rods having $b_2/R_0 = 0.021$ and b_1/R_0 ranging from 0.22 to 0.29.

During migration a steady drift of the spherical elliptical orbit constant C occurs, until the limiting values $C = \infty$ for rods and $C = 0$ for discs are reached. Thus, as shown in Fig. 61, at equilibrium the long axes of the rods and the faces of the discs are oriented in planes passing through the axis of the tube. The same drift in C is observed in Couette flow, in which these limiting values correspond to maximum energy dissipation in the flow.[2]

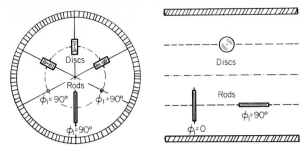

FIG. 61. Projections of the orbits of rods and discs viewed along the X_3 axis (left) and in the median $(X_2 X_3)$ plane (right) at equilibrium. The particles have migrated to a radial position halfway between tube axis and wall and rotate in orbits of $C = \infty$ for rods and $C = 0$ for discs. (After Karnis et al.[31a])

The rate of change in C with time from an initial value C_0 in Couette flow, due to inertia of the fluid, has been calculated by Saffman[36] to be

$$\ln \frac{C}{C_0} = - \frac{0.24\gamma^2 b_1 (b_1 - b_2)\rho}{\eta_0} t \qquad \text{(VII2)}$$

Although a linear dependence of log C with time is found, and C increases with increasing γ and particle size,[31a] equation (VII2) predicts limiting orbits having $C = 0$ for a rod $(b_1 > b_2)$ and $C = \infty$ for a disc $(b_1 < b_2)$, the opposite of what is observed.

Over a period of time in which C remains appreciably constant the measured $\dot{\theta}_1$ and $\dot{\phi}_1$ for the cylindrical particles and the measured angular velocities of rigid spheres in Couette flow are in agreement with Jeffery's theory.[31a] Similarly, measurements of the angular velocities of rigid spheres at $Re_p > 10^{-3}$ in the tube[120] show that they are close to the predicted value of $2RU_{3M}/R_0^2$, the small discrepancies found being attributed to wall effects, as in the case of the translational velocities,[31a,120] where the spheres lag the fluid, that is, $(U_3 - u_3) > 1$, at all radial positions.[119]

c. Nonneutrally Buoyant Spheres

The effect of buoyancy on the radial migration of freely rotating spheres is to displace the equilibrium position toward the wall or the tube axis, depending on whether the sedimentation velocity $u_{3\infty}$ is of the same sign or of opposite sign to that of the fluid velocity U_3. Thus, if the sphere is more dense than the fluid and flow is downward, there is an increase in $R^\dagger/R_0$, whereas when a less dense sphere is in a downward-flowing liquid, or a denser sphere is in an upward-flowing liquid, the equilibrium position is displaced toward the tube axis. With density differences in the range of

0.9 to $10\%^{120,166}$ it was reported that the particles always either migrated to the tube wall or axis. However, with very slight density differences ($\Delta\rho = 0.04 - 0.06\%$) migration in the tube[101] or in a rectangular duct[119] (essentially a two-dimensional Poiseuille flow) ceased before the extreme positions had been reached.

Mention should be made of a simple but ingeniously devised experiment, in which both the drag and lift forces on a single freely suspended sphere were measured by suspending the particle in a less dense fluid flowing upward through an inclined tube.[167] In this manner both the drag and lift forces were balanced by the corresponding components of the net gravitational force on the sphere. In all cases the radial force was found to be inwardly directed.

Finally, in the case of a sphere settling near the wall of a tube in an otherwise stagnant fluid inward migration has been observed[101] and the particle followed all the way to the axis.[31a]

3. THEORETICAL INTERPRETATION OF RESULTS

a. Unbounded Flows

The creeping or Stokes flows may be regarded as the leading terms in an asymptotic solution of the Navier–Stokes equation for small Reynolds numbers. For solutions at higher Reynolds numbers various perturbation schemes have been used. For a sphere rotating about its own axis in an infinite fluid the Stokes' solution is the zeroth-order approximation to the Navier–Stokes equation and is valid throughout the whole fluid domain. Higher-order terms can then be obtained by ordinary perturbation methods using Re as the small perturbation parameter. However, for flow streaming past a sphere in an unbounded fluid the Stokes solution becomes invalid at distances of the radius vector r from the sphere, such that $\text{Re}\, r/b = 0(1)$.[172,173] No matter how small Re, the assumption that inertial terms are negligible compared to viscous terms becomes invalid at large r/b. Hence, Oseen[173] suggested that the problem could be solved by replacing the nonlinear inertial term $\mathbf{V} \cdot \nabla \mathbf{V}$ in the Navier–Stokes equation by the linearized approximation $\mathbf{V}^0 \cdot \nabla \mathbf{V}$, where $\mathbf{V}^0$ is the uniform stream velocity field at large distances from the sphere. But Oseen's solution only provides a uniformly valid zeroth-order solution of the Navier–Stokes equation throughout all space.

To evaluate the higher-order terms it has been found necessary to use

[172] A. N. Whitehead, Quart. J. Math. 23, 78 and 1143 (1889).
[173] C. W. Oseen, "Neue Methoden und Ergebnisse in der Hydrodynamik." Akad. Verlagsges., Leipzig, 1927.

singular perturbation schemes[174-176] and to use two expansions of the exact solution, each of which is only locally valid in a different region of the fluid. The "inner" expansion is an ordinary perturbation in the parameter Re, considered small, and leads to linear differential equations that are not required to satisfy the boundary conditions at infinity. The "outer" expansion, valid only at great distances from the sphere, uses "stretched" coordinates $\tilde{r} = \text{Re} \cdot r$ and results in the removal of the Reynolds number from the differential equations of motion, making the viscous and inertial terms of comparable order. However, the equations of the outer expansion are not valid near the particle and are not unique without the specification of an inner boundary condition, just as the equations of the "inner" expansion lack an outer boundary condition.

Both sets of conditions are simultaneously met by use of the matching principle, according to which two asymptotic solutions of the same differential equations and boundary conditions must be asymptotically equal in their common domain of validity. The overlap domain, in which both solutions are valid, occurs when $\text{Re} \cdot r/b = 0(1)$. This dimensionless distance from the sphere is small in terms of the stretched coordinate $\tilde{r}/b = \text{Re} \cdot r/b$ but is very large in terms of the unstretched inner coordinate r/b.

Proudman and Pearson,[176] using the matching technique, found the force acting on a rigid sphere correct to the first order in the Reynolds number. Rubinow and Keller[177] applied the technique to find the force and torque on a sphere spinning about any axis through an unbounded fluid at rest (or in uniform flow) at infinity. Correct to first order in the Reynolds number, it was found that the hydrodynamic force consists of a drag force $\mathbf{P}_D$ antiparallel to the motion of the sphere,

$$\mathbf{P}_D = -6\pi\eta_0 b\mathbf{v}[1 + \tfrac{3}{8}\text{Re} + 0(\text{Re})] \qquad \text{(VII3)}$$

as well as of a lift force $\mathbf{P}_L$ at right angles to this direction, given by

$$\mathbf{P}_L = \pi b^3 \rho \boldsymbol{\omega}^0 \times \mathbf{v}[1 + 0(\text{Re})] \qquad \text{(VII4)}$$

where $\text{Re} = b|v|/\nu$, ν being the kinematic viscosity. This force, which arises from "slip spin," is comparable to the Magnus force[178] arising at very high Re, used to explain phenomena such as the curving of a spinning ball.

[174] P. A. Lagerstrom and J. D. Cole, *J. Rat. Mech. Anal.* **4**, 817 (1955).

[175] S. Kaplun and P. A. Lagerstrom, *J. Math. Mech.* **6**, 585 (1957).

[176] I. Proudman and J. R. A. Pearson, *J. Fluid Mech.* **2**, 237 (1957).

[177] S. I. Rubinow and J. B. Keller, *J. Fluid Mech.* **11**, 447 (1961).

[178] S. Goldstein (ed.), "Modern Developments in Fluid Dynamics," Vol. 1, p. 83. Oxford Univ. Press, London and New York, 1938.

The experimental results in Poiseuille flow clearly indicate that the presence of the walls is a cardinal feature of radial migration and therefore, as has been pointed out,[102,122] any treatment that does not explicitly consider inertia effects in their presence is inadequate. Despite this, much of the experimental data in tube flow has been interpreted on the basis of equation (VII4). It is assumed that the particle spin $\omega_1^0 = \gamma/2 = 2U_{3M}R/R_0^2$ and v is identified with an axial slip velocity $u_3 - U_3$, which in the case of a neutrally buoyant particle in the absence of wall effects is given by equation (IV11). The lift force is then

$$P_2 = -2\pi b^5 \rho U_{3M}^2 R/3R_0^4$$

acting to move the sphere inward all the way to the tube axis with a velocity $\dot{R} = u_2$, which may be calculated by means of Stokes' law, $P_2 = 6\pi\eta_0 b\dot{R}$, and is

$$u_2 = -\frac{b^4 \rho U_{3M}^2 R}{9\eta_0 R_0^4} \tag{VII5}$$

$$= -\tfrac{2}{9}\mathrm{Re}_t\left(\frac{b}{R_0}\right)^4 \frac{R}{R_0} U_{3M} \tag{VII6}$$

Recognizing that this conclusion was at odds with the experimental facts and finding the labor of calculating the additional terms in P_L in a shear flow prohibitive, Rubinow and Keller[177] suggested multiplying equation (VII6) by a factor $-(R^\dagger - R)/R$ to bring it into qualitative agreement with experiment, but quite apart from the failure of the theory to explain the existence of an equilibrium value of R it has been shown by Saffman[179] and Cox and Brenner[180] to be inapplicable to Poiseuille flow.

Saffman[36] has calculated the lateral velocity due to inertia of a small sphere in an unbounded parabolic velocity profile by iteration of the Navier–Stokes equations and obtained an expression similar to equation (VII6). In a later, more rigorous treatment Saffman[179] calculated the lift force on a small sphere in a uniform, simple shear flow $(V_3 = \gamma X_2)$ with translational velocity v_3 parallel to the streamlines. The treatment assumes that the particle (Re_p), shear (Re_s), and rotational (Re_ω), Reynolds numbers, defined by

$$\mathrm{Re}_p = \frac{2b|V_3 - v_3|}{v}, \qquad \mathrm{Re}_s = \frac{(2b)^2\gamma}{v}, \qquad \mathrm{Re}_\omega = \frac{(2b)^2|\omega|}{v}$$

[179] P. G. Saffman, J. Fluid Mech. 22, 385 (1965).
[180] R. G. Cox and H. Brenner, Forthcoming Publication.

are very small. Provided that $\mathrm{Re}_s \gg (\mathrm{Re}_p)^2$, using singular perturbation techniques leads to the result that the lift force P_2 is

$$P_2 = 81.2\eta b^2(v_3 - V_3)(\gamma/v)^{1/2} + 0(v^{-1/2}) \qquad \text{(VII7)}$$

and that in the absence of an external torque the sphere spins with angular velocity ($=\gamma/2$) of the undisturbed fluid. Thus, unlike the Rubinow–Keller theory, the lift force was shown to be independent of the angular velocity and to arise from "slip-shear." The theory was thus able to account for the results obtained with the nonrotating sphere.[101,171] Although Saffman's analysis demonstrated the lack of universal applicability of the Rubinow–Keller theory, it could be argued that had the opposite case, in which $\mathrm{Re}_p \gg \mathrm{Re}_s$, been analyzed, the lift force due to slip spin might have dominated or been comparable to that due to slip shear. However, Brenner[108] has pointed out that, at least in the case of the neutrally buoyant sphere, in which the axial slip velocity is of the order of $\frac{2}{3}U_{3M}(b/R_0)^2$ when b/R_0 is small, and in which it can be shown that

$$\frac{\mathrm{Re}_s}{(\mathrm{Re}_p)^2} = 0\left(\frac{1}{\mathrm{Re}_p}\right)$$

that is, as $\mathrm{Re}_p \to 0$ the ratio becomes infinite, Saffman's condition for $\mathbf{P}_L$ is always met and the Rubinow–Keller theory is inapplicable.

b. Poiseuille Flow

The theories given above, while each able to predict qualitatively certain of the observed features of the radial migration in the tube, are unable to account for the two-way migration of neutrally buoyant particles that lag the fluid at all radial positions.[31a] A full treatment of the problem of a freely rotating rigid sphere in a tube of finite radius has, however, been attempted by Cox and Brenner,[180] who obtained a first-order solution of the Navier–Stokes equation. The sphere was assumed to move parallel to the tube axis, and the lateral force required to maintain it at a fixed R was calculated and converted into an equivalent radial migration velocity by the application of Stokes' law.

The use of singular perturbation techniques resulted in a solution in the form of a double expansion in the two parameters Re_t and b/R_0, each assumed to be $\ll 1$. The only restriction imposed was that the sphere be not too close to the wall, i.e., $b/(R_0 - R) \ll 1$. Solutions in five separate cases ranging from the neutrally buoyant particle to that of the sphere settling with a velocity $u_{3\infty}$ in an otherwise quiescent fluid were given, and three are presented here.

(1) Neutrally buoyant sphere, $|u_{3\infty}/U_{3M}| \ll (b/R_0)^2$:

$$\frac{u_2}{U_{3M}} = \mathrm{Re}_t \left(\frac{b}{R_0}\right)^3 F_1(R/R_0) \tag{VII8}$$

(2) Nonneutrally buoyant sphere, $(b/R_0)^2 \ll |u_{3\infty}/U_{3M}| \ll 1$:

$$\frac{u_2}{U_{3M}} = \mathrm{Re}_t \frac{b}{R_0} \cdot \frac{u_{3\infty}}{U_{3M}} F_{II}(R/R_0) \tag{VII9}$$

(3) No net flow in the tube, provided that in addition $\mathrm{Re}_\infty = 2b|u_{3\infty}|/v \ll 1$:

$$\frac{u_2}{u_{3\infty}} = \frac{2bu_{3\infty}}{v} F_{III}(R/R_0) \tag{VII10}$$

The other two cases are intermediate between (1) and (2) and between (2) and (3).

Although the calculations of the three functions of R/R_0 are incomplete, equations (VII8) to (VII10) show qualitative agreement with experiment. Thus equation (VII8) is of the same form as equation (VII1) empirically found by Segré and Silberberg.[102] Further, since the equilibrium position must correspond to $F_1(R^\dagger/R_0) = 0$, this implies that $R^\dagger/R_0$ is independent of Re_t and b/R_0, provided these are small. This also is borne out by experiment. Finally, equation (VII9) indicates, as found experimentally, that the direction of lateral migration depends on the algebraic sign of $u_{3\infty}/U_{3M}$.

In the intermediate case between a neutrally buoyant and nonneutrally buoyant sphere, that is, $|u_{3\infty}/U_{3M}| = O(b/R_0)^2$, the theory of Cox and Brenner indicates that the range of density differences required to prevent migration of the particle all the way to the tube axis or wall is remarkably narrow.[108] This may account for the failure of some investigators to observe eccentric equilibrium positions in the nonneutrally buoyant case.

4. RADIAL MIGRATION IN OSCILLATORY AND PULSATILE FLOWS

Migration of rigid spheres, discs, and rods both toward and away from tube axis has been observed in oscillatory flow[15] and in pulsatile flow, that is, oscillatory flow superimposed on steady flow.[168] However, the phenomena are somewhat more complex and are best understood by considering the variation in the velocity distribution in the tube with time during each cycle of the flow.

a. Equations of Fluid Motion

Exact solutions of the Navier–Stokes equations for oscillatory flow of a homogeneous liquid through a circular tube have been obtained by Sexl,[181] Lambossy,[181a] and Womersley.[182] We consider the case of a pulsatile flow in which a simple harmonic pressure gradient is superimposed on a steady gradient:

$$-\frac{\partial p}{\partial X_3} = \mathscr{K}'\eta_0 \exp(iwt) + 2\mathscr{K}\eta_0 \tag{VII11}$$

where $\mathscr{K} = 4Q_0/\pi R_0^4$ (Q_0 being the steady volume flow rate), $\mathscr{K}'$ is a real positive constant, and w is the angular frequency. The Navier–Stokes equations in cylindrical coordinates assuming laminar incompressible flow is then

$$\frac{\partial^2 U_3}{\partial R^2} + \frac{1}{R}\cdot\frac{\partial U_3}{\partial R} - \frac{\rho}{\eta_0}\frac{\partial U_3}{\partial t} + \mathscr{K}'\exp(iwt) + 2\mathscr{K} = 0 \tag{VII12}$$

where $U_3(R, t)$ is the fluid velocity at a radial distance R and time t. The solution of equation (VII12) is of the form

$$U_3(R, t) = \mathscr{U}_3(R)\exp(iwt) + U_3(R) \tag{VII13}$$

where $U_3(R)$ is given by equation (IV3) and $\mathscr{U}_3(R)$ is the solution of Bessel's equation,

$$\frac{d^2\mathscr{U}_3}{dR^2} + \frac{1}{R}\cdot\frac{d\mathscr{U}_3}{dR} + \frac{i^3 w\rho}{\eta_0}\mathscr{U}_3 = -\mathscr{K}'$$

The fluid velocity is given by[182]

$$U_3(R, t) = \frac{\mathscr{K}'\eta_0}{i\rho w}\left(1 - \frac{J_0(\mu R/R_0 i^{3/2})}{J_0(\mu i^{3/2})}\right)\exp(iwt) + \frac{\mathscr{K}}{2}(R_0^2 - R^2) \tag{VII14}$$

where μ is a dimensionless parameter defined by

$$\mu = R_0(w\rho/\eta_0)^{1/2}$$

By expressing the zero-order Bessel functions in equation (VII14) in terms of phase and modulus,

$$J_0(\mu R/R_0, i^{3/2}) = M_0(\mu R/R_0)\exp[i\Theta_0(\mu R/R_0)]$$

$$J_0(\mu i^{3/2}) = M_0(\mu)\exp[i\Theta_0(\mu)]$$

[181] T. Sexl, Z. Physik **61**, 349 (1930).

[181a] P. Lambossy, Helv. Phys. Acta **25**, 371 (1952).

[182] J. R. Womersley, J. Physiol. (London) **127**, 553 (1955).

and taking the real part of $\mathscr{K}'\eta_0 \exp(iwt)$ as $\mathscr{K}'\eta_0 \cos wt$ Womersley[182] obtained the expression

$$U_3(R, t) = \frac{\mathscr{K}'\eta_0 M_0'}{\rho w} \sin[wt + \xi_0(R)] + \frac{\mathscr{K}}{2}(R_0^2 - R^2) \quad \text{(VII15)}$$

The quantity M_0' is defined as

$$M_0' = (1 + H_0^2 - 2H_0 \cos d_0)^{1/2} \quad \text{(VII16a)}$$

where

$$H_0 = \frac{M_0(\mu R/R_0)}{M_0(\mu)} \quad \text{(VII16b)}$$

$$d_0 = \Theta_0(\mu) - \Theta_0(\mu R/R_0) \quad \text{(VII16c)}$$

and the phase angle is given by

$$\xi_0(R) = \tan^{-1}\left(\frac{H_0 \sin d_0}{1 - H_0 \cos d_0}\right) \quad \text{(VII17)}$$

Similarly, the volume flow rate calculated from equation (VII14) was shown to be[182]

$$Q(t) = Q_M \sin(wt + \xi_{10}) + Q_0 \quad \text{(VII18)}$$

where

$$Q_M = \pi R_0^4 \mathscr{K}' M_{10}'/\mu^2$$

and M_{10}' and ξ_{10} are functions of μ and have been tabulated.[182]

Finally, the constant $\mathscr{K}'$ in equation (VII14) may be calculated from the volume displacement per half cycle,

$$\Delta\mathscr{V} = \int_{-\xi_{10}/w}^{(\pi - \xi_{10})/w} Q(t)\, dt$$

and shown to be

$$\mathscr{K}' = \Delta\mathscr{V}\mu^2 w/2\pi R_0^4 M_{10}' \quad \text{(VII19)}$$

The theory of Womersley has been tested for both oscillatory and pulsatile flow[168] by using the integrated form of equation (VII15), giving the displacement profile. For example, for oscillatory flow

$$X_3(R, t) = -\frac{\mathscr{K}'\eta_0 M_0'}{\rho w^2} \cos[wt + \xi_0(R)] \quad \text{(VII20)}$$

the total amplitude of oscillation $a(R)$ being

$$a(R) = 2\mathcal{K}'\eta_0 M_0'/\rho w^2$$
$$= \Delta\mathcal{V} M_0'/\pi R^2 M'_{10} \qquad\qquad (VII21)$$

The paths of neutrally buoyant, small, rigid tracer spheres $(b/R_0 < 0.05)$ were followed in a liquid of $\eta_0 = 0.14$ poise subjected to oscillatory flow by driving the piston of a reciprocating pump in simple harmonic motion.[15,183] Good agreement with equations (VII20) and (VII21) was found over a range of values of μ from 5 to 11, as illustrated for $\mu = 7.11$ in Fig. 62. The flattening of the profile in the center of the tube, characteristic of oscillatory flow at sufficiently high μ, increases with increasing μ. The maximum value of the displacement amplitude a occurs at a radial position away from the tube center (Fig. 62), and fluid motion near the wall is out of phase with that in the center. This illustrates the fact that flow close to the wall is dominated by viscous forces (and is nearly in phase with the negative pressure gradient), whereas that in the central portion of the tube is dominated by inertial forces (and lags the pressure gradient).

Linford and Ryan[184] investigated Womersley's equations for oscillatory flow by measuring both the pressure gradient and the velocity profile, the latter being recorded by photographing a thin filament of ink introduced into the fluid. Good agreement with theory was found over a range of μ of 1.8 to 21. At values of $\mu \approx 20$, where Re_t, calculated from the maximum mean linear flow rate during the cycle, was in excess of 10,000, turbulent flow was observed when $\Delta\mathcal{V}$ was high.

Mention should also be made of a theory by Atabek and Chang[185] for laminar oscillatory flow in the inlet length of a circular tube which was confirmed by measurements of velocity profiles.[185a]

b. Radial Migration

The radial migration of rigid spheres, discs, and rods suspended in viscous fluids undergoing oscillatory flow has been studied at various μ, $\Delta\mathcal{V}$, and b/R_0.[15,168] As in steady flow, both inward and outward migration is observed, and the rotational orbits of rods and discs assume limiting values that correspond to maximum dissipation of energy in Couette flow.

The inward migration rates of rigid spheres initially introduced at the wall increase with increasing μ and $\Delta\mathcal{V}$, with an attendant shift of the equilibrium radial position toward the wall. The rates also increase with increasing b/R_0 and, as in steady flow, this results in a decrease in $R^\dagger/R_0$.

[183] M. Takano, H. L. Goldsmith, and S. G. Mason, *J. Colloid Interface Sci.* **23**, 248 (1967).
[184] R. G. Linford and N. W. Ryan, *J. Appl. Physiol.* **20**, 1078 (1965).
[185] H. B. Atabek and C. C. Chang, *Z. Angew. Math. Phys.* **12**, 185 (1961).
[185a] H. B. Atabek, C. C. Chang, and L. M. Fingerson, *Phys. Med. Biol.* **9**, 219 (1964).

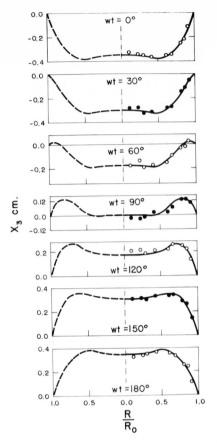

FIG. 62. Displacement profiles at 30° intervals over a half-cycle obtained by following the motion of tracer spheres in the liquid undergoing oscillatory flow at $\mu = 7.11$. The lines drawn are calculated from equation (VII20) using the measured $\Delta \mathscr{V}$, w, and phase angle $\xi_0(R)$ computed from equation (VII17). (After Shizgal et al.[15])

Whereas at a given Re_p and b/R_0 there is only one equilibrium position in steady flow, in oscillatory flow at values of $\mu > 5$ more than one equilibrium position could be demonstrated. The measured $R^{\dagger}/R_0$ correspond approximately to radial distances at which the calculated time-average velocity gradient is zero. These equilibrium positions change with increasing μ and depend also on b/R_0.

The velocity gradient in the fluid is given by differentiating equation (VII14) with respect to R and is[182]

$$\gamma(R, t) = \frac{\mathscr{K}' R_0 H_1}{\mu} \cos(wt - d_1) - \mathscr{K} R \qquad \text{(VII22)}$$

where

$$d_1 = -\tfrac{1}{4}\pi - \Theta_1(\mu R/R_0) + \Theta_0(\mu), \qquad H_1 = M_1(\mu R/R_0)/M_0(\mu)$$

and $\Theta_1(\mu R/R_0)$ is the phase of the first-order Bessel function $J_1(i^{3/2}\mu R/R_0)$ having modulus $M_1(\mu R/R_0)$.

The net time-average velocity gradient is defined by

$$\langle \gamma(R)\rangle = \frac{\displaystyle\int_{t_1}^{t_2} \gamma(R,t)\,dt}{\displaystyle\int_{t_1}^{t_2} dt} + \mathscr{K}R$$

which upon substituting $\gamma(R,t)$ from equation (VII22) and integrating from $t_1 = -\xi_0(R)/w$ to $t_2 = [\pi - \xi_0(R)]/w$ yields

$$\langle \gamma(R,t)\rangle = \frac{2\mathscr{K}'R_0 H_1}{\pi}\sin[\xi_0(R) + d_1] \qquad \text{(VII23)}$$

Thus, the net time-average velocity gradient is zero when $(\xi_0 + d_1) = n\pi$. Below $\mu = 5$, $\langle\gamma(R)\rangle = 0$ at $R = 0$, but above $\mu = 5$ first two, and then three, positions of zero $\langle\gamma(R)\rangle$ are found.

Liquid drops in systems of viscosity ratio $\lambda < 1$ undergo oscillating deformation and orientation in good agreement with the theory for steady flow when the oscillatory value of the velocity gradient, given by the first term on the right-hand side of equation (VII22), is inserted into equations (III32) and (III33).[183] Unlike rigid spheres and cylinders, however, the drops always migrate to the tube axis.[168] The time average mean migration rates in oscillatory flow are given by

$$\langle -\dot{R}\rangle = \frac{\displaystyle\int_{d_1/w}^{(2\pi + d_1)/w} (-\dot{R})\,dt}{\displaystyle\int_{d_1/w}^{(2\pi + d_1)/w} dt}$$

and when $\dot{R}$ is substituted from equation (IV16) with $\gamma(R,t)$ from equation (VII22) this yields

$$\langle -\dot{R}\rangle = \left(\frac{\mathscr{K}'H_1}{\mu}\right)^2 \cdot \frac{b^4\eta_0}{\sigma} \cdot \frac{R_0^2}{(R_0 - R)^2} \cdot \frac{99(9\lambda^2 + 17\lambda + 9)(19\lambda + 16)}{280(\lambda + 1)^3} \qquad \text{(VII24)}$$

The experimental results agree with equation (VII24).[168] However, in systems of $\lambda > 1$ the measured deformations are observed to lag the calculated values and the deformation amplitudes are appreciably lower than those computed from the oscillating γ. This suggests that a part of the normal stress acting at the surface is consumed for viscous extensional flow within the drop, and calculations show[183] that this will result in a decrease in

$\langle - \dot{R} \rangle$ for viscous drops over that given by equation (VII24), an effect which has been observed.[168] When λ reaches 50, however, the drops behave as rigid spheres in oscillatory and pulsatile flow.

5. DILUTE AND CONCENTRATED SUSPENSIONS

a Necklace Formation

At dilutions at which rigid spheres still behave independently, Segré and Silberberg[102,186] and Jeffrey[120] have found that in steady flow particles that have become concentrated in a thin cylindrical layer and are all moving with the same velocity align themselves into regular columns or necklaces parallel to the tube axis. These constellations of particles appear to progress along the tube without changing relative positions.

However, this steady state is only apparent since, if followed sufficiently far along the tube, particle interactions may be observed: "Two particles which happen to move along the same line attract each other, until they reach a minimum (not touching) distance. Then repulsion predominates, until by damped oscillations an equilibrium distance is found, and the particles form a stable pair. If the two particles were not initially aligned, a transverse displacement concomitant with this longitudinal rocking motion brings the particles into the same final configuration. This, too, however, is final only if no other particle is found nearby; otherwise this also tends to align itself with the first two, and a group of three, equidistant particles is formed. And the process goes on, by capture of isolated spheres or by fusion of already formed groups, until all the suspension is nothing more than a collection of nice, regular, long necklaces."[186]

b. Two-Phase Flow

When suspensions of rigid spheres at volume concentrations from 5 to 30% undergo steady or oscillatory flow in a rigid tube at high Reynolds numbers, a particle-free zone develops at the wall.[15,31a] At a given instant the width of this zone is not constant along the tube length but fluctuates about a mean value, which increases with time of flow until equilibrium is reached. At a given b/R_0 the rate of formation of the particle-free layer increases with increasing Re_t but its thickness decreases with increasing c.

As a result of the formation of the particle-free layer and two-phase flow of the suspension there is a drop in the pressure gradient at a given Q and a marked change in the velocity profile in steady flow. This is illustrated in Fig. 63, where Δp may be seen to decrease with time to a limiting value

[186] G. Segré, Proc. 4th Intern. Congr. Rheol., 1963 Pt. 4, p. 103. Wiley (Interscience), New York, 1965.

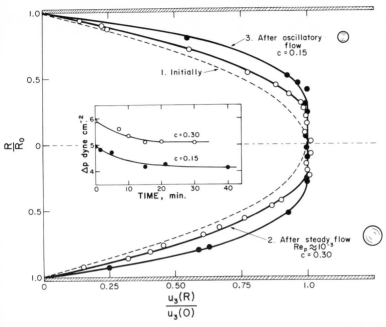

FIG. 63. The effect of a particle-free zone at the tube wall on the particle velocity distribution and pressure drop (inset graph) in concentrated suspensions of rigid spheres. The dashed curve corresponds to the initial velocity profile observed with two suspensions, one of which ($c = 10\%$) was then subjected to steady flow at $\text{Re}_p = 7 \times 10^{-3}$, until at equilibrium the profile shown in curve 2 (open circles) resulted. The other, $c = 22\%$, was subjected to oscillatory flow at $\mu = 5.1$ and $\Delta\mathscr{V} = 1.1$ cm^3. After reaching equilibrium the profile in steady flow at $\text{Re}_p \approx 10^{-5}$ is that shown in curve 3 (closed circles). (After Karnis et al.[31a])

when the velocity distribution in the tube is blunted and exhibits a region of partial plug flow in the center. Unlike the partial plug flow observed with concentrated suspensions at low Re_p and described in Section V, these velocity distributions result from two-phase flow, that is, having a central core of high viscosity surrounded by a particle-free layer of low viscosity.

The presence of a particle-free zone at the wall, which acts as a lubricating layer, may in the limit produce plug flow of the suspension when, because of the absence of a velocity gradient, the particles cease to rotate. In certain situations as, for example, in the pipeline transportation of wood chips and other particles, in which rotation might conceivably cause mechanical attrition, or again in the flow of pulp suspensions, in which rotation can cause fiber aggregation,[187] stopping particle rotation can be an advantage.

---

[187] O. L. Forgacs, A. A. Robertson, and S. G. Mason, "Fundamentals of Papermaking Fibres," p. 447. British Paper & Board Makers' Assoc., Kenley, Surrey, England, 1958.

There will also be a change, possibly a reduction, in the total power required to maintain flow.[15]

VIII. Applications to Hemorheology

1. GENERAL

Rheologically, vertebrate blood may be described as a concentrated suspension of deformable particles that circulates in distensible vessels whose sizes range from less than 1 to 3000 particle diameters. In man the suspended cells occupy from 40 to 50 % of the suspension volume, and of these about 93 % by number (5 × 10⁶ cells/cm³) are red blood cells (erythrocytes). The remainder are made up of various categories of white cells (leucocytes, 5 to 8 × 10³/cm³), and a large number of platelets (2.5 to 5.0 × 10⁵/cm³), which because of their small size amount to only $\frac{1}{20}$ of the red cell suspension volume.

The human red cell is a biconcave disc (Fig. 64) having a diameter[188] of 8.5 ± 0.41 microns and respective maximum and minimum thickness of 2.4 ± 0.13 microns and 1.0 ± 0.08 microns, the mean cell volume being 87 (microns)³. The suspending liquid or plasma is an aqueous solution of 7 % by weight of various proteins and about 1 % each of inorganic and other organic substances. Among the proteins is fibrinogen, which is removed when a clot is formed, leaving behind a straw-yellow liquid known as serum.

The ease with which such a concentrated suspension circulates in the body is surprising, especially in the microcirculatory system, where it is frequently necessary for the cells to deform in order to travel through sections

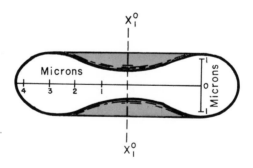

FIG. 64. The shape of the normal human red cell with the mean dimensions given by Ponder[188] and showing the axis of revolution $X_1{}^0$.

[188] The value of the diameter and thickness of the red cell are taken from E. Ponder, "Hemolysis and Related Phenomena." Grune & Stratton, New York, 1948. The maximum diameter of the red cell is often given as 7.5 microns, but this is considered to be due to a shrinkage artifact.

of the capillary bed. By contrast, a 30% model suspension of neutrally buoyant nonflexible rubber discs ($r_p = 0.18$) in aqueous glycerol could only be made to flow with great difficulty through a tube $R_0 = 0.75$ cm, $b/R_0 = 0.20$.[94]

As shown in Fig. 65, suspensions of rigid discs and spheres, in which $\langle b \rangle = 0.05$ microns and 7.2 microns,[189,190] and which obey Mooney's equation, equation (VI4), have a considerably higher apparent relative viscosity[190] than that of normal blood at hematocrits[191] of 25 to 50%. Even at

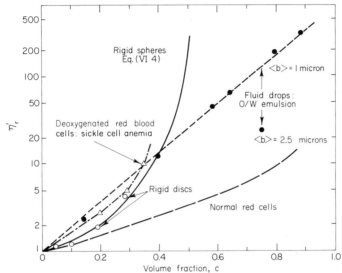

FIG. 65. The apparent relative viscosities of a normal red cell suspension (from the data of Bayliss[87]) as a function of concentration compared with that of suspensions of small rigid spheres,[189] fluid drops (from the data of Richardson[194,195] for an oil in water emulsion, closed circles), and rigid discs (from the data of Müller,[94] open circles). Also shown are the points obtained with sickled red blood cells[192] (open triangles), demonstrating the effect of shape and rigidity on η'_r.

[189] J. G. Brodnyan and E. L. Kelley, *J. Colloid Sci.* **20**, 7 (1965). Agreement with Mooney's equation was obtained when the effect of charged double layers surrounding the spheres was taken into account.

[190] H. M. Taylor, S. Chien, M. I. Gregerson, and J. L. Lundberg, *Nature* **207**, 77 (1965). The authors made a comparison between the viscosities of blood and a latex as a function of particle concentration. The measured η_r appear to obey equation (VI4).

[191] The term *hematocrit* refers to the volume concentration of the red cells, as obtained by centrifuging the blood in a capillary tube and measuring the fraction of the suspension length occupied by the red-cell column. This value of c is slightly too high because of the presence of trapped plasma.

[192] J. W. Harris, "The Red Cell. Production, Metabolism, Destruction: Normal and Abnormal." Harvard Univ. Press, Cambridge, Massachusetts, 1963.

$c = 0.8$ the apparent relative viscosity of blood is still only about 11 (the viscosity of plasma relative to water is about 1.8), and people suffering from the disease polycythemia vera may have hematocrits as high as 80%. If the deformability of the red cells is responsible for the relative ease of blood flow, its viscosity should be compared with that of emulsions, as illustrated in Fig. 65. At a given c the limiting apparent viscosities of emulsions at high γ vary with the type and amount of emulsifying agent[193] and globule size,[194] η_r' decreasing with increasing b. However, even for a relatively coarse oil-in-water emulsion ($\langle b \rangle = 2.5$ microns), the values of η_r' as measured in a Couette viscometer[194] over a range of concentrations are appreciably higher than those found for blood. Richardson[195] has pointed out that at concentrations approaching 50% "work must be done in compressing and relaxing some globules as they jostle each other in relative motion. Round about 75%, the emulsion will take on a quasi-solid consistency like margarine and flow can only take place by the small globules jumping from one hole between more or less immobile large ones to the next, in a process of 'successive dislocation.' " At the very highest concentration, that of a packed column of erythrocytes that has been spun down in a centrifuge and has been estimated to contain about 2% trapped plasma,[196] blood never reaches a quasisolid consistency, although there must be deformation of the cells in flow (and even at rest).[197]

The effect of flexibility besides that of the shape of the red cell on viscosity is strikingly shown in a pathological condition known as sickle cell anemia. Here, when the oxygen tension is sufficiently low, the cell appears to lose its deformability and changes its shape to the form of a sickle blade. As shown in Fig. 65, the viscosity of the sickled blood as measured in a capillary viscometer is considerably higher than that of normal blood,[198,199] η_r' being greater than that for the suspension of rigid spheres. When sickling occurs in the body, flow is arrested in the smaller vessels, causing the vascular stasis that underlies the clinical features of the condition. Again, if red cells are hardened by treatment with formalin, acetaldehyde, or tannic acid, the viscosity of a suspension of such cells may increase by a factor of 10.[200]

[193] J. Broughton and L. Squires, *J. Phys. Chem.* **42**, 253 (1938).

[194] E. G. Richardson, *J. Colloid Sci.* **5**, 404 (1950).

[195] E. G. Richardson, in "Flow Properties of Disperse Systems" (J. J. Hermans, ed.), p. 39. North-Holland Publ., Amsterdam, 1953.

[196] H. W. Thomas, *Lab. Pract.* **10**, 771 (1961).

[197] A. C. Burton, "Physiology and Biophysics of the Circulation." Year Book Med. Publ., Chicago, Illinois, 1965. It is shown that with the closest (hexagonal) packing of the biconcave cells in sheets, the maximum hematocrit before onset of deformation is 58%.

[198] J. W. Harris, H. A. Brewster, T. H. Ham, and W. B. Castle, *A.M.A. Arch. Intern. Med.* **97**, 145 (1956).

[199] L. Dintenfass, *J. Lab. Clin. Med.* **64**, 594 (1964).

[200] K. Kuroda, Y. Mishiro, and I. Wada, *Tokushima J. Exptl. Med.* **4**, 73 (1958).

Several reviews and general discussions of blood rheology have been given[87,201-204a] which together cover the work done on blood flow *in vivo* and *in vitro* as well as some relevant experiments with model suspensions. The present discussion is limited to those aspects of the flow of dilute and concentrated model suspensions described in the previous sections which may be important to an understanding of blood flow. It should be emphasized, however, that hemorheology is a relatively new field of research and that much remains to be done before the mechanism of blood flow in both large and small vessels can be elucidated.

2. FLOW PROPERTIES IN LARGE VESSELS

a. Rotations

Cinefilms of flowing dilute suspensions of erythrocytes ($c < 2\%$) in rigid tubes[205] show that there is a distribution in the rotational orbits of the cells. At velocity gradients up to $20 \sec^{-1}$, the measured angular velocities $\dot{\phi}_1$ of single human red cells in plasma are in accord with equation (III13).[205] This is illustrated in Fig. 66(a) for a corpuscle having $r_e = 0.26$. As in the case of rigid discs, there are two limiting orbits: that of the particle spinning about its axis of revolution [$C = 0$, equation (II20)] and that of rotation of the axis of revolution in the X_2X_3 plane and with the cell in an end-on position ($C = \infty$). The mean effective or particle orbit volume is therefore higher than the particle volume and, if cell crowding and interaction effects are neglected, the observations reported for rigid discs of $r_p = \frac{1}{4}$ in Table IV suggest a critical concentration for impeded rotation of 40%. Some observations made in fixed, microtomed sections of a quick-frozen rabbit femoral artery,[197,206] in which the cells tend to be oriented such that the biconcave surfaces are perpendicular to planes passing through the axis of the tube (cf. Fig. 50), suggest that in concentrated suspensions the orbit distribution may change toward $C = \infty$.

b. Multiplets

In addition to collision aggregates of red cells, in which particles are brought together by the velocity gradient, and which exist for a certain time

[201] A. C. Burton, *in* "Medical Physiology and Biophysics" (T. C. Ruch and J. F. Fulton, eds.). Saunders, Philadelphia, Pennsylvania, 1960.

[202] R. E. Wells, *New Engl. J. Med.* **270**, 832, 889 (1964).

[203] R. L. Whitmore, *Biorheology* **1**, 201 (1963).

[204] A. L. Copley and G. Stainsby (eds.), "Flow Properties of Blood and Other Biological Systems". Pergamon Press, New York, 1960.

[204a] E. O. Attinger (ed.), "Pulsatile Blood Flow." McGraw-Hill, New York, 1964.

[205] H. L. Goldsmith, *Science* **153**, 1406 (1966).

[206] R. H. Phibbs and A. C. Burton, *Federation Proc. Abstr.* **24**, 154 (1965).

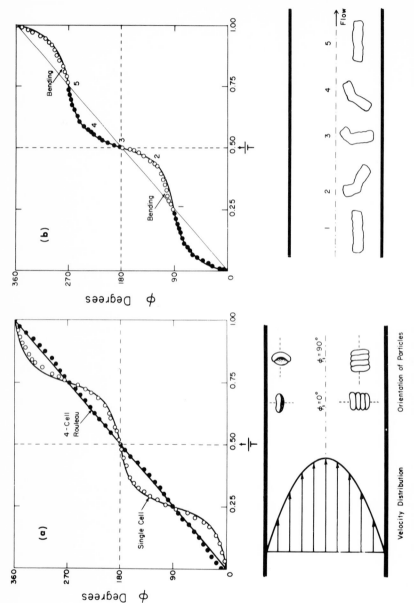

FIG. 66. The measured ϕ_1 plotted against t/T for a single human erthrocyte and rouleaux during an orbit in Poiseuille flow. (a) Single cell having $r_e = 0.26$ and a four-cell linear aggregate having r_e close to 1.0. (b) A 15-cell rouleau which bent in the second and fourth quadrants (open circles) as shown by the tracings in the lower portion. The lines were calculated from equation (III13) using the experimentally measured r_e. (After Goldsmith.[205])

before separation occurs, the erythrocytes form more permanent aggregates called rouleaux. In these the cells are arranged like stacked poker chips (Fig. 67). Fahraeus[207] has described their formation under a microscope slide: "When two previously solitary corpuscles come into contact with their broadsides, we get the impression that they are pressed together by invisible forces. If the first point of contact is situated at the edge of one or both corpuscles, the corpuscles perform a quick motion, the one gliding up against the other, broadside against broadside. The motion does not stop until the peripheral edges of the corpuscles are brought to a position corresponding to the surface of a cylinder. Rouleaux formation is thus a mechanism manifestly proving the existence of a force which tends to diminish the interface between corpuscles and plasma. If stickiness, which has played a great role in theories of agglutination, gave rise to the aggregation here in question the corpuscles would remain in the position of their first contact and we should not obtain the peculiar architecture of the aggregates, the rouleaux formation."

In Poiseuille flow, rouleaux containing n cells in linear array, of regular shape and not deformed by the velocity gradient rotate as rigid discs ($n = 2, 3$) or rods ($n > 5$),[205] as shown by the plots of equation (III13) in Fig. 66. Similar results have been obtained in Couette flow with physical models of rouleaux consisting of linear stacks of rigid discs.[75] However, the ease of deformation of the red cell rouleaux while rotating is quite remarkable. Depending on the magnitude of γ and the particle length, buckling occurs in a manner quite analogous to that described for dacron and nylon filaments and pulp fibers [Fig. 66(b)]. At a given r_p, the value of $(\gamma\eta_0)_{crit.}$ [equation (III65)] at which shear induced bending sets in is only 10^{-7} that found for dacron filaments[29] of the same diameter. Also in common with flexible filaments,[81] wood pulp fibers[82] and chains of spheres in which a liquid phase meniscus bridges adjacent particles,[75] long linear aggregates of erythrocytes (> 20 cells) are observed to execute snake orbits similar to those illustrated in Figs. 23b and 24.

It has been shown[208] that the structural model developed for kaolin suspensions[209] may be applied to blood settling in a tube. In this model the floc, or hard, aggregate (consisting of a rouleau of about four cells in blood) is distinguished from the loose aggregate of flocs that breaks down under shear. The loose aggregates may extend in a continuous network (cf. Fig. 67) that possesses structural strength. The yield stress of the red-cell suspension in a given tube, determined in the settling experiments, is of the same order as that obtained from plots of shear stress versus shear rate,[208] which have

[207] R. Fahraeus, *Physiol. Rev.* **9**, 241 (1929).
[208] S. E. Charm, W. McComis, and G. Kurland, *J. Appl. Physiol.* **19**, 127 (1964).
[209] A. S. Michaels and J. C. Bolger, *Ind. Eng. Chem. Fundamentals* **1**, 24 (1962).

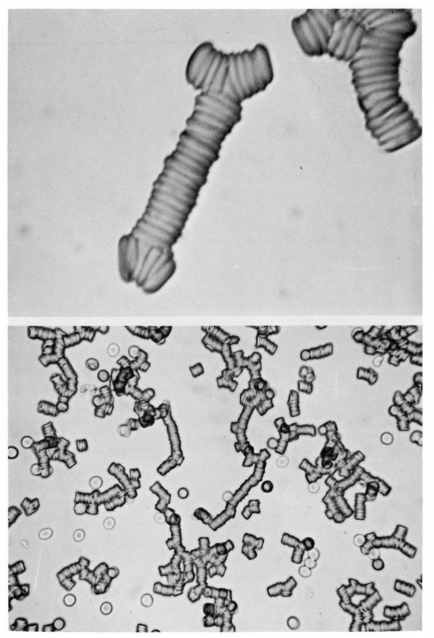

FIG. 67. Photomicrographs of rouleaux of human erythrocytes, showing two separate aggregates (upper) and networks of rouleaux (lower). The upper photograph also illustrates the variation in cell diameter which is found in blood.

been found, at low γ in both capillary[210,211] and Couette instruments,[212] to follow Casson's equation,[213] which was derived for printing inks. In Casson's model mutually attractive particles suspended in a Newtonian medium at low shear rates are considered to form rigid rodlike aggregates, whose length varies inversely with the shear rate. With blood it is presumed that the aggregates are rouleaux whose mean length decreases with increasing shear rate. It also appears that in the absence of fibrinogen the blood does not exhibit a yield stress.[214] Thus, suspensions of red cells in physiological saline show Newtonian behavior at all shear rates, and it has been claimed[211] that in this medium the cells do not form rouleaux.

c. Deformation

The ready deformation of erythrocytes in vessels of diameter the same as or smaller than the cells (cf. Fig. 68) has been frequently seen and reported.[215–217] However, it is only with the advent of high-speed cinematographic techniques that it has been possible to demonstrate deformation in flow in vessels larger than the cells. Bloch[218] has described the complex nature of the flow in arterioles 2 to 5 times the cell diameter: "erythrocytes spin about their long and short axes, deform readily and travel across the 'lamina' in an irregular helical spiral."

Applying the technique of Mitchison and Swann,[219] Rand and Burton[220] determined the stiffness of the red-cell membrane by measuring the pressure required to suck the cells into a micropipette. It was shown that the membrane tension σ was only 0.02 dyne cm^{-1} and that a red cell could be forced out of a micropipette having a tip of only 2 microns internal diameter, with a pressure drop of less than 1 mm of water. During outflow the erythrocyte folds over; apparently, the membrane can withstand large bending strains but only limited tangential strains. Thus, if a long tongue of a cell is pulled into a micropipette, the membrane within the tube spontaneously collapses

[210] S. E. Charm and G. Kurland, *Am. J. Physiol.* **203,** 417 (1962).

[211] E. W. Merrill, A. M. Benis, E. R. Gilliland, T. K. Sherwood, and E. W. Salzman, *J. Appl. Physiol.* **20,** 954 (1965).

[212] E. W. Merrill, E. R. Gilliland, G. Cokelet, H. Shin, A. Britten, and R. E. Wells, Jr., *J. Appl. Physiol.* **18,** 255 (1963).

[213] N. Casson, *in* "Rheology of Disperse Systems" (C. C. Mill, ed.). Pergamon, London, 1959.

[214] E. W. Merrill, G. R. Cokelet, A. Britten, and R. E. Wells, Jr., *Circulation Res.* **13,** 48 (1963).

[215] A. Krogh, "The Anatomy and Physiology of Capillaries," pp. 5, 7–12. Hafner, New York, 1959.

[216] P. I. Bränemark and J. Lindström, *Biorheology* **1,** 139 (1963).

[217] M. M. Guest, T. P. Bond, R. G. Cooper, and J. R. Derrick, *Science* **142,** 1310 (1963).

[218] E. H. Bloch, *Am. J. Anat.* **110,** 125 (1962).

[219] J. M. Mitchison and M. M. Swann, *J. Exptl. Biol.* **31,** 443 (1954).

[220] R. P. Rand and A. C. Burton, *Biophys. J.* **4,** 115 (1964).

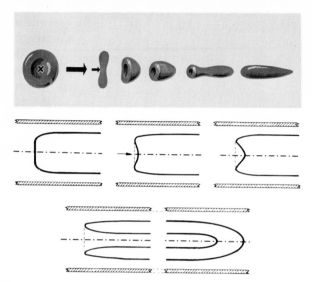

FIG. 68. The deformation of red cells into thimble-shaped forms during flow in capillaries (upper part, after Bränemark and Lindström[216]) compared with the development of reentrant cavities at the trailing end of flowing bubbles suspended in liquids undergoing Poiseuille flow (lower part).

on itself.[221] This is similar to the breakup of a flowing long, cylindrical bubble in a tube after being brought to rest.[138,222]

If the value of σ given above may be used as a measure of the deformability of the red cell, inward migration rates of particles close to the walls of vessels can be calculated[121] from equation (IV16). Table V gives the calculated $\dot{R}$ in the larger vessels of the mesenteric vascular bed of the dog, for which Mall[223] has given data on vessel size and mean linear flow rates. Assuming, for simplicity, a spherical particle of radius 4 microns and $\lambda = 1$, the table shows that the migration rates would be significant in the arterial branches. Some authors[224–226] have speculated on the importance of the tubular-pinch effect in blood flow. The table shows, however, that in the smaller vessels the inward migration rate due to inertia, calculated from equation (VIII), is much smaller than that due to deformation.[121]

[221] R. P. Rand, Biophys. J. 4, 304 (1964).

[222] S. Goren, J. Fluid Mech. 12, 309 (1962).

[223] J. P. Mall, modified in J. Schleier, Arch. Anat. Physiol. Leipzig 173, 172 (1919).

[224] S. I. Rubinow, Biorheology 2, 117 (1964).

[225] G. Vejlens, 2nd European Conf. Microcirculation, Pavia, 1962, Bibliotheca Anat. Fasc. 4, p. 46 (1964).

[226] G. Segré and A. Silberberg, 2nd European Conf. Microcirculation, Pavia, 1962. Bibliotheca Anat. Fasc. 4, p. 83 (1964).

TABLE V

ESTIMATED MIGRATION RATES IN THE MESENTERIC VASCULAR SYSTEM OF THE DOG

Vessel	Vessel length,* cm	R_0,* microns	b/R_0†	U_{3M},* cm sec⁻¹	R/R_0	$\dot{R}$ (deformation),‡ microns sec⁻¹	$\dot{R}$ (deform.)/$\dot{R}$ (inertia)§
Mesenteric artery	6.0	1500	2.7×10^{-3}	33	0.95	−0.03	3.4
Main branches	4.5	500	8.0×10^{-3}	20	0.95	−9.7	30
End branches	9.3	300	1.3×10^{-2}	18	0.95	−60.7	84
					0.90	−13.6	24
					0.85	−5.4	17

* From the data of Mall, modified by Schleier.[223]

† Where b is taken to be 4 microns.

‡ Calculated from equation (IV16) with $\eta_0 = 0.012$ poise, $\sigma = 0.02$ dyne cm⁻¹,[220] $\rho = 1.03$ gm cm⁻³, and $\lambda = 1$.

§ Calculated from equation (VII1), assuming R† = $0.6R_0$.

In connection with axial migration it should be mentioned that Fahraeus[207] observed that *in vivo* the larger white blood corpuscles are transported more centrally than the smaller erythrocytes, but in the case of diminished suspension stability (and as found by Vejlens in slow flow[103]) the reverse occurred.

d. Two-Phase Flow

The establishment of a lubricating cell-free layer at the wall of the vessel through the action of inwardly directed forces, with the accompanying drop in pressure gradient,[31a] may be of importance in the circulation of the blood. However, little direct visual evidence of the existence of such a layer has been adduced. Copley and Staple[227] observed a marginal cell-free zone in the arterial and venal flow in hamster cheek pouches and also found that injected graphite particles 1 to 2 microns in diameter behaved in a fashion similar to the red cells, concentrating in an axial core. Bloch,[218] however, has pointed out that what appears to the naked eye (or what is seen on a cinefilm at a speed of 16 to 24 frames/sec) as a well-defined clear zone at the vessel wall proves to be irregular in width and to fluctuate with time and distance along the tube (cf. suspensions of rigid spheres[31a]), when seen on a high-speed cinefilm.

Indirect evidence of the existence of a particle-free layer has been obtained (i) from the reduction of the apparent viscosity, at a given wall shear stress, with decreasing tube radius in capillary viscometers, as mentioned in Section VI (Fahraeus–Linquist effect[152,228–230]), (ii) from a reduction in particle concentration within the tube from that in the reservoir,[196,207] and (iii) from a difference in the mean velocities of the cells and plasma *in vivo*[231,232] and *in vitro*.[233] However, effects similar to (i) and (ii) are observed with suspensions of rigid spheres[93,99,100] at values of $Re_p < 10^{-5}$ and are

[227] A. L. Copley and P. H. Staple, *Biorheology* **1**, 3 (1962).

[228] R. Fahraeus and T. Lindquist, *Am. J. Physiol.* **96**, 562 (1931).

[229] K. Kümin, Inaugural dissertation, University of Bern, 1948, Paulus Press, Fribourg, Switzerland (1949).

[230] Some doubt has been expressed about the existence of the Fahraeus–Lindquist effect.[211] The apparent viscosity of blood has been reported to decrease in tubes of $R_0 < 0.05$ cm. However, Merrill *et al.*[211] find that at a given wall shear stress ($= \Delta pR/2\Delta X_3$) the limiting apparent viscosities at high flow rates are independent of tube radius over a range $R_0 = 0.013$ to 0.085 cm. Similar results were obtained by L. C. Cerny, F. B. Cook, and C. C. Walker, *Am. J. Physiol.* **202**, 1188 (1962).

[231] P. Dow, P. F. Hahn, and W. F. Hamilton, *Am. J. Physiol.* **147**, 493 (1946).

[232] A. C. Groom, W. B. Morris, and S. Rowlands, *J. Physiol.* (*London*) **136**, 218 (1957).

[233] H. W. Thomas, R. J. French, A. C. Groom, and S. Rowlands, *Proc. 4th Intern. Congr. Rheol. 1963* Pt. 4, p. 381. Wiley (Interscience), New York, 1965.

due to wall effects (as discussed in Sections IV and VI) rather than to particle migration. A detailed analysis of the problem made by Thomas[234] and also the measured differences in the circulation time between red cells and plasma obtained *in vitro*[233] suggest that, although wall effects are undoubtedly present, they are supplemented by an effect that is shear-dependent and hydrodynamic in origin.

A peripheral cell-free layer has been seen on high-speed cinefilms of flowing blood in rigid tubes[235] (R_0 = 20 to 50 microns). As observed *in vivo*,[218] the layer thickness fluctuated considerably, so that at a hemotocrit of 40% the standard deviation was approximately equal to half the thickness of the layer. The measured velocity profiles[235] at low flow rates and hematocrit of 40% were blunted without a region of partial plug flow; with increasing flow rate the velocity distribution became more and more parabolic. This fact, besides the observed plug flow of whole blood in tubes in the region close to zero flow,[211] suggests that two-phase flow with a central core of aggregates of cells, in which the velocity gradient is zero or close to zero, may occur at low shear rates. At higher γ the aggregates presumably break down, and gradients are established throughout the core of the suspension. This contrasts with the flow behavior of wood pulp suspensions, in which the flexible fibers form a continuous network that moves as a plug surrounded by clear suspending fluid and that is progressively compressed as the flow rate increases, until flow in the peripheral layer becomes turbulent and the plug eventually breaks up.[104,187]

3. FLOW PROPERTIES IN SMALL VESSELS

The smallest blood vessels or capillaries are comparable in diameter to the red cells and often even smaller. The picture that has emerged from visual studies of the capillary bed[216,218,236,237] is one of intermittent flow[236,237] with groups of erythrocytes passing through, separated by plasma. With the use of improved optical techniques[218,238] and high-speed cinemicrography[216–218] it has become possible to observe the flow behavior of the individual cells. It appears that the cells are most commonly oriented such that their faces are perpendicular to the axis of the vessel. At zero flow they are seen edge on but, as flow increases, deformation sets in, and the erythrocytes often become thimble-shaped, as shown in Fig. 68, the degree of deformation increasing with increasing flow rate. By contrast, rigid discs

[234] H. W. Thomas, *Biorheology* 1, 41 (1962).
[235] G. Bugliarello, C. Kapur, and G. Hsia, *Proc. 4th Intern. Congr. Rheol. 1963* Pt. 4, p. 351. Wiley (Interscience), New York, 1965.
[236] P. A. G. Monro, *Biorheology* 1, 239 (1964).
[237] A. A. Palmer, *Quart. J. Exptl. Physiol.* 44, 149 (1959).
[238] P. I. Bränemark and I. Jonsson, *J. Roy. Microscop. Soc.* 82, 245 (1963).

having diameters slightly less than that of a rigid tube do not orient themselves broadside on, when suspended in a fluid undergoing Poiseuille flow,[239,239a] but rotate in complex orbits.

The erythrocytes or erythrocyte columns move axisymmetrically down the capillaries, surrounded by a thin sleeve of plasma and, when deformed, more of the cell surface is in close proximity to the capillary endothelium, where the exchange of respiratory gases occurs.

a. Flowing Bubbles

A physical model of the flow of red blood cells in capillaries has been provided by the flow of large bubbles suspended in wetting liquids undergoing laminar flow through rigid tubes.[138,240] The bubbles assume a cylindrical shape and travel axisymmetrically down the tube, surrounded by a film of suspending liquid. This film is of constant thickness between the bubble ends and increases in width as the bubble velocity increases.[138,241] The leading ends of the bubbles are prolate spheroidal; the trailing ends, generally oblate. The velocity profiles in the bubble and suspending phase depend on the viscosity ratio λ and may be calculated by assuming parallel creeping motion in the film surrounding the cylindrical portion of the bubble.[138]

When the bubble has negligible viscosity compared to the suspending liquid, there is no flow in the film.[138] The ratio of the pressure drop across the bubble ends to the pressure drop across an equal length of pure suspending fluid, the "bubble resistance factor," has been shown for short bubbles (bubble length from 2 to 4 tube diameters) in systems of $\lambda = 0$ to be only slightly greater than unity.[242] In this case the additional pressure drop due to the presence of the bubble arises from effects associated with the bubble ends.

When $\lambda > 0$, velocity gradients develop in the film and internal circulation can be seen inside the bubble, showing that the gradient has been transmitted across the interface (Fig. 69). As λ increases, the gradients in the film increase, and those in the bubble decrease, until at high λ the system approaches plug flow. In this case the bubble resistance factor may have values as high as 10 to 20,[243] and in a system where a train of bubbles follow each other down a tube a large increase in the pressure gradient at a given flow rate would be expected.

[239] H. L. Goldsmith and S. G. Mason, unpublished work, 1963.
[239a] S. P. Sutera and R. M. Hochmuth, Division of Engineering, Brown University Report No. 1.
[240] J. W. Prothero and A. C. Burton, Biophys. J. 1, 565 (1961).
[241] F. Fairbrother and A. E. Stubbs, J. Chem. Soc. 1, 527 (1935).
[242] H. L. Goldsmith, Ph.D. thesis, McGill University, Montreal, Canada, 1961.
[243] R. H. Marchessault and S. G. Mason, Ind. Eng. Chem. 52, 79 (1960).

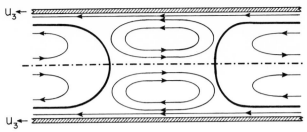

FIG. 69. Bolus flow: schematic representation of streamline patterns in the liquid between the rear end of one bubble (right) and the leading end of the following bubble (left). The motions of the fluid elements are relative to those of the bubbles which are moving with a velocity u_3. Also shown are the internal circulation patterns in the bubbles. (After Goldsmith and Mason.[105])

It is interesting, therefore, that the measured apparent viscosity of blood in tiny glass capillaries having a tip of 5 microns bore is only 5% greater than that of plasma.[244] Further work will be necessary, however, before the significance of this result is understood.

Flowing bubbles in systems of $\lambda > 0$ are deformed at their trailing ends in a manner resembling that of the red cell in the capillary. Thus, as shown in Fig. 68, at high flow rates the ends develop reentrant cavities of continuous-phase liquid, whose volume increases with increasing bubble velocity. The cavity volume also increases with time and, as it penetrates the bubble, is pulled out into a thread of liquid, which travels the length of the bubble along its axis.[138]

b. Bolus Flow

The flow pattern developed when a train of bubbles follow each other down a tube has been termed "bolus" flow,[240] since it is observed[105,240] that within an axial core of the liquid between the rear end of one bubble and the leading end of the next circulation patterns are set up, as shown schematically in Fig. 69. The effect of such flow is that a bolus of suspending liquid between the bubbles is carried down the tube with the bubble velocity. Prothero and Burton[240] have carried out model experiments using a thermal analogue to determine how much more effective the mixing motion of bolus flow is in relation to gaseous exchange. It was found that at high flow rates the heat transfer in bolus flow is approximately twice that in Poiseuille flow for the same volume of fluid; at low flow rates the difference is less and tends to zero at zero flow rate. The increase in resistance to flow was also measured[244] and found to be about 30%, as would be expected in the system air–water, in which λ is effectively zero.

[244] J. W. Prothero and A. C. Burton, *Biophys. J.* **2,** 199 (1962).

Nomenclature

ENGLISH LETTERS

$a(R)$	Amplitude of oscillation of fluid at a radial distance R from tube axis
A_{23}, A_{31}, etc.; $\langle A_{23} \rangle$	Respective $X_2 X_3$ and $X_3 X_1$ projections, etc., of unit area of equatorial plane of cylindrical particle; average value of A_{23}
dA	Area element
b_1, b_2, b_3	Principal semiaxes of ellipsoid
$b_1, b_2 (= b_3)$	Semiaxis of revolution and semiequatorial axis of spheroid
b; b_A, b_B	Radius of sphere and undistorted drop; radii of type A and type B spheres
$b_{(1)}, b_{(2)}$	Radii of curvature of deformed drop
$b_1'(\phi_1), b_2'(\phi_1)$	Projected lengths of principal semiaxes of spheroid on $X_2 X_3$ and $X_1 X_3$ planes
B_1, B_2, B_3	Components of Cartesian tensor $\mathbf{B}$ for ellipsoid
B	Value of $\mathbf{B}$ for spheroid
c; c_A, c_B	Volume fraction of singlet spheres; volume fraction of type A and type B spheres
c'	Volume fraction of doublet spheres
c_0	Surface concentration
c^*	Critical particle concentration
C; C_0	Spherical elliptical orbit constant; initial value of C
d_0, d_1	Function of $\Theta_0(z)$ and $\Theta_1(z)$, respectively
D_{11}^0, D_{23}^0; D_{11}, D_{23}	Components of fluid dilatation referred to the X_1^0, X_2^0, X_3^0 and X_1, X_2, X_3 coordinate systems, respectively
e_1^0, e_1	Respective first components of the internal electric field vectors $\mathbf{e}^0$ and $\mathbf{e}$ referred to the particle and Couette field coordinate systems
E_b	Bending modulus of rod
f; f^+, f^-	Two-body collision frequency; frequencies in regions of positive and negative particle flux, respectively
$F_I(R/R_0), F_{II}(R/R_0)$	Functions of R/R_0 in equations (VII8) to (VII10)
$g(z), G(z)$	Differential and integral distribution functions with respect to z; $z = \phi_1, \theta_1, \tau$, etc.
$g_C(\phi_1), G_C(\phi_1)$, etc.	Differential and integral distribution functions with respect to ϕ_1, etc., evaluated at a given C
h	Film thickness: separating colliding spheres [equations (V6) to (V8)]; at wall of Couette or capillary viscometer [equations (V12) and (V13)]
h_0	Value of h for colliding spheres at $\phi_1 = 0$
H	$\gamma/(\mathfrak{r}_e^2 - 1)$
i	$\sqrt{-1}$
I	Moment of inertia
j_{11}, j_{23}, etc.	Constants involving the spheroid integrals $\alpha_1, \alpha_1', \alpha_1''$, etc., and D_{11}^0, D_{23}^0, etc. [equation (III51)]
J	Perpendicular distance of tangent plane at the point (X_1^0, X_2^0, X_3^0) from spheroid center
$J_0(z), J_1(z)$	Bessel functions of order 0 and 1, respectively, and argument z
k	Modulus (of elliptic integral, etc.)

k_s', k_s; K_s', K_s	Respective internal streamline constants in hyperbolic and Couette flows; external streamline constants
$K(k)$	Elliptic integral of first kind with modulus k
l	Length, X_2 distance of particle center from wall in Couette flow
$\langle l \rangle$; $\langle\!\langle l \rangle\!\rangle$	Mean free path traveled by reference sphere between collision in Couette or Poiseuille flow; mean free path for all particles in the tube
l_1, l_{23}; $\langle l_1 \rangle$, $\langle l_{23} \rangle$	Projections of unit length of the axis of revolution of cylindrical particle on the X_1 axis and X_{23} plane, respectively; mean values
L_2; L_3	$b_2'(\phi_2) + b_1'(\phi_1)$; $b_2'(\phi_3) + b_1'(\phi_3)$
m	b/R^*
$M_0(z)$, $M_1(z)$	Modulus of Bessel functions of order 0 and 1, respectively, and argument z
M_0'	Function of d_0 and M_0
M_{10}'	Function of dimensionless ratio μ
n	Integer
n; n_A, n_B	Number of particles per cubic centimeter of suspension; number of type A and type B spheres per cubic centimeter of suspension
n'	Number of doublets of spheres per cubic centimeter of suspension
p_1, p_2, p_3	Components of force per unit area along $X_1{}^0$, $X_2{}^0$, $X_3{}^0$ axes of spheroid
Δp_{rr}	Difference in normal stresses at interface
Δp_0	Difference in surface pressure in interface
P_1, P_2, P_3	Net components of force acting on portion of surface of spheroid
$P_1^{(c)}$	Electrostatic force of attraction acting along line joining particle centers
P_1'	Approximate value of P_1 for large r_p
P_ϕ	Shear force per unit width of drop surface
p	Pressure field
$\mathbf{P}_D$, $\mathbf{P}_L$	Respective drag force and lift force on spinning sphere
q	Dielectric constant ratio of suspended phase to suspending phase
Q; Q_M	Volume flow rate; maximal oscillatory volume rate of flow
Q_0	Volume flow rate of steady component in pulsatile flow
r	Radial coordinate in plane polar, spherical polar, or cylindrical polar coordinate system with origin at the particle center
$\tilde{r}$	Re $\cdot r$, stretched radial coordinate
r_c	Radius of contact of drops
R; R^*	Radial coordinate from center of Couette apparatus and from tube axis; value of R at stationary layer in Couette flow and for reference sphere in Poiseuille flow
$R^\dagger$	Equilibrium radial position of particle in tube flow
R'	Initial radial position
$\langle\!\langle |\Delta R| \rangle\!\rangle$	Mean time-average radial displacement of spheres normal to wall in Pouseuille flow
R_c	Radius of suspension core in which u_3 is constant
R_0	Tube radius
R_I, R_{II}; ΔR	Radii of inner and outer cylinders of Couette apparatus; $(R_{II} - R_I)$
Re	Generalized particle Reynolds number
Re_t	Tube Reynolds number
Re_p, Re_s, Re_ω	Respective particle, shear, and rotational Reynolds numbers

Re_∞	Sedimentation Reynolds number
$s_{r1}, s'_{r1}; S_{r1}, S'_{r1}$	First components of the radial stress vector inside fluid drop in Couette and plane hyperbolic flow; components outside the drop
$s_{r\phi}, s_{rr}; S_{r\phi}, S_{rr}$	Respective tangential and normal components of the radial stress vector inside the drop in Couette flow; components outside the drop
s'_{rr}, S'_{rr}	Normal components of the radial stress vector in plane hyperbolic flow
S_{11}, S_{23}, etc.	Components of the stress $\mathbf{S}$ acting on surface of ellipsoid
$S_{rr}^{(e)}$	Normal component of the electric radial stress vector
t, t_0	Time, initial time
T	Period of rotation through $\Delta\phi_1 = 2\pi$
u_2, u_3	Respective particle velocities along X_2 and X_3 axes in tube flow
$u_{3\infty}$	Sedimentation velocity of particle in tube with suspending fluid at rest
$U_3(R), U_{3M}$	Fluid velocity in the X_3 direction at a radial distance R from tube axis in Poiseuille flow, maximum fluid velocity at $R = 0$
U_3'	$U_3(R) - U_3(R^*)$
v_1, v_2, v_3	Particle velocities along X_1, X_2, and X_3 axes in Couette flow
v_1', v_2', v_3'	Particle velocities along respective X_1', X_2', X_3' axes in plane hyperbolic flow
$V_1, V_2, V_3; V_1', V_2', V_3'$	Respective components of fluid velocity along X_1, X_2, X_3 axes; components in plane hyperbolic flow
$\mathbf{V}^{(0)}, \mathbf{V}^{(1)}, \mathbf{V}^{(2)}$, etc.	Zero, first, second, etc., reflected velocity fields
V_e	Electric potential in suspending phase
w	Angular velocity of oscillatory flow
W	Rate of energy dissipation
$x_1, x_2, x_3; x_1', x_2', x_3'$	Coordinates inside fluid drop in Couette flow; coordinates in plane hyperbolic flow
X_1^0, X_2^0, X_3^0	Particle Cartesian coordinate system
$X_1, X_2, X_3; X_1', X_2', X_3'$	Fixed coordinate system of external Couette flow field; coordinate system of plane hyperbolic flow field
$\langle|\Delta X_2|\rangle, \langle\!\langle|\Delta X_2|\rangle\!\rangle$	Time-average displacement of sphere for a given collision; mean time-average displacement over all collisions
y	Distance between particle centers
Y	$X_2 + l$
Z_1^0, Z_1	First component of fluid rotation referred to the particle and external field coordinate system, respectively
$Z(\beta k)$	Jacobian zeta function with parameter β and modulus k

GREEK LETTERS

$\alpha_1, \alpha_2, \alpha_3, \alpha_1', \alpha_1''$, etc.	Ellipsoid integrals
$\alpha_1, \alpha_2, \alpha_1', \alpha_1''$, etc.	Spheroid integrals ($\alpha_2 = \alpha_3, \alpha_2' = \alpha_3'$, etc.)
β	Parameter of Jacobian zeta function
$\gamma, \gamma_{crit.}$	Rate of shear, critical value of γ
$\Gamma_1, \Gamma_2, \Gamma_3$	Components of hydrodynamic torque Γ
$\Gamma_1^{(e)}$	First component of electric torque $\Gamma^{(e)}$
δ	$\tan^{-1}(\mathscr{E}_1/\mathscr{E}_2)$
Δ	Difference operator
ε	Dielectric constant of suspending phase
ζ	Variable of integration in equation (II2) and definition of $K(k)$

$\eta, \eta_0; \eta_s$	Viscosity of suspension, suspending phase; two-dimensional shear and area viscosity
η_r, η_r'	Respective relative and apparent relative viscosities
$[\eta]$	Intrinsic viscosity
θ, ϕ	Polar coordinates of spheroid
$\theta_1, \phi_1; \theta_1', \phi_1'$	Spherical polar coordinates referred to the polar axis X_1 of the external flow field; coordinates in plane hyperbolic flow
$\theta_2, \phi_2; \theta_3, \phi_3$	Spherical polar coordinates with X_2 and X_3 as the polar axis
$\theta_1^0, \phi_1^0, \phi_2^0$	Values of θ_1, ϕ_1, and ϕ_2 at apparent contact during collision of spheres
$\Theta_0(z), \Theta_1(z)$	Phase of Bessel function of order 0 and 1, respectively, and argument z
κ	Surface compressibility
λ	Viscosity ratio of suspended phase to suspending phase
$\Lambda_0(\Phi, k)$	Heuman lambda function with parameter Φ and modulus k
μ	$R(w\rho/\eta)^{1/2}$
ν	Kinematic viscosity
$\xi_0(R)$	Phase angle between pressure gradient and linear flow velocity in oscillatory flow
ξ_{10}	Phase angle between pressure gradient and volume flow rate in oscillatory flow
$\Xi(b/R_0)$	Function of b/R_0 (Section IV p. 153)
ρ	Density
σ	Interfacial or surface tension
$\tau, \langle\tau\rangle, \tau_M$	Lifetime of collision doublet, mean and maximum lifetimes
Υ	Particle-crowding factor [equation (VI4)]
$\phi_1^{0\prime}, \phi_2^{0\prime}$	Values of ϕ_1 and ϕ_2 at apparent separation of spheres
$\phi_{crit.}, \phi_i$	Respective steady-state values of ϕ_1 at $\mathscr{E}_{0.\,crit.}$ and $\mathscr{E}_0 > \mathscr{E}_{0.\,crit.}$
ϕ_m	ϕ_1 orientation of principal axis of deformed drop in shear flow
ϕ_n	ϕ_1 orientation of principal axis of deformed drop in combined electric and shear field
ϕ_r	Rectilinear collision angle
ϕ_{2M}, ϕ_{3M}	Maximum values of ϕ_2, ϕ_3
Φ	Parameter of Heuman lambda function
χ	Factor defined by equation (II49); reciprocal gives increase in T for rotation in shear and electric field
χ'	χ/i
ψ	$(1 + \frac{2}{5}\lambda)\mathfrak{D}$
$\Psi^{(s)}$	$4\gamma\eta_0(19\lambda + 16)/(16\lambda + 16)$
$\Psi^{(e)}$	$9\varepsilon\mathscr{E}_0^2(q - 1)^2/16\pi(q + 2)^2$
$\Psi^{(n)}$	$\Psi^{(e)2} + 2\Psi^{(e)}\Psi^{(s)} \sin 2\psi + \Psi^{(s)2}$
ω_1	$\dot\phi_1$, angular velocity of fluid about X_1 axis
$\omega_1^0, \omega_2^0, \omega_3^0$	Respective particle spins about X_1^0, X_2^0, and X_3^0 axes
ω_D	Value of ω_1 for doublet axis of interacting cylinders

Ω_U, Ω_L Respective angular velocities of upper and lower disc in counter-rotating-disc Couette device

Ω_I, Ω_{II} Respective angular velocities of inner and outer cylinders of Couette apparatus

$\Omega(R)$ Angular velocity of fluid at a radial distance R from center of Couette apparatus

ENGLISH SCRIPT

ℓ $\cosh^{-1}(y/2b)$

$\mathcal{D}_\phi, \mathcal{D}_r$ Respective tangential and normal electrical displacements

$\mathscr{E}_1^{\,0}, \mathscr{E}_1$ First components of the external electric field vector referred to particle and Couette field coordinate system, respectively

$\mathscr{E}_r, \mathscr{E}_\phi$ Radial and tangential components of external electric-field vector

$\mathscr{E}_0, \mathscr{E}_{0,\,\mathrm{crit.}}$ Value of electric-field strength, critical value for impeded rotation

$\mathscr{F}$ $f(q, r_e)\varepsilon\mathscr{E}_0^{\,2}/\eta_0$

$\mathscr{K}$ $d\gamma/dR = 4Q/\pi R_0^{\,4}$ in Poiseuille flow; $(\Omega_U + \Omega_L)/\Delta X_2$ in Couette flow between parallel counterrotating discs

$\mathscr{K}'$ Constant in oscillatory flow defined by equation (VII19)

$\mathscr{R}$ Ratio of resistivities of suspending phase to suspended phase

$\mathscr{S}$ $2\sinh^2 \ell \sum\limits_{1}^{\infty} \dfrac{n}{\sinh 2n\ell + n \sinh 2\ell}$

$\mathscr{U}_3$ Maximum oscillatory fluid velocity

$\mathscr{V}$ Volume

$\mathscr{V}_p$ Particle volume

$\mathscr{V}_E(C), \langle \mathscr{V}_E \rangle$ Effective particle volume, mean value of $\mathscr{V}_E$

GERMAN LETTERS

$\mathfrak{a}$ Interaction coefficient

$\mathfrak{A}$ Function of r_e defined by equation (II43b)

$\mathfrak{D}_1, \mathfrak{D}_2$ Respective deformations of liquid drop in X_2X_3 and X_1X_3 planes

$\mathfrak{D}; \mathfrak{D}_B, \mathfrak{D}'$ Ratio of viscous forces to surface-tension forces (deformation), given by equation (III32); value at burst, limiting value

$\mathfrak{D}_1^{(e)}, \mathfrak{D}_B^{(e)}$ Deformation of drop in X_2X_3 plane in electric field; deformation at burst

$\mathfrak{D}^{(n)}$ Deformation of drop in X_2X_3 plane in combined shear and electric field

$\mathfrak{f}(q, r_e)$ Function of q and r_e, defined by equation (II43)

$\mathfrak{f}(q, \lambda, \mathscr{R})$ Function of q, λ, and $\mathscr{R}$, defined by equation (III39)

$\mathfrak{m}$ Circulation number on streamline inside drop $= T\gamma/4\pi$

$\mathfrak{n}$ Number of axial spins during one complete rotation of spheroid

$\mathfrak{P}(q, r_e)$ Function of $\mathfrak{f}(q, r_e)$ and $\mathfrak{A}$, defined by equation (II42)

$r_e; r_p, r_{\mathrm{crit.}}$ Equivalent ellipsoidal axis ratio; particle axis ratio, critical value of r_p for shear-induced buckling

CHAPTER 3

HIGH-SHEAR VISCOMETRY

Arie Ram

I. Introduction

High-shear viscometry deals with the flow properties of liquids under rates of shear of 10^4 to $10^6 \, \text{sec}^{-1}$. In some non-Newtonian systems—in particular, dilute polymer solutions—this is the domain of the ultimate (second) Newtonian region. The fact that the limiting value of viscosity, η_∞, is a constant (shear-independent) parameter is of great importance in polymer science. The behavior of particles in suspension and the conformation of polymer chains in solution or as melts stimulate much theoretical and practical curiosity. In many flow problems the limiting viscosity parameter represents the state of greater order, wherein most of the disturbing effects responsible for non-Newtonism have been suppressed. It is therefore called the "bone" viscosity parameter of the system.

Polymer characterization has been extensively utilized in low-shear viscometry. However, whenever the larger macromolecules are examined, the shear sensitivity of the apparent viscosity becomes increasingly significant even at low rates of shear. Because of this, high-shear viscometry may furnish a new tool for polymer characterization. Intrinsic viscosity is used for characterizing polymers in dilute solutions, and melt viscosity has been found to be related to polymer chain length. A special practical tool is the melt indexer.[1,2] Unfortunately, its use, mainly in polyethylene production, is unsound from a rheological point of view, because it provides a single-point value of viscosity for a definitely non-Newtonian system. In fact, the melt-index method has been recently widely criticized.[3-5] A possible means of improvement is working in the second Newtonian region.

High shear prevails under various polymer processing and fabrication conditions, usually when the methods are based on extrusion and injection. Any attempt at predicting the flow properties therefore depends on high-shear information. A direct correlation between the index of moldability, represented by the spiral mold length, and the high-shear viscosity parameter has been suggested.[6] Thus, predicting the workability of plastics on the basis of extensive melt-rheology information is becoming a necessity.

Another practical field in which high-shear viscometry plays an important role is lubrication. Admixing polymers (so-called viscosity index improvers) with a view to modifying the temperature sensitivity of lubricant viscosity leads to the formation of a pseudoplastic system. Conditions of higher shear also prevail in working motors or cranks. The rheological properties of the modified oils under high-shear conditions have accordingly been explored.[7-9]

[1] J. P. Tordella and R. E. Jolly, *Mode. Plastics* **31**, 146 (1953).
[2] A.S.T.M. Designation: D 1238–57 T.
[3] S. L. Aggarwal, L. Marker, and M. J. Carreno, *J. Appl. Polymer Sci.* **3**, 77 (1960).
[4] H. P. Schreiber, *SPE (Soc. Plastics Engrs.) Trans.* **1**, 86 (1961).
[5] A. Rudin and H. P. Schreiber, *SPE (Soc. Plastics Engrs.) J.* **20**, 533 (1964).
[6] G. Penzin, *SPE (Soc. Plastics Engrs.) Trans.* **3**, 260 (1963).
[7] R. S. Porter and J. F. Johnson, *J. Am. Soc. Naval Engrs.* **73**, 511 (1961).
[8] C. W. Georgi, *World Petrol. Congr., Proc. 4th Congr., Rome, 1955* Sect. VI, p. 211. Izd. AN SSSR, Moscow 1955.
[9] H. H. Horowitz, *Ind. Eng. Chem.* **50**, 1089 (1958).

Regular data on low-shear viscometry and even the use of the viscosity index scale are therefore of questionable value for polymer-modified oils.

Dispersions and related heterogeneous systems are increasingly dealt with in industry. Here the concept of the limiting viscosity parameter under high shear has acquired popularity. Calculated or extrapolated values of this parameter are being used for empirical correlations with various flow systems.[10,11]

To the polymer scientist the prediction of the physical behavior of macromolecular chains in the shear field is extremely important. In contrast to low-shear viscometry, the status of research in high-shear viscometry and the amount of accumulated data are still unsatisfactory. Even in the absence of a clear-cut boundary the difference in flow character between the two regions is remarkable. Many theoretical approaches have been modified to include the limiting viscosity parameter, but the usefulness of most of the derivations is still questionable.

Although represented as a unique branch in rheology, high-shear viscometry constitutes an important link within general viscometry. Any progress toward a self-consistent rheological equation of state promises the eventual achievement of a unified approach. In the following paragraphs the various aspects of high-shear viscometry will be discussed on the basis of the contemporary knowledge.

The nomenclature used in this chapter is listed in the last section.

II. Analysis of Flow Behavior at High Shear Rates

1. FLOW BEHAVIOR OF PURE LIQUIDS AND DISPERSED SYSTEMS

Flow behavior at high shear rates should be described with the aid of general flow theories. However, the actual circumstances make this difficult. Starting with "simple" homogeneous liquids, Newtonian behavior is characterized by a *constant* ratio between shear stress and rate of shear:

$$\tau = \eta \cdot \frac{dU}{dy} = \eta \cdot D \qquad \text{or} \qquad \eta = \frac{\tau}{D} \tag{1}$$

It will be shown later that Newton's law is generally obeyed by low-molecular-weight (so called Newtonian) liquids only when high shear rates are excluded.

By contrast, colloid suspensions and polymer melts or solutions show remarkable deviations from Newton's law; these deviations set in at relatively

---

[10] D. G. Thomas, *A.I.Ch.E. J.* **8**, 266 (1962).

[11] D. M. Eissenberg, "Developments in Theoretical and Applied Mechanics," Vol. 1, pp. 277–290. Plenum Press, New York, 1963.

low shear rates. All such systems are grouped together under the name of non-Newtonians,[12] their most general flow equation being

$$\tau = f(D) \tag{2}$$

The type of function depends mainly on the system and its flow conditions. The single parameter of viscosity may be replaced with two or more parameters. In some cases the effects of time (shear history) cannot be disregarded.

This chapter is devoted mainly to a special group of non-Newtonians known as pseudoplastics, by which term most polymer melts or solutions may be described. In studying them it is preferable to use the apparent viscosity, the varying ratio between shear stress and rate of shear:

$$\tau = \eta_a \cdot D \tag{3}$$

The apparent viscosity of pseudoplastic materials is shear thinning, i.e., decreases with increasing shear rate (or stress). Both equations (2) and (3) are qualitative and, as such, useless for any mathematical derivation. Many attempts have been made to derive exact equations for pseudoplastic liquids, but most of the derivations are still limited in application and no basic simple equation has been found to cover the flow behavior of pseudoplasticity.

Among the empirical relationships between shear stress and rate of shear the so-called power law is very popular in engineering:

$$\tau = KD^n \tag{4}$$

where $n < 1$ for pseudoplastics. This law may also be written

$$\eta_a = \tau/D = KD^{n-1} \tag{5}$$

Its advantage lies in the replacement of the single material property η by two parameters, K and n, which are conveniently applicable in the treatment of most aspects of flow analysis with only a slight complication of the mathematical derivation.[13] Fortunately, the power law is frequently obeyed by polymer melts and solutions in the normal working range. The flow curve representing the relationship between shear stress and rate of shear is still a straight line in a logarithmic coordinate system and thus permits extrapolation and interpolation of data. It should, however, be borne in mind that this law cannot be basically exact. It predicts, for pseudoplastics, zero viscosity at high shear rates, ($\eta_\infty = 0$) and an infinite viscosity at zero shear rate ($\eta_0 = \infty$). Zero shear rates prevail, as a rule, at the center of a tube, so that this "boundary condition" is not covered by the power law. It has also been

[12] E. W. Merrill, in "Modern Chemical Engineering Series" (A. Acrivos, ed.), Vol. 1, Chapt. 4. Reinhold, New York, 1964.

[13] A. B. Metzner, in "Processing of Thermoplastic Materials" (E. C. Bernhardt, ed.), Chapt. 1. Reinhold, New York, 1962.

noted that the exponent n in equation (4) may vary with the flow conditions of one material.[14,15] It is, moreover, known that a Newtonian region prevails at low shear rates. As for the second extreme, the high shear region, it has been known that a limiting value of viscosity exists. To account for the first Newtonian viscosity η_0 and the second Newtonian viscosity η_∞ additional constants are included:

$$\eta_a = \eta_0[1 - B(D)^n + \cdots +] \qquad (6)$$

Theory[16,17] predicts, for Eq. (6), $n = 2$.

$$\frac{1}{\eta_a} = \frac{1}{\eta_0}(1 + a\tau) \qquad \text{(Ferry}^{18}\text{)} \qquad (7)$$

$$\eta_a = \eta_\infty + \frac{\eta_0 - \eta_\infty}{1 + \tau^2/b^2} \qquad \text{(Reiner-Philippoff)} \qquad (8)$$

$$\eta_a = \eta_\infty + (\eta_0 - \eta_\infty)\exp(-KD) \qquad \text{(Golub}^{19}\text{)} \qquad (9)$$

$$\eta_a = \eta_\infty + bD^{n-1} \qquad \text{(Sisko}^{20}\text{)} \qquad (10)$$

$$\eta_a = \eta_0 \frac{\tau/B}{\sinh(\tau/B)} \qquad \text{(Powell-Eyring}^{21}\text{)} \qquad (11)$$

2. EXISTENCE OF SECOND NEWTONIAN REGION AT HIGH SHEAR RATES

According to Reiner,[22,23] a "generalized Newtonian liquid" is actually a pseudoplastic liquid obeying Newton's law at both extremes. Such a complete flow curve is shown in Fig. 1. The second Newtonian viscosity η_∞ is the limiting value at relatively high shear conditions.

The earliest reference to the second Newtonian viscosity is made by Ostwald.[24] It deals with flow curves of various suspensions, representing pseudoplasticity at low rates of shear and straightening out into a Newtonian line at high shear rates. According to Ostwald, η_∞ is the only constant value

[14] R. G. Shaver, and E. W. Merrill, *A.I.Ch.E. J.* **5**, 181 (1959).
[15] A. P. Metzger and R. S. Brodkey, *J. Appl. Polymer Sci.* **7**, 399 (1963).
[16] G. M. Guzman and J. M. G. Fatoue, *J. Colloid. Sci.* **35**, 441 (1959).
[17] A. Peterlin and M. Copic, *J. Appl. Phys.* **27**, 434 (1956).
[18] J. D. Ferry, *J. Am. Chem. Soc.* **64**, 1330 (1947).
[19] M. A. Golub, *J. Polymer Sci.* **18**, 27 (1955).
[20] A. W. Sisko, *Ind. Eng. Chem.* **50**, 1789 (1958).
[21] R. E. Powell and H. Eyring, *Nature* **154**, 427 (1944).
[22] M. Reiner, "Deformation, Strain and Flow." Lewis, London, 1960.
[23] M. Reiner "Lectures on Theoretical Rheology." North-Holland Pub., Amsterdam, 1960.
[24] W. Ostwald, *Kolloid. Z.* **36**, 99 (1925).

detected in experiments. Philippoff and Hess[25] investigated polymer solutions and described them by a flow curve similar to Fig. 1. The flow curve thus comprises four distinct regions: a first Newtonian region at low rates of shear (initial slope representing η_0), a non-Newtonian pseudoplastic region with a decreasing slope (point ratios representing η_a), a second Newtonian region at high rates of shear, about 10^5 to 10^6 sec^{-1} and a region of steepening slope identified as the onset of turbulence. The last region is of little value in pure rheological treatment because of the intervention of "eddy viscosity." The second Newtonian region is represented by a straight line, extrapolable to the origin in rectilinear coordinates, or by a unity slope in log–log coordinates. The whole flow curve is characterized by an S shape.

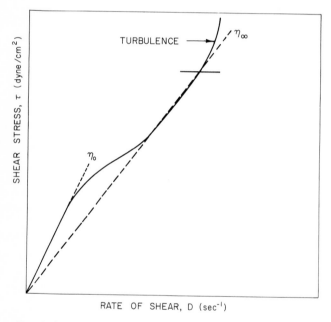

FIG. 1. A complete flow curve of pseudoplastic, shear thinning, liquid.

3. THEORETICAL ANALYSIS

It is well known that no current atomistic theory predicts the viscosity values of even simple fluids completely. Only semiempirical approaches have been successful for analyzing the changes in viscosity with temperature

[25] W. Philippoff and K. Hess, Z. Physik. Chem. B31, 237 (1936).

of the viscosity of a mixture of simple liquids. Thus, no real rheological equation of state exists. For the most promising results see Chapter 1. We follow Eyring and associates,[26-29] whose rate-process theory treats the flow of simple homogeneous liquids as another such process with respect to the jumping of molecules over the energy barriers between equilibrium positions. Eyring's general flow equation reads

$$\eta = \frac{\lambda_1 RT}{J\lambda\delta} \bigg/ \left[1 + \frac{1}{3!}\left(\frac{\tau\delta}{2RT}\right)^2 + \frac{1}{5!}\left(\frac{\tau\delta}{2RT}\right)^4 + \cdots + \right] \tag{12}$$

where J is the jump frequency "at rest" (no shear) and λ_1, λ, and δ are related to molecular dimensions and jump distances. Equation (12) states that every liquid is basically non-Newtonian in nature. Shear-independent viscosity only prevails at low shear rates, where the τ-containing terms of the polynomial vanish. In practice the range of this initial Newtonian viscosity is fairly broad in simple liquids, quite often prevailing throughout the laminar flow regime. It is rather limited, however, in more complicated systems. Eyring's general equation does not predict the vanishing effect of the shear stress on the variable viscosity under high shear.

In order to cover the second Newtonian region Ree and Eyring[30,31] have modified the basic theoretical system to one of heterogeneous "flow groups" of polymer segments. Each group is characterized by its surface fraction X_n and relaxation time β_n. In the case of a polymer solution, starting from the pure solvent, the relaxation time increases with increasing complexity of the flow unit:

$$\eta = \sum_0^n \frac{X_n\beta_n}{\alpha_n} \frac{\sinh^{-1}(\beta_n D)}{\beta_n D} \tag{13}$$

$$\alpha = \delta/2RT, \qquad \beta_1 D \ll 1, \qquad \beta_2 D \geqslant 1, \qquad \beta_3 D \gg 1$$

for

$$\beta D \to 0 \qquad \lim \frac{\sinh^{-1}(\beta D)}{\beta D} = 1$$

[26] H. Eyring, J. Chem. Phys. 4, 283 (1936).
[27] R. H. Ewell and H. Eyring, J. Chem. Phys. 5, 726 (1937).
[28] R. H. Ewell, J. Appl. Phys. 9, 252 (1938).
[29] S. Glasstone, K. L. Laidler, and H. Eyring, "The Theory of Rate Processes." McGraw-Hill, New York, 1941.
[30] T. Ree and H. Eyring, J. Appl. Phys. 26, 793 (1955).
[31] T. Ree and H. Eyring in "Rheology" (F. R. Eirich, ed.), Vol. 2, Chapt. 3, p. 83. Academic Press, New York, 1958.

Therefore

$$\frac{\ln \eta_{r0}}{c} = \frac{1}{c} \ln \left[X_0 + \frac{X_1 \beta_1}{\alpha_1 \eta_s} + \frac{X_2 \beta_2}{\alpha_2 \eta_s} + \cdots + \right], \tag{14}$$

where $\eta_r = \eta/\eta_s$ = reduced viscosity; for

$$\beta D \to \infty, \qquad \lim \frac{\sinh^{-1}(\beta D)}{\beta D} = 0$$

Therefore

$$\frac{\ln \eta_{r\infty}}{c} = \frac{1}{c} \ln \left[X_0 + \frac{X_1 \beta_1}{\alpha_1 \eta_s} \right] \tag{15}$$

where X_0 and η_s refer to the solvent proper. When very dilute solutions are used, equation (14) predicts the low-shear intrinsic viscosity, and equation (15) predicts its high-shear counterpart as shown by (16) and (17):

$$[\eta]_0 = \lim_{\substack{c \to 0 \\ D \to 0}} \frac{\eta_0 - \eta_s}{\eta_s c} = \lim_{\substack{c \to 0 \\ D \to 0}} \frac{\ln \eta_{r0}}{c} \tag{16}$$

$$[\eta]_\infty = \lim_{\substack{c \to 0 \\ D \to \infty}} \frac{\eta_\infty - \eta_s}{\eta_s c} = \lim_{\substack{c \to 0 \\ D \to \infty}} \frac{\ln \eta_{r\infty}}{c} \tag{17}$$

The general equation, (13), describes η_0 and η_∞ for any heterogeneous system, polymer solutions and dispersions, and polymer melts which normally consist of different flow groups. The use of (13) is however restricted, for it comprises the various unknown relaxation times β_n and the stress area fraction X_n. Some of these values have been calculated by Ree and Eyring through an approximate fitting of curves to existing flow data short of the second Newtonian region. A superposition of three curves represents the contribution of distinct flow groups. As there is no means of assuming the values of the various flow parameters, no prediction of η_∞ is possible without actual data or extrapolation. It should also be mentioned that, by equation (15), only the groups having the lowest relaxation times contribute to the high-shear intrinsic viscosity; in such case $[\eta]_\infty$ should not depend on the length of the chain. This possibility will be discussed later in connection with the relationships between molecular weight and high-shear intrinsic viscosity. It will be shown that this is not the common rule.

Another complete flow theory has been suggested by Bueche.[32,33] The motion of the solvent exerts a torque on the polymer chains, which rotate

[32] F. Bueche, "Physical Properties of Polymers." Wiley (Interscience), New York, 1962.
[33] F. Bueche, J. Chem. Phys. 22, 1570 (1954).

with an angular velocity of $D/2$. The friction losses decrease with increasing frequency of rotation. Bueche suggested the following equation for the complete flow curve with the two Newtonian viscosities at the extremes:

$$\frac{\eta - \eta_s}{\eta_0 - \eta_s} = 1 - \frac{6}{\pi^2} \sum_{n=1}^{N} \frac{D^2\beta^2}{n^2(n^4 + D^2\beta^2)} \left[2 - \frac{D^2\beta^2}{n^4 + D^2\beta^2} \right] \tag{18}$$

Here the relaxation time β is related to the molecular weight of the whole polymer molecule which acts as an imbedded spring. The summation term vanishes at shear rates of low value and is independent of shear rates of high value; thus

$$\frac{\eta_\infty - \eta_s}{\eta_0 - \eta_s} = 1 - \frac{6}{\pi^2} \sum_{n=1}^{N} \frac{1}{n^2} \tag{19}$$

In these derivations n is an integer referring to the mode of vibration of the spring system in flow and N is the number of segments in a single chain. As N increases, equation (19) will predict the limiting case $\eta_\infty = \eta_s$. Such a total vanishing of the polymer's specific viscosity contradicts most experimental results. In any case, equation (18) is too cumbersome for practical use.

The physical causes of non-Newtonian viscosity are often analyzed. The following factors are believed to favor reduction of the viscosity with increasing shear[22,34]: progressive breakdown, under shear, of the polymeric association arising from intermolecular attraction or entanglement, increasing orientation of anisometric rigid particles in the direction of flow, and increasing deformation of flexible molecules.

Besides being changed in shape and dimensions, the trapped solvent, immobilized within the swollen coil, is expelled. This is an additional contribution to the drop in viscosity. Only reversible and time-independent effects are dealt with here; the irreversible decrease in viscosity, such as that caused by chain degradation, is not considered for the moment.

4. ROLE OF ENTANGLEMENT OF POLYMER CHAINS IN MELTS AND SOLUTIONS

The role of entanglement–disentanglement is essential in synthetic polymers, which exhibit, when long enough, "hopelessly entangled" chains. This is supported by the existence of a critical molecular weight, above which non-Newtonian flow dominates the behavior in melts and concentrated solutions.[35–38] The change is apparent on plots of melt viscosity versus

[34] L. J. Sharman, R. H. Jones, and L. H. Cragg, *J. Appl. Phys.* **24**, 703 (1953).

[35] F. Bueche, *J. Appl. Phys.* **24**, 423 (1953).

[36] F. Bueche, *J. Appl. Phys.* **26**, 738 (1955).

[37] T. G. Fox and P. J. Flory, *J. Phys. Colloid. Chem.* **55**, 221 (1951).

[38] R. Signer, *Trans. Faraday Soc.* **32**, 296 (1936).

molecular weight. For various polymers the following relationship holds:

$$\eta_0 = AM^b \qquad (20)$$

The exponent b is about 1 at low molecular weight and rises sharply to 3.4 above the critical chain length. The critical molecular weight indicates the threshold value for entanglement control. It is generally agreed that increasing shear will disentangle polymer coils. Porter and Johnson[39] have shown that b drops under high shear from 3.4 toward unity; that is, it approaches the value b of unentangled polymer. The disentanglement efficiency depends on the range of molecular weight and on the level of concentration (in the case of a solution). Generally, viscosity drops as a result of disentanglement. Complete disentanglement may be predicted only in the case of very dilute solutions or melts of relatively short chain length. Moreover, the reversibility of the entanglement–disentanglement process is doubtful. Scott Blair[40] states that there is no obvious cause for reentanglement at rest to be as severe as in the polymerization process. This author[41] noted in the course of his experimental work that in some cases a mild absolute drop in the viscosity of polymer dilute solutions might result from irreversible disentanglement. Such a phenomenon may often be associated with the concept of polymer degradation. It should be emphasized, however, that an irreversible reduction in the viscosity of a polymer solution, either on standing or through shear, is not necessarily evidence of polymer degradation. It is always safer to check the intrinsic viscosity. Even in the range of very dilute solutions residual entanglements of very long polymer chains may survive by unsuitable treatment. Disentanglement through the slipping of physical loops is also related to the flexibility of the chain. Thus viscometric data represent the joint contribution of several effects. The association processes are best described either by molecular (secondary) valences or by electrical dipole forces. Their concentration dependence is often widely different from that of physical entanglement. The connection between disentanglement and degradation processes will be discussed later.

5. Role of Particle Orientation under Shear

Orientation is the main contribution to the pseudoplasticity of most colloid suspensions and polymer solutions. The dispersed particles, mostly asymmetric in shape, tend to orient their major axis in the direction of flow under shear. Thus shear conditions affect the competing processes of

[39] R. S. Porter and J. F. Johnson, *J. Appl. Polymer Sci.* **3**, 107 (1960).

[40] G. W. Scott Blair, "A Survey of General and Applied Rheology." Pitman, New York, 1944.

[41] A. Ram, Sc.D.Thesis, M.I.T., Cambridge, Massachusetts, 1961.

random Brownian movement (which may be intensified at higher temperatures) and of alignment with the flow direction. Whereas flexible chains may change their shape, rigid particles (such as nitrocellulose[42]) acquire high sensitivity to shear as the result of progressive orientation. Since the continuous phase (the solvent) is disturbed by the dispersed particles, it is obvious that an alignment will effectively reduce the viscosity of the whole system.[42a] The intrinsic viscosity of rigid long chains completely aligned in the flow direction will no longer depend on the length of the chain. Such an extreme result has also been predicted by Ree and Eyring on the basis of the heterogeneous-flow theory. However, no polymer material is perfectly rigid. Furthermore, alignment is only statistically feasible. Long chains actually lie in different shear fields and therefore undergo rotational as well as translatory displacement. According to Bueche,[33] this rotational movement, controlled by the rate of shear, comprises stages of stretching and folding. Bondi[43] analyzed the effect of orientation on pure liquids with the aid of Eyring's flow theories and found a decrease in viscosity with increasing shear up to a limiting value at infinite shear stress. The ratio of the viscosity in the unaligned state to that at full alignment equals the ratio of the average random chain length to the length of a fully stretched chain. As to the intrinsic viscosity, Kirkwood and Auer[44] showed that with completely oriented rigid rods it decreases to one fourth of its value at full Brownian movement (Kuhn,[45] however, claims a ratio of one half). In the case of long macromolecules that are neither rigid nor completely flexible, the alignment, rotation, deformation, folding, and unfolding of coils all jointly affect the flow resistance. They are controlled by the rates of shear and the relaxation times of the flow units. The orientation of particles under shear, as in streaming birefringence, may be successfully followed optically.[46] Thus, the physics of sheared particles may be independently observed. In the case of polymer melts residual "frozen" orientation is frequently observed in the final solid state. It entails marked effects on mechanical strength and is extensively utilized in spinning.

6. DEFORMATION OF POLYMER CHAINS UNDER SHEAR

The role of deformation under shear is the subject of widespread controversy. On the one hand, flexible or deformable chains may exhibit reduced

[42] E. H. Immergut and F. R. Eirich, Ind. Eng. Chem. 45, 2500 (1953).

[42a] See Chapter 2.

[43] A. Bondi, J. Appl. Phys. 16, 539 (1945).

[44] J. G. Kirkwood and P. L. Auer, J. Chem. Phys. 19, 281 (1951).

[45] W. Kuhn, J. Polymer Sci. 12, 14 (1954).

[46] A. Peterlin, in "Rheology" (F. R. Eirich, ed.), Vol. 1, Chapt. 15, p. 615. Academic Press, New York, 1956.

dimensions under shear at finite concentrations;[34,47] on the other, shearing of very dilute solutions (region of intrinsic viscosity) may result in the extension and elongation of readily deformable chains.[48-53] The resistance to deformation, sometimes referred to as "internal viscosity," decreases with increasing chain length. Peterlin[17,54-60] claims that, whereas the intrinsic viscosity of rigid molecules is extremely shear-sensitive, that of completely flexible chains may be almost shear-independent. On the other hand, with increasing shear the intrinsic viscosity may even increase when the effects of elongation outweigh those of orientation. An uncoiling of macroscopic particles under shear conditions has been demonstrated by Forgacs and Mason.[61]

In low-shear viscometry Einstein's[62] original equation,

$$\eta_r = \eta/\eta_s = 1 + 2.5\phi \tag{21}$$

where ϕ is the volume fraction of particles, may be modified so as to relate intrinsic viscosity to molecular dimensions and weight. Deformation of the particle and noncreeping flow change the situation completely. A substitution of a real nonspherical random coil by an equivalent hydrodynamic sphere results in a variety of relationships according to the specific geometry involved:

$$[\eta] = 2.5\frac{4}{3}\pi R^3 \frac{1}{M} \tag{22}$$

N_A is missing here!

where R is the equivalent radius. The relationship under high shear is dictated by the relative change in radius. Shear deformation may elongate flexible chains in dilute solutions or, by contrast, cause crowding and a reduction in the apparent volume at higher concentrations.

[47] E. Edelman, *Proc. 2nd Intern. Congr. Rheol., Oxford, 1953* p. 107. Butterworth, London and Washington, D.C., 1954.

[48] H. Leaderman, "Physics of High Polymers." Utrecht Univ. Press, Utrecht, 1951.

[49] W. Kuhn and H. Kuhn, *Helv. Chim. Acta* **28,** 1533 (1945).

[50] W. Kuhn and H. Kuhn, *Helv. Chim. Acta* **29,** 609 (1946).

[51] J. G. Kirkwood and J. Riseman, *J. Chem. Phys.* **16,** 565 (1948).

[52] P. E. Rouse, *J. Chem. Phys.* **21,** 1272 (1955).

[53] B. H. Zimm, *J. Chem. Phys.* **24,** 269 (1956).

[54] A. Peterlin, *J. Polymer Sci.* **8,** 621 (1952).

[55] A. Peterlin, *J. Chem. Phys.* **33,** 1799 (1960).

[56] A. Peterlin, Symposium on Non-Newtonian Viscometry, A.S.T.M., Washington, Oct. 1960, p. 115. A.S.T.M., Philadelphia, 1962.

[57] A. Peterlin and D. T. Turner, *Nature* **197,** 488 (1963).

[58] A. Peterlin and D. T. Turner, *J. Chem. Phys.* **38,** 2315 (1963).

[59] S. Burow, A. Peterlin, and D. T. Turner, *J. Polymer Sci.* **2,** 67 (1964).

[60] A. Peterlin, International Symposium on Second-Order Effects in Elasticity, Plasticity and Fluid Dynamics, Haifa, Israel, April, 1962. Pergamon, London, 1964.

[61] O. L. Forgacs and S. E. Mason, *J. Colloid Sci.* **14,** 473 (1959).

[62] A. Einstein, *Ann. Physik.* [4] **19,** 289 (1906).

7. BEHAVIOR OF SIMPLE LIQUIDS AT HIGH SHEAR RATES

Simple liquids are tacitly assumed to be Newtonian. The assumption is usually correct for gases and "thin" liquids of low viscosity in the working range. At higher shear rates not involving turbulence simple liquids of relatively low molecular weight have shown non-Newtonian features. There is conclusive evidence[12,41,63-67] that glycerol and pure mineral oils become pseudoplastic at high shear rates (above 60,000 sec^{-1}) or shear stresses of about 5000 to 8000 dynes/cm^2 conforming to Eq. (12) which predicts pseudoplasticity on increasing shear. Grunberg and Nissan[68] calculated a departure of pentane from Newtonian to pseudoplastic flow at a critical shear rate of 5×10^6 sec^{-1} (shear stress of 10^4 dynes/cm^2). They predicted that the critical shear stress is inversely proportional to the molecular weight. Thus, dilute solutions of relatively small particles in low-viscosity liquids show only a mild shear effect. It has, however, been mentioned that the relative change in flow resistance encountered in heterogeneous systems is sometimes more predictable than the flow behavior of simple liquids.

III. High-Shear Viscometry Data

1. ULTIMATE (SECOND NEWTONIAN) VISCOSITY

The ultimate viscosity, η_∞, under high shear conditions of polymer solutions has been studied by several investigators.[12,19,22,25,39,41,47,69-77a]

[63] A. Ram and A. Tamir, *Ing. Eng. Chem.* **56**, 47 (1964).

[64] I. A. Dumanskii and L. V. Khailenk, *Kolloidn. Zh.* **22**, 277 (1960).

[65] W. W. Hagerty, *J. Appl. Mech.* **17**, 54 (1950).

[66] S. M. Neale, *Chem. Ind.* (*London*) **56**, 140 (1937).

[67] R. N. Weltmann, *Ing. Eng. Chem.* **40**, 272 (1948).

[68] L. Grunberg and A. H. Nissan, *Nature* **156**, 241 (1945).

[69] E. W. Merrill, *J. Polymer Sci.* **38**, 539 (1959).

[70] E. W. Merrill, H. S. Mickley, A. Ram, and G. Perkinson, *Trans. Soc. Rheol.* **5**, 237 (1961).

[71] E. W. Merrill, H. S. Mickley, A. Ram, and W. H. Stockmayer, *Trans. Soc. Rheol.* **6**, 119 (1962).

[72] E. W. Merrill, A. Ram, H. S. Mickley, and W. H. Stockmayer, *J. Polymer Sci. Pt. A* **1**, 1201 (1963).

[73] J. G. Brodnyan, F. H. Gaskins, and W. Philippoff, Symposium on Non-Newtonian Viscometry, A.S.T.M., Washington, Oct. 1960, p. 14. A.S.T.M., Philadelphia, 1962.

[74] J. G. Brodnyan and E. L. Kelley, *Trans. Soc. Rheol.* **5**, 205 (1961).

[75] W. Philippoff, F. H. Gaskins, and J. G. Brodnyan, *Trans. Soc. Rheol.* **1**, 109 (1957).

[76] S. N. Chinai and W. C. Schneider, *J. Appl. Polymer Sci.* **7**, 909 (1963).

[77] S. Claesson and U. Lohmander, *Makromol. Chem.* **44**, 461 (1961).

[77a] E. H. Merz and R. E. Colwell, *A.S.T.M.* (*Am. Soc. Testing Materials*) *Bull.* **232**, 63 (1958).

Others[30,78] have calculated it by extrapolation from medium shear rates. Philippoff[75] was the first to show complete flow curves that clearly represented the first and second Newtonian viscosities of polyisobutylene solutions and also of nitrocellulose solutions over a wide range of concentrations.[25,73] The ultimate viscosity was determined at shear rates of about 10^5 to 10^6 sec^{-1}. The higher concentrations necessitated higher shear rates. Reiner[22] analyzed the ultimate viscosity of rubber solutions. The concentration dependence of the ultimate viscosity is most interesting, and will be discussed in the next section.

2. ULTIMATE VISCOSITY NUMBER OF DILUTE POLYMER SOLUTIONS

The ultimate viscosity number (UVN) is defined as

$$UVN = \frac{\eta_\infty - \eta_s}{c\eta_s} \tag{23}$$

with c = concentration in g/dl. It is well known that the ratio η_∞/η_0 is largely dependent on the concentration level. Philippoff[75] showed a five-thousand-fold change in this ratio through a mere eighteenfold change in polyisobutylene concentration. Surprisingly, however, the UVN is essentially independent of concentration over this entire range. In most other studies it is noted that the typical UVN of various polymer solutions is either concentration-independent or shows only a mild dependence. The range of extremely dilute solutions (region of intrinsic viscosity) under high shear was studied by Claesson and Lohmander.[77] It has been claimed that structural pseudoplasticity only manifests itself above a certain minimal concentration.[22,79,80] That is the reason for Immergut and Eirich's[42] statement that only data at finite concentration have full rheological significance. This problem should not be overlooked. To determine the intrinsic contribution one has to extrapolate to zero concentration. On the other hand, viscometric measurements at extreme dilution will hardly show shear effects, unless extreme accuracy is ensured.

The effect of the type of solvent on the UVN has been shown to be similar to its effect on intrinsic viscosity at low shear, namely higher values in better solvents, except that the differences are much smaller. Quantitative comparison is difficult, because most data are scattered and the types of polymer and ranges of concentration vary considerably.

[78] E. Passaglia, J. T. Yang, and N. J. Wegemer, *J. Polymer Sci.* **27**, 333 (1960).

[79] A. Gemant, *J. Appl. Phys.* **13**, 210 (1942).

[80] C. M. Conrad, V. W. Tripp, and T. Mares, *J. Phys. Colloid Chem.* **55**, 1474 (1951).

3. EMPIRICAL CORRELATIONS BETWEEN ULTIMATE VISCOSITY NUMBER AND MOLECULAR WEIGHT OF POLYMERIC MATERIALS

The author made an attempt,[41] reported in part by Merrill and colleagues,[12,69–72] to find direct correlations between the UVN and the molecular weight of different polymer groups. The following is a summary of the results.

1. Flexible polymers in dilute solutions (chiefly polyisobutylene and polystyrene) clearly show a second Newtonian region at high shear rates of 50,000 to 10^5 sec^{-1}.

2. These polymers show a constant (concentration-independent) UVN above a critical molecular weight of about one million. The UVN is higher in better solvents; see Fig. 2.

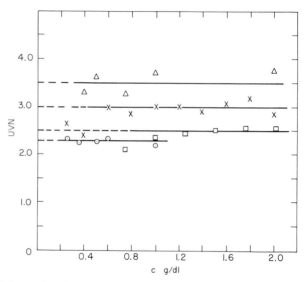

FIG. 2. Ultimate viscosity number versus concentration: polyisobutylene, molecular weight 1.1×10^6, at 30°C. △, in cyclohexane; ×, in toluene; □, in benzene; ⊙, in decaline.

3. The constant UVN may be successfully related to the molecular weight of the polymer samples, as in low-shear intrinsic viscosity:

$$\text{UVN} = K_1 M^{a_1} \qquad (24)$$

The values of a_1 lie between 0.3 and 0.5. The concentration level is 0.2 to 2.0 g per 100 cm^3. The constant UVN may serve empirical characterization. Typical results are shown in Fig. 3.

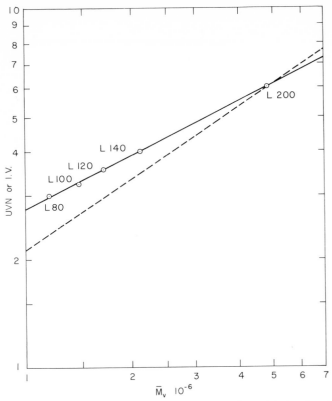

FIG. 3. Ultimate viscosity number versus molecular weight: polyisobutylene in toluene at 30°C. ——, high shear (UVN); – – –, low shear (I.V. = intrinsic viscosity).

4. Below a molecular weight of about 10^6 both polyisobutylene and polystyrene in various solvents exhibit values of the UVN that increase linearly with concentration. In the same range of concentration low-shear viscosity numbers increase exponentially. On the other hand, values of the UVN extrapolated to zero concentration roughly approximate the corresponding low-shear intrinsic viscosity.

5. When polyisobutylene of unusually high molecular weight (5×10^6 to 12×10^6) is dissolved in a viscous solvent (decalin), extrapolation to zero concentration indicates an increase in the UVN in accordance with Peterlin's theory.[54-60] A schematic description of the UVN shown by three different groups of polymers is given in Fig. 4.

The shear dependence of small particles is still in dispute. In solutions the effects of shear manifest themselves at medium concentrations. The drop in viscosity with increasing shear is quite small. Sharman and colleagues[34]

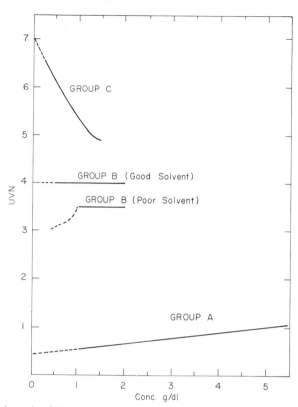

FIG. 4. Schematic of dependence of ultimate viscosity number on concentration. Group A, $\overline{M}_v < 10^6$; group B, $\overline{M}_v > 10^6$; group C, $\overline{M}_v > 5 \times 10^6$.

found that polystyrene exhibits a "threshold" molecular weight of about 700,000, below which intrinsic viscosity is not affected by shear. A minimal critical molecular weight of about 10^6 for the shear sensitivity of intrinsic viscosity has been claimed by Pao[81] and Peterlin and Copic.[17] When the small particles are approximately spherical and relatively rigid, the shear effect in dilute solutions is negligible, except where interaction occurs. The effect of the solvent itself, however, should not be disregarded. Viscous lubricants modified with polymer additives show shear effects even at low concentrations. This problem has been discussed by Philippoff,[82] who measured the viscosities and streaming birefringences of polymers in various solvents. His conclusions are as follows: "The experiments have shown that the main influence of the viscosity of the solvent is to allow one to obtain

[81] Y. H. Pao, J. Chem. Phys. 25, 1294 (1956).
[82] W. Philippoff, J. Polymer Sci. 57, 141 (1962).

larger shear stresses with polymer solutions. With low viscosity solvents and with a small intrinsic viscosity (polyisobutylene of about 200,000) no non-Newtonian effects are usually encountered. This is not due to the fact that they are non-existent, but is due to the great difficulty in applying sufficient high shear stresses to the solutions." It is obvious that the difference is due to experimental conditions only.

The empirical relationship between the UVN and molecular weight has been described, but unfortunately there are no comparable published data. It is, however, of interest to compare extrapolated data. Ree and Eyring's[30,31] calculation yields

$$\frac{\ln \eta_{r\infty}}{c} = \text{UVN} = K_1 M^{0.4} \tag{25}$$

for polystyrene in benzene. Infinite shear [Eq. (15)], on the other hand, eliminates the effect of molecular weight. Jeffery[85] arrived at a minimal intrinsic viscosity of fully oriented long rods. This value is a constant independent of chain length. For Philippoff's experiments with nitrocellulose solutions[25] the author determined the relationship between the UVN and molecular weight and found a very low exponent (UVN $\propto M^{0.14}$). Porter and Johnson[83,84] found that their calculated activation energies for polymer solutions verified a sharp drop from around 9 kcal/mole at low shear to about 3.5 kcal/mole at high shear—about the same as for the pure solvent.

In conclusion, there are two extreme expectations:

1. Fully oriented, rigid, long chains should show vanishing effects of molecular weight on the UVN (Jeffery).

2. Completely flexible polymer chains should show vanishing effects of shear on intrinsic viscosity (Peterlin). In this case the UVN and also $[\eta]$ will vary as $M^{0.7}$ to $M^{0.8}$ in good solvent. Polymer chains neither completely rigid nor totally flexible should show a lower exponent, say $M^{0.4}$. The data thus far available are insufficient for a more quantitative conclusion.

4. UTILIZATION OF HIGH-SHEAR VISCOSITY DATA FOR POLYMER MELTS

The problem of the ultimate viscosity of polymer melts has been studied by Porter and Johnson,[39,83,84] Philippoff,[86] Marker and co-workers,[87] and Bagley and West.[88] The first-named have reached very high shear rates

[83] R. S. Porter and J. F. Johnson, *J. Appl. Polymer Sci.* **3**, 200 (1960).

[84] R. S. Porter and J. F. Johnson, *J. Appl. Phys.* **32**, 2326 (1961).

[85] G. B. Jeffery, *Proc. Roy. Soc.* **A102**, 175 (1922).

[86] W. Philippoff and F. H. Gaskins, *J. Polymer Sci.* **21**, 205 (1956).

[87] L. Marker, R. Early, and S. L. Aggarwall, *J. Polymer Sci.* **38**, 381 (1959).

[88] E. B. Bagley and D. C. West, *J. Appl. Phys.* **29**, 1511 (1958).

(about 10^6 sec^{-1}) and found that molten polyethylene is shear-dependent only above a critical molecular weight indicated by melt swell and elasticity which may be interpreted as the onset of entanglement. An analysis of high-shear viscosity of melts, published by Malkin and Vinogradov,[88a] shows that $\eta_\infty \propto M$, as found experimentally. It is obvious that in polymer melts the role of entanglement is controlling. At high shear the dependence of η_∞ on M above the critical molecular weight illustrates the importance of the disentanglement process. As for the lower range, the region described by Porter and Johnson may essentially be a second Newtonian region, possibly very close to the first one. Philippoff and Gaskins[86] succeeded in establishing the first Newtonian line at lower shear rates. Recently Ballman[88b] determined the ultimate viscosity of polystyrene melts at high shear rates and found a tendency of the viscosity to increase with further increasing shear.

5. INFLECTION POINT ON THE FLOW CURVE OF POLYMER SOLUTIONS

A completely different approach to high-shear viscometry has been adopted by Umstätter[89] and Schurz.[90,91] They related the inflection point on the logarithmic plot of the flow curve, characterized by its S shape, to the molecular weight of the polymer in solution. By using the value of the shear rate at the inflection point, $\hat{D}$, they postulated an empirical equation:

$$\hat{D} = aM^{-b} \qquad (26)$$

In fact, the inflection point is often concentration-dependent. Its physical significance is still insufficiently clear. When compared with high-shear data, it does not represent the second Newtonian parameter.

IV. Instrumentation in High-Shear Viscometry

1. APPLICABILITY AND LIMITATIONS OF CONVENTIONAL EQUIPMENT

Most high-shear data have been determined with either capillary or rotational concentric-cylinder viscometers. The crucial problem in high-shear viscometry is to provide a wide range of working conditions without impairing accuracy. Any side effects should be either eliminated or carefully taken into account by means of suitable correction terms. Although the

[88a] A. Y. Malkin and G. V. Vinogradov, *J. Polymer Sci. Pt. B* **2**, 671 (1964).
[88b] R. L. Ballman, *Nature* **202**, 288 (1964).
[89] H. Umstätter, *Kolloid-Z.* **139**, 120 (1924).
[90] J. Schurz, *Kolloid-Z.* **154**, 97 (1957); **155**, 45 (1957).
[91] J. Schurz, *Intern. Symp. Second-Order Effects in Elasticity, Plasticity and Fluid Dynamics, Haifa, Israel, April, 1962.* Pergamon, London, 1964.

cone-and-plate viscometers yield accurate viscosity data, their use is limited at shear rates of more than about $20,000 \, \text{sec}^{-1}$. The concentric-cylinder viscometers yield reliable data, provided the gap is very small compared with the cylinder radius. At high shear rates various disturbances may interfere; they will be discussed later. Capillary viscometers are by far the simplest to construct, but an exact calculation of viscosity from them involves a variety of correction terms.

2. Specific Instruments for High-Shear Viscometry

Capillary high-pressure viscometers are often popular because of their relative simplicity and flexibility. The procedure and calculations involved are described elsewhere.[63,74–77a] The classical mathematical derivation of Rabinowitsch[92] permits the calculation of shear rates at the capillary wall without recourse to an assumed flow function. The shear stress is controlled by varying the external pressure and tube geometry. The main corrections, particularly important under high-shear conditions, refer to the kinetic energy, entrance effects, and heat dissipation. The first-named correction represents the equivalent pressure energy converted into kinetic energy in the flowing fluid:

$$\Delta P \, \text{eq.} = m\rho \overline{U}^2 \tag{27}$$

The constant m, named after Hagenbach, includes friction losses on contraction. In view of the high speeds involved, the kinetic energy correction is of marked significance at high shear rates.

The entrance effect, expressed by a fictitious equivalent tube length $(L + NR)$, is due to an extra pressure drop at the inlet prior to establishment of the stream profile and described as follows:

$$\tau = \frac{\Delta P}{2(L/R + N)} \tag{28}$$

where L and R are tube length and radius, ΔP applied pressure, and N a constant. It has been shown that the entrance effect also becomes more pronounced at higher shear rates. This is particularly so in polymer melts, obviously owing to elasticoviscous phenomena.

Temperature control is always critical in viscometry. At high shear heat production through friction increases correspondingly. Its rate per unit volume equals the product of shear stress and rate of shear. The possible rise in temperature should always be allowed for and corrected. A good control system and small residence time are beneficial.

Special high-shear rotary viscometers for research were designed by

[92] B. Rabinowitsch, Z. Physik. Chem. A145, 1 (1929).

Merrill[93,94] and Barber and associates.[95] They consist of concentric cylinders with extremely narrow gaps. The Barber viscometer, having several exchangeable rotors, permits variation of the gap. In work with large macromolecules the ratio of gap width to chain length must be considered. The accuracy of the point value of the viscosity decreases with increasing gap width, and an average value of shear rate must be introduced. Rotational viscometers permit direct readings and automatic recording. Because of technical considerations the inner cylinder is normally the rotor. This may entail a considerable disadvantage in the form of Taylor vortices,[96] in which case the onset of an "early transition" in simple liquids is dictated by the geometry of the viscometer. The critical Reynolds number is related to the cylinder radius R and the gap width h as follows:

$$(N_{Re})_c = 41.8(R/h)^{1/2} \qquad (29)$$

If a viscometer has a bottom, a special correction is called for, and heat generation through the dissipation process is intensified by accumulation. For this reason the rotational viscometers of the above-described type have no bottoms. Another problem in rotational viscometers is the interference of normal stresses, reflected in the climbing of elasticoviscous polymers up through the gap, which is known as the Weissenberg effect.[97] The normal stresses do not affect the shear stress reading, but they entail difficulties during operation of the viscometer.[15] A fuller description of viscometric instrumentation is given by Oka,[98] and a recent book[99] covers most of the available commercial instruments for both the low and high shear rates.

V. Structural Turbulence

1. DEFINITION AND COMPARISON WITH REYNOLDS TURBULENCE

The onset of turbulence in simple liquids in pipe flow is well defined and controlled by a lower limit of the Reynolds number:

$$N_{Re} = \frac{\bar{U}d\rho}{\eta} \qquad (30)$$

[93] E. W. Merrill, J. Colloid Sci. 9, 7 (1954).

[94] E. W. Merrill, ISA (Instr. Soc. Am.) J. 3, 173 (1956).

[95] E. M. Barber, J. R. Muenger, and F. J. Villforth, Anal. Chem. 27, 425 (1955).

[96] G. I. Taylor, Phil. Trans. Roy. Soc. A223, 289 (1923).

[97] K. Weissenberg, Nature 159, 310 (1947).

[98] S. Oka, in "Rheology" (F. R. Eirich, ed.), Vol. 3, Chapt. 2, p. 18. Academic Press, New York, 1960.

[99] J. R. Van Wazer, J. W. Lyons, K. Y. Kim, and R. E. Colwell; "Viscosity and Flow Measurement." Wiley (Interscience), New York, 1963.

The critical value for flow in pipes is about 2100. The breakdown of the laminar flow regime is due to purely hydrodynamic causes and is controlled by the ratio between inertia and viscous forces. The onset of "Reynolds turbulence" is thus universal for all fluids and should not depend on their specific nature. A simple indication of the appearance of turbulent vortices is a sudden increase in the shear stress or apparent viscosity. This may be defined by an additional viscosity term, the eddy viscosity, E, whence:

$$\tau_t = (\eta + E)\frac{dU}{dy} \tag{31}$$

where τ_t = shear stress at turbulent flow. Polymeric systems, however, exhibited often a different type of turbulence at relatively low Reynolds numbers. It was named "structural turbulence" by Ostwald and Auerbech[100] in view of its association with the structure of the system and was successfully analyzed by Reiner;[22,101] other early reports of it were made.[102–105] Although vortices were actually observed,[103] the onset of turbulence was generally established through the sharp increase in apparent viscosity. More recent data are reported by Bestul and Bryant[106] and Brodnyan and Kelley.[107] Porter and Johnson[108] observed structural turbulence in capillary jets only, whereas Ram[41,63] and Merrill and co-workers[109] observed it both in capillary and rotational viscometers. In contrast to Reynolds turbulence, structural turbulence is governed by the type of material and not by the Reynolds number, which may be relatively small. Characteristic flow curves are shown in Fig. 5.

2. CRITICAL SHEAR STRESS

Structural turbulence has been found to be characterized by a critical shear stress, which varies with the type of polymer, its molecular weight, and its concentration in solution.[63]

[100] W. Ostwald and P. Auerbech, *Kolloid-Z.* **38**, 261 (1926).
[101] M. Reiner, *Kolloid-Z.* **39**, 314 (1926).
[102] E. Hatschek and R. S. Jane, *Kolloid-Z.* **38**, 33 (1926).
[103] E. N. C. Andrade and J. W. Lewis, *Kolloid-Z.* **38**, 260 (1926).
[104] R. Schnurman, *Proc. 1st Intern Congr. Rheol., Holland, 1948* Vol. 2, p. 142. North-Holland Publ., Amsterdam, 1949.
[105] C. N. Davies, *Proc. 1st Intern. Congr. Rheol., Holland, 1948* Vol. 2, p. 152. North-Holland Publ., Amsterdam, 1949.
[106] A. B. Bestul and C. B. Bryant, *J. Polymer Sci.* **19**, 255 (1958).
[107] J. G. Brodnyan and E. L. Kelley, *Trans. Soc. Rheol.* **9**(2), 371 (1965).
[108] R. S. Porter and J. F. Johnson, *Rheol. Acta* **2**, 82 (1962).
[109] E. W. Merrill, H. S. Mickley, and A. Ram, *J. Fluid Mech.* **13**, 86 (1962).

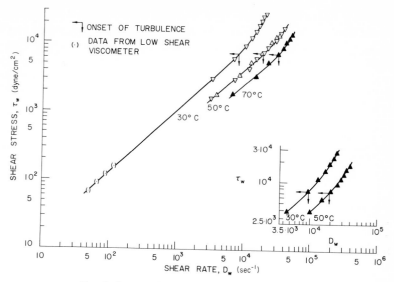

FIG. 5. Structural turbulence in polymer solutions.

A criterion of this behavior was suggested by Reiner.[22,101] If there exists a critical shear stress, its value at the wall is

$$(\tau_w)_c = \frac{\Delta P R}{2L} = \frac{4\overline{U}_c \eta}{R} = \frac{4\overline{U}_c \eta_a}{R} \frac{3n' + 1}{4n'} \tag{32}$$

where

$$n' = \frac{d \ln \tau_w}{d \ln(4\overline{U}/R)}$$

Therefore, at the onset of turbulence the critical velocity $\overline{U}_c$ is proportional to the capillary radius R:

$$\overline{U}_c \propto R$$

$$\overline{U}_c \propto 1/\eta$$

The critical Reynolds number is strongly dependent on the capillary radius and is not represented by a constant value.

$$(N_{Re})_c \propto U_c R \propto R^2$$

On the other hand, Reynolds turbulence is characterized by a critical Reynolds number,

$$(N_{Re})_c = \frac{\overline{U}_c 2R\rho}{\eta} \tag{33}$$

274   ARIE RAM

whence

$$U_c \propto 1/R$$

$$U_c \propto \eta$$

The relationship between critical velocity and capillary radii for various polyisobutylene solutions and also for the pure solvent are shown in Fig. 6. It is clear that structural turbulence is controlled by the shear stress, whose critical value was found in the author's experiments to decrease with increasing molecular weight or dilution.

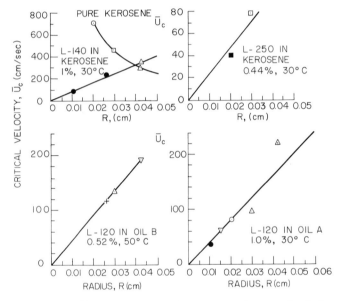

FIG. 6. Critical velocity $\overline{U}_c$ versus capillary radius R at onset of structural turbulence: criterion for structural turbulence or Reynolds turbulence. Solutions of polyisobutylene.

3. PHYSICAL INTERPRETATION OF STRUCTURAL TURBULENCE

The physical significance of structural turbulence is not yet clear. If it were a regular type of turbulence, it should entail eddies, irregular flow and increased momentum transfer. In most cases, however, only an increase in momentum transfer is observed. An analysis of flow data[63] indicates that structural turbulence most probably occurs in dilute solutions of long flexible chains under increasing shear stress, the entangled macromolecular structure being gradually disrupted when the chains tend to uncoil and stretch. Vortices, too, may be caused by elastic cycles of the springlike rotating coils, as has been mentioned. Thus, a study relating flow elasticity to turbulence is called for. Direct physical proof of the existence of vortices is still unavailable. The changes in flow profile provide little evidence, because the profiles of both laminar-pseudoplastic and general turbulent flow are flat.

4. Hydrodynamic Instability in Flow of Pseudoplastic Liquids

The Reynolds number is introduced either through a rigorous solution of the basic flow equations (Navier–Stokes) or through application of the engineering approach of dimensional analysis. Although the importance of a single flow-regime criterion cannot be underestimated, it is very difficult to predict the numerical value of the critical N_{Re} of even simple liquids.

Hydrodynamic stability has been discussed by Lin,[110] and several instability criteria of non-Newtonians have been suggested by various authors.[111-116] These are essentially modifications of the regular N_{Re}. Dodge and Metzner[116] have suggested the following N_{Re} for non-Newtonians:

$$N'_{Re} = \frac{d^{n'} \overline{U}^{2-n'} \rho}{g_c K'(8)^{n'-1}}$$

(34)

assuming f (Fanning friction coefficient) is $16/N'_{Re}$ and τ_w is $K'(8\overline{U}/d)^{n'}$. Other investigators favor a direct replacement of the viscosity term in N_{Re} by its proper value, either η_a or η_∞ (Thomas[117]). Ryan and Johnson[115] have adopted another stability criterion by referring to local disturbances. This criterion goes to zero both at the center of the tube and at the wall:

$$Z = \frac{R\rho U}{\tau_w} \cdot \frac{dU}{dy}$$

(35)

The maximal value of Z becomes $(4/27)^{1/2}(d\overline{U}\rho/\eta)$ and is therefore related to the critical Reynolds number. A similar group of criteria, more general in nature, has been suggested by Hanks.[111,112] These stability criteria have been utilized with non-Newtonians with some success. It is safe to say,

[110] C. Lin, "Theory of Hydrodynamics Stability." Cambridge Univ. Press, London and New York, 1955.
[111] R. W. Hanks, *A.I.Ch.E. J.* **9**, 45 (1963).
[112] R. W. Hanks and E. B. Christiansen, *A.I.Ch.E. J.* **8**, 467 (1962).
[113] D. C. Bogue, *Ind. Eng. Chem.* **51**, 874 (1959).
[114] J. N. Kapur and S. Goel, *Appl. Sci. Res. Sect. A* **11**, 304 (1963).
[115] N. W. Ryan and M. M. Johnson, *A.I.Ch.E. J.* **5**, 433 (1959).
[116] W. W. Dodge and A. B. Metzner, *A.I.Ch.E. J.* **5**, 189 (1959).
[117] D. G. Thomas, *A.I.Ch.E. J.* **6**, 631 (1960).

however, that hydrodynamic criteria will predict flow instability only if they consider local ratios of inertia to viscous forces.

5. FLOW BEHAVIOR OF POLYMER SOLUTIONS UNDER REYNOLDS TURBULENCE

The onset of Reynolds turbulence has been claimed to occur at values of modified N_{Re} either close to or exceeding the general critical level. This "delayed" Reynolds turbulence presents a contrast to the "early" structural turbulence at low N_{Re} and high shear rates. The idea of a delayed turbulence is associated with a pronounced decrease in the Fanning friction co-efficient f on the addition of small portions of a polymer to simple liquids (usually water) under turbulent flow. The reason for this phenomenon (which is of large practical importance) is still in dispute.[14,118–125] According-ing to Thomas,[118] it is well established that non-Newtonians flowing in round pipes always show greater turbulent-friction coefficients than do non-Newtonians in laminar flow, but they show smaller ones than do Newtonians in turbulent flow. The basis of comparison usually is some modified Reynolds number. Beyond the last general statement there is little agreement among investigators, and it has been suggested that at least one additional parameter is required for a complete definition of a non-Newtonian liquid in turbulent motion. Shaver and Merrill[14] and Metzner and associates[119,120] have made extensive use of N'_{Re} described by equation (34) and have succeeded in predicting a general equation and in making correlations between N'_{Re} and the friction coefficient in the turbulent flow regime by means of the parameter n', identical with the exponent n of the power law whenever the latter holds. New findings on ultraefficient drag reducers[121,125–126a] show that the existing hydrodynamic parameters no longer suffice for such predictions. Elasticoviscous materials have been shown to be most efficient in drag reduction; hence, the latest approaches postulate the change of kinetic energy of turbulence into forms of elastic energy. It is thus interesting that both two phenomena, structural turbulence under high shear conditions (combined with low Reynolds numbers) and a reduced friction coefficient at high Reynolds numbers (under ordinary

[118] D. G. Thomas, *A.I.Ch.E. J.* **8,** 266 (1962).
[119] A. B. Metzner and J. C. Reed, *A.I.Ch.E. J.* **1,** 434 (1955).
[120] A. B. Metzner, *Ind. Eng. Chem.* **49,** 9 (1957).
[121] A. B. Metzner and M. G. Park, *J. Fluid Mech.* **20,** 291 (1964).
[122] A. T. Ippen and C. Elata, M.I.T. Hydrodynamics Lab. Tech. Rept. No. 45 (1960).
[123] D. C. Bogue and A. B. Metzner, *Ind. Eng. Chem. Fundamentals* **2,** 143 (1963).
[124] D. M. Eissenberg and D. C. Bogue, *A.I.Ch.E. J.* **10,** 723 (1964).
[125] J. G. Savins, *Soc. Petrol. Engrs. J.* **4,** 203 (1964).
[126] C. Elata, J. Lehrer, and A. Kahanovitz, *Israel J. Tech.* **4,** 87 (1966).
[126a] A. Ram, E. Finkelstein, and C. Elata, *Ind. Eng. Chem.* in press.

shear conditions), are supposed to be both controlled by elasticoviscous behavior.

6. ROLE OF ELASTICITY

In view of the shortcomings of the simple and modified Reynolds numbers for predicting structural turbulence or other anomalies (melt fracture, drag coefficient reduction) in polymer flow the concept of "elastic turbulence" has been introduced.[127,128] This phenomenon occurs when the increase in elastic energy is faster than the decrease in viscous loss for elasticoviscous fluids. Some ratio between the elastic and viscous forces will provide an elastic Reynolds number characterized, for example, by the ratio between normal stress $P_{11} - P_{22}$ and shear stress τ; this will involve the shear modulus and the relaxation times of the polymer segments. Though Metzner[121] succeeded in measuring normal stresses in polymer melts and concentrated solutions, considerable difficulties are still encountered with dilute solutions. Thomas and Walters[129] point out that elasticity of the liquid reduces the critical Taylor number at which instability sets in. The role of elasticity has acquired importance in the correction of entrance phenomena in capillary flow for which an elastic term becomes predominant with increasing shear.[130,131] Vinogradov and colleagues[128] described the elastic turbulence criterion as determining the stability of flow. An elastic Reynolds number $N_{Re(e)}$ is defined, which depends on the interrelation between the elastic force and viscous friction:

$$N_{Re(e)} = \frac{\beta R}{\bar{U}\lambda^2} \tag{36}$$

where $1/\lambda$ is equal to D and β is the relaxation time. Therefore,

$$N_{Re(e)} = \frac{\beta R D^2}{\bar{U}} \tag{37}$$

Relaxation times may be calculated with the aid of Rouse's theory[132]

$$\beta = \frac{6(\eta_0 - \eta_s)}{\pi^2 n' k T} \tag{38}$$

where n' is the polymer concentration in molecules per unit volume and k is Boltzmann's constant. Equation (37) may be used to predict the critical elastic Reynolds number for the onset of structural turbulence. It may also

[127] W. F. Busse, *Phys. Today* **17**, 32 (1964).
[128] G. V. Vinogradov, A. Y. Malkin, and A. I. Leonov, *Kolloid-Z.* **191**, 25 (1963).
[129] R. H. Thomas and K. Walters, *J. Fluid Mech.* **18**, 33 (1964).
[130] E. B. Bagley, *J. Appl. Phys.* **28**, 624 (1957).
[131] D. L. Johnson and E. Baer, *J. Appl. Polymer Sci.* **7**, 1359 (1963).
[132] P. E. Rouse, *J. Chem. Phys.* **21**, 1272 (1953).

be used to predict the dependence of the critical conditions on molecular weight and concentration.

7. MELT FRACTURE OF POLYMERS UNDER HIGH-SHEAR STRESSES

Melt fracture is directly related to structural turbulence. Extensive analyses of the phenomenon are available elsewhere,[133–139] and a special chapter by Tordella will be included in Volume V. It begins in extruded melts with roughness and waviness at some critical shear stress and ends at higher stresses in a total fracture of the extrudate. Apart from its theoretical interest, it obviously presents a major difficulty in plastics processing. Because of its resemblance to the "frozen profile" of turbulence, it has been associated with a state of onset of instability. The critical N_{Re} involved are, however, low to the point of insignificance, which rules out hydrodynamic turbulence. Elastic interference is so closely involved in these flow cases that an elastic turbulence, somewhat like structural turbulence in dilute polymer solutions, has been proposed. It should be emphasized that, whereas structural turbulence is characterized by an increase in apparent viscosity, melt fracture shows no such increase but rather gross changes in the shape of the extrudate. Philippoff and Gaskins[140] and Bagley[141] analyzed the elastic component S_R, called the "recoverable elastic shear strain." When Hooke's law of shear holds, then

$$\tau = GS_R \tag{39}$$

where G is the shear modulus. Bagley found that under high shear stress Hooke's law is exceeded at some critical value $(S_R)_c = 7$; this level of shear strain leads to the onset of fracture. Elastic turbulence is caused by uncoiling and stretching of polymer chains, and the similarity between melt fracture and structural turbulence lies in the fact that both phenomena are controlled by critical shear stresses (or, possibly, normal stresses) that are inversely proportional to the molecular weight of the polymer. This constant relationship between τ_c and molecular weight has been utilized for polymer characterization.

[133] J. P. Tordella, *Trans. Soc. Rheol.* **1**, 203 (1957).

[134] J. P. Tordella, *J. Appl. Phys.* **7**, 454 (1956).

[135] E. R. Howells and J. J. Benbow, *Plastic Inst.* (*London*) *Trans. J.* **30**, 240 (1962).

[136] J. J. Benbow and P. Lamb, *SPE* (*Soc. Plastics Engrs.*) *Trans.* **3**, 7 (1963).

[137] A. P. Metzger, C. W. Hamilton, and E. H. Merz, *SPE* (*Soc. Plastics Engrs.*) *Trans.* **3**, 21 (1963).

[138] E. B. Bagley, *Trans. Soc. Rheol.* **5**, 203 (1960).

[139] J. F. Hutton, *Nature* **200**, 646 (1963).

[140] W. Philippoff and F. H. Gaskins, *Trans. Soc. Rheology* **2**, 263 (1958).

[141] E. G. Bagley, *Trans. Soc. Rheol.* **5**, 355 (1961).

The critical elastic Reynolds number proposed by Vinogradov and colleagues[128] in equation (36), may be modified with the aid of $\beta = \eta/G$ (relaxation time for a Maxwell body). In that case

$$(\mathrm{Re})_e = D\eta/G = \tau/G = S_R \qquad (40)$$

Replacing D in equation (40) with $\bar{U}/R$, one has, under critical conditions,

$$\bar{U}_c \propto R$$

Since $G \propto 1/M$, where M is molecular weight, then $\tau_c \propto 1/M$. The ratio of normal stress to shear stress has not, however, been found to have a critical constant.

White[142] has introduced a new dimensionless number, named the Weissenberg number N_{We}. This dimensionless ratio is significant in fluids in which a nonlinear relationship between shear stress and shear rate is observed. These fluids possess a "memory" of their deformation history and exhibit recoil on an abrupt arrest of flow. The Weissenberg number is defined by the ratio of elasticoviscous forces to viscous forces as follows:

$$N_{\mathrm{We}} = \mathscr{J}\eta \, \bar{U}/L \qquad (41)$$

where $\mathscr{J}$ is the steady-state shear compliance of linear elasticoviscosity. It can be easily shown that this equation is also reducible to the critical elastic strain S_R.

Another type of instability was analyzed by Hutton,[139] who used silicone fluids and cone-and-plate viscometers. He observed a pronounced Weissenberg effect, that is, a climb-up of the liquid through normal stresses under high shear. The critical conditions for fracture are

$$fGS_R^2 V/2 > 2\pi R^2 \sigma \qquad (42)$$

where f = fraction of elastic energy converted into surface energy
V = volume of sample
R = radius of cone
σ = surface tension of liquid

The shear modulus G and the recoverable elastic shear strain S_R are related to the shear stress τ and normal stress $P_{11} - P_{22}$ as follows:

$$G = \frac{\tau^2}{P_{11} - P_{22}}, \qquad S_R = \frac{P_{11} - P_{22}}{\tau} \qquad (43)$$

[142] J. L. White, J. Appl. Polymer Sci. 8, 2339 (1964).

They predict a critical normal stress in a cone-and-plate viscometer when

$$(P_{11} - P_{22})_c \geqslant (6/f)\sigma/R\varphi \tag{44}$$

where φ is the cone angle. This criterion depends strongly on the geometry of the viscometer and very slightly on the molecular weight of the polymer (dictated by almost constant surface tension). The uncertainty with regard to the normal-stress data limits the applicability of this simple criterion.

8. PSEUDODILATANCY OF DILUTE POLYMER SOLUTIONS

The possibility of an increase in the intrinsic viscosity of flexible polymer chains with increasing shear was extensively discussed by Peterlin and co-workers.[17,54–60] Selby[143] calls the increase "pseudodilatancy." The flow curve shows initial shear thinning at low rates of shear and then an upward turn to "pseudodilatancy." Hitherto dilatancy has been associated with the volume dilatation of concentrated solutions, which "dry" on shearing. Metzner and Whitlock[144] analyzed the rheological dilatancy of concentrated solutions and found it may be independent of volume dilation.[144a] Most data on dilatancy, however, refer to highly concentrated solutions, whereas Peterlin's work deals with the flow of extremely dilute solutions. His theoretical and experimental results may be summarized as follows.[145] The hydrodynamic interaction between the polymer chain and its environment leads to rotation of the chains which with increasing shear rates is accompanied by periodical extension and folding. Changing relative distances between the chain segments the hydrodynamic resistance coefficient may decrease by orientation and energy storage, but the positive short-range contribution of adjoining segments becomes eventually more significant when the deformation process favored over the alignment by higher shear rates. With the use of highly viscous solvents high shear stresses are obtainable at lower shear rates. The conditions for pseudodilatancy thus are dilute solutions, long chains (complete flexible), and viscous solvents. There is considerable similarity between the flow curves of structural turbulence and those of pseudodilatancy. In the former case the increase in viscosity was not confined to viscous solvents but was also observed in kerosene solutions of polyisobutylene. Peterlin's theory is inapplicable in that case, but the increase may be due to a different competition between orientation and deformation in the shear of flexible chains.

[143] T. W. Selby, *Am. Soc. Testing Mater. Spec. Tech. Publ.* **299** (1964).
[144] A. B. Metzner and M. Whitlock, *Trans. Soc. Rheol.* **2,** 239 (1958).
[144a] See also Chapter 5.
[145] S. P. Burow, A. Peterlin, and D. T. Turner, Reprint No. 46, Camille Dreyfus Laboratory, Durham, North Carolina, March, 1964.

VI. Polymer Degradation under High-Shear Conditions

1. IRREVERSIBLE VISCOSITY DROP IN POLYMER MELTS AND SOLUTIONS

The irreversible decrease in polymer viscosity after exposure to high-shear conditions is usually associated with chain scission. As stated before, this decrease may as well be due to other causes, such as irreversible disentanglement or dissociation. It is therefore preferable to establish degradation by checking for a change in the intrinsic viscosity. Moreover, an absolute method, such as light-scattering, may yield better information regarding the change in average dimensions of polymer chains. Further, chemical analysis may reveal the presence of fresh degradation products, such as free radicals, monomers, or other short chains. A thorough checking for polymer degradation remains therefore indispensable.

2. SHEAR CONDITIONS AFFECTING DEGRADATION

In general, degradation is regarded as the reverse reaction of polymerization, that is, depolymerization. As such it appears as a rate process, controlled by polymer concentration and energy state. Like every chemical reaction, the resultant state depends on the duration of the process before equilibrium is established. Theories of random degradation[146,147] presume equal strength of, and accessibility to, all chain links, irrespective of their length. The rate of chain degradation, however, depends on the number of links, that is, on the molecular weight. Among other effects inducing degradation are severe flow conditions. Degradation due to beating and high-speed stirring has been quoted.[148–152] In the experiments conducted with dilute polymer solutions turbulent conditions were assumed to concentrate the strain within the polymer chain by intensive vibration or collapse of cavities. The degradation of polymer melts or concentrated solutions under conditions of purely laminar flow and high shear also has been reported.[145,153–157] In view of the high energy required for disrupting

[146] N. Grassie, "Chemistry of High Polymer Degradation Process." Wiley (Interscience), New York, 1956.
[147] H. H. G. Jellinek, "Degradation of Vinyl Polymers." Academic Press, New York, 1955.
[148] E. W. Merrill, H. S. Mickley, and A. Ram, J. Polymer Sci. 62, S109 (1962).
[149] P. Alexander and M. Fox, J. Polymer Sci. 12, 533 (1954).
[150] M. R. Fenske and E. E. Klaus, Lubrication Eng. 2, 101 (1955).
[151] W. R. Johnson and C. C. Price, J. Polymer Sci. 45, 217 (1960).
[152] F. Rodriguez and C. C. Winding, Ind. Eng. Chem. 51, 1281 (1959).
[153] A. Tamir, M.S. thesis, Israel Institute of Technology, Haifa, 1963.
[154] R. S. Porter and J. F. Johnson, J. Phys. Chem. 63, 202 (1959).
[155] A. B. Bestul, J. Appl. Phys. 25, 1069 (1954).
[156] A. B. Bestul and H. V. Belcher, J. Appl. Phys. 24, 1011 (1953).
[157] P. Goodman, J. Polymer Sci. 25, 325 (1957).

primary valence bonds (of the order of 80 kcal/mole) "weak spots" in a polymer chain are often postulated. The flowing medium exerts a drag on the polymer chain, which in turn depends on the shear stress and extended chain surface and is increased by instability and turbulence. Long chains, being spread over varying flow regions, are stretched and twisted according to local velocity gradients; the weak spots, being the junctions of entangled segments, are regarded as the loci of strain concentration. The response of the polymer chain to increasing shear depends on the rigidity of the structure. The entangled segments may either slip, one over the other, or tear apart. Thus, disentanglement and degradation are competing reactions to shearing forces, the final reaction depending on the ratio of residence time to relaxation time of the polymer segment. Most evidence favors the presence of points of entanglement as premise for effective degradation. As shown in Fig. 7,[148] shear degradation occurs in very high molecular-weight species and is arrested approximately at an asymptotic level independent of the initial value. It is believed that at the final stage no more weak spots of entanglement exist. Pohl[158] has proposed the following

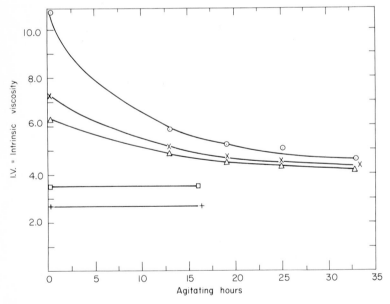

FIG 7. Degradation of polymer solutions upon agitation: polyisobutylene in toluene at 30°C. L300, ⊙; L250, × ; L200, △; L140, □; L100, +.

[158] H. A. Pohl and J. K. Lund, *SPE* (*Soc. Plastics Engrs.*) *J.* **15,** 390 (1959).

index for chain breakdown:

$$B = \tau M^2 \tag{45}$$

where M is the molecular weight and B denotes the critical conditions for degradation. This relationship has been confirmed by Bueche.[32]

The effect of temperature on the degradation process is also of interest. Although disentanglement should be favored at elevated temperatures, degradation has been found to increase with decreasing temperature presumably by way of increased stress concentrations.[158] Another finding is that the heat generated by dissipation does not initiate degradation. The different effects of various solvents are also of importance.[152] The viscosity of the solvent affects the shear stress, and the interaction between solvent and polymer ("goodness" of the solvent) affects both chain dimension and disentanglement under shear. An additional problem is the chemical nature of the solvent and its ability to participate in chain transfer with the new active chains generated by the scission process. These latter aspects of polymer degradation under shear have not yet been analyzed. For these reasons degradation is one of the main difficulties encountered in high-shear viscometry. Any viscometric data under high-shear conditions are useless, unless degradation is avoided.

Nomenclature

c	Concentration	R	Radius
d	Diameter of tube	R	Constant of ideal gas
D	Rate of shear	S	Strain
E	Eddy viscosity	T	Temperature
f	Fanning friction coefficient	U	Velocity
G	Shear modulus	V	Volume
h	Width of the gap between concentric cylinders	Z	Number of segments
		Z	Stability criterion
J	Frequency of jump	β	Relaxation time
$\mathscr{J}$	Shear compliance	δ	Shear volume of jumping unit
k	Constant of Boltzmann		
K	Parameter of power law	η	Viscosity (Newtonian)
L	Length	η_a	Apparent viscosity
m	Coefficient of Hagenbach	η_0	Viscosity in the first Newtonian region
M	Molecular weight		
n	Exponent of power law	η_∞	Viscosity in the second Newtonian region
N	Avagadro's number		
N	Entrance correction	η_r	Reduced viscosity
P	Pressure	η_s	Viscosity of pure solvent
$P_{11} - P_{22}$	Normal stress	$[\eta]$	Intrinsic viscosity
r	Distance	λ	Distance of jump

$\lambda_1, \lambda_2, \lambda_3$	Dimensions of jumping unit	σ	Surface tension
ρ	Density	τ	Shear stress
		ϕ	Volume fraction
SUBSCRIPTS		φ	Angle of cone
c	Critical	t	Turbulent
e	Elastic	w	Wall

Chapter 4

RHEOLOGICAL ASPECTS OF THE MIXING OF PLASTICS COMPOUNDS

James T. Bergen

I. Introduction

1. Objective of the Mixing Operation

The mixing of plastics compounds is recognized as a process operation of great importance in the commercial production of useful articles from high polymers. The effectiveness of succeeding process operations, the properties of the final product, and its cost often depend upon the achievement of a suitable mixture of the components. The mixing consists of incorporating or blending into a polymer matrix of ingredients such as colorants, reinforcing fillers, inert fillers, chemical additives, and liquid plasticizers.

Gases may be dispersed in a polymer by mixing, as in the production of a plastic foam. Not infrequently several polymers may be massed or blended together by mixing.

Ideally, the objective of these many mixing operations is to achieve a uniform spatial distribution of one or more additives within the three-dimensional matrix of the polymeric material. The plastics compounds may consist of many different ingredients (rubber compounds, for example, may contain as many as fifteen). It is more convenient to consider from a rheological point of view that the operation amounts to a concurrent mixing of each component with the remainder of the system. Any mixing process may thus be pictured as a two-component operation.[1]

2. BACKGROUND CONSIDERATIONS

Commercial mixing operations have developed largely as an art; the mixing of rubber compounds was practiced about a century ago. Only in the past twenty years have serious attempts been made to investigate the fundamentals. A very large body of literature on the subject exists, in which all phases of the general topic of mixing are treated. The most extensive systematization of this literature has taken place within the field of chemical engineering, where mixing is treated as a unit process. Regular reviews (see, for example, Oldshue[2]) indicate that rheological considerations are widely employed in these discussions and that various levels of technological accomplishment have been reached. The development of a theory of the mixing of plastics compounds is a special area of this effort and one in which progress has lagged considerably behind practice, so that technology is frequently found to be lacking in fundamental basis.

The purpose of this discussion is to describe the theory of the ideal mixing process for plastics as it exists today and to point out some areas of technology in which basic work still remains to be done. It will become evident that many of these problems must depend upon advances in rheology for their solution.

II. Definition of a Mixture

1. DISCRETE PARTICLES OF TWO COMPONENTS

To envision a mixture conveniently, let us consider the simple representation of a two-component mixture of discrete particles, Fig. 1. Figure 1(a) is a tray containing equal numbers of black and white marbles in a complete state of segregation.

---

[1] W. D. Mohr, in "Processing of Thermoplastic Materials" (E. C. Bernhardt, ed.), Chapt. 3. Reinhold, New York, 1959.

[2] J. Y. Oldshue, Ind. Eng. Chem. **57**, No. 11, 115 (1965).

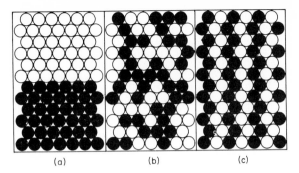

FIG. 1. Simple mixing of solids: (a) segregated components; (b) example of random mixture; (c) example of ideal, or special, mixture.

a. Random Mixture

If the marbles are stirred or shaken sufficiently and then poured back onto the tray, an arrangement as in Fig. 1(b) might be observed. This is only one of many similar configurations that would have approximately equal probability of occurrence. Hence, such arrangements may be referred to as a random mixture or an arrangement corresponding to a maximum "entropy" of the system. Local variations in composition are evident; so a random mixture is seen to result often in local clustering. Triangular blocks of 10, say, will show a variable ratio of black to white marbles.

b. Ideal Mixture

When triangular blocks of 10 include the same proportion of black and white marbles as is present in the overall mixture, the arrangement represents an "ideal" mixture in the sense that the spatial distribution of black and white marbles is uniform. Such a special mixing is depicted in Fig. 1(c) and cannot readily be achieved by a random mixing process, since it is clear that the probability of its occurrence in a random process is very small.

2. BASIC CONCEPTS IN DESCRIBING A MIXTURE

a. Random versus Systematic Mixtures

The random mixture exemplified by Fig. 1(b) resulted from a random process, in this case random collisions between marbles. The ideal mixture of Fig. 1(c) could (except for a highly improbable random event) be expected to result only from a systematic process. As will be seen later, this distinction forms a basic concept in the consideration of the mixing of plastics.

b. Scale of Scrutiny

The basis upon which the uniformity of spatial distribution of components in Fig. 1 is seen to depend is the dimension of the sample or region to be included in the examination of the mixture. This dimension has been designated the "scale of scrutiny" by Danckwerts[3] and is a characteristic length representing the minimal size of regions of examination that would cause the mixture to be imperfect in terms of the purpose for which it is intended. In Fig. 1, for example, a scale of scrutiny involving triangular areas of less than 4 marble diameters on a side represents a limit, below which no arrangement would be found to represent an ideal mixture.

c. Scale of Segregation

Most simply, the scale of segregation is a dimension that characterizes the size of clumps of an unmixed component in the mixture. In a random mixture this dimension would be a random variable that could be expressed either as a mean or as an upper limit occurring with some small but finite probability. In a systematic mixture, such as that in Fig. 1(c), it may be conceived of as the distance between the alternate layers of black or white marbles. This quantity serves as one parameter by which a mixture may be defined.

Another parametric quantity that serves to define a mixture is the "intensity of segregation," proposed by Danckwerts.[3] It is a measure of the deviation from the mean of all samples in the mixture and may be estimated statistically from a smaller number of samples chosen at random from the mixture. A convenient measure of the intensity of segregation is the coefficient of variation, or ratio of standard deviation to the mean, which is described more fully by Mohr[1] and by Moore.[4] For the purpose of this discussion, as will be developed later, it is assumed that the intensity of segregation is unchanged in the ideal mixing process of plastics.

These concepts are basic to the description of a mixture in the discussion to follow. They apply in general to all mixtures, and their special significance in the mixing of plastics will be developed in relation to the theory of the ideal mixing process.

III. Laminar-Flow Mixing Theory

1. DEFINITION OF LAMINAR-FLOW MIXING

The process of mixing in the field of chemical and engineering technology, as well as the theories related to it, depends primarily on randomizing

---

[3] P. V. Danckwerts, *Research (London)* **6**, 355 (1953).
[4] W. R. Moore, *Plastics Inst. (London) Trans. J.* **32**, 247 (1964).

mechanisms to achieve a suitable state of mixture. The random collision of solid particles or gas molecules, turbulent flow of gases and liquids, molecular diffusion, thermal convection, and Brownian motion of colloidal particles are examples. Even in the case of rather viscous fluids ($\eta \doteq 200$ poise)* local turbulence accounts for ultimate mixing,[5] and the mixing of glass in glass tanks and open channels is attributed mainly to diffusion and thermal convection.[6] In the latter case the viscosity of the glass mixture is extremely high, and the time period for mixing extends over many hours. These mechanisms of mass transport are more or less well understood and are treated at length in the literature.[2]

Current theories of the mixing of plastics differ from other mixing theories in one important respect: the ultimate mechanism of mass transport is restricted to the laminar streamline flow of a substantially incompressible fluid. It is assumed that any flow in a polymeric plastics mixture is far removed from the turbulent region, such that any vortex motion is absent and all inertial effects are inconsequential. Random collisions of solid particles are effective only over negligibly small distances; diffusion transport is assumed to be so slow or to be confined to such small effective distance that it can be disregarded.

Thus, the question with which the theoretical treatment of the mixing of plastics compounds is concerned is how to achieve a mixture of initially segregated components solely through the application of laminar streamline flow. It is immediately clear that, since the flow is specifically laminar or systematic, any mixture resulting from it must be systematic rather than random. This important concept is fundamental to all discussions of the mixing of plastics compounds.

Recalling the basic concepts employed to describe any mixture (see Section II,2,a,b), the objective of laminar-flow mixing can be defined as the reduction of the initial scale of segregation of the mixture to a measure comparable to the scale of scrutiny applied, by a process of laminar-flow deformation alone.

2. CRITERIA OF LAMINAR-FLOW MIXING

a. Increase of Interfacial Area

In considering the mixing of solid particles Brothman et al.[7] observed that the increase in degree of mixedness was related to the increase in interfacial area between the components of the mixture. Spencer and Wiley[8] reached the

---

* Here $\doteq$ signifies "of the order of."

[5] J. P. Gray, Chem. Eng. Progr. 59, No. 3, 55 (1963).
[6] A. R. Cooper, J. Am. Ceram. Soc. 42, 93 (1959).
[7] A. Brothman, G. N. Wollman, and S. M. Fellman, Chem. Met. Eng. 52, No. 4, 102 (1945).
[8] R. S. Spencer and R. M. Wiley, J. Colloid Sci. 6, 133 (1951).

same conclusion regarding the mixing of very viscous liquids by laminar-flow deformation. They stated that the mixing process must create sufficient new interfacial surface to satisfy the "condition of apparent continuity," which is equivalent to stating that the scale of segregation must be reduced to a measure less than that of the scale of scrutiny. They pictured the total mixture as divided into equal cells or cubes of edge dimension h; the mixing process must cause some of the final interface between components to pass through each cube. Furthermore, if h is one half of the scale of scrutiny, the condition of apparent continuity is fulfilled. Suppose that initially segregated material, Fig. 2(a), is being "mixed" by a laminar flow process to yield the configuration shown in Fig. 2(b). Nowhere in the mass is it now possible to distinguish segregated material in terms of the scale of observation.

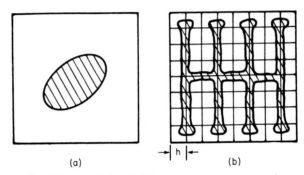

(a)   (b)

FIG. 2. Increase in interfacial area between components.[1]

Spencer and Wiley show that for unidirectional shear, or laminar flow, in the direction x_1 of the coordinate system (x_1, x_2, x_3) the material may be considered to be deformed into more or less plane, parallel sheets during the mixing process. The interfacial surface area, s, will increase in relation to the original surface area. Thus according to the equation[8]

$$(s/s_0)^2 = 1 - 2(\partial u_1/\partial x_2)\cos \alpha_1 \cos \alpha_2 + (\partial u_1/\partial x_2)\cos^2 \alpha_1 \qquad (1)$$

where $\cos \alpha_i$ is the direction cosine to coordinate i of the normal to the original surface at point (x_1, x_2, x_3), u_1 is the x_1 component of the displacement vector at the point (x_1, x_2, x_3), and $\partial u_1/\partial x_2$ is the gradient of the x_1 component of the displacement vector in the x_2 direction, a measure of the shearing strain, and

$$u_2 = u_3 = 0$$

$$\partial u_1/\partial x_1 = \partial u_1/\partial x_3 = 0$$

The application of equation (1) can be visualized by considering a system consisting of two very viscous liquids (of equal viscosity), in which the minor,

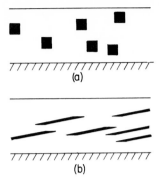

FIG. 3. Mixing action of fluid shear. (After Mohr et al.[9]).

dark-colored component is present as randomly distributed, discrete cubes; see Fig. 3(a). If the upper boundary is displaced to the right, the cubes will be deformed, as in Fig. 3(b), with a resultant increase in interfacial surface area. If, for the sake of simplicity, the cubes are oriented with edges along the axes (x_1, x_2, x_3), the direction cosines will take the numerical values

$$\cos \alpha_i = 1, \qquad i \neq j, k$$
$$\cos \alpha_i = 0, \qquad i = j, k \tag{2}$$

On substitution of equation (2) into equation (1) the following values of interfacial surface area, s, of the faces of the primer cubes with respect to the initial values, s_0, will hold:

$$x_1\text{-}x_2 \text{ plane:} \qquad (s_{1,2}/s_0)^2 = 1$$
$$x_2\text{-}x_3 \text{ plane:} \qquad (s_{2,3}/s_0)^2 = 1 + (\partial u_1/\partial x_2)^2 \tag{3}$$
$$x_1\text{-}x_3 \text{ plane:} \qquad (s_{1,3}/s_0)^2 = 1$$

Equations (3) state that the deformed surface area is proportional to the undeformeo surface area; the proportionality depends upon the orientation of the initial interface with respect to the direction of shear. The maximal increase in interfacial area and the maximal degree of mixing will occur when the interface cuts perpendicularly across the shear displacement vectors.[8]

b. Reduction of Striation Thickness

In Fig. 4(a) the initial interfaces are oriented normal to the shear displacement vectors, and the distances between interfaces is r_0. If the upper surface is displaced by a distance L, the angle ϕ diminishes from its original value,

[9] W. D. Mohr, R. L. Saxton, and C. H. Jepson, Ind. Eng. Chem. 49, 1857 (1957).

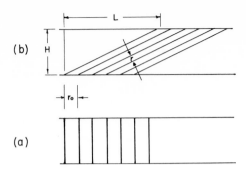

Fɪɢ. 4. Reduction in striation thickness, r, from initial striation thickness, r_0. (After McKelvey.[10])

and it may be seen in Fig. 4(b) that

$$r = r_0 \sin \phi \qquad (5)$$

Since $\tan \phi = H/L$, then, when $L \gg H$,

$$\tan \phi \doteq \sin \phi \doteq \phi$$

so that

$$r = r_0 \phi = r_0(H/L) \qquad (6)$$

Thus, in any practical case involving large deformation the distance between interfaces, r, varies inversely with the amount of deformation or strain.

The dimension r was termed "striation thickness" by Mohr et al.[9] and was given the physical significance of being the characteristic distance between lamellae or striations of laminar-flow mixtures. It is thus a physical measure of Danckwerts' scale of segregation. Furthermore, Spencer and Wiley[8] showed that the average striation thickness may be computed from the ratio of total interfacial area between the components, S, and the total volume of the system, V:

$$r = \frac{2}{S/V} \qquad (7)$$

The average striation thickness may be determined by visual observation of samples of a mixture at various stages of mixing. This provides a more convenient measure of the scale of segregation than does the interfacial area between components. The mixing process is thus seen by equation (7) to result in a reduction of the striation thickness concurrent with the increase in interfacial area, as discussed in Section III,2,a.

[10] J. M. McKelvey, "Polymer Processing," Chapt. 12. Wiley, New York, 1962.

c. Deformation by Simple Shear

An example of flow in simple shear occurs between rotating concentric cylinders, often termed Couette flow. It is well known that the displacement vectors in this case form concentric circles in the radial cross section of the annulus. In Fig. 5(a) the initially unmixed system is represented within the annulus as a white fluid containing a black fluid as a minor component, represented by the radial line. If the inner cylinder is rotated in a counterclockwise direction by n revolutions, the result is as pictured schematically for $n = 1$, 2, 3, and 1000 revolutions successively. The interfacial area is seen to be increasing and the striation thickness, or thickness separating adjacent lines, is decreasing, until at $n = 1000$ it has diminished below the resolving power of the eye and would appear to be uniformly grey.

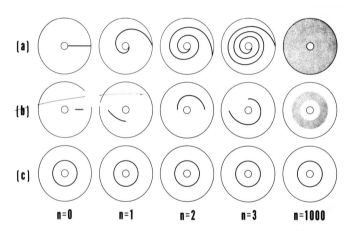

FIG. 5. Representation of laminar, nondiffusive mixing.[10]

In Fig. 5(b) the dark material is confined to the center portion of the annulus. Mixing proceeds in this region as in Fig. 5(a) but, since no mass transport can occur radially, no mixing occurs in the outer or inner areas. In Fig. 5(c) the dark material is distributed completely parallel to the shear displacement vectors or streamlines, and no mixing can occur, because the dark particles now move with substantially the same velocity.

By following reasoning similar to that employed by Schrenk *et al.*[11] the decrease in striation thickness for the case shown in Fig. 5(a) may be calculated as follows:

Figure 6 illustrates a portion of the annulus in which the initial distance between interfaces is r_0, shown at position A. After a displacement of the

[11] W. J. Schrenk, K. J. Cleereman, and T. Alfrey, *SPE (Soc. Plastics Engrs.) Trans.* **3**, 192 (1963).

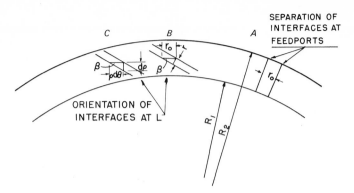

F$_{IG}$. 6. Calculation of striation thickness after mixing between concentric cylinders. (After Schrenk et al.[11])

inner cylinder through an angle θ, the striation thickness will have been reduced to the value r, indicated at position B. From the triangle at B it is clear that

$$r_0/r = \csc \beta \qquad (8)$$

At C (actually superposed on B, but displaced for convenience of representation) the following relationship is seen to hold:

$$\cot \beta = \rho \frac{d\theta}{d\rho} \qquad (9)$$

where ρ is the annular distance having limiting values of R_1 and R_2.

Substituting equation (9) into equation (8) by means of the relation

$$\csc^2 \beta = 1 + \cot^2 \beta \qquad (10)$$

we have

$$(r_0/r)^2 = 1 + \left(\rho \frac{d\theta}{d\rho} \right)^2 \qquad (11)$$

When θ becomes very large, and $d\theta/d\rho \gg 1$, then equation (11) may be simplified to

$$(r_0/r)^2 = \left(\rho \frac{d\theta}{d\rho} \right)^2 \qquad (12)$$

The factor $d\theta/d\rho$ is the design criterion for a laminar-flow mixer based on the Couette-flow principle. As $d\theta/d\rho$ becomes very large, improved mixing results, since the striation thickness, r, becomes very small.

The shear rate, $\dot{\gamma}$, for Couette flow is well known (see, for example, Oka[12]):

$$\dot{\gamma}(\rho) = \frac{2/\rho^2}{1/R_1^2 - 1/R_2^2}\Omega = \frac{d}{dt}\left(\rho\frac{d\theta}{d\rho}\right) \tag{13}$$

where Ω is the constant angular velocity of the inner cylinder. Assuming conditions of steady flow, for which acceleration terms may be ignored, equation (13) may be integrated between time $t = 0$ and $t = T$ to yield

$$\rho\frac{d\theta}{d\rho} = \frac{2/\rho^2}{1/R_1^2 - 1/R_2^2}\Omega T \tag{14}$$

$$\rho\frac{d\theta}{d\rho} = \frac{2/\rho^2}{1/R_1^2 - 1/R_2^2}(2\pi n) \tag{15}$$

where $\Omega T = 2\pi n$, and n is the number of revolutions of the inner cylinder. Substituting equation (15) into equation (12) gives

$$r_0/r = \frac{4\pi n}{\rho^2(1/R_1^2 - 1/R_2^2)} \tag{16}$$

or inversely,

$$r/r_0 = \frac{\rho^2(1/R_1^2 - 1/R_2^2)}{4\pi n} \tag{17}$$

Equation (17) states that the striation thickness, r, decreases with increasing number of revolutions, that r decreases more rapidly as R_1 approaches R_2, and that r is minimal at $\rho = R_1$, the inner cylinder. Referring to Fig. 5(a): if the thickness of the annulus, ρ, is 1 cm and the inner cylinder R_1 is 1 mm, the maximal striation thickness at the outer cylinder would be about 50 μ after 1000 revolutions, or about the limit of resolving power of the eye. Near the inner boundary of the annulus, R_1, the value of r would be about 0.5 μ.

d. Shear-Rate Criterion

From equation (13) it is seen that the shear rate is greatest at the inner cylinder, R_1, where likewise the reduction of striation thickness is greatest. Thus it is apparent that by the time the material at the outer cylinder, R_2, is adequately mixed, the material elsewhere is "overmixed." This leads to the conclusion that for the ideal mixer all the streamlines should lead through the region of maximal shear rate, if mixing is to be achieved with a minimum

[12] S. Oka, in "Rheology: Theory and Applications" (F. R. Eirich, ed.), Vol. 3, p. 32. Academic Press, New York, 1960.

of shear deformation. This ideal is approached as R_1 is made to approach R_2, that is, as the annulus is made to be increasingly thin, for the Couette-type mixer.

e. Composition along All Streamlines

Referring to Figs. 5b and 5c, it is evident that adequate mixing would never be achieved, even if n became extremely large. This observation leads to the conclusion that, because the composition averaged along any streamline is not roughly equal to that of the mixture as a whole, the mixing process is less than ideal. In the cases cited, closed streamlines exist along which the average composition consists entirely of one component, whereas in the case of Fig. 5a the average composition along any streamline is seen to be roughly equal to that of the overall mixture.

f. Summary of Laminar-Flow Mixing Theory

The considerations discussed thus far are intended to portray the background from which the present theory of the ideal laminar-flow mixing process was developed. Although in this discussion particular rheological situations were used as examples, a more rigorous analysis of generalized shear deformation has been presented.[13] A summary of the criteria is presented here for the convenience of the reader:

1. The total shear deformation effected must be sufficient to reduce the striation thickness to a desired value.

2. The initial interfaces between components should be oriented normal to streamlines of shear flow.

3. The average composition along any streamline should approximate that of the overall mixture.

4. All flow paths should lead through the region of maximal rate of shear.

5. The velocity distribution must be described completely throughout the system to be mixed.

These requirements are not likely to be fulfilled completely in any real situation, and the extent to which they are not fulfilled is a measure of the departure from the ideal mixing process.

IV. Experimental Evaluation of Laminar-Flow Mixing Theory

In a recent paper Schrenk et al.[11] present what seems to be the only published account of an effort to evaluate quantitatively the theory of laminar-flow mixing discussed in the previous section. Since this work bears so directly on the topic of this discussion, it will be described in some detail.

[13] J. T. Bergen, J. A. Krumhansl, and G. F. Carrier, SPE (Soc. Plastics Engrs.) Tech. Papers **4**, 987 (1958).

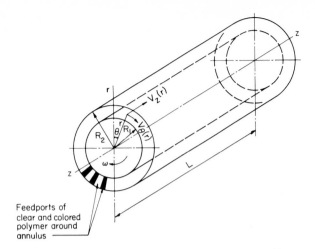

FIG. 7. Diagram showing geometry of continuous mixer.[11]

1. DESCRIPTION OF THE MIXER

The mixing device is basically a continuous Couette-type mixer, within which the flow is both tangential and axial. A schematic representation is shown in Fig. 7. The mixing action is carried out in the annulus of coaxial cylinders, the inner cylinder of which may be rotated clockwise or held fixed, and the outer cylinder of which may be rotated clockwise or counterclockwise or held fixed. Molten polymers, suitably colored, are pumped from an extruder through feedports at the inlet end of the annulus at fixed rates and at the same temperature. Two components to be mixed are introduced through alternate feedports, so that, upon entering the annulus, the interfaces between these components are oriented normal to the tangential velocity components associated with the rotational motion of the cylinders. In this respect the flow paths are the same as those discussed in Section III,2,c. Superposed upon this motion is an axial component resulting from the axial flow through the annulus. The resultant motion of the fluids causes the initial striation thickness, equal to the distance between feedports, to be reduced as the material flows axially through the mixer. The amount of reduction in striation thickness could be predicted from the velocity distribution as a function of annular position ρ and axial position L. Experimental observations of the actual striation thickness could then be compared with those predicted from theory. The dimensions of the mixer elements are as follows:

Length of annulus:	6.75 inches
Internal diameter of outer cylinder:	4.00 inches
Outer diameter of inner cylinder:	3.33 inches

Both the outer and inner cylinder are oil-heated.

2. HYDRODYNAMIC EQUATIONS OF FLOW

The fluids are assumed to be Newtonian, and each fluid is assumed to be pumped at a fixed rate through the annulus. The latter assumption allows the ratio of components to be set throughout the experiment.

The axial velocity of a particle, V_Z, due to the pressure drop in the annular channel is given as a function of radial position, ρ, in the annulus by the well-known equation for axial flow through the annulus. This velocity profile is very nearly parabolic for the particular geometry employed. The tangential displacement, θ, of a given particle is related to the rate of volumetric extrusion, Q, and to the length of travel along the axis, L. For example, if the inner cylinder is rotated and the outer cylinder held fixed, the expression for θ becomes

$$\theta = \frac{TR_2^4\left[\frac{(1 - \alpha^2)^2}{\ln 1/\alpha} + \alpha^4 - 1\right]\left[\rho^2/R_2^2 - 1\right]}{8\eta Q\left\{\rho^4 - \rho^2 R_2^2\left[1 - R_2^2\frac{1 - \alpha^2}{\ln 1/\alpha}\ln (R_2/\rho)\right]\right\}}$$

where T is the torque required to produce rotation, η is viscosity, Q is the volumetric extrusion rate, and α is equal to R_1/R_2.

In this expression the axial length L is contained in the term T. It is a linear function beginning at $L = 0$ at the annulus; hence, θ is a linear function of L.

Similar equations are developed for differing experimental conditions of rotation of the inner and outer cylinders. Differentiation of these equations provides the expressions for $d\theta/d\rho$, from which the striation thickness may be calculated as a function of ρ and L by means of equation (12).

3. DESCRIPTION OF EXPERIMENTS

Four different experiments are described, corresponding to the four choices of rotation of the inner and outer cylinders; see Fig. 8.

Case I. The inner cylinder is rotated at constant angular velocity, while the outer cylinder is held fixed. The tangential velocity distribution, $V_\theta(\rho)$, may be computed from equation (13) to yield

$$V_\theta(\rho) = \rho\frac{d\theta}{dt} = \frac{\Omega}{1/R_1^2 - 1/R_2^2}\left(\frac{\rho}{R_2^2} - \frac{1}{\rho}\right)$$

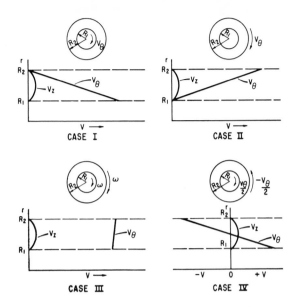

FIG. 8. Characteristics of four basic mixers.[11]

In Fig. 8 for this case the distribution $V_\theta(\rho)$ is seen to be very nearly linear. Superposition of the tangential velocity, $V_\theta(\rho)$, and the axial velocity, $V_z(\rho)$, leads to a helical flow path for any particle as it moves through the mixer. The helix angle is a function of ρ, the radial position in the annulus. Hence, the interfaces between components are caused to increase as the mixture flows through the annulus and, in turn, the striation thickness, originally the distance between feedports, is reduced.

Case II. The outer cylinder is rotated in a counterclockwise direction, while the inner cylinder is held fixed. The velocity distribution $V_\theta(\rho)$ is now seen to be the inverse of that of Case I.

Case III. Both cylinders are rotated at the same angular velocity. The velocity distribution, $V_\theta(\rho)$, is here rather flat; the resulting difference in helix angle over the radius, ρ, is small, and the displacement gradient, $d\theta/d\rho$, is found to be zero at a point near the mid-radius. At this point the mixing action is zero, and the striation thickness should be unchanged from that at the inlet.

Case IV. The inner cylinder is rotated in a clockwise direction, while the outer cylinder is rotated counterclockwise at the same velocity. This experiment is not similar to Case I or Case II, since now both cylinders are moving in relation to the feedports.

As indicated in Section IV,1, functions are derived for these four cases, from which the striation thickness, r, may be predicted. At a given point along the axis r is a function of the radial position, ρ, and the displacement gradient, $d\theta/d\rho$; see equation (12). Substituting into equation (12) the appropriate functions for $d\theta/d\rho$ yields expressions for r corresponding to the four cases described above.

Therefore, the critical design criterion is the distribution of $d\theta/d\rho$ as a function of ρ; these distributions are shown graphically in Fig. 9. In Cases I and II of the value $d\theta/d\rho$ varies from a small but finite number depending upon which cylinder is rotated; in Case III it passes through zero near the mid-radius, and in Case IV it is everywhere considerably greater than zero.

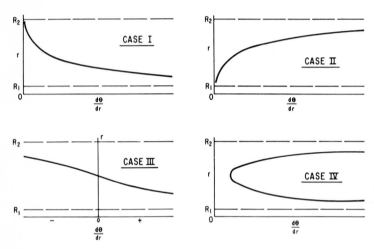

FIG. 9. Graphical illustration of $d\theta/d\rho$ for the four basic mixers.[11]

4. EXPERIMENTAL RESULTS

Experiments were carried out by pumping black and white polyethylene melt from two extruders at relative rates of 1 and 9, respectively. Both the extruders and the mixer were maintained at 450°F. When steady-flow conditions were obtained, the extruders and rotor drives were simultaneously stopped and the mixer was cooled rapidly and removed. By sectioning the mixture and measuring the striation thickness, r, in the frozen-in patterns a quantitative measure of the degree of mixing could be obtained. The results are shown in Fig. 10, where the ratio r_0/r is presented for the Case II mixer (Figs. 8 and 9) at the mid-channel point, $\rho = \frac{1}{2}(R_1 + R_2)$, as a function of axial distance from the feedports. The theoretical values of r_0/r are indicated by the broken line. Similar agreement between theory and experiment was obtained for the remaining cases.

This work demonstrates clearly that it is possible to design accurately a laminar-flow mixer that fulfills substantially all the criteria of the ideal laminar-flow mixing process. As was pointed out previously, however, the mixing devices actually in use in the plastics industry today evolved as a part of the art of mixing practice and as a result of empirical experience. Some of the principles of laminar-flow mixing theory can be found in the operating characteristics of these machines. The following discussion is a brief review of the literature on the characteristics of the more common industrial plastics mixers as they relate to the theory and rheology of the mixing of plastics.

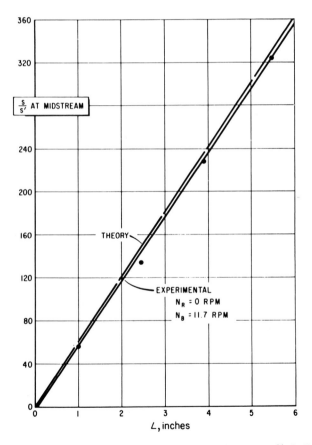

FIG. 10. Comparison of measured and calculated mixing for Case II.[11] Ordinate: ratio of initial striation thickness, S, to observed striation thickness, S'. Abscissa: distance, L, from feedport N_R and N_B denote speed of inner and outer cylinders, respectively.

V. Mixers in the Plastics Industry

The most important mixing devices used at present in the processing of plastics materials are of three general types: the extruder, the internal mixer, and the two-roll mill. Alterations of these devices that have appeared from time to time have improved their effectiveness as mixers, but the basic principles remain the same. A general review of the technology of these mixers is given by Rushton et al.[14] and in a series of articles in Simonds.[15] In the sections to follow the roles of the basic principles of laminar-flow mixing in the functioning of industrial plastics mixers will be discussed. The discussion is necessarily brief because the literature on the subject is limited, and only in the case of the simple screw extruder has a broad attempt been made to deal with the basic aspects of mixing; in general, practice has far outstripped theory.

1. THE SCREW EXTRUDER

The flow conditions in the screw extruder have been studied extensively,[15-18] and the use of the screw extruder as a mixer has been analyzed.[9] A review of mixing of thermoplastics is presented by McKelvey;[10] it includes a discussion of mixing in the extruder. The complete description of the three-dimensional flow appears to be mathematically intractable. The simplified two-dimensional theory that has been developed for the present has many useful technological applications. By appealing to this description of the flow the mixing action of the extruder can be visualized reasonably well. Figure 11 illustrates schematically the velocity components of flow in the helical channel, where u_z is the component in the direction of the helical channel of the screw, and u_y is the cross-channel component. In this figure the coordinate axes are fixed with respect to the screw, so that the barrel is taken to be moving instead of the screw. The shear rate is seen to vary with the radial position in the channel (the radial component of flow at the walls of the channel is ignored through the assumption of two-dimensional flow); near the mid-radius of the channel the resultant axial velocity is greatest, while the shear rate is least in this region. Therefore, it is to be expected that unmixed material would appear first in a rod-shaped extrudate

[14] J. A. Rushton, R. D. Boutros, and C. W. Selheimer, "Encyclopedia of Chemical Technology," Vol. 9, p. 133. Wiley (Interscience), New York, 1952.

[15] H. R. Simonds, "Encyclopedia of Plastics Equipment," pp. 37, 201, 353, 355, 360, 365. Reinhold, New York, 1964.

[16] J. F. Carley, R. S. Mallouk, and J. M. McKelvey, Ind. Eng. Chem. **45**, 97 (1953).

[17] W. L. Gore and J. M. McKelvey, "Rheology: Theory and Applications" (F. R. Eirich, ed.), Vol. 3, p. 589. Academic Press, New York, 1960.

[18] R. M. Griffith, Ind. Eng. Chem., Fundamentals **1**, No. 3, 181 (1962).

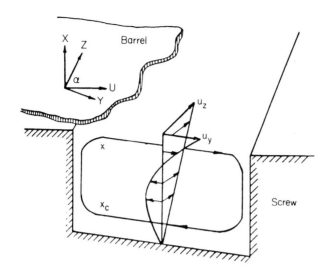

FIG. 11. Velocity components in extruder channel.[18]

as an annular band near the mid-radius, as has been observed experimentally.[9] This effect may be suppressed by increasing the length of the screw, increasing the width-to-depth ratio of the channel, or operating with a high ratio u_y/u_z resulting from a restriction of the extrudate flow at the exit of the extruder. Various methods have been suggested for overcoming this difficulty by modifying the screw design or by providing suitable mixing heads at the downstream end of the extruder screw.[19] The purpose of these modifications is to apply shear vectors in directions other than that of the helical axis.

Other efforts to improve the mixing characteristics of the extruder while retaining its basic simplicity of design and capacity for large volume throughput have been made. The variations include a meshed, double-screw mixer, an independent twin-screw mixer, and an interrupted-flight single-screw mixer, as examples.[15] In the case of the latter device the interrupted flights of the rotor mesh with the fixed blades of the barrel, and the rotor is also caused to oscillate axially as it turns. This oscillation leads to fore-and-aft mixing that is superposed on the complex motion experienced by the mixture as it passes between the fixed and rotating blades.

[19] F. W. Kroessner and S. Middleman, *Polymer Eng. Sci.* **5,** No. 4, 230 (1965).

2. THE INTERNAL MIXER

Another type of widely used mixing devices is the internal mixer, in which the mixture is contained within a closed chamber. Descriptions of the internal mixer are given in several sources, such as pages 37 and 365 of Simonds[15] and in Riegel.[20]

The schematic representation of one such machine is shown in cross section in Fig. 12. The mixture occupies the space within the chamber, being confined by a hydraulic ram. Two rotors turn in opposite directions within parallel troughs. During part of each revolution a portion of the mixture is sheared between the blades projecting from the rotors and the walls of the troughs. Here the shear rate is large, owing to the small clearance between the blade tip and the wall. An analysis of the flow conditions existing in this region has been reported.[21] Elsewhere the differential rate of rotation of the rotors results in a complicated motion that rapidly transports the material back and forth between blades. Furthermore, the projecting blades form helices in the axial direction; this leads to axial transport of the material from end to end of the chamber.

Mixing devices of this type are very effective in mixing plastics rapidly and in large amounts; however, they are usually very massive and, therefore, costly, and in addition they consume a great deal of power per unit of material output.[13]

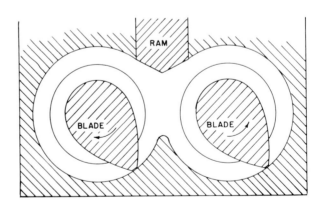

FIG. 12. Schematic radial cross section of Banbury internal mixer.

[20] E. R. Riegel, "Chemical Process Machinery," pp. 293–299. Reinhold, New York, 1953.

[21] J. T. Bergen, *in* "Processing of Thermoplastic Materials" (E. C. Bernhardt, ed.), p. 405. Reinhold, New York, 1959.

3. THE TWO-ROLL MIXING MILL

The two-roll mill is widely used in plastics mixing operations. It consists of two steel rollers mounted parallel, their axes in a horizontal plane, and rotated in opposite directions by a suitable drive motor and gear train. The axis of one roll is fixed; the other is adjustable through jack screws. This allows the clearance, or "nip," between the rolls to be adjusted. The mixture is caused to pass through the nip, where it undergoes an intense shearing action. A description of the mixing mill is given on page 353 of Simonds[15] and by Riegel.[22]

The velocity distribution within the region of the nip (the only region of a mixing mill where any shear or mixing occurs) has been established by Gaskell;[23] a representative plot of velocity profiles is shown in Fig. 13.

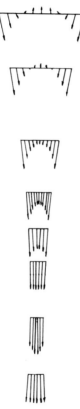

FIG. 13. Velocity profiles at several points in two-roll mill.

[22] E. R. Riegel, "Chemical Processing Machinery," p. 301. Reinhold, New York, 1953.
[23] R. E. Gaskell, J. Appl. Mech. 17, 334 (1950).

The shear rate is seen to be greatest along the surfaces of the rolls and to vanish everywhere along the plane of symmetry between the curved boundaries. Thus, there is little mixing effect in the region of this plane. It is usual practice to allow the material to form a continuous band on one roll and so pass through the nip repeatedly. However, the plane of symmetry between rolls tends to lie along a closed streamline, so that the presence of unmixed material may persist near the center of the banded sheet of material for a surprisingly long time.

Further, there is little axial flow within the calender nip, so that mixing in the axial direction is nil, if the milling process is left to itself. For this reason, and to facilitate mixing of material near the center of the sheet, it is almost universal practice to cut the band of mixture away from the roll surface and to fold it and transport it axially either manually or by means of mechanical devices.

4. SUMMARY OF LAMINAR-FLOW MIXING PRACTICE

It is evident that during the past fifteen years considerable progress has been made toward the goal of formulating a quantitative theory for the mixing of plastics by laminar-flow shear deformation. Conceptual criteria and a quantitative description of the completeness of mixing have been established. Direct experiments confirm the validity of the theory, as long as the conditions of the theory are fulfilled. Although progress has been slow, the importance of the theory should not be overlooked, since it represents a beginning toward a clearer understanding of an important process that has heretofore developed largely as an art.

It is likewise evident that the present theory of laminar-flow mixing has rather limited technological application. The restriction to idealized systems, to isothermal, rheologically simple, flows, and to Newtonian fluids does not result so much from limitations in the theory of laminar-flow mixing as it does from limitations on current theories of the flow of fluids in general. Efforts have been made to describe the adiabatic flow of non-Newtonian fluids in the extruder[18,19] and in a simplified internal mixer.[24] The temperature distribution in the nip of a two-roll mill has been treated by Finston,[25] and Gaskell[23] discusses the flow of a Bingham material for the same case.

Plastics materials, however, and especially thermoplastics, are recognized to be often viscoelastic under conditions of processing and mixing, and quantitative theories of the flow of viscoelastic fluids are still in a state of development. It is clear that any successful treatment of the extremely

[24] W. R. Bolen and R. E. Colwell, SPE (Soc. Plastics Engrs.) J. **14**, No. 8, 24 (1958).
[25] M. Finston, J. Appl. Mech. **18**, 12 (1951).

complex flow conditions that exist in many industrial plastics processes (cf. the studies of Maddock[26] on the extrusion of thermoplastics) will depend on further progress in the fundamentals of viscoelastic-fluid flow.

An interesting technological development is the "elastic melt extruder," first described by Maxwell and Scalora[27] and later extended by Henry and Plymale.[28] This device employs the principle of a parallel-plate viscometer, in which one plate is rotated parallel to a second plate, which is held fixed. The mixture is introduced at the periphery of the heated plates, where it is subjected to simple shear in the gap between the moving plate and the fixed plate. Owing to the Weissenberg effect (a result of the normal stresses that occur in a viscoelastic fluid being sheared) a pressure gradient appears that is directed inward along the radius of the plates. Thus, the molten polymer flows inward toward an exit hole in the center of the fixed plate, being subjected to considerable shear deformation as a consequence. In both the papers cited the authors remark on the good mixing achieved with pigmented plastics in even a single pass. The shear displacement and shear rate are fairly uniform across the gap between the plates, so that more uniform mixing of the extrudate compared with that in the screw extruder is to be expected. No further work of this type has been reported; one can imagine other devices with different geometries that would utilize normal stresses and secondary flows specifically to benefit the mixing process.

VI. Dispersive Mixing

In the preceding discussions the shearing forces developed in the course of laminar-flow mixing are presumed to be sufficient to cause individual particles to move freely in the mixture along the direction of the shear displacement vectors; that is, interparticle forces have been assumed to be negligible. In practice this assumption has been found to be incorrect in many cases. The forces necessary to deagglomerate clumps of particles, to displace air or other molecules that initially may surround particles, and to bring about wetting of the particles by the polymer matrix frequently are not inconsequential. The practice of plastics mixing thus is often concerned with efforts at achieving shearing stresses of sufficient magnitude to separate particles from one another or to provide for adequate wetting of individual particles, so that dispersive mixing will result.

[26] B. H. Maddock, *SPE* (*Soc. Plastics Engrs.*) *J.* **15,** 383 (1959).
[27] B. Maxwell and A. J. Scalora, *Mod. Plastics* **18,** No. 8, 107 (1959).
[28] J. E. Henry and C. E. Plymale, *SPE* (*Soc. Plastics Engrs.*) *J.* **21,** No. 4, 391 (1965).

1. COMMINUTION OF PARTICLES AND AGGLOMERATES

As a result of the familiar techniques of dry grinding of solid particles initially coarse particles or agglomerates are reduced to varying degrees of fineness in the dry state, and the resultant powder is introduced into the mixer together with the polymer in the hope of achieving a random mixture of the ultimate particles in the polymer matrix. Experience has shown that such mixtures often tend to contain larger clumps or agglomerates of the comminuted solid. One means of facilitating the dispersion of these agglomerates is to carry out the grinding or comminution process in a suitable liquid medium and in the presence of surface-active agents that promote wetting of the solid and that minimize recombination or agglomeration of the solid particles once they are fractured. Recent discussions of milling of solids with surface-active agents prior to mixing in plastics and of the comminution of pigments in organic solvents and plasticizers are given by Reeve and Zabel;[29] discussion of the aqueous dispersion of insoluble organic materials is given by Harper and Seaman.[30]

2. SHEAR DISPERSION OF PARTICLES

During the process of mixing of solids in plastics it is necessary that the shearing stresses at some point exceed the interparticle forces present in the solid agglomerate. These shearing stresses are a function of the viscous properties of the polymer matrix and the rate of shear achieved in the mixer itself. To predict the effectiveness of the dispersion process it is necessary to know the magnitude of the forces between the particles constituting the agglomerate. One such analysis is given by McKelvey[10] for a two-particle agglomerate in a field of uniform shear stress. He concludes that high shear stresses promote dispersion, that a critical shear stress exists, below which dispersion will not occur, and that initial orientation of the axis between particles with respect to the direction of shear is important when the shear stress is only slightly larger than the critical shear stress.

Further discussion of the mechanics of dispersion of solid particles in plastics lies beyond the scope of this chapter and would properly constitute a separate effort.

VII. Conclusion

1. DISTRIBUTIVE MIXING

Laminar-flow mixing, as it is discussed by various workers, can be seen to depend upon the existence of continuous, uninterrupted flow throughout

[29] T. B. Reeve and R. H. Zabel, *Color Eng.* **3**, No. 6, 12 (1965).
[30] H. R. Harper and D. Seaman, *Kolloid-Z.* **204**, No. 1, 283 (1965).

a system. Indeed, the greatest difficulty in the description of plastics mixing is encountered in attempting to obtain the rheological description of the flow within the system. Consequently, progress in the theory of laminar-flow mixing will follow any advances in theoretical aspects of the rheology of fluid flow.

Another aspect of plastics mixing was considered some time ago by Spencer and Wiley,[8] being denoted by them as "distributive mixing." This mixing process involves streamline flow, but it differs from laminar-flow mixing in that the streamlines are interrupted, or cut, by cutting or folding of the mass being mixed. Segments or cells of the mixture may thus be transported in toto with respect to one another, so that gross randomization of the spatial position of cells of the mixture occurs. This procedure affects the *distribution* of the components within the mass, a consideration that is somewhat different from that of creating new interfacial areas along stream-lines which, needless to say, do not cross one another.

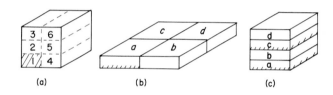

FIG. 14. Example of distributive mixing. (After Spencer and Wiley.[8])

An example of this problem is given in Spencer and Wiley's paper as follows. Consider a mass to be mixed which is subdivided into N discrete cells numbered 1 through N. In Fig. 14(a) the mass is represented as a cube for which $N = 6$, all of the minor component being contained in cell 1. The cube is compressed to one fourth of its original height (it may likewise be so deformed by stretching normal to the direction of compression) and becomes the form shown in Fig. 14(b). Now cut the mass into equal quarters and reassemble it as shown in Fig. 14(c). One can see intuitively that the distribution is now more uniform than in Fig. 14(a); this procedure will be called a mixing cycle, which can be repeated at will until the desired degree of uniformity of distribution of the minor component is achieved.

Spencer and Wiley represent this process (or similar ones involving cutting or folding of the mixture, as in the alternate rolling and folding over of biscuit dough) in mathematical terms as follows. Let $a_j{}^n$ be the concentration of the minor component [shaded, Fig. 14(a)] in the jth cell after n mixing cycles. Further, let the distribution matrix $[D_{ij}]$ represent the fraction of material originally in the ith cell that will be found in the jth cell after one mixing cycle. If $a_i^{(0)}$ is the initial concentration in the ith cell, the concentrations in the cells after n mixing cycles will be

$$a_j^{(n)} = [a_i^{(0)}][D_{ij}]^n \tag{18}$$

In the simple example shown in Fig. 14(a), all elements $a_i^{(0)}$ are zero except a_1 which is given a density value of 1.0; hence,

$$[a_i^{(0)}] = [1 \quad 0 \quad 0 \quad 0 \quad 0 \quad 0] \tag{19}$$

After the single mixing cycle, Fig. 14(c), the distribution matrix $[D_{ij}]$ will be found to be

$$[D_{ij}] = \tfrac{1}{4} \begin{bmatrix} 1 & 1 & 0 & 1 & 1 & 0 \\ 1 & 1 & 0 & 1 & 1 & 0 \\ 1 & 0 & 1 & 1 & 0 & 1 \\ 1 & 0 & 1 & 1 & 0 & 1 \\ 0 & 1 & 1 & 0 & 1 & 1 \\ 0 & 1 & 1 & 0 & 1 & 1 \end{bmatrix} \tag{20}$$

Equation (20) states that the present concentration of material, originally in cell 1 (the minor component), in the "new" cell 1 is $\tfrac{1}{4}$ ($i = 1, j = 1$), and the concentration in the "new" cell 3 is 0 ($i = 1, j = 3$), and so forth.

Successive mixing cycles will produce distribution matrices represented as successive powers of the matrix in equation (20):

$$[D_{ij}]^2 = \tfrac{1}{16} \begin{bmatrix} 3 & 3 & 2 & 3 & 3 & 2 \\ 3 & 3 & 2 & 3 & 3 & 2 \\ 3 & 2 & 3 & 3 & 2 & 3 \\ 3 & 2 & 3 & 3 & 2 & 3 \\ 2 & 3 & 3 & 2 & 3 & 3 \\ 2 & 3 & 3 & 2 & 3 & 3 \end{bmatrix}$$

$$[D_{ij}]^3 = \tfrac{1}{64} \begin{bmatrix} 11 & 11 & 10 & 11 & 11 & 10 \\ 11 & 11 & 10 & 11 & 11 & 10 \\ 11 & 10 & 11 & 11 & 10 & 11 \\ 11 & 10 & 11 & 11 & 10 & 11 \\ 10 & 11 & 11 & 10 & 11 & 11 \\ 10 & 11 & 11 & 10 & 11 & 11 \end{bmatrix}$$

Thus, successive mixing cycles smooth out the original fluctuations quite rapidly. It is important to observe that, if the mechanism of mass transport were to be continuous streamline flow, by compressing the cube in Fig. 14(a) without cutting and reassembling, mixing would never be complete, because none of the material in cell 1 would ever be transported to cells 4, 5, and 6. Other examples of distributive mixing by repetitive deformation and cutting are given by Spencer and Wiley, who conclude by pointing out that in practice rather large and unwieldy matrices would be encountered, requiring the use of high-speed digital computation methods for solution.

Unfortunately, this initial effort has never been pursued, and investigation into this aspect of plastics mixing has been neglected. For the three general types of industrial plastics mixers mentioned in Section V it is evident that distributive mixing is an essential part of the function of the mixer, because in each case the design includes provision for interrupting other-wise continuous streamlines and for folding or "shuffling" gross quantities of the mixture.

2. MIXING OF LIQUIDS IN PLASTICS

a. Droplet Formation

One very common objective in plastics mixing is to obtain a suitable mixture of a liquid component in a very viscous or viscoelastic polymer. The liquid in many instances is a low-vapor-pressure organic liquid that is intended to soften or plasticize a thermoplastic polymer system. McKelvey[10] considers the problem of mixing immiscible Newtonian liquids by laminar flow, the viscosity of one of the two components differing widely from that of the other. He concluded that, when the viscosity, η, of the minor compo-nent (plasticizer) is very much smaller than the viscosity, η', of the major component (polymer), the shear rate within the minor component is deter-mined by the velocity gradient through the layer of plasticizer; that is, the very much more viscous polymer tends to behave as if it were a rigid bound-ary. This conclusion is open to some question, since Taylor,[31] studying the

[31] G. I. Taylor, *Proc. Roy. Soc.* (*London*) A**146**, 501 (1934).

distortion of drops of an immiscible fluid by shearing within a medium of much higher viscosity ($\eta/\eta' = 0.0003$), observed that the drops tended to form infinitely long threads at low shear rate but developed instability or varicosity, eventually forming many small droplets as the thread ruptured owing to varicose instability. These results were confirmed by Tomotika,[32,33] who found that the unstable threads of liquid broke up into droplets of volume about $\frac{1}{1000}$ that of the original drop. The possible existence of such hydrodynamic instability during the mixing of low-viscosity plasticizers in polymers should be investigated, because it implies a mechanism of mixing that is markedly different from that associated with stable laminar shear.

b. Molecular Diffusion

The question of molecular diffusion as a mechanism of the mixing of plasticizers in polymers and plastics has been considered by various workers,[1,8,10] but it has been generally concluded that diffusion in polymers is too slow or acts over too small distances to be considered an effective transport mechanism. Significantly, no estimate of the rate of diffusion of plasticizers in polymers under the conditions of process mixing has been reported.

Experimental data on the sorption of plasticizers by polymers are very limited; considerably more work is reported on the desorption of plasticizers from the polymer–plasticizer system, but a review of the literature indicates that these data are not applicable to sorption. A further shortcoming in available data is due to the limited temperature range covered in many experiments; almost all observations are confined to temperatures below the glass transition temperature, which is far below the range of temperature at which thermoplastics are processed.

On the other hand, the specific effect of different plasticizers on the mixing characteristics of plastics compounds has long been recognized,[34] in that the time to achieve a suitable mixture is known to depend upon the chemical structure of the plasticizer and upon its molecular weight. A more recent study[35] exemplifies the technological evaluation of plasticizing action in terms of "fusion rates" for poly(vinyl chloride). The polymer, introduced in the form of finely divided solid particles a few hundred microns in diameter, is mixed with the plasticizer in the cold to form a "wet" blend. Upon heating of this wet blend in a suitable mixer it is observed that at a fairly definite threshold temperature the mixture, originally a loose, granular mass, rapidly "fuses" into a coherent, viscoelastic solid. This phenomenon is

[32] S. Tomotika, *Proc. Roy. Soc.* A**150,** 322 (1935).
[33] S. Tomotika, *Proc. Roy. Soc.* A**153,** 302 (1936).
[34] H. S. Bergen and J. R. Darby, *Ind. Eng. Chem.* **43,** 2404 (1951).
[35] N. W. Touchette, H. J. Seppala, and J. R. Darby, *Plastics Technol.* **10,** 33 (1964).

accompanied by a very large and rapid rise in torque of the mixer blades and is followed by an expected asymptotic decrease as the temperature is further increased. The nature of this phenomenon strongly suggests that some fundamental physical change in the polymer–plasticizer system occurs over a very narrow temperature range; the nature of this change is not at all well understood.

It is clear, too, that this process does not take place by simple laminar-flow mixing. The plasticizing effect requires that the plasticizer be distributed uniformly throughout the polymer at the molecular level. Furthermore, the "fusion" temperature is observed to range from about 20°C below the glass transition temperature of the polymer to as much as 50°C above it, depending on the structure of the plasticizer. Consequently, some physical process of mass transport at the molecular level must be invoked to explain these facts.

One such process is diffusion of the plasticizer molecules into the polymer mass. Observations of the diffusion of di-ethylhexylphthalate, commonly known as di-octylphthalate, into poly(vinyl chloride) granules are reported by Dannis,[36] who used dielectric constant measurements to follow the sorption of the polymer. His results were later confirmed by Grotz,[37] who used weight increase of disks of poly(vinyl chloride) to follow the diffusion of dioctylphthalate at temperatures between 65 and 85°C. The conventional Arrhenius plot of the data of Grotz is shown in Fig. 15, where an extrapolation is indicated by the dashed line. Although such an extrapolation is open to considerable question, it allows an estimate of the order of magnitude of the coefficient of diffusion, D, in the region 100 to 125°C, which represents the lower region of usual poly(vinyl chloride) mixing temperature ranges (the error in extrapolated values of D will, if anything, lead to too low a value). By using such values of D for corresponding temperatures, the diffusion of di-octylphthalate into a plane sheet of poly(vinyl chloride) was calculated with the appropriate equation.[38] The results are shown in Table I as the thickness of a plane sheet into which 90% of the equilibrium plasticizer content would diffuse in 1 minute if the sheet were immersed in liquid plasticizer.

For the lower region of poly(vinyl chloride) processing temperatures, 100 to 125°C, the effective diffusion thickness is estimated to be 0.1 to 1.0 mm. This distance may be taken to represent the order of magnitude of the striation thickness that the mixing process must achieve for diffusion to be effective as a means of mass transport of plasticizer in poly(vinyl

[36] M. L Dannis, *J. Appl. Phys.* **21,** 510 (1950).

[37] L. C. Grotz, *J. Appl. Polymer Sci.* **9,** 209 (1965).

[38] P. Meares, "Polymers: Structure and Bulk Properties," pp. 319–320. Van Nostrand, Princeton, New Jersey, 1965.

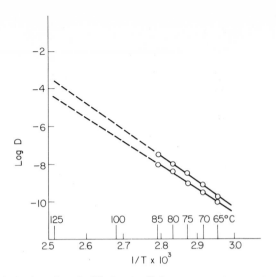

FIG. 15. Arrhenius plot of diffusion coefficient versus temperature: dioctylphthalate in poly(vinyl chloride). (Data of Grotz.[37])

TABLE I

ESTIMATED SHEET THICKNESS OF POLY(VINYL CHLORIDE) CORRESPONDING to 90 %
EQUILIBRIUM DIFFUSION OF DIOCTYLPHTHALATE IN 1 MINUTE

$T, °C$	Sheet thickness, μ
85	17
100	83
125	870

chloride). It is several orders of magnitude greater than that assumed by previous workers. Therefore, molecular diffusion can be assumed to be a valid mechanism of mass transport in this aspect of the mixing plastics; indeed, it may well be the overriding mechanism in many instances.

Another study, carried out by McKinney,[39] illustrates this point. Granular poly(vinyl chloride) was mixed with dioctylphthalate in the cold and then stored at 60°C for a period of several days. Samples of this mixture taken at successively longer times were subjected to differential thermal analysis (as described by McKinney[40]), so that the change in glass transition temperature might be observed as the mixture was allowed to age at 60°C. In Fig. 16 the results of the analysis are displayed; the temperature differential,

[39] P. V. McKinney, unpublished data, 1965.
[40] P. V. McKinney, J. Appl. Polymer Sci. 9, 3359 (1965).

ΔT, between a reference standard (pentaerythritol) and the sample is plotted on the ordinate against the actual temperature of the apparatus, the temperature being increased at about 3°C per minute. The initial untreated mixture exhibits an exothermic rise in ΔT in the region of 72°C and then a decrease in ΔT to a new level at about 80°C. Similarly, for the sample "aged" 30 minutes at 60°C the exothermic peak appears in the region of 74°C, but it is more broad and less sharp than that of the initial sample. Also of importance is the onset of a broad endothermic dip along the left-hand portion of the curve, which in successive curves can be seen to develop into a well-defined endothermic minimum after 3 days' aging at 60°C. Meanwhile, the initial exothermic peak is seen to shift to slightly higher temperatures and finally to disappear.

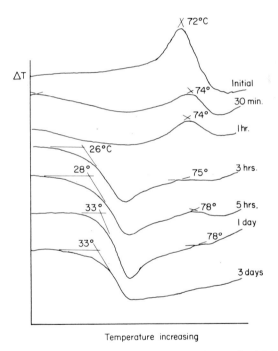

FIG. 16. Differential thermal analysis results on aging of mixture of poly(vinyl chloride) and dioctylphthalate "wet" blend. (Data of McKinney.[39])

The significance of these results may be recognized from the fact that the exothermic rise, signifying the glass transition of the unplasticized poly-(vinyl chloride), was found to occur independently at approximately 78°C; the glass transition of the complete mixture of polymer and plasticizer was likewise found to occur at approximately 32°C. Therefore, the experimental curves in the figure appear to demonstrate absorption by diffusion of plasticizer in the polymer particles.

As an additional check of this conclusion the foll~~ing calculation was performed. By means of the diffusion coefficient at 60°C. from Fig. 15 the equation for diffusion into a sphere[38] was solved for the effective radius of the sphere that would have yielded 99% equilibrium diffusion in 3 days, corresponding to the lower curve in Fig. 16. This estimated particle size was found to be about 150 μ, which agrees surprisingly well with the range of particle size reported by the manufacturer, 75 to 400 μ.

This example of diffusion transport of plasticizer over significantly large dimensions, even at temperatures well below those encountered in plastics mixing, points out that considerable further work is needed if we are to understand the basic mechanisms involved. As Moore[4] has shown, it is possible to predict solubility equilibria or compatibility from molecular weight and structural formula data on the plasticizer and polymer. It should likewise be possible to predict the kinetics of the mixing of liquids in polymers from similar data. At the very least, the need of further information on the role of diffusion in this aspect of the mixing of plastics is quite obvious.

Note added in proof: In Section IV, the velocity profile is based on the assumption that the fluid is Newtonian. Recently, the velocity profile for the same (helical) flow of Section IV is derived theoretically by Savins and Wallick[41] for a viscoelastic (shear-thinning) fluid, and experimental results for helical flow of a power-law fluid are given by Rea and Schowalter.[42] These results would allow the degree of mixing in helical flow to be predicted for such non-Newtonian liquids.

[41] J. G. Savins and G. C. Wallick, *A.I.Ch.E.J.* **12**, 357 (1966).

[42] D. R. Rea and W. R. Schowalter, Presented at Fall Meeting, Society of Rheology, Atlantic City (Nov. 1966).

Nomenclature

a_i	Local concentration	V	Volume
r	Striation thickness	V_i	Directional velocity
s, S	Interfacial area	X, Y, Z	Coordinates
t, T	Time	α_i	Angle of vector to x_i
u_i	Displacement	α	Ratio of radii
x_i	Coordinate directions	$\dot{\gamma}$	Shear rate
H	Thickness of layer	$\partial u_i / \partial x_j$	Shear strain
L	Layer displacement	η	Viscosity
n	Number of revolutions	θ	Angular rotation
Q	Volume rate of flow	μ	Micron
R	Radius	ρ	Annular radial position
T	Torque	Ω	Angular velocity

Chapter 5

THE RHEOLOGY OF LIQUID CRYSTALS

Roger S. Porter and Julian F. Johnson

I. Introduction

1. Definition of Liquid Crystals

A surprising number of organic compounds—more than three thousand —are known to exhibit liquid-crystal behavior. Liquid crystals are systems that are molecularly ordered yet possess mechanical properties resembling those of fluids. The ordered, or aggregate, domains in liquid crystals contain typically about 10^5 molecules and exhibit longer relaxation times, about 10^{-5} sec, than do simple isotropic liquids, whose relaxation times are about 10^{-11} sec. Isotropic liquids exhibit much shorter range ordering, in the form of swarms, which occur most prominently near mesophase and solid-phase transitions.

Liquid-crystal aggregates can occur in solution. They can also be formed within pure compounds. On being heated the compounds go through a first-order transition to the liquid-crystal state, or mesophase, and at a higher temperature go through another first-order transition to the normal, or isotropic, liquid state. The weakest associations are broken first, giving the molecules certain degrees of freedom in the liquid-crystal state. Mesophase transitions can be reversible on cooling, sometimes with supercooling. The transitions on heating and cooling are generally accompanied by preeffects and posteffects to the first-order transformation. Some liquid-crystal phases are metastable or monotropic; these terms mean that the phase is observed only on solvent evaporation or on cooling from the isotropic liquid.[1] A single compound can exhibit more than one type of liquid crystal. The liquid-crystal state is perhaps more satisfactorily referred to as a mesophase;[2] it has also been referred to as anisotropic or para-crystalline phase.[3] The liquid-crystal terminology is descriptive and widely used.

2. RHEOLOGY BY LIQUID-CRYSTAL TYPE

Lehmann reported in 1890 that cholesteryl benzoate, when heated, first formed a turbid system that flowed as readily as oil but still retained many of the characteristics of crystals.[4] A qualitative viscosity classification of liquid crystals, developed by Lehmann, has recently been restated;[5] meso-phases with viscosities near 1 cp are called liquid crystals, and flowing crystals have viscosities of about 10 cp.

It is perhaps most useful to examine liquid crystals on the basis of structure. The three basic structural types are the nematic, smectic, and cholesteric.[6] The nematic (threadlike) type of liquid crystal is the simplest mesophase and in pure compounds consists totally of rigid bundles of needlelike structures; see Fig. 1.[6,7] The only structural requirement of nematic aggregates is that the molecules present a parallel or nearly parallel orientation within a bundle.

Molecular aggregates of the nematic type are readily oriented. Active surfaces and electric and magnetic fields can cause nematic mesophase orientation. Shear is also influential in inducing orientation of the generally large and anisotropic nematic structures. The viscosity of these liquid

[1] G. W. Gray, "Molecular Structure and the Properties of Liquid Crystals." Academic Press, New York, 1962.

[2] G. H. Brown and W. G. Shaw, Chem. Rev. 57, 1049 (1957).

[3] F. B. Rosevear, J. Am. Oil Chemists' Soc. 31, 628 (1954).

[4] O. Lehmann, Z. Physik. Chem. 5, 427 (1890).

[5] V. A. Usol'tseva and I. G. Chistyakov, Russ. Chem. Rev. (English Transl.) 32, 495 (1963).

[6] J. L. Fergason, Sci. Am. 211, No. 2, 77 (1964).

[7] H. Mark and A. V. Tobolsky, "Physical Chemistry of High Polymeric Systems," 2nd ed., pp. 229–239. Wiley (Interscience), New York, 1950.

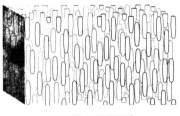

NEMATIC MESOPHASE

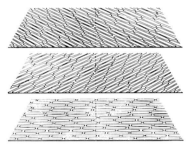

CHOLESTERIC MESOPHASE

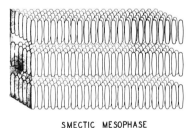

SMECTIC MESOPHASE

FIG. 1. Schematic representation of liquid crystals.

crystals thus depends not only on aggregate shape and size but on shear rate, temperature, and direction of viscosity measurement relative to electric and magnetic fields.

In the smectic (greaselike) mesophase, molecules are arranged side by side in a series of stratified layers; see Fig. 1.[6,7] Molecules in the layers may be arranged in rows or at random. In general, the soaplike smectic mesophase has greater ordering than the nematic mesophase. In some smectic mesophases molecules are further ordered by being arranged in paired rows within individual layers.[2,6] The long axes of molecules are parallel to one another and essentially perpendicular to the plane of the layer, which can be one or more molecules thick.[6] The smectic mesophase can exist in Grandjean terraces, which serve as slip-planes for flow.[2]

The cholesteric mesophase is formed principally by derivatives of cholesterol, but not by cholesterol itself. It resembles the smectic mesophase in that the molecules are arranged in layers. Within each layer, however, the parallel arrangement of molecules is reminiscent of the nematic mesophase.[6] The cholesteric layers are thin, and the long axis of the essentially flat molecules are parallel to the plane of the layers.[6] The repeating unit for a cholesteric mesophase is shown in Fig. 1. The scattering of light, giving vivid colors, is a striking feature of the cholesteric mesophase. Like the smectic and unlike the nematic, this mesophase is reportedly inactive in electric and magnetic fields;[1,2] the cholesteric mesophase of at least the nonanoate ester, however, is said to be orientable in an electric field.[8]

3. SCOPE OF THE REVIEW

The very nature of liquid crystals invites the development and correlation of rheological properties. The first viscosity studies were made before 1900.[4,9,10] Several books,[1,10] reviews,[2,5,11] and symposia[11a,12,13] on liquid crystals have been published. None have emphasized rheology and, indeed, some have led, as will be pointed out, to errors in the citation and interpretation of liquid-crystal rheology. Some general flow characteristics of mesophase types have been reported.[13a]

The best and also the most recent publications on the general properties of liquid crystals are by Brown,[2] by Fergason,[6] and by Gray.[1] This chapter covers the rheology and related phenomena for all types of liquid crystals and closely related systems. A few of the publications prior to 1910 are considered in detail. Others not cited are lacking in definition with respect to system purity and in instrumental detail. Purity is extremely important

[8] J. H. Muller, Z. Naturforsch. 20a, 849 (1965).

[9] R. Schenck, Z. Physik. Chem. 27, 167 (1898).

[10] R. Schenck, "Kristallinische Flüssigkeiten und flüssige Kristalle." Engelmann, Leipzig, 1905.

[11] G. H. Brown, Ind. Res. p. 53, May (1966).

[11a] J. D. Bernal and D. Crowfoot, Trans. Faraday Soc. 29, 1032 (1933).

[12] L. C. Flowers and D. Berg, presented at the Symposium on Ordered Fluids and Liquid Crystals, Division of Colloid and Surface Chemistry, National ACS Meeting, Atlantic City, New Jersey, September, 1965.

[13] R. S. Porter and J. F. Johnson, eds., "Ordered Fluids and Liquid Crystals." Advan. Chem. Series No. 63, Am. Chem. Soc., 1967.

[13a] R. S. Porter, E. M. Barrall II, and J. F. Johnson, J. Chem. Phys. 45, 1452 (1966).

in dealing with liquid crystals. The crystals generally exist over short temperature intervals defined by small heats of transition, and these features can lead to discrepancies and errors in mesophase identification and in the properties reported.

4. SIGNIFICANCE OF LIQUID-CRYSTAL RHEOLOGY

A defined rheology of liquid crystals is of prime scientific interest. Nematic mesophase viscosities reportedly change fourfold with mesophase orientation in either electric or magnetic fields. The viscosity of smectic and cholesteric phases can change reversibly by decades due to changes in simple shear. Liquid-crystal behavior is also of practical importance, for mesophases in a myriad of commercial and biological systems have been reported. The order in many colloids is analogous to that of liquid crystals.[14,15] Extensive work has been done to show that liquid crystals are similar to emulsions;[15a–15d] mesophase behavior in systems of synthetic detergents, soaps, and dyes has been reported.[3,16,17] Liquid crystals have also been found in synthetic polymers[18] and copolymers[19] and in gels of microcrystalline cellulose.[20]

In biological systems[17,21] the flow of ordered systems of rods, or tactoids, is analogous to the nematic mesophase, and has been reported to exist in polypeptides,[22] tobacco mosaic virus,[15] collagen,[23] paramyosin, and bovine serum albumin.[24,24a] The reports of "blood dust" in the human body[25] and the transport of cholesterol derivatives are probably related to liquid-crystal properties.[21] Haemoglobin, α-keratin, and many other biologically important substances are also able to form liquid crystals.[5]

[14] W. Ostwald and H. Malss, *Kolloid-Z.* **63**, 61 (1933).

[15] C. Robinson, *Trans. Faraday Soc.* **52**, 571 (1956).

[15a] E. Bose, *Physik.-Z.* **8**, 347 (1907).

[15b] E. Bose and F. Conrat, *Physik.-Z.* **9**, 169 (1908).

[15c] E. Bose, *Physik-Z.* **9**, 707 (1908).

[15d] E. Bose, *Physik.-Z.* **10**, 32 (1909).

[16] E. E. Jelley, *Nature* **139**, 631 (1937).

[17] C. Robinson, J. C. Ward, and R. B. Beevers, *Discussions Faraday Soc.* **25**, 29 (1958).

[18] A. Skoulios and G. Finaz, *J. Chim. Phys.* **59**, 473 (1962).

[19] C. Sadron, *Chem. Ind.* (*London*) p. 1230 (1962).

[20] J. Hermans, Jr., *J. Polymer Sci.* **C2**, 129 (1963).

[21] G. T. Stewart, *Nature* **192**, 624 (1961).

[22] C. Robinson and J. C. Ward, *Nature* **180**, 1183 (1957).

[23] E. Fukada and M. Date, *Biorheology* **1**, 101 (1963).

[24] J. W. Allis and J. D. Ferry, *J. Am. Chem. Soc.* **87**, 4681 (1965).

[24a] J. W. Allis and J. D. Ferry, *Proc. Natl. Acad. Sci.* **54**, 369 (1965).

[25] F. Rinne, *Trans. Faraday Soc.* **29**, 1016 (1933).

II. The Nematic Mesophase and Systems of Rods

1. SYSTEMS OF RODS

Many solutions may be represented approximately as systems of rigid rods. The solutions thus are a prototype of the nematic mesophase. The theory of separation into ordered and disordered phases for systems of tactoids or rigid rods is well established.[26] Spontaneously ordered aniso-tropic phases are to be expected, according to the modified lattice theory of Flory, for an assembly of rods without diluent when the length-to-diameter ratio is $\geqslant 2e \geqslant 5.44$. The ratio required for the separation of an anisotropic phase increases with dilution of the rods. The anisotropic state is achieved at the sacrifice of disordering entropy. On the other hand, in the ordered state competition for sequences of sites suitable for occupying rods is diminished.[27] When small brass rods are shaken in a flat dish, a virtually random arrangement is obtained, as long as only a few particles are present, but when the density is increased, assemblies of rods are observed; these assemblies probably are the simple result of the requirements of space filling.[7]

Viscosity studies of solutions simulated by systems of rods, such as sodium desoxyribonucleate, have generally been conducted at low con-centrations in attempts at seeking information about molecular shape or the effect of electric charges. Helders and Ferry have studied sodium desoxyribonucleate at relatively high concentrations, 10^{-4} to 10^{-2} g/ml.[28] Solution viscosities of this compound, which had a molecular weight of 5.8×10^6, were measured with three different types of instrument and indicate considerable overlapping of molecular domains but no liquid-crystal formation. The prominent non-Newtonian flow observed was successfully treated by reduced variables. The results were consistent with a rather stiff coil configuration of the desoxyribonucleate.[28] Darskus and colleagues have studied the concentration effects on the sedimentation velocity and viscosity of desoxyribonucleic acid solutions,[29] and Wada has reported on their chain regularity and flow dichroism.[30]

A study of DNA samples with intrinsic viscosities of $\geqslant 90$ dl/g showed a discrepancy in the viscosity data, which suggests that a shear-sensitive structure persists in solutions with DNA concentrations of only 5×10^{-5} g/ml and less.[31] The optical rotatory power of the DNA polymer helix and

[26] P. J. Flory, *Proc. Roy. Soc.* **A234,** 73 (1956).
[27] J. Hermans, Jr., *J. Colloid Sci.* **17,** 638 (1962).
[28] F. E. Helders and J. D. Ferry, *J. Phys. Chem.* **60,** 1536 (1956).
[29] R. L. Darskus, D. O. Jordan, and T. Kurucsev, *Nature* **201,** 1215 (1964).
[30] A. Wada, *Biopolymers* **2,** 361 (1964).
[31] A. B. Robins, *Trans. Faraday Soc.* **60,** 1344 (1964).

of its complex with acridine orange under streaming conditions has also been studied.[32] Trace impurities have been shown to greatly influence solution viscosities of DNA.[32a] Robins has reviewed the rheology of DNA solutions.[32b] Low-gradient capillary[33] and rotational[34] viscometers applicable to solutions of RNA and related systems have been described.

The molecules of tobacco mosaic virus are also rodlike and have diameters of 150 Å and lengths of 2700 Å and longer.[5,23] In aqueous solutions they associate in several different stages of aggregation and form a gel structure at concentrations greater than 4%. Spindle-shaped tactoids also have been observed; in these the tobacco mosaic molecules are in parallel association.[5,35,35a]

The solution properties of collagen are thought to be similar to those of tobacco mosaic virus.[23] Fukada has studied the dynamic viscosity, rigidity, and steady-flow viscosity of collagen solutions in dilute hydrochloric acid at concentrations of 0.5 to 2.0 g/100 ml.[23] Even at these relatively low concentrations the rheology is that of a gel structure. The observed variation of viscosity with shear rate is similar to the relationship between dynamic viscosity and frequency. Identifying 2π times the frequency with shear rate, the dynamic viscosity is slightly lower than the steady-flow viscosity at high shear rate.[23] A concentration–time reduction scheme superimposes the viscosity data quite well for moderately concentrated solutions of collagen. Zero-frequency viscosity is proportional to the sixth power, and rigidity to the 3.3th power, of concentration. The results appear to confirm that the rodlike molecules of collagen associate with each other, forming a network structure in solution.[23]

The flow properties of cellulose gels have been discussed by Hermans[20] in terms of a particle network model, in which the microcrystals have an axial ratio of about 10. The measurement of two values for yield stress, depending on previous shear rate, is explained by assuming that the elongated particles are oriented by shearing with their long axes almost parallel. A liquid crystal phase is formed in this way.[20] Viscosity anomalies reported for quite a different system, solutions of the dye 1 : 1'-diethyl-ψ-cyanine, have been attributed to rodlike aggregates of dye molecules.[16] Blakeney has also measured the effect of concentration on viscosity using suspensions of straight rigid rods composed of nylon fibers.[35b]

[32] S. F. Mason and A. J. McCaffery, *Nature* **204**, 468 (1964).

[32a] R. L. Scruggs and P. D. Ross, *Biopolymers* **2**, 593 (1964).

[32b] A. B. Robins, *Biorheology* **3**(3), 153 (1966).

[33] Y. P. Vinetskii, *Zh. Fiz. Khim.* **37**, 1512 and 2790 (1963).

[34] E. V. Frisman, L. V. Shchagina, and V. I. Vorob'ev, *Biorheology* **2**, 189 (1965).

[35] J. D. Ferry, *Advan. Protein Chem.* **4**, 1 (1948).

[35a] E. Wada, *J. Polymer Sci.* **14**, 305 (1954).

[35b] W. R. Blakeney, *J. Colloid Sci.* **22**, 324 (1966).

2. SOLUTIONS OF POLY-γ-BENZYL-L-GLUTAMATE AND POLY-γ-BENZYL-D-GLUTAMATE

The liquid-crystal behavior of solutions of polypeptides has been widely reported, most commonly that of poly-γ-benzyl-L-glutamate.[22,27,36,37,38] In certain solvents the molecules of both the L and the D enantiomorphs are in the α helical configuration, intramolecular hydrogen bonding giving considerable rigidity. This rigid-rod configuration is responsible for the stability of liquid-crystal phases in these systems over a considerable concentration range. The large helical twist, which varies with solvent, is visible if observed at right angles to the axis of the helix.[22]

The rodlike α helices of poly-γ-benzyl glutamates show a remarkable viscosity dependence on concentration and shear. The rheological behavior of compounds with the rod configuration in solvents such as water,[36] dioxane,[15] and m-cresol[27,37,38] is notably different from that of the same compounds in such solvents as dichloroacetic acid.[37] In the latter poly-γ-benzyl-L-glutamate is a solvated random coil. The viscosity of poly-γ-benzyl-D-glutamate solutions has been shown to depend markedly on the ionic state of the α-carboxyl side-chain group.[36] Sedimentation, diffusion, and viscosity measurements of solutions of poly-γ-benzyl-L-glutamate have been reported.[39] The Kirkwood–Auer theory[40] gives a good qualitative description of the α helix within poly-γ-methyl-D-glutamates and poly-γ-methyl-L-glutamates in dilute solution.[41] Preliminary results for the D,L copolymer indicate that a small amount of D residues make some imperfection in the α helix.[41]

Figure 2 shows the reduced viscosity (η_{sp}/C) dependence on shear stress for m-cresol solutions of poly-γ-benzyl-L-glutamate.[37] The data were obtained by Yang from a polymer with an M_w of 208,000 and an M_w/M_n of 1.3. The effect of shear stress on the rod configuration is remarkable and is most pronounced at higher concentrations. The viscosity of the coil configuration, by contrast, is only mildly reduced by shear stress over the same range. Figure 2 indicates that the curves of reduced viscosity versus shear stress for the two different forms of the compound (possibly) cross over at comparable concentrations. Over the range studied by Yang the intrinsic viscosity of the rod form changed at constant stress from 2.9 to about 0.2;

[36] H. Edelhoch and R. E. Lippoldt, *Biochim. Biophys. Acta* **45**, 205 (1960).

[37] J. T. Yang, *J. Am. Chem. Soc.* **80**, 1783 (1958).

[38] J. T. Yang, *J. Am. Chem. Soc.* **81**, 3902 (1959).

[39] V. N. Tsvetkov, Y. V. Mitin, I. N. Shtennikova, V. R. Glushenkova, O. V. Tarasova, V. S. Skazka, and N. A. Nikitin, *Vysokomolekul. Soedin.* **7**, No. 6, 1098 (1965).

[40] J. G. Kirkwood and P. L. Auer, *J. Chem. Phys.* **19**, 281 (1951).

[41] H. Tanaka, A. Sakanishi, M. Kaneko, and J. Furuichi, presented at the 14th Annual Rheology Symposium (Japan), Sendai, November, 1965.

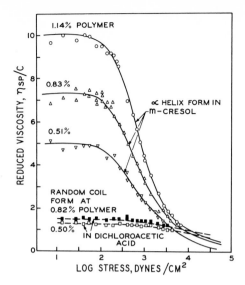

FIG. 2. Non-Newtonian-flow characteristics for poly-γ-benzyl-L-glutamate solutions.

that of the coil form, from 1.2 to about 0.8. The α helix of the rod form remains stable even when subjected to stresses of 5×10^4 dynes/cm². Molecular-weight polydispersity broadens the shear region for non-Newtonian flow.[38] An excellent fit of viscosity versus shear for the coil form in dilute solution has been obtained, however, without consideration of polydispersity.[42] From the theoretical fit an axial ratio and polymer dimensions may be calculated.

An exceptional rheological study of m-cresol solutions of poly-γ-benzyl-L-glutamate has been reported by Hermans.[27] Four different molecular weights were studied in capillaries at concentrations as high as 18 wt-% and at shear stresses of 50 to 4×10^4 dynes/cm². Below a critical concentration the viscosity of the rod form in m-cresol increased with concentration. Because of the formation of an anisotropic (liquid-crystal) phase, however, the viscosity decreased with concentration above a critical concentration. The critical concentration varied inversely with polymer molecular weight, as shown in Fig. 3. When an axial ratio of about 150 was used for molecular weights of nearly 270,000, results are consistent with the theory of Flory.[26,27] The high-concentration liquid-crystal phase is caused by restrictions on rotation, which in turn are caused by volume exclusion, for solutes of high axial ratios. Robinson has shown the same type of result with a viscosity

---

[42] H. A. Scheraga, *J. Chem. Phys.* **23**, 1526 (1955); R. Cert and H. A. Scheraga, *Chem. Rev.* **51**, 185 (1951).

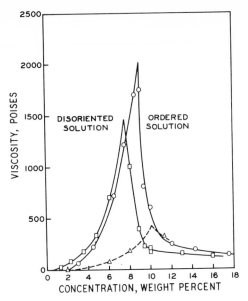

FIG. 3. Viscosity change with concentration for poly-γ-benzyl-L-glutamate in *m*-cresol. Molecular weights: □, 342,000; ○, 270,000; △, 220,000. Stress < 100 dynes/cm².

reduction of 60% on increasing the concentration of poly-γ-benzyl-L-glutamate from 19 to 30 g per 100 g in dioxane.[15] Solutions of lowest concentration were isotropic, whereas more concentrated solutions were highly birefringent. The high optical rotatory power is reportedly reminiscent of the cholesteric mesophase.[15]

Hermans has shown that the curves in Fig. 3 are shear-sensitive and that all viscosities are lower at higher shear, the viscosity maximum disappearing completely in the high-shear Newtonian limit.[27] This means that the ordering produced by statistical interference is not complete. Maximal order in the liquid-crystal phase can, however, be achieved by hydrodynamic shear forces or by further increases in concentration.[43] A complete theoretical interpretation of the rheology of the anisotropic phase has not yet been achieved. The question may also exist whether the α helix structure of poly-γ-benzyl glutamate is the same in dilute and in concentrated solution.

3. ONE-COMPONENT SYSTEMS

a. General Behavior

Most of the reported rheology of liquid crystals has concerned the nematic mesophase. In turn, studies of the nematic mesophase have been dominated

[43] S. H. Bastow and F. P. Bowden, *Proc. Roy. Soc.* **151**, 220 (1935).

by research on a single compound, *p*-azoxyanisole. Its crystal structure reveals the origin of the rodlike structures that exist in the nematic mesophase. Figure 4 shows the stable crystal form as determined by Bernal and Crowfoot.[11] The X and Y coordinates of the molecule are fixed, but the Z coordinate is variable. This is almost certainly due to translation along the favored Z direction.[11] This translation may be due to kinetic energy or may represent a statistical uncertainty in the Z coordinates of the molecule.

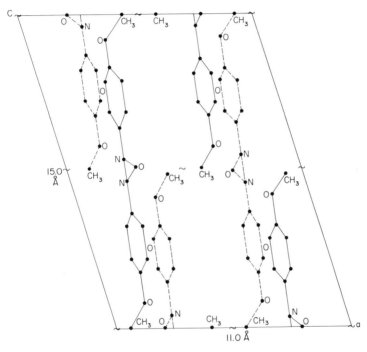

FIG. 4. Stable crystal structure of *p*-azoxyanisole determined by x-ray.

The molecules do not lie in layers but are imbricated, so that the ether group of one molecule is nearly in contact with the azoxy group of its neighbor.[11] These features lead to stabilization of the rod structures in the nematic mesophase, which is formed on heating of the stable crystalline form of the pure compound shown in Fig. 4. Other compounds that form nematic mesophases show similar structural relationships. This type of mesophase is most commonly formed by essentially linear molecules involving aromatic rings with polar groups at intervals along the molecule.[2]

The viscosity of the nematic mesophase is characteristic and unique. The general flow behavior of the nematic mesophase is illustrated in Fig. 5 with

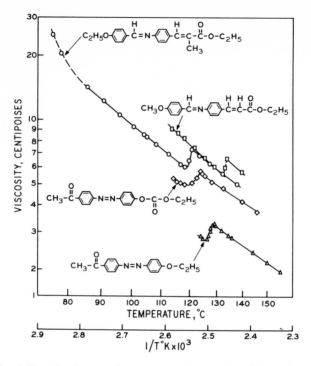

FIG. 5. Viscosity change with temperature for nematic and isotropic states.

some old and previously uncorrelated data, which has been tabulated.[44] Many other pure compounds that form nematic mesophases exhibit equivalent flow behavior.[4,9,10,15b–15d,45–48] Notably, the higher-temperature isotropic phase has a characteristically higher viscosity than the nematic mesophase at lower adjacent temperatures. This is because the mesophase is readily and highly oriented in the direction of flow. The nematic–isotropic transition and its associated visual clarification occur at the steepest viscosity increase; see Fig. 5. Vorlander has independently studied the two most viscous compounds represented in Fig. 5. Higher transition temperatures were reported for these compounds, implying a purity difference.[49] On

[44] Landolt-Börstein, "Physikalisch-chemische Tabellen," Vol. I, p. 165. Springer, Berlin, 1923.

[45] F. Kurger, Physik.-Z. 14, 651 (1913).

[46] M. W. Neufeld, Physik.-Z. 14, 646 (1913).

[47] R. S. Porter and J. F. Johnson, J. Appl. Phys. 34, 51 (1963).

[48] R. S. Porter and J. F. Johnson, J. Phys. Chem. 66, 1826 (1962).

[49] D. Vorlander, Physik.-Z. 31, 428 (1930).

cooling to temperatures lower than those given in Fig. 5 the two most viscous compounds form smectic mesophases. This change is characterized by an additional abrupt viscosity increase at lower transition temperatures.[49]

Different investigators, using viscometers of different dimensions and types, have generally obtained equivalent nematic mesophase viscosities in studies of the same compound. Certain early capillary viscosity data, however, are unusually high near the nematic–isotropic transition.[15d,45,46] The anomalously high values were measured at overpressures in capillary viscometers and were not adjusted with any of the now common capillary corrections. Further uncertainty may be due to the fact that the capillary residence times approached the relaxation times of nematic aggregates. Bose suggested in 1909 that the higher nematic viscosities were due to a Reynolds-type turbulence,[15d] but he was probably wrong.[47] It is also unlikely that the higher nematic viscosities measured at overpressures are due to a breaking up of the nematic structures, as has been suggested,[45,46] because tests in a rotational viscometer show no such viscosity increase of nematic structures even at extreme shear rates, such as $\geqslant 10^5 \, \text{sec}^{-1}$.[47,50,51] Other capillary viscosity data on nematic mesophases do not show an increase with overpressures.[44] It appears that the anomalously high viscosities observed in the early capillary experiments at overpressures are artifacts.

b. Dependence on Purity

Nematic–isotropic transitions are known to involve generally transformation heats that are only a small percentage of normal melting heats.[52] Thus, impurities may strongly suppress nematic–isotropic transitions and have a concomitantly large influence on rheological characteristics. A practical consequence has been observed by Peter and Peters, who report that the viscosity of p-azoxyanisole decreased about 4% in the nematic range on repeated temperature cycling of the pure compound.[51]

Figure 6 shows viscosity measurements, made by Ubbelohde and his associates, of the nematic and isotropic states of mixtures of p-azoxyanisole and phenanthrene. At concentrations of 95% and greater p-azoxyanisole displayed the general and characteristic nematic flow behavior, only the temperature axis being shifted regularly with concentration. Under the conditions shown by the dotted lines in Fig. 6 measurements could not be made, because two liquid phases were present.[53] The viscosity maximum of the pure compound is several degrees lower than those generally reported.[24,48] At concentrations of p-azoxyanisole in phenanthrene of 93%

[50] S. Peter, *Angew. Chem.* **67,** 112 (1955).
[51] S. Peter and H. Peters, *Z. Physik. Chem.* (*Frankfurt*) [N.S.] **3,** 103 (1955).
[52] E. M. Barrall, II, R. S. Porter, and J. F. Johnson, *J. Phys. Chem.* **68,** 2810 (1964).

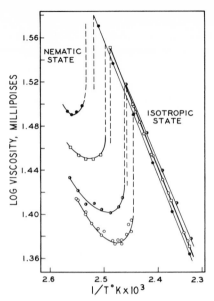

FIG. 6. Viscosity change with temperature: *p*-azoxyanisole plus phenanthrene. Percentage of compound: ●, 95.21; □, 97.16; ◑, 99.15; ○, 100.

and less neither viscosity minima nor maxima were observed; indeed, the nematic mesophase is not observed in this concentration range.[53] The temperature range and the properties of mesophases are more affected by added components than are isotropic liquids. The viscosity and flow activation energy for the isotropic state of *p*-azoxyanisole plus phenanthrene varies in only a small and regular way with composition. Flow rates of about 0.05 to 0.15 ml/min in an 0.2-mm bore capillary indicated a Newtonian flow for all compositions and temperatures covering the ranges of isotropic and mesophase states.[53]

c. Shear Orientation

An extensive study of the shear dependence of viscosity in the nematic and isotropic states of *p*-azoxyanisole has been reported by Peter and Peters.[50,51] An Ubbelohde glass capillary viscometer was used at shear rates of 60 to 18,000 sec⁻¹, calculated at the wall. For the nematic mesophase at low shear rates the non-Newtonian flow shown in Fig. 7 was observed.[51] The lines in this figure accurately represent the data points

---

[53] E. McLaughlin, M. A. Shakespeare, and A. R. Ubbelohde, *Trans. Faraday Soc.* **60**, 25 (1964).

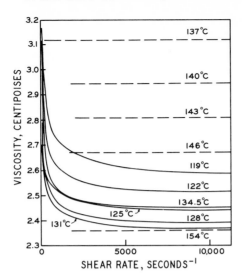

FIG. 7. Viscosity change with capillary shear rate: nematic and isotropic states of *p*-azoxy-anisole. Nematic, ——; isotropic, – – –.

originally given.[51] Newtonian behavior was observed at several temperatures for the isotropic liquid and at shear rates greater than $10,000 \, \text{sec}^{-1}$ in the mesophase.

The highest viscosities at the lowest shear rates, measured by Peter and Peters in the non-Newtonian range of the nematic mesophase, correspond closely to the values of Becherer and Kast, who used the method of Helmholtz.[54] This method involves following the damping of an oscillating sphere filled with the test fluid. This measuring of apparent viscosities should be one of the best ways of studying randomly oriented mesophases.[54] The only mesophase viscosities measured by the Helmholtz method are shown in Fig. 8. This figure also gives some orientation data on *p*-azoxy-anisole, which will be described.

The rate and magnitude of nematic mesophase orientation and the associated change in apparent viscosity can be markedly influenced by adhesion and orientation at surfaces and by the effect of oriented layers on the bulk nematic mesophase. The depth and direction of this orientation can depend on the composition and history of the surface. Such effects have been widely noted and may be responsible for the apparently non-Newtonian-flow results with *p*-azoxyanisole.[15c,47,51,55]

[54] G. Becherer and W. Kast, *Ann. Physik.* [5] **41**, 355 (1942).
[55] G. Foex, *Trans. Faraday Soc.* **29**, 958 (1933).

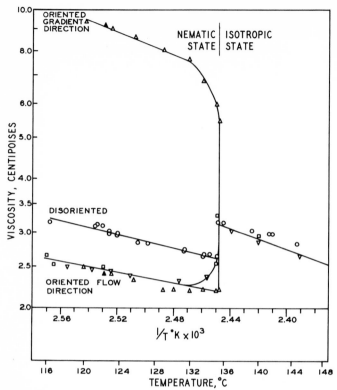

FIG. 8. Viscosity change with magnetic-field orientation: p-azoxyanisole. △, Meisowicz;[59] ▲, Meisowicz;[58] ○, Becherer and Kast; Porter and Johnson, □, rotational viscometry; ▽, capillary viscometry.

d. Magnetic-Field Orientation

As a complement to their studies of viscosity versus shear rate, Peter and Peters measured the light-transparency of the nematic mesophase of p-azoxyanisole in a rotational instrument.[51] They conclude that the shear orientation of nematic aggregates is of a different type from that caused by a magnetic field. Neufeld also has reported magnetic and shear orientation studies of the nematic mesophase of p-azoxyanisole.[46] Orientation caused by shear and by the wall of the capillary, only 0.09 mm in diameter, led to the result that magnetic fields as strong as 8795 gauss across the capillary could not alter maximal mesophase orientation in the flow direction; anisaldazene and p-azoxyanisole were the compounds studied.[46] Tsvetkov and Mikhailov have attempted to orient magnetically the flowing nematic mesophase in a capillary with a rectangular cross section.[56] They found

[56] V. N. Tsvetkov and G. M. Mikhailov, *Acta Physicochim. U.R.S.S.* **8,** 77 (1938).

that the highest viscosity corresponded to the application of a magnetic field perpendicular to the molecular axis, which is also the capillary axis. At high flow rates there was no influence of the magnetic field. At low flow rates the orientation induced by a magnetic field perpendicular to the flow did not reach saturation and was temperature-dependent for fields of almost 10,000 gauss. The difference in viscosity between mesophase flow oriented with, and that oriented without, a perpendicular field for *p*-azoxyanisole was found to be severalfold.[56,57]

The most definitive orientation viscosity studies of nematic mesophases have been published by Miesowicz.[58–60] Orientation was achieved either by magnetic fields or by temperature gradients in the cooling mesophase.[60] Thick layers of liquid were exposed to a low oscillating shear by a plate that could be oriented along the axes relative to a magnetic field. This was accomplished by immersing a glass plate in test fluid contained in a thermostated vessel with parallel walls 6 mm apart. The plate was hung vertically from the beam of an analytical balance, and the damping of plate oscillations was the measure of viscosity. The mesophase viscosities of *p*-azoxyanisole and *p*-azoxyphenetol have been obtained in this way.[58–60] Values have been given for maximal magnetic-field orientation in the direction of (a) flow, (b) velocity gradient, and (c) perpendicular to these. The maximal and minimal viscosities, shown by flow and velocity gradients, respectively, and the viscosities of the isotropic liquid of *p*-azoxyanisole have been measured as a function of temperature by Miesowicz.[59] A reinterpretation of his data is given in Fig. 8 in combination with other viscosities.[48] It is of note that flow-orientation viscosities cannot be reduced further by a magnetic field except near the isotropic transition region. In this region the coefficient of expansion is known to increase markedly. The nematic aggregates probably change in shape and size near the transition. The nature of the viscosity–temperature curve near the isotropic transition was in dispute in the early literature.[15b] The curvature was incorrectly considered an artifact. The temperature dependence of viscosity in the disoriented nematic mesophase is certainly not the same as that in the isotropic phase, as Fig. 8 shows; this has been incorrectly remarked upon[54] and quoted.[1] Viscosities in the disordered state of the nematic mesophase of *p*-azoxyanisole were measured by the method of Helmholtz, described earlier.[54] The viscosity of the nematic mesophase of *p*-azoxyanisole has also been measured by Tsvetkov

[57] V. N. Tsvetkov, *Akad. Nauk. SSSR, Otd. Tekhn. Nauk. Inst. Mashinoved., Soveshch. Vyazkosti Zhidkistei i Kolloidn. Rastvorov, 3rd Conf. 1941* (1945); *Chem. Abstr.* **40**, 3033 (1946).

[58] M. Miesowicz, *Nature* **136**, 261 (1935).

[59] M. Miesowicz, *Bull. Intern. Acad. Polon. Sci. Classe Sci. Math. Nat. Ser. A*, p. 228 (1936); *Chem. Abstr.* **31**, 3354 (1937).

[60] M. Miesowicz, *Nature* **158**, 27 (1946).

in a rotating magnetic field.[61] Results consistent with related experiments suggest that the mesophase exhibited an angular deflection of about 1° in the rotating field. The cross section of the nematic aggregate was calculated to be about 7.1×10^{-5} cm.[61]

Orientation effects are a general property of the nematic mesophase but not of the corresponding higher-temperature isotropic, or true liquid phase. Magnetic and flow-birefringence measurements suggest, however, that there is considerable order at the lowest temperatures in the true liquid state.[62] Non-Newtonian viscosities of the isotropic state have not been measured. However, changes in specific heat and volume expansion suggest, as does the birefringence, a real and progressive molecular association on cooling of the true liquid in a temperature region within a few degrees of the nematic transition.[62] The occurrence of pretransition effects on both sides of nematic–isotropic transitions can be at least qualitatively explained by the heterophase fluctuation theory of Frenkel.[63] Ericksen has considered some magnetohydrodynamic effects in the nematic type of liquid crystals.[63a]

e. *Electric-Field Orientation*

In 1935 Bjornstahl showed the influence of an electric field on apparent nematic mesophase viscosity.[64] A saturation value for the increase in viscosity was postulated at high field strength. The viscosity was derived from the decrement of the damping of a disk, which was immersed in the test fluids, *p*-azoxyanisole and *p*-azoxyphenetol. An increase in decrement was found when the electric field was applied perpendicular to the disk, that is, parallel to the velocity gradient. Later Bjornstahl described his first electric-field orientation experiments as rather primitive.[65] In his later experiments he used an electric field on the same two compounds during shear in a Couette viscometer. When the direct-current field was increased at 145°C., the mesophase viscosity of *p*-azoxyphenetol started to change at about 200 volts/cm and increased rapidly with increasing field strength. Viscosity measurements were made at several speeds of Couette viscometer rotation. Only at the lowest shear rate, at a speed of $\Omega/2\pi = 0.63$, was an apparent saturation viscosity reached at the maximal field of about 900 volts/cm ; see Fig. 9. Experiments were also made at these same conditions but with an alternating current. The alternating current had less effect on

[61] V. N. Tsvetkov, *Acta Physicochim. URSS* **11**, 97 (1939).

[62] V. N. Tsvetkov, *Acta Physicochim. URSS* **19**, 96 (1944).

[63] J. Frenkel, "Kinetic Theory of Liquids." Oxford Univ. Press (Clarendon), London and New York, 1946.

[63a] J. L. Ericksen, *Arch. Rational Mech. Anal.* **23**, 266 (1966).

[64] Y. Bjornstahl, *Physics* **6**, 257 (1935).

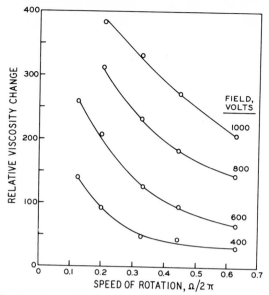

FIG. 9. Viscosity change with shear rate in an electric field: p-azoxyphenetol at 145°C.

nematic mesophase viscosity than the direct current, and the effect further decreases with increasing frequency.[65]

The electric conductivity of the nematic mesophase reportedly decreases in the direction of the shear gradient for flow.[64] The electric field apparently counteracts the orientation due to shear and works toward an orientation parallel to the field. These conclusions are purportedly in agreement with certain other orientation studies.[65] However, Mikhailov and Tsvetkov have reported contradictory results for nematic mesophase orientation in an electric field.[66] They measured the mesophase capillary flow of p-azoxyanisole in an alternating electric field of 0 to 5×10^5 hz. The field could be maintained either parallel or perpendicular to the direction of flow. It was concluded that the molecules tend to arrange themselves at all frequencies perpendicular to the lines of force. That is, longitudinal electric fields decrease flow time, and transverse low-frequency fields increase it.[66] At the temperature for the transition to the isotropic phase, and presumably at higher temperatures, no effect of electric field was observed. The orientation observed by Tsvetkov has been postulated to be due to the current generated in the nematic phase rather than the orienting effect of the electric field itself.[66]

More recently some interesting information concerning the orientation of nematic mesophases in direct-current and alternating-current electric

[65] Y. Bjornstahl and O. Snellman, *Kolloid-Z.* **86**, 223 (1939).
[66] G. M. Mikhailov and V. N. Tsvetkov, *Acta Physicochim. URSS* **10**, 415 (1939).

fields has been published by Carr.[67] Microwave techniques involving dielectric measurements of high-purity p-azoxyanisole indicate a mesophase alignment of the long axes of the molecules perpendicular to the externally applied field. Alignment parallel to the field, which has also been reported, appears to be due to extremely small amounts of impurity.[67] This leads to a conductivity which has a maximum in the direction of the field. At modest electric-field strengths a "stirring action" due to orientation of the nematic mesophase has been observed with the unaided eye in studies of p-azoxyanisole.[68]

III. The Cholesteric Mesophase

1. DESCRIPTIVE RESULTS

The flow properties of cholesteric mesophases of several esters of cholesterol have been observed. The viscosity data have concerned the acetate, [69,69a,70] propionate,[13,69,69a,71] butyrate,[69,69a] hexanoate,[12] nonanoate,[12] palmitate,[70] stearate,[70] benzoate,[7,9,10,15a] and the carbonate ethyl ester of cholesterol.[13,14,72]

The intense colors of cholesteric mesophases reportedly change with mechanical stress.[6] The flow of this type of mesophase probably influences many biological processes.[21] The high optical rotatory power[15] and the even structure[73] of poly-γ-benzyl-L-glutamate solutions also have been considered to be akin to the cholesteric mesophase, although glutamate derivatives have been claimed by the various investigators to exhibit typical nematic mesophase behavior.[22]

Figure 10 is a rare pictorial description of flow orientation in a cholesteric mesophase. The figure shows that molecular aggregates in the mesophase of cholesteryl myristate readily orient in the direction of flow.* Gaubert obtained a similar photograph in 1913 with crossed nicols and a magnification of 60 to 100.[74] Optical observations of cholesteryl bromide at room temperature also indicate that uniaxial orientation can be easily achieved mechanically.[75,76] Aggregates in cholesteric mesophases are reportedly

---

* This light microscope picture is reproduced by the courtesy of Dr. Edward M. Barrall II, Chevron Research Company, Richmond, California.

[67] E. F. Carr, J. Chem. Phys. 39, 1979 (1963).
[68] R. Williams, J. Chem. Phys. 39, 384 (1963).
[69] W. Ostwald and H. Malss, Kolloid-Z. 63, 192 (1933).
[69a] W. Ostwald, Trans. Faraday Soc. 29, 1002 and 1022 (1933).
[70] R. S. Porter and J. F. Johnson, J. Appl. Phys. 34, 55 (1963).
[71] A. S. C. Lawrence, Trans. Faraday Soc. 29, 1033 and 1080 (1933).
[72] D. Vorlander, Z. Krist. 79, 61 (1931).
[73] C. Robinson, Tetrahedron 13, 219 (1961).
[74] P. Gaubert, Bull. Soc. Franc. Mineral. 35, 64 (1913).
[75] J. Fischer, Z. Physik. Chem. A160, 110 (1932).
[76] D. Vorlander, Trans. Faraday Soc. 29, 90 (1933).

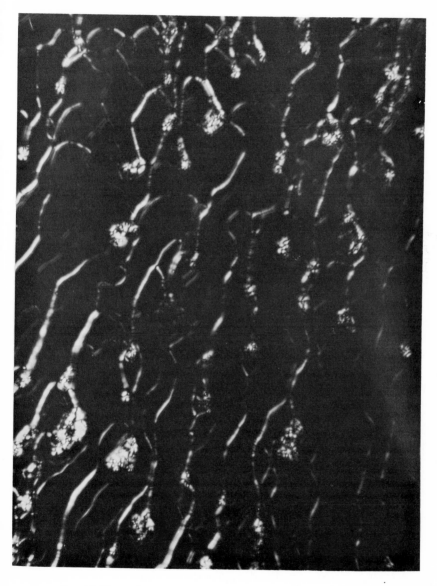

Fig. 10. Flow orientation in the mesophase of cholesteryl myristate.

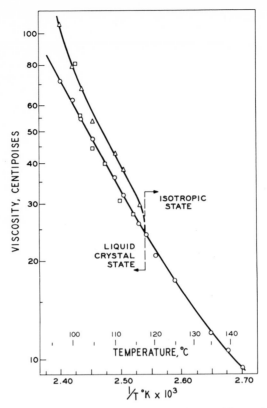

FIG. 11. Viscosity change with temperature for mesophase and isotropic states of cholesteryl acetate. △, $10^4 \sec^{-1}$; ○, isotropic and high-shear limit; □, Ostwald's highest viscosities.

smaller in size than in the nematic mesophases.[70] The shapes and relative dimensions of nematic and cholesteric aggregates are also grossly different.

2. MESOPHASE RHEOLOGY

Figure 11 shows how the viscosity of cholesteryl acetate changes with temperature and shear. The results typify the data on other esters of cholesterol. In the isotropic state concentric cylinder and capillary viscosity measurements indicate entirely Newtonian flow at shear rates to $3 \times 10^5 \sec^{-1}$. Viscosity–shear anomalies have been reported at temperatures for the isotropic melt of propionate and butyrate esters after conditioning near the temperature of transition to the mesophase.[69] Non-Newtonian behavior is uniformly observed in the temperature range for the mesophase, where the magnitude of viscosity change with shear increases with decreasing temperature. The apparent viscosities of cholesteric mesophases undoubtedly can change many tens of times as a function of shear. Limiting mesophase viscosities at low shear have not been measured, because the esters are generally too viscous and may exhibit a yield stress.[13,69,70]

Limiting mesophase viscosities of cholesteryl esters at high shear can generally be found, although stresses to 10^4 to 10^5 dynes/cm^2 are required. Limiting high-shear Newtonian viscosities of a mesophase form a continuous series with higher-temperature viscosity measurements of the isotropic liquid; see Fig. 11. This is in contrast to the viscosity maximum at mesophase–isotropic transitions, which is characteristic of nematic liquid crystals. Such behavior of the cholesteric mesophase does not necessarily imply that mesophase aggregates are completely destroyed by shear. Aggregate orientation can also contribute markedly to viscosity decreases due to shear.

A curve of the general shape given in Fig. 11 was reported by Bose in 1907 for cholesteryl benzoate in terms of relative scales for viscosity and temperature.[15a] Anomalies corresponding to "indefinitely" high viscosities of several esters at cholesteric–isotropic transitions have been reported[14,69,72] and republished.[1,5,71] The effect varied with shear rate and the direction of temperature from which the transition was approached.[72] Indeed, this observation of an indefinitely high viscosity is not characteristic of the cholesteric mesophase and has been said to be fictitious,[71] possibly due to turbulence, density fluctuation, or other characteristics inherent at critical condition. Measurements of the mesophases of several cholesteryl esters made by capillary and by rotational viscometers indicate that the apparent viscosity maximum at the isotropic transition decreases and disappears completely at increasing rates of shear.[70,71]

3. INFLUENCE OF PURITY

Purity has a marked effect on the viscosity of cholesteric mesophases. For example, a recrystallized acetate ester at 100°C. and a shear rate of 10^5 sec^{-1} had a viscosity of 64 cp; an unpurified commercial sample had a value of 20 cp less under the same conditions.[70] This difference was probably not due to the large reduction of the cholesteric–isotropic transition temperature by impurity because the comparison was made well within the range of non-Newtonian flow and visually observed liquid crystals.[70] The viscosity drop likely occurs through the properties of aggregates formed in the mesophases after exclusion of impurities, as happens in normal crystallization. This also is the likely reason for the low cholesteric mesophase viscosities of the acetate ester reported by Ostwald.[13,69] These results are included in Fig. 11. The relative values given by Ostwald have been established on an absolute scale by comparison with other ester data, reported in centipoises, for the isotropic state of the acetate ester. In the isotropic temperature range cholesteryl ester purity should have only a minor influence on viscosity. The Ostwald capillary pressure unit of 1.0 approximates a shear stress of 5.6×10^3 dynes/cm^2. This is a correction of a previously stated conversion factor.[70]

4. COMPARISON AMONG ESTERS

An interesting comparison may be made among the available data on cholesteryl esters. Table I contains a compilation of ester viscosities obtained at 117°C., which is a temperature within the isotropic-liquid range for each of the five esters. All values are quite similar. The acetate ester is slightly higher, possibly because the comparison is made within a degree of its cholesteric mesophase transition. Indeed, at temperatures higher than 140°C. the acetate has the lowest viscosity of the five esters by a slight amount, indicating a higher flow activation energy. Table I summarizes the available flow activation energies of the esters.

TABLE I

VISCOSITIES AND FLOW ACTIVATION ENERGIES OF CHOLESTERYL ESTERS

Cholesteryl ester	Viscosity at 117°C, cp	Flow activation energy, kcal/mole	
		Liquid crystal[a]	Isotropic liquid at 125°C
Acetate	25.8	15.5	11.9
Propionate	22.1	—	—
Butyrate	22.3	—	—
Palmitate	21.4	~10.0	8.1
Stearate	22.0	11.9	8.0

[a] High-shear limit, average over liquid-crystal range.

It is important to note that viscosity anomalies were not observed when the temperature ranges for smectic–cholesteric transitions of the palmitate and stearate esters were crossed.[70] If both mesophases were present, the viscosity difference between them was small for these test conditions. The cholesteric mesophase has been considered a variation of the smectic phase, although differing in optical properties.[2]

IV. The Smectic Mesophase and Soap Systems

1. SOAP SYSTEMS

One-component smectic mesophase and soap systems generally possess a high viscosity and considerable rigidity.[15a,43,69] It has been reported that normal liquid flow does not occur in the smectic mesophase, the move-

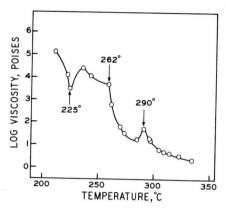

FIG. 12. "Residual" viscosity change with temperature for anhydrous sodium stearate.

ment being one of gliding in one plane.[77] A common example of a smectic mesophase is the double layers of inner and outer surfaces of a soap bubble.[6] *Smectic* means soaplike, and the mesophases of soap compounds are nominally considered smectic.[43,78,79]

The mesophase viscosities of a number of pure soaps and soap solutions have been reported. The viscosity and rigidity of the mesophase formed by a 1 % solution of ammonium oleate in water at 20°C. has been reported.[43] Extrusion plastomer measurements at 140°C. on a molecular-weight series of pure straight-chain sodium soaps indicate an abrupt increase in the yield value when the number of carbon atoms in the soap drops below eight.[79] The flow of pure and technical anhydrous sodium stearate from 45 to 336°C. has been studied as a function of shear by Puddington.[78,79] Soap blocks were melted in vacuo for removal of traces of water. Large changes in soap mobility were observed at 68 and 105°C. The compound is generally non-Newtonian below, and Newtonian above, 298°C. This is the isotropic transition temperature observed viscometrically. The "residual" viscosity reported for "infinite rate of shear" for pure sodium stearate is shown in Fig. 12 as a function of temperature from 211 to 336°C. These viscosities were calculated in an unconventional manner, by extrapolating to infinite flow rate on a plot of viscosity versus reciprocal flow rate. The correlation shown in the figure shows the general change to increasing viscosity at lower temperatures for pure sodium stearate with discontinuities at 225, 262, and 290°C. The two higher temperature transitions correspond to the subneat-neat and isotropic transitions as elucidated by other measurement

[77] S. Glasstone, "Textbook of Physical Chemistry," 2nd ed., p. 515. Van Nostrand, Princeton, New Jersey, 1946.
[78] B. D. Powell and I. E. Puddington, *Can. J. Chem.* **31,** 828 (1953).
[79] F. W. Southam and I. E. Puddington, *Can. J. Res.* **B-25,** 125 (1947).

techniques. The pure sodium stearate reportedly is thixotropic, and therefore may account for an apparent 8°C. variation in the measurement of isotropic–mesophase transitions.[78]

2. Other Compounds

The rheological characteristics of mesophases of the smectic type have been investigated with ethyl p-azoxybenzoate and only a few other pure compounds such as ethyl p-azoxycinnamate.[7] This type of compound has compositions and structures similar to materials that form nematic mesophases.[2,7,14,69] The viscosity of smectic mesophases, however, is several times that of nematic phases, because the molecules in the smectic state are not only parallel to each other, but their ends lie in the same plane and so produce a layered structure.[6]

The most extensive work on smectic mesophase rheology exists for ethyl p-azoxybenzoate, which was first studied by E. Eichwald (*Inaugural-dissertation*, Marburg 1905).[10,76] Above its isotropic transition temperature, this compound exhibits entirely Newtonian behavior over a wide shear range. The mesophase viscosity is high and prominently non-Newtonian. This is illustrated in Fig. 13 with the data of Ostwald.[13,69] The solid lines accurately describe the data originally given. A table of viscosities as a function of capillary pressure, besides a plot of the data, was published twice identically

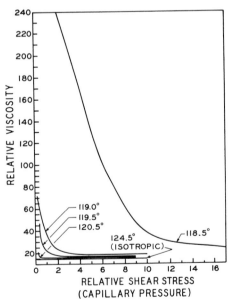

Fig. 13. Viscosity change with shear stress: smectic and isotropic state of ethyl p-azoxybenzoate.

in 1933.[13,69] Viscosity measurements were not readily obtained at temperatures near the transition to the true solid phase. The capillary meniscus could not be seen well. At the lower temperatures in the mesophase range the highly viscous dark-red melt showed properties of a "jelly"; that is, it was distinctly elastic.[69]

Smectic mesophase viscosities are generally higher than isotropic ones at about the same temperatures. The shear range of smectic mesophase viscosities may be defined by the same method as cholesteric mesophases; see Fig. 11. Rheological measurements thus represent a simple and definitive method of distinguishing smectic and cholesteric mesophases from nematic ones. Present data, however, indicate that cholesteric and smectic mesophases, which may be classified by optical and other means, cannot be distinguished rheologically. As for soap systems, viscosity measurements in certain cases can distinguish transitions between mesophases; note Figure 12. Multiple mesophases of the smectic and nematic type are formed by ethyl p-(4-ethoxybenzylideneamino)-α-methylcinnamate and by ethyl p-(4-methoxybenzylideneamino)cinnamate.[1,49] Vorlander has shown that abrupt changes in the viscosities of these compounds can be observed at each of the mesophase–mesophase transitions and also at the mesophase–isotropic transition.[1,49]

V. Discussion

1. APPLICATION OF FLOW THEORIES

An adequate, complete rheological theory of mesophase types has yet to be developed. Even the mechanism responsible for non-Newtonian flow of cholesteric and smectic mesophases is yet to be defined. Our understanding of the nematic mesophase is on a considerably sounder foundation.[26,40] Some related discussion is included in Section II,1. The condition for organization and orientation of assemblies of rods is well established, essentially in terms of excluded volume for rotation, and the theory of shear orientation in hydrodynamic flow of dilute systems of rods and of other defined shapes also has been generally confirmed experimentally.[42,80,*] For example, the theoretical interpretation of the non-Newtonian viscosity of solutions of ellipsoidal particles has been given by Kuhn and Kuhn and by Saito, who used Peterlin's distribution function for the orientation of particle axes in a streaming liquid.[42] The energy dissipated through both hydrodynamic orientation and Brownian motion has been calculated on the basis

---

* Cf. Chapter 2 of this volume and also Chapter 14 by Frisch and Simha, Volume 1 of this series.

[80] J. G. Kirkwood and R. J. Plock, *J. Chem. Phys.* **24**, 665 (1956).

of Jeffrey's general relations.[81,82] Raasch has also considered the dynamics in viscous shear flow of single infinitely long particles in planar flow, near a wall, and on interacting with other particles.[82] Finally, formulas have been derived for the rate of dissipation of energy due to the motion of a rigid rotational ellipsoid in a uniaxially drawn liquid.[83] A theory for the non-Newtonian viscosity of solutions of rodlike particles has been developed by Kirkwood,[40,80] and calculations and discussion based on this work have been given by Scheraga.[42] Extensive data on solutions of rodlike polypeptides confirm the theory[38,41] and suggest that the lengths of rigid particles may be estimated from their intrinsic viscosity at zero shear.[84] The effect of polydispersity on non-Newtonian flow has been theoretically treated with respect to solutions of prolate ellipsoids with axial ratios of 300.[85]

A fragmentary interpretation suggests that changes in specific volume may be used for expressing the viscosity changes of liquid crystals.[86] Continuum theories of the mechanical behavior of liquid crystals are still unsettled. A solution for the nematic mesophase has been offered.[86a] Ericksen also suggests that a partial explanation is available in the hydrostatic theory summarized by Frank.[87,88] It has been shown that the commonly used forms of the laws of conservation of linear momentum, moment of momentum, and energy seem inadequate for describing liquid crystals. Revised forms, however, have been given,[87] and special boundary conditions for anisotropic flow have been treated by continuum mechanics.[89,90] Wang has reported on the symmetries of certain hyperelastic simple liquid crystals,[90a] and Coleman has reported mathematical examples of simple materials having isotropy groups appropriate to liquid crystals.[90b]

2. VISCOELASTICITY

A complete interpretation of the frequency dependence of shear storage and loss moduli has been given for dilute poly-γ-benzyl-L-glutamate in a

[81] G. B. Jeffery, *Proc. Roy. Soc.* **A102**, 161 (1922).

[82] J. Raasch, Doctor of Engineering Thesis, Karlsruhe Technical University, Germany, 1961.

[83] R. Takserman-Krozer and A. Ziabicki, *J. Polymer Sci.* **A1**, 507 (1963).

[84] J. T. Yang, *J. Polymer Sci.* **54**, 514 (1961).

[85] M. E. Reichmann, *J. Phys. Chem.* **63**, 638 (1959).

[86] R. O. Herzog and H. Kudar, *Trans. Faraday Soc.* **29**, 1006 (1933).

[86a] J. L. Ericksen, *Phys. Fluids* **9**, 1205 (1966).

[87] J. L. Ericksen, *Trans. Soc. Rheol.* **5**, 23 (1961); *ibid* to be published.

[88] F. C. Frank, *Discussions Faraday Soc.* **25**, 19 (1958).

[89] F. M. Leslie, *Proc. Cambridge Phil. Soc.* **60**, 949 (1964).

[90] F. M. Leslie, *J. Fluid Mech.* **18**, 595 (1964).

[90a] C. C. Wang, *Proc. Natl. Acad. Sci.* **55**, 468 (1966).

[90b] B. D. Coleman, *Arch. Rat. Mech. Anal.* **20**, 41 (1965).

helicogenic solvent.[41,91] The results were intermediate between the predictions of the Kirkwood–Auer theory for rigid rods[40] and the Zimm theory* for flexible random coils with dominant hydrodynamic interactions. This indicated that behavior at low frequency is rodlike but that at high frequency the helix displays some flexibility.[91] Comparable measurements have been reported on poly-γ-methyl-D-glutamate.[91a] Measurements of the same type, as well as steady-flow viscosities, have been reported for dilute solutions of collagen.[23] At a concentration of 1.0 g per 100 ml in dilute HCl the rod-like molecules associate and form a network structure. The dynamic viscosities were found to be lower at high and equivalent shear rates than the apparent steady-flow viscosities.[23] The results of some similar tests are given in Section II,1.

As predicted by the Kirkwood–Auer theory,[40] elasticity measured by transverse-wave propagation has been observed in dilute solutions of rod-like bovine fibrinogen.[92] The microelasticity and structure of keratin and collagen also have been discussed.[93] Normal stress[93a] and qualitative elastic effects have been reported for a variety of liquid-crystal type of systems. By visual and optical means the elasticities of solutions of polypeptides,[15] the smectic mesophase,[69] and a variety of other pure compounds that form liquid crystals have been estimated.[94] Unfortunately, the theory of nonlinear flow as discussed in the next chapter is not yet adequately adapted to strongly anisotropic fluids.

* See Chapter 1 of Volume III of this series.

[91] N. W. Tschoegl and J. D. Ferry, *J. Am. Chem. Soc.* **86**, 1474 (1964).

[91a] H. Tanaka, A. Sakanishi, and J. Furuichi, *Rept. Progr. Polymer Phys. Japan* **8**, 193 (1965).

[92] J. D. Ferry and F. E. Helders, *Biochim. Biophys. Acta* **23**, 569 (1957).

[93] P. Mason, *Kolloid-Z.* **202**, 139 (1965).

[93a] T. Kotaki, *J. Chem. Phys.* **30**, 1566 (1959).

[94] V. Frederiks and V. Zolina, *Trans. Faraday Soc.* **29**, 919 (1933).

ACKNOWLEDGMENTS

The authors are pleased to acknowledge the courtesy of the following copyright owners in permitting them to reprint some of the figures reproduced here: the American Chemical Society for Fig. 2 (from the *Journal of the American Chemical Society*) and Fig. 8 (from the *Journal of Physical Chemistry*), the Faraday Society for Figs. 4, 6, and 13 (from *Transactions of the Faraday Society*), the American Institute of Physics for Fig. 11 (from the *Journal of Applied Physics*), Academic Press, Inc., for Fig. 3 (from the *Journal of Colloid Science*), *Kolloid-Zeitschrift* for Fig. 9, the *Canadian Journal of Chemistry* for Fig. 12, *Zeitschrift für physikalische Chemie* for Fig. 7, and *Scientific American* for Fig. 1.

CHAPTER 6

NONLINEAR STEADY-FLOW BEHAVIOR

Hershel Markovitz

I. Introduction

Rheology, in general, is the study of the deformation and flow of matter. The actual field of interest and activity of most rheologists is more restricted. It is more accurately described in a statement which appears in a brochure of the Society of Rheology (U.S.): "Rheology, a branch of mechanics, is the study of those properties of materials which determine their response to mechanical force." The rheologist, who attempts to isolate this particular aspect of the total mechanics problem, deals as much as possible with relatively simple deformations of materials that exhibit complex mechanical behavior. He leaves the investigation of the complex deformations of simple materials to the students of classical hydrodynamics, fluid dynamics, and elasticity. In this chapter we shall deal with the phenomenological aspects

347

of these material properties which determine the flow behavior of non-Newtonian fluids. We shall make only brief allusions to molecular theories which have been proposed to explain these properties.

There are several phenomena familiar to rheologists that may be used to typify the complex mechanical behavior of fluids in simple deformations. One of these is *shear-dependent viscosity*, the dependence of the viscosity on the rate of shear in steady flow. This effect shows up in a variety of experiments used to determine the viscosity: lack of proportionality between rate of flow and applied pressure head in flow through tubes, or between applied torque and angular velocity of rotation in Couette and cone–plate viscometers.

Another phenomenon is the existence of *normal stress effects*. In steady Couette flow, fluids sometimes exert a greater normal thrust on the inner cylinder than on the outer—just the opposite from what is expected on the basis of centrifugal effects alone. In cone–plate flow a greater normal thrust is exerted on the plate near the center than at the edge. Similar phenomena occur in other types of flow.

In addition to this steady-flow behavior there are time-dependent effects that are usually associated with experiments in viscoelasticity: creep, stress-relaxation, etc.

Our main emphasis in this chapter will be the normal stress effect, but we shall find that nonlinear viscosity is intimately associated with it. In fact, we shall find that these steady-flow phenomena cannot be properly described unless the viscoelastic behavior also is taken into account in formulating the theory.

In the next section we deal with the general theory of our subject. To put the problem in its proper framework we first review the general approach to the solution of problems in the mechanics of continua. An important task here is the mathematization of the "properties of materials which determine their response to mechanical force," that is, the formulation of a *constitutive equation* which tells how the force and motion are related for the material. This question has received the attention of many theoreticians in recent decades. The exposition given here includes a discussion of the motivation of some of the theories, their inadequacies, and their relation to the more classical (perfect, Newtonian, and linear viscoelastic) fluids. Even for the most general of these theories, that of the memory fluid, it is possible to obtain solutions for a class of steady flows (viscometric flows) in terms of three functions that are characteristic of the material. In Section III both the theoretical and experimental aspects of some viscometric flows are presented. These include Couette flow, axial flow between coaxial cylinders, cone–plate flow, and torsional flow. In Section IV we discuss the viscometric functions that have been determined experimentally. We also briefly mention some mechanistic theories which include normal stress effects. In the final section we indicate the behavior of the memory fluid in some flows that are not viscometric.

II. General Theory

In this section we discuss the mathematical theories of the continuum mechanics of fluids. This is not intended to be a mathematical treatise; nor do we seek the utmost generality. The purpose of this discussion is to give an indication of the concepts that are employed, the assumptions on which the theories are based, the mathematical methods that are used, and the results that are obtained. We thus proceed with a minimum of mathematics (and with a consequent loss of exactitude) to give an outline of what has been an exciting and productive theoretical endeavor. We hope to convey the physical concepts behind the mathematical expressions, even though we gloss over a considerable part of the theoretical structure and difficulties involved.

Readers who are dissatisfied with the weaknesses in this presentation and who seek a more rigorous approach are urged to consult the original works and the more scholarly reviews of the subject. Even those of us who cannot comprehend those writings completely are rewarded, for such efforts as we exert, by obtaining a deeper appreciation of the beauty and accomplishments of the subject. This field is blessed with authors who write with clarity, who examine difficulties with care rather than gloss over them, and who are not above spicing their weighty tomes with dashes of humor. C. Truesdell has recently completed, in collaboration with Toupin and Noll, comprehensive reviews of the field.[1,2] A less formal presentation may be found in a series of lectures that he delivered at an industrial research laboratory, and that was subsequently published.[3] The theory discussed in this chapter, based mainly on these works and on the exposition in a recent monograph,[4] follows the approach of Coleman and Noll as given therein.

Parts of the theory or related theories have been developed independently by others, frequently with different emphasis and from other points of view.

[1] C. Truesdell and R. Toupin, in "Handbuch der Physik" (S. Flügge, ed.), Vol. 3, Pt. I, Springer, Berlin, 1960.

[2] C. Truesdell and W. Noll, in "Handbuch der Physik" (S. Flügge, ed.), Vol. 3, Pt. Springer, Berlin, 1965.

[3] C. Truesdell, "The Principles of Continuum Mechanics." Socony-Mobil Oil Co., Texas, 1961.

[4] B. D. Coleman, H. Markovitz, and W. Noll, "The Viscometric Flows of Non-Ne Fluids." Springer, Berlin, 1966.

Some of them have been reviewed by other authors in this treatise[5-7] and elsewhere.[8-13]

Some of the concepts used in this subject date back many centuries, but most of the development has taken place within the last two decades or so. Neither in this section nor in the remainder of this chapter do we attempt to assign precedence for various advances. Considerations of these sometimes controversial questions have appeared in numerous historical treatments, to which the interested reader is referred.[1,2,4,9,10,13-15] References given in this chapter are intended to be further sources of information for the reader rather than necessarily being assignments of priority.

The natural mathematical language for our discussion is that of vectors and tensors; it is assumed that our reader has at least a slight acquaintance with these concepts. An appendix to this article gives some useful formulas and explains some of the notation and terminology.

We shall limit our discussion here to purely mechanical phenomena: forces and motions. Thus, not only do we ignore certain classes of experiments (for example, those involving electric and magnetic fields), but we ignore an important aspect of all processes, thermodynamics.

1. MOTION

Here we are concerned only with those motions during which the body is being strained, that is, during which there are changes in the distances between the material points of which the body is composed. There are several ways in which the strain may be characterized.

[5] M. Reiner, in "Rheology: Theory and Applications" (F. R. Eirich, ed.), Vol. 1, Chapt. 2. Academic Press, New York, 1956.

[6] J. G. Oldroyd, in "Rheology: Theory and Applications" (F. R. Eirich, ed.), Vol. 1, Chapt. 16. Academic Press, New York, 1956.

[7] S. Oka, in "Rheology: Theory and Applications" (F. R. Eirich, ed.), Vol. 3, Chapt. 2. Academic Press, New York, 1958.

[8] R. S. Rivlin, J. Rational Mech. Anal. 5, 179 (1956).

[9] R. S. Rivlin, in "Phénomènes de Relaxation et du Fluage en Rhéologie Non-Linéaire," p. 83. C.N.R.S., Paris, 1961.

[10] R. S. Rivlin, in "Research Frontiers in Fluid Dynamics" (R. J. Seeger and G. Temple, eds.), p. 144. Wiley, New York, 1965.

[11] A. C. Eringen, "Nonlinear Theory of Continuous Media." McGraw-Hill, New York, 1962.

[12] A. G. Fredrickson, "Principles and Applications of Rheology." Prentice-Hall, Englewood Cliffs, New Jersey, 1964.

[13] A. S. Lodge, "Elastic Liquids." Academic Press, New York, 1964.

[14] R. S. Rivlin, Phys. Today 12, 32 (1959).

[15] A. Jobling and J. E. Roberts, in "Rheology: Theory and Applications" (F. R. Eirich, ed.), Vol. 2, Chapt. 13. Academic Press, New York, 1958.

Let us refer to the (not necessarily unstrained) configuration of the body at a time t as the *reference configuration*. Let x_k be the coordinates of a particular material point in the reference configuration with respect to a specified Cartesian coordinate system.[16] The material point may thus be labeled by stating the values of x_k and t. As the material deforms, the position of the material point changes. We refer to the coordinates of the same material point at some other time τ as $\xi_i = \xi_i(\tau)$. When we wish to remind ourselves of the dependence of ξ_i on τ and on the material point, we write

$$\xi_i = \xi_i(x_k, \tau; t) \tag{1}$$

The gradient F_{ij} of ξ_i with respect to x_k, defined by the equation

$$F_{ij} = F_{ij}(\tau) = F_{ij}(x_k, \tau; t) = \partial \xi_i / \partial x_k \tag{2}$$

is a measure of the deformation called the *relative deformation gradient*. When $t = \tau$ and therefore no motion has taken place with respect to the reference configuration, then

$$F_{ij} = F_{ij}(x_k, t; t) = \delta_{ij} \tag{3}$$

where δ_{ij}, the Kronecker delta, is 0 when $i \neq j$ and 1 when $i = j$.

However, equation (3) does not hold for all motions when the material is not being strained (that is, when distances between points are not being changed). A better (but not the only) measure of the actual strain is the symmetric tensor formed from the F_{ij} according to the relation

$$J_{ij} = \sum_k F_{ki} F_{kj} - \delta_{ij} = \sum_k \frac{\partial \xi_k}{\partial x_i} \frac{\partial \xi_k}{\partial x_j} - \delta_{ij} \tag{4}$$

For any rigid motion it can be shown that

$$J_{ij} = 0 \tag{5}$$

If $J_{ij} \neq 0$, the material is strained relative to the reference configuration. In the limiting case of small deformations the *infinitesimal strain tensor* E_{ij} of the classical theory of elasticity is related to J_{ij} by the expression

$$J_{ij} = 2[E_{ij}(\tau) - E_{ij}(t)] \tag{6}$$

If the reference configuration happens to be one that is unstrained (that is, $E_{ij}(t) = 0$), then J_{ij} is twice the infinitesimal strain.

A measure of the rate of straining at time t may be obtained from the gradient L_{ij} of the velocity $\dot{x}_i$ of the material particle. It can be shown that

---

[16] To avoid complicating the equations more than is necessary for our purpose we employ Cartesian coordinate systems through most of this section. Thus, the distinction between covariant, mixed, and contravariant components is unnecessary here.

L_{ij} is related to $F_{ij}(x_k, \tau; t)$ by the equation

$$L_{ij} = L_{ij}(t) = \frac{\partial \dot{x}_i}{\partial x_j} = \frac{d}{d\tau} F_{ij} \bigg|_{\tau = t} \qquad (7)$$

The symmetric tensor D_{ij} formed from L_{ij},

$$D_{ij} = \frac{1}{2}(L_{ij} + L_{ji}) = \frac{1}{2}\left(\frac{\partial \dot{x}_i}{\partial x_j} + \frac{\partial \dot{x}_j}{\partial x_i}\right) \qquad (8)$$

is called the *rate of deformation tensor*. The antisymmetric part, W_{ij}, of L_{ij}, defined by the equation

$$W_{ij} = \frac{1}{2}(L_{ij} - L_{ji}) = \frac{1}{2}\left(\frac{\partial \dot{x}_i}{\partial x_j} - \frac{\partial \dot{x}_j}{\partial x_i}\right) \qquad (9)$$

is called the *spin* or *vorticity*. It is a measure of the rate of rotation of the fluid. These two tensors are familiar from the classical theories of hydrodynamics.

2. FORCES

It is assumed that there are two types of force acting in the body. One of these is the *external body force*, which is regarded as coming from outside the body. It can be characterized by a force density per unit mass b_i. The body force acting on an infinitesimal volume dv is then given by $b_i \rho dv$, where ρ is the density. If b_i is conservative, as it is when the body force is due to gravity, it can be derived from a scalar potential ψ. We then write

$$b_i = \frac{\partial \psi}{\partial x_i} \qquad (10)$$

The only body force that will be of interest here is that due to gravity.

The other type of force is that exerted by one part of the body on the other across the boundary surface separating the parts. It is assumed that this contact force can be characterized by a vector field s_i, the contact force per unit area, which depends not only on the position x_j but also on the orientation of the boundary surface at that point. Indicating by n_k the external unit normal vector on the surface at x_j, we write

$$s_i = s_i(x_j, n_k) \qquad (11)$$

and call s_i the *stress vector*. It can be shown that there exists a *stress tensor field* $s_{ij}(x_k)$, in terms of which the stress vector may be expressed as

$$s_i(x_j, n_k) = \sum_p s_{ip}(x_j) n_p \qquad (12)$$

The contact forces will be discussed in terms of the stress tensor s_{ij}.

3. GENERAL LAWS OF MECHANICS

There are two laws of mechanics that are of fundamental importance here. The first of these is the *law of balance of linear momentum*. It is expressed by the differential equations

$$\sum_j \frac{\partial s_{ij}}{\partial x_j} + \rho b_i = \rho \ddot{x}_i \tag{13}$$

or, when b_i has a potential ψ, by

$$\sum_j \frac{\partial s_{ij}}{\partial x_j} + \rho \frac{\partial \psi}{\partial x_i} = \rho \ddot{x}_i \tag{14}$$

where the $\ddot{x}_i$ are the components of the acceleration and ρ is the density. These equations are also called the *dynamical equations* or the *equations of motion*.

With equation (13) satisfied the second law, the *balance of moment of momentum*, may be stated as

$$s_{ij} = s_{ji} \tag{15}$$

That is, the stress tensor is symmetric. Equations (13) and (15) are called *Cauchy's first and second laws of motion*, respectively.

4. CONSTITUTIVE EQUATIONS

The general laws of mechanics are not sufficient to determine the forces and motions in a body. The properties of the material of which the body is composed must also be known. Mathematically, the properties of an *ideal material* are specified through a *constitutive equation*, which relates the stress to the motion.

The proposal and investigation of constitutive equations has been an area of great research activity in theoretical continuum mechanics in recent years. The mathematicians realized that a basic understanding of deviations in the behavior of actual fluids from that expected from the classical ideal fluids could not be achieved by "patching up" some solutions to special problems as, for example, by allowing some of their constants to vary. The source of the failure is at the very base of the classical theories: the inadequacy of the constitutive equations and the assumptions behind them in reflecting the properties of the materials of interest. Although stopgap measures might produce solutions to special problems,[17] they could not

---

[17] We do not intend here to be contemptuous of problem-solving. On the contrary, it is all the more remarkable that the engineers and scientists who deal with nonclassical fluids are able to develop processes and design plants.

lead to real insight, and they failed to capture unexpected aspects of the problem under investigation.

The theoretician is guided in his assignment of properties to the ideal material by the behavior of some actual materials over a range of experimental conditions. He begins with a decision concerning the general nature of the stress–deformation relation. The form of the relation is then simplified by demanding compliance with a set of physically reasonable principles and by taking advantage of the properties of the tensors that occur in the relation.

In this subsection we begin with a list of some of the principles that guide the formulation of constitutive equations. We then illustrate how the procedure outlined above works out for several ideal materials of special interest.

a. Basic Principles

Truesdell and Toupin[1] have tabulated a set of principles that should be used in forming constitutive equations. Some of these are summarized below.

(1) *Coordinate invariance.* Every constitutive equation must be stated in such a way that it is unaffected by the choice of the coordinate system; that is, the body should be "unaware" of the particular coordinate system which we happen to employ. If the equations are stated in tensorial form, this rule will be satisfied.

(2) *Isotropy.* The constitutive equation must reflect the symmetry of the material represented. If, for example, the material is *isotropic* (that is, if it has no preferred directions of response when it is in an undistorted state), then any rotation of the material coordinates must leave the constitutive equations unchanged when the reference configuration is taken to be an undistorted state. In this chapter we adopt that definition of a fluid which is used in much of the recent work in continuun mechanics and which is based on assigning to a fluid the highest degree of symmetry. Not only is a fluid isotropic, but any configuration can be taken as the reference configuration— an undistorted state. If the fluid is held in any configuration for a long time, the stress must eventually reduce to a hydrostatic pressure. Materials with yield stresses, for example, are not considered fluids, because they can support shear stresses less than the yield stress indefinitely.

(3) *Material indifference.*[18] To be consistent with our notion that the response of a material is independent of the observer the constitutive equation must have a form such that it is not changed by an arbitrary rigid rotation of the frame of reference. In other words, if both the material and

[18] This principle, implicitly used by earlier investigators, was explicitly formulated by Noll.[19]

[19] W. Noll, *Arch. Rational Mech. Anal.* **2,** 197 (1958).

reference system are rotated at the same time, the constitutive equation remains the same. Even though the constitutive equation must satisfy the requirement of material indifference, it is to be noted that the laws of motion do not; apparent forces must be invoked in comparisons of the experimental results of observers who are rotating with respect to each other.

b. Examples

Here we shall give brief discussions of several constitutive equations essentially in the order of their historical development. It will be readily seen how the complexity of these equations increases as more phenomena are included within their scope.

Because many more problems can be solved for incompressible fluids than for compressible ones, the recent literature of continuum mechanics deals almost exclusively with the former. In this chapter the discussion is limited to incompressible fluids. For these fluids, the motion determines the stress only to within an isotropic pressure. Moreover, only those motions which occur without change of volume, *isochoric motions*, are possible.

1. The *perfect*, or *inviscid*, fluid has the simplest constitutive relation. The stress is always hydrostatic. This property is expressed by the constitutive equation

$$s_{ij} = -p\delta_{ij}, \tag{16}$$

where p is a scalar.

2. The constitutive equation of the *Newtonian*, or *Navier–Stokes, fluid* can be based on the assumption that the stress is a linear function of the velocity and the velocity gradient,[1,20] that is,

$$s_{ij} = \lambda_{ij}\left(\dot{x}_k, \frac{\partial \dot{x}_l}{\partial x_m}\right) \tag{17}$$

where the λ_{ij} are *linear functions*. The principle of material indifference does not allow any dependence on the velocity nor an unrestricted dependence on $\partial \dot{x}_i/\partial x_k$, because observers rotating with respect to each other would measure different values for these quantities. The symmetric tensor formed from $\partial \dot{x}_i/\partial x_j$, the rate of deformation tensor D_{ij}, defined in (8) is, however, the same for all observers. Thus, (17) reduces to

$$s_{ij} = -p\delta_{ij} + 2\eta D_{ij} \tag{18}$$

where η is a constant called the *shear viscosity* or, most frequently, the *viscosity*. Substitution of (18) into Cauchy's first law (13) yields the *Navier–*

[20] Actually, the stress is usually assumed to be a linear function of the velocity gradients only. The velocity is inserted here to show the power of the principle of material indifference which rejects such a dependence even if it is hypothesized originally.

Stokes equations. At equilibrium, where there is no relative motion, (18) reduces to (16), the equation for the perfect fluid. Except in extreme circumstances, the behavior of fluids consisting entirely of small molecules such as water are very well represented by (18). In fact, in a number of situations of importance in engineering the perfect fluid defined by the simpler (16) suffices. The study of these two ideal materials, the perfect and Newtonian fluids, forms the bulk of the subject matter discussed in the fields of fluid dynamics and hydrodynamics.

3. A constitutive equation that can describe nonlinear viscosity and some normal stress effects is that of the *viscoinelastic,* or *Reiner–Rivlin, fluid.* This equation can be based on the assumption that the stress depends on the velocity and the velocity gradient but that, in contrast to the Newtonian fluid, the dependence need not be linear; that is,

$$s_{ij} + p\delta_{ij} = r_{ij}\left(\dot{x}_k, \frac{\partial \dot{x}_l}{\partial x_m}\right) \tag{19}$$

where r_{ij} is not necessarily linear. To satisfy the principle of material indifference, however, (19) cannot be so general; it must reduce to

$$s_{ij} + p\delta_{ij} = f_{ij}(D_{kl}) \tag{20}$$

Another consequence of this principle is that f_{ij} is an isotropic function of D_{lk}.[21] A further simplification of (20) can be achieved by taking advantage of the fact that the tensors $(s_{ij} + p\delta_{ij})$ and D_{ij} are both symmetric and that the function f_{ij} is isotropic. A theorem of tensor functions then asserts that equation (20) for the incompressible viscoinelastic fluid, can be represented by

$$s_{ij} + p\delta_{ij} = \aleph_1 D_{ij} + \aleph_2 \sum_k D_{ik}D_{kj} \tag{21}$$

where $\aleph_1$ and $\aleph_2$ are functions of II and III, two of the principal invariants of D_{ij}, defined by the equations

$$\mathrm{II} = \mathrm{trace}\, D_{ij}^2 = \sum_i \sum_j D_{ij}D_{ji} \tag{22}$$

$$\mathrm{III} = \det D_{ij} \tag{23}$$

It is to be noted that (21) is a much simpler tensor relation than (20). The former contains, at most, products of two tensors, whereas the latter could have a much greater complexity. Of course, a great deal of arbitrariness remains, in that the coefficients in (21) are unspecified scalar functions of II and III.

---

[21] By isotropic function, we mean, for example, that the components f_{ij} are the same in all Cartesian coordinate systems. Thus they are not affected by any rotation of the coordinate system.

Because of the nature of the basic assumption, equation (19), underlying the viscoinelastic fluid, the stress depends only on the instantaneous value of the rate of deformation. Thus it is not to be expected that this fluid can represent the time-dependent behavior characteristics of viscoelastic fluids. It also turns out, as we shall show below, that it does not correctly describe the normal stress effect in Couette flow of those fluids that have been investigated.

4. A constitutive equation that is able to account for some time-dependent characteristics and all experimentally investigated normal stress effects is that of the *fluid of differential type*, or *Rivlin–Ericksen fluid*, which can be based on the assumption that the stress depends not only on the velocity and velocity gradient but also on higher time-derivatives of these; that is,

$$s_{ij} + p\delta_{ij} = f_{ij}(\dot{x}_k, \partial\dot{x}_l/\partial x_m, \ddot{x}_n, \partial\ddot{x}_p/\partial x_q, \ldots, x_r^{(N)}, \partial x_s^{(N)}/\partial x_t) \qquad (24)$$

where, again, f_{ij} is not restricted to being linear and the superscript (N) here refers to the Nth time derivative. The principle of material indifference eliminates all dependence on the velocity $\dot{x}_i$ and its higher time derivatives and forces the dependence on the gradients to simplify to the form

$$s_{ij} + p\delta_{ij} = f_{ij}(A_{lk}^{(1)}, A_{nm}^{(2)}, \ldots, A_{rs}^{(N)}) \qquad (25)$$

where the symmetric tensors $A_{ij}^{(n)}$ are called the *n*th *Rivlin–Ericksen tensors*, defined by

$$A_{ij}^{(1)} = 2D_{ij} \qquad (26)$$

$$A_{ij}^{(n+1)} = \dot{A}_{ij}^{(n)} + \sum_k \left(\frac{\partial \dot{x}_k}{\partial x_j} A_{ik}^{(n)} + \frac{\partial \dot{x}_k}{\partial x_i} A_{kj}^{(n)} \right) \qquad (27)$$

and where f_{ij} is an isotropic function. The $A_{ij}^{(n)}$ have the property that they are the same for all observers. Again advantage can be taken of the isotropy of f_{ij} and the symmetry of the $A_{ij}^{(n)}$ and s_{ij} to restrict further the generality of f_{ij} but, since we shall not have use of the complicated result, we shall not reproduce it here. From equation (24) it is seen that the stress is constant whenever the fluid is motionless, and thus fluids of the differential type cannot describe the phenomenon of gradual stress relaxation in which, although the fluid is motionless, the stress changes with time because of some previous deformation.

---

[22] This, perhaps surprising, name for the most complex fluid that we shall treat here was employed by Noll[19] because that was the simplest case of the general class of materials that he considered, materials in which the stress at a given particle at a given time is determined by the history of the motion in the immediate neighborhood of that particle. In his simple fluid not the entire motion, but only the deformation gradient, is relevant. As it turned out, however, the simple fluid was adequate for the phenomena of interest, and more complex fluids were not investigated in detail.

5. All the models listed above are special cases of a more general ideal fluid, which is able to explain all the types of behavior that have been mentioned. This is the *memory fluid*, also called the *simple fluid*.[22] Here it is supposed that the stress at a given particle and at a given time, $s_{ij}(x_k, t)$ is determined by the history of the relative deformation gradient $F_{ik}(\tau)$, defined in equation (2) This can be stated mathematically as

$$s_{ij}(x_l, t) = -p\delta_{ij} + \mathop{\mathfrak{H}}_{-\infty}^{t}\,_{ij}\,[F_{km}(\tau)] \tag{28}$$

where $\mathfrak{H}_{ij}$ is a functional; that is, its value is determined when its argument, the relative deformation gradient F_{km}, is known for all times previous to the current time t.[23] If the material is to obey the principle of material objectivity and to have the symmetry demanded of a fluid, equation (28) can be shown to reduce to

$$s_{ij}(t) = -p\delta_{ij} + \mathop{\mathfrak{G}}_{s=0}^{\infty}\,_{ij}\,[J_{kl}(t - s)] \tag{29}$$

where the dummy parameter s measures the time backward from the current time t, where $J_{kl}(t - s)$ is the symmetric strain tensor, defined in equation (4), evaluated at $\tau = t - s$, and the reference configuration is taken as the configuration of the fluid at the current time t. The functional $\mathfrak{G}_{ij}$ is isotropic; that is, equation (29) must remain unchanged when the frame of reference undergoes any rigid motion.

For the incompressible memory fluid, which is the only kind we are considering, the motion determines the stress only up to an isotropic pressure. This indeterminacy is commonly removed by adopting the normalization

$$\sum_k (s_{kk}(t) + p\delta_{kk}) = \sum_k \mathop{\mathfrak{G}}_{s=0}^{\infty}\,_{kk}\,[J_{ij}(t - s)] = 0 \tag{30}$$

or

$$p = -\tfrac{1}{3}\sum_k s_{kk} \tag{31}$$

It is sometimes convenient to write expressions in terms of the *stress deviator* t_{ij}, defined by the equation

$$t_{ij} = s_{ij} + p\delta_{ij} \tag{32}$$

Thus, equation (29) may be written as

$$t_{ij}(t) = \mathop{\mathfrak{G}}_{s=0}^{\infty}\,_{ij}\,[J_{kl}(t - s)] \tag{33}$$

---

[23] Perhaps the simplest commonly used functional is the definite integral, whose value is determined by a function, its integrand.

and the normalization condition, equation (30), as

$$\sum_k t_{kk}(t) = 0 \tag{34}$$

It can be shown that, if a memory fluid has always been at rest, that is, $J_{ij}(t - s) = 0$ for all s, then the stress is just an isotropic pressure,

$$s_{ij} = b\delta_{ij} \tag{35}$$

and, because of the normalization [equation (30)],

$$\overset{\infty}{\underset{s=0}{\mathfrak{G}}}_{ij}(0) = 0 \tag{36}$$

Other properties of the memory fluid will be discussed later.

Despite its generality the memory fluid does not include all ideal fluids that have been discussed in continuum mechanics in recent years. For example, a dependence of the stress on higher spatial gradients could have been assumed. Some of the materials called *anisotropic fluids*[24] also are not memory fluids. In these fluids the stress is assumed to depend, not only on the deformation, but also on a vector at the material point.[25] We shall not deal further with these fluids, since for the purposes of this chapter the complexity of the memory fluid suffices.

In addition to the above constitutive equations other relations, many of which are special cases of the memory fluid, have been proposed in the literature. Many of them have arisen from extensions of various equations of linear viscoelasticity to nonlinear problems in such a way as to ensure compliance with the principle of material objectivity.[26-29] A difficulty that arises here is that generalizations made in this way are not unique: there are many correct nonlinear equations that reduce to the same linear one. A choice among the possible generalizations presumably, then, rests on the adequacy of the resulting equation for a given real fluid (or class of fluids) or for a given type of deformation. If one of these special fluids should turn out to be a good representation, a considerable simplification could result. A justification of the use of simpler special equations is that they can qualitatively indicate behavior in the more complicated flow problems, in which solutions can be obtained with the special fluids but not with the general

[24] J. L. Ericksen, *Arch. Rational Mech. Anal.* **4**, 231 (1960); *Trans. Soc. Rheol.* **4**, 29 (1960).

[25] An ideal fluid of this type may be appropriate for the description of the mechanical behavior of liquid crystals.

[26] J. G. Oldroyd, *Proc. Roy. Soc.* **A200**, 523 (1950).

[27] T. W. DeWitt, *J. Appl. Phys.* **26**, 889 (1955).

[28] M. Yamamoto, *Progr. Theoret. Phys.* (*Kyoto*) Suppl. **10**, 19 (1959).

[29] H. Markovitz, *Trans. Soc. Rheol.* **6**, 349 (1962).

memory fluid. An increasing number of papers dealing with problems of this type are being published. We do not discuss that literature here.* Here we limit our attention to those simpler equations which result as mathematically correct, well-stated, limiting cases derived from the equation of the memory fluid. These will be discussed now.

Although the current stress in a memory fluid depends on the history of the deformation, it is not expected that this dependence should be as strong for ancient deformations as it is for recent ones. That is, if two deformation histories of a given viscoelastic material have been the same in the recent past, the stresses in the two cases will be practically the same, even though deformations may have differed greatly in the distant past. These vague statements have been made mathematically exact in the theory of materials with *fading memory*.[30] For such materials it is possible to approximate the equation of the memory fluid by expansions that are useful in special flow situations. Two approximations have been treated in the literature. These may be called the *slow flow* and the *recently small deformation* approximations.

In the *slow flow approximation*[30] a sequence of deformation histories is considered. Each history differs from a reference history in that the time scale is slowed down by a factor α, where $0 < \alpha < 1$; that is, the deformation that occurs at a time t is the same as that which occurred in the reference history at a time αt. The stress can then be expressed as a power series in α. If the series is truncated at the term in α^n, the resulting equation for the stress is said to be the constitutive equation of an *nth-order fluid*. It then turns out that the *zeroth-order fluid* is the perfect fluid [equation (16)] and the *first-order fluid* is the Newtonian fluid [equation (18)]; thus, these classical fluids may be considered to have the status of approximations to the general fluid with fading memory in the limit of slow flows. The next approximation, the *second-order fluid*, has the constitutive equation

$$s_{ij} + p\delta_{ij} = \eta_0 A_{ij}^{(1)} + \beta \sum_k A_{ik}^{(1)} A_{kj}^{(1)} + \gamma A_{ij}^{(2)} \tag{37}$$

where the $A_{ij}^{(n)}$ are the Rivlin–Ericksen tensors defined in equations (26) and (27), where η_0 is the viscosity, and where the constants β and γ are called normal stress coefficients (for reasons to be made clear below). We can think of equation (37) as representing the first deviations from the Newtonian fluid as the flow is speeded up. Higher-order fluids have also been investigated,[31] but we shall not reproduce the more complicated equations here.

* *Note added in proof*: Recent survey articles on this subject are: T. W. Spriggs, J. D. Huppler, and R. B. Bird, *Trans. Soc. Rheol.* **10**:1, 191 (1966); D. C. Bogue and J. D. Doughty, *Ind. Eng. Fund.* **5**, 243 (1966).

[30] B. D. Coleman and W. Noll, *Arch. Rational Mech. Anal.* **6**, 355 (1960).
[31] W. E. Langlois and R. S. Rivlin, *Rend. Mat. Roma* **22**, 169 (1963); W. E. Langlois, *Trans. Soc. Rheol.* **7**, 75 (1963); **8**, 33 (1964).

The *recently small deformation* approximation arises upon consideration of deformation histories, in which the deformation has been small, at least in the recent past. It is possible to make this statement mathematically exact[32] and to develop approximations of various orders. These are called equations of *nth order viscoelasticity*. In the special case of first-order viscoelasticity where the deformation has always been infinitesimal, the incompressible memory fluid may be approximated by Boltzmann's equation of infinitesimal linear viscoelasticity:[33]

$$s_{ij}(t) + p\delta_{ij} = 2G(0)E_{ij}(t) + 2 \int_0^\infty \frac{dG(s)}{ds} E_{ij}(t - s)\, ds \tag{38}$$

where E_{ij} is the infinitesimal strain tensor of the classical theory of elasticity [cf. equation (6)] and $G(t)$ is the *stress-relaxation modulus*. This equation is the basis of much of the work on viscoelasticity.[33]

The approximation correct to second order (with no limitation to infinitesimal deformation) is

$$s_{ij} + p\delta_{ij} = \int_0^\infty \frac{dG(s)}{ds} J_{ij}(s)\, ds$$

$$+ \sum_k \int_0^\infty \int_0^\infty [a(s, r)J_{ik}(s)J_{kj}(r) + b(s, r)J_{kk}(s)J_{ij}(r)]\, ds\, dr \tag{39}$$

where m, a, and b are material functions determined by the functional $\mathfrak{G}_{ij}$ of equation (29), and J_{ij} is the measure of strain defined in equation (4).

It can be shown that a slow flow is also an example of a recently small deformation; that is, the motion is so slow that the configurations that were much different from the present one occurred such a long time ago that they have already been forgotten by the material. As a result, the material constants that appear in equation (37), the constitutive equation for the second-order fluid, are determined by the material functions that occur in equation (39), the equation for second-order viscoelasticity. These relations are given by the expressions[34]

$$\eta_0 = \int_0^\infty G(s)\, ds \tag{40}$$

$$-\gamma = \int_0^\infty sG(s)\, ds \tag{41}$$

$$\beta = \int_0^\infty \int_0^\infty sra(s, r)\, ds\, dr \tag{42}$$

[32] B. D. Coleman and W. Noll, *Rev. Mod. Phys.* **33**, 239 (1961).

[33] H. Leaderman, *in* "Rheology: Theory and Applications" (F. R. Eirich, ed.), Vol. 2, Chapt. 1. Academic Press, New York, 1958; B. Gross, "Mathematical Structure of the Theories of Viscoelasticity." Hermann, Paris, 1953.

[34] B. D. Coleman and H. Markovitz, *J. Appl. Phys.* **39**, 1 (1964).

Equation (40) is a result long known in linear viscoelasticity.[35] The integral on the right side of equation (41) has been shown[35] to be equal to $\eta_0{}^2 J_e$, where J_e is the steady-state compliance. Furthermore, since $G(t) > 0$ for all t, then η_0 and γ must obey the inequalities

$$\eta_0 > 0, \qquad \gamma < 0 \tag{43}$$

The first of these inequalities is to be expected, but that the theory of the memory fluid, general as it is, should lead to such a limitation on a normal-stress coefficient is rather surprising.

5. VISCOMETRIC FLOWS

In this section we first consider the simple shearing flow of the memory fluid. It turns out that three functions are required to characterize the material in this flow. The same three functions are also sufficient for a wider class of flows, the viscometric flows, which include many commonly used in rheological investigations.

Rather than discuss the flow of each of the ideal fluids mentioned in the previous section individually, we shall treat the most general fluid, the memory fluid, first and then indicate the result for the special fluids. It is rather amazing that such an approach is possible. Equation (29) appears to be so general and vague that it seems unlikely that anything definite could be deduced concerning a specific type of deformation. However, it turns out that the viscometric flows are so simple that definite results can be obtained.[36]

a. Simple Shearing Flow[37,38]

We consider a flow such that a Cartesian coordinate system, x_1, x_2, x_3 (or x, y, z respectively), can be so chosen that the components of the velocity field $\dot{x}_i(x_k, t)$ are

$$\dot{x}_1 = 0, \qquad \dot{x}_2 = \kappa x_1, \qquad \dot{x}_3 = 0 \tag{44}$$

where κ is a constant, *the rate of shear*. Such a motion is called a *simple shearing flow*. See Fig. 1.

The position ξ_i of a material point at any time τ can be found by solving the differential equation

$$\dot{\xi}_1 = 0, \qquad \dot{\xi}_2 = \kappa \xi_1, \qquad \dot{\xi}_3 = 0 \tag{45}$$

[35] J. D. Ferry, "Viscoelastic Properties of Polymers." Wiley, New York, 1961.
[36] The treatment in this section is based to a great extent on Refs. 37 and 38.
[37] B. D. Coleman and W. Noll, *Arch. Rational Mech. Anal.* **3**, 289 (1959).
[38] B. D. Coleman and W. Noll, *Ann. N.Y. Acad. Sci.* **89**, 672 (1961).

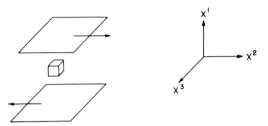

FIG. 1. Simple shearing flow.

We take our reference configuration as that which occurs at the present time $\tau = t$ (here we are using the notation defined in Section II1). Thus, the particle which has the coordinates x_1, x_2, and x_3 at time t, had coordinates

$$\xi_1 = x_1, \qquad \xi_2 = x_2 + \kappa x_1(\tau - t), \qquad \xi_3 = x_3 \qquad (46)$$

at time τ. The components of the relative deformation gradient $F_{ij} = \partial\xi_i/\partial x_j$ are readily calculated,

$$\|F_{ij}(\tau)\| = \begin{Vmatrix} 1 & 0 & 0 \\ \kappa(\tau - t) & 1 & 0 \\ 0 & 0 & 1 \end{Vmatrix} \qquad (47)$$

and therefore the matrix of J_{ij} can be calculated from its definition, equation (4):

$$\|J_{ij}(t - s)\| = \begin{Vmatrix} \kappa^2 s^2 & -s\kappa & 0 \\ -s\kappa & 0 & 0 \\ 0 & 0 & 0 \end{Vmatrix} \qquad (48)$$

where $t - s = \tau$. Thus the tensor $J_{ij}(t - s)$ has the form

$$J_{ij}(t - s) = -sA_{ij}^{(1)} + \tfrac{1}{2}s^2 A_{ij}^{(2)} \qquad (49)$$

where, as the reader can verify from equations (26) and (27), $A_{ij}^{(1)}$ and $A_{ij}^{(2)}$, which are given by

$$\|A_{ij}^{(1)}\| = \kappa \begin{Vmatrix} 0 & 1 & 0 \\ 1 & 0 & 0 \\ 0 & 0 & 0 \end{Vmatrix}, \qquad \|A_{ij}^{(2)}\| = 2\kappa^2 \begin{Vmatrix} 1 & 0 & 0 \\ 0 & 0 & 0 \\ 0 & 0 & 0 \end{Vmatrix} \qquad (50)$$

are Rivlin–Ericksen tensors. They are independent of t.

Since, by equation (49), $J_{ij}(t - s)$ is determined by $A_{ij}^{(1)}$ and $A_{ij}^{(2)}$ in a steady simple shearing flow, and since $t_{ij}(t)$ is determined by $J_{ij}(t - s)$ according to equation (33), then $t_{ij}(t)$ is independent of t and depends only on $A_{ij}^{(1)}$ and

$A_{ij}^{(2)}$. The functional $\mathfrak{G}_{ij}$ reduces to a function q_{ij}. This is expressed by the equations

$$t_{ij}(t) = \mathop{\mathfrak{G}}_{\substack{ij \\ s=0}}^{\infty} [J_{kl}(t-s)] = q_{ij}(A_{kl}^{(1)}, A_{mn}^{(2)}) \tag{51}$$

where, to satisfy the principle of material indifference, q_{ij} must be an isotropic function. It can then be shown by a somewhat lengthy mathematical argument that the matrix of t_{ij} has the form

$$\|t_{ij}(t)\| = \begin{Vmatrix} t_{11} & t_{12} & 0 \\ t_{12} & t_{22} & 0 \\ 0 & 0 & t_{33} \end{Vmatrix} \tag{52}$$

where the t_{ij} are functions only of κ, the rate of shear. All the t_{ij} are not independent, however, because of the normalization, equation (34). Thus, the properties of the memory fluid in simple shearing flow can be expressed by the three independent functions of κ, the *viscometric functions* τ, σ_1, and σ_2, defined by the equations[39]

$$t_{12} = \tau(\kappa), \qquad t_{11} - t_{33} = \sigma_1(\kappa), \qquad t_{22} - t_{33} = \sigma_2(\kappa) \tag{53}$$

We refer to τ as the *shear stress function* and to σ_1 and σ_2 as the *normal stress functions*. It is clear from equation (32) that, in terms of the stress components s_{ij}, we may similarly write

$$s_{12} = \tau(\kappa), \qquad s_{11} - s_{33} = \sigma_1(\kappa), \qquad s_{22} - s_{33} = \sigma_2(\kappa) \tag{54}$$

It can further be proved that τ must be an odd function of κ, and σ_1 and σ_2 even functions,

$$\tau(-\kappa) = -\tau(\kappa), \qquad \sigma_1(\kappa) = \sigma_1(-\kappa), \qquad \sigma_2(\kappa) = \sigma_2(-\kappa) \tag{55}$$

and that

$$\sigma_1(0) = \sigma_2(0) = 0 \tag{56}$$

The *viscosity function* η, defined by

$$\eta(\kappa) = \tau(\kappa)/\kappa \tag{57}$$

is the familiar function used by rheologists to express the shear dependence of the viscosity.

We adopt the symbol S for values of the shear stress when we wish to distinguish them from the function; thus we write

$$S = \tau(\kappa) \tag{58}$$

[39] The choice of the stress differences σ_1 and σ_2 as the material function is somewhat arbitrary; other normal stress differences or the stress deviators can also serve the purpose.

It is also convenient to have a name for the inverse of the function τ; we write

$$\kappa = \lambda(S) \tag{59}$$

Assuming that λ is single-valued, there is no problem in writing the normal stress function as functions of the shear stress; we indicate this dependence by using the symbol

$$\hat{\sigma}_i(S) = \sigma_i[\lambda(S)] = \sigma_i(\kappa), \qquad i = 1, 2 \tag{60}$$

In this subsection we have thus far discussed simple shearing flow without considering whether such a flow is compatible with any physical situation. It is known, of course, that such a flow can arise in the case of a Newtonian fluid placed between two parallel infinite plates, one of which is moving with constant linear velocity in its own plane. To verify whether the same holds true for a memory fluid it is necessary to set up the appropriate boundary value problem. By using the results of this section in conjunction with the balance of momentum equation (13) and the appropriate boundary conditions it is possible to show that simple shearing flow is indeed compatible with all requirements. As in the case of the Newtonian fluid, κ is given by

$$\kappa = V/d \tag{61}$$

where V is the velocity of the moving plate and d is the distance between the plates. The new feature that arises for the memory fluid is that the normal stresses s_{11}, s_{22}, and s_{33} are not equal to each other, as they are for the Newtonian fluid in simple shearing flow; their differences are determined by the σ_i according to equation (54). We shall not give the details[40] of the analysis here. We indicate how such an analysis proceeds in connection with a more interesting type of deformation, Couette flow.

It is gratifying that it is possible to solve a flow for the memory fluid, even if a very simple flow. However, the result would have rather limited usefulness for the experimenter, if this were the only flow for which results could be obtained, or if it turned out that a new set of functions was required for each flow. Fortunately, a class of flows of great interest to rheologists can be discussed successfully in terms of the same three functions, τ, σ_1, and σ_2, which we met in simple shearing flow.

b. Simple Viscometric Flows[38]

Most of the flows used by rheologists (such as Poiseuille, Couette, cone–plate) in their study of the steady-flow behavior of fluids are examples of

[40] The interested reader may find them in Ref. 38 or 4.

simple viscometric flows. These flows are characterized by a velocity field whose contravarient components have the form

$$\dot{x}^1 = 0, \qquad \dot{x}^2 = v(x^1), \qquad \dot{x}^3 = 0 \tag{62}$$

in some orthogonal coordinate system,[41] in which the covariant components of the metric tensor g_{ij} are independent of the x^2 coordinate. The simple shearing flow, equation (44), is a special case of equation (62) when the co-ordinate system is Cartesian. By an extension of the argument used for simple shearing flow in the previous subsection it can be shown that the strain tensor $J_j{}^i(t - s)$ is given by the equation

$$J_j{}^i(t - s) = -sA_j^{(1)i} + \tfrac{1}{2}s^2 A_j^{(2)i} \tag{63}$$

similar to equation (49), for simple viscometric flows. It then follows that the physical components of the stress, $s_{\langle ij\rangle}$, have a matrix similar to equation (52),

$$\|s_{\langle ij\rangle}\| = \begin{Vmatrix} s_{\langle 11\rangle} & s_{\langle 12\rangle} & 0 \\ s_{\langle 12\rangle} & s_{\langle 22\rangle} & 0 \\ 0 & 0 & s_{\langle 33\rangle} \end{Vmatrix} \tag{64}$$

and that

$$t_{\langle 13\rangle} = t_{\langle 23\rangle} = s_{\langle 13\rangle} = s_{\langle 23\rangle} = 0 \tag{65}$$

$$t_{\langle 12\rangle} = s_{\langle 12\rangle} = \tau(\kappa) \tag{66}$$

$$t_{\langle 11\rangle} - t_{\langle 33\rangle} = s_{\langle 11\rangle} - s_{\langle 33\rangle} = \sigma_1(\kappa) = \hat{\sigma}_1(S) \tag{67}$$

$$t_{\langle 22\rangle} - t_{\langle 33\rangle} = s_{\langle 22\rangle} - s_{\langle 33\rangle} = \sigma_2(\kappa) = \hat{\sigma}_2(S) \tag{68}$$

where the rate of shear is

$$\kappa = \sqrt{\frac{g_{22}}{g_{11}}} \frac{dv}{dx^1} \tag{69}$$

g_{ij} is the metric tensor (see the Appendix) and where the functions τ, σ_1, and σ_2 are the same as those which appeared in the expressions for the stress components in simple shearing flow. Specific examples of simple viscometric flows will be discussed below.

The whole class of *viscometric flows* consists of those for which $J_j{}^i(t - s)$ has the form of equation (63), and the Rivlin–Ericksen tensors can be expressed in the form given in equation (50). Here κ, the rate of shear, is independent of t but can depend on the material point. This class includes deformations which are more general than the simple viscometric flow

[41] Up to this point we have employed only Cartesian coordinate systems. In the discussio of viscometric flows it becomes convenient to allow the use of other orthogonal coordin- systems. We now must distinguish between covariant, contravariant, and mixed compone-

equation (62); for example, it includes flows characterized by a velocity field whose contravarient components have the form

$$\dot{x}^1 = 0, \qquad \dot{x}^2 = u(x^1), \qquad \dot{x}^3 = w(x^1) \tag{70}$$

We shall not discuss these more complicated flows here; the interested reader is referred to the literature.[2,8,42,43]

We note here that all viscometric flows occur without change of volume and are thus possible flows for an incompressible fluid.

c. Special Constitutive Equations

For the memory fluid, aside from the limitation implied by the equations (55) and inequalities (43), the viscometric functions are quite arbitrary. For some of the less general fluids discussed in Section II4, the material functions take on special values. We list them here.

For the perfect fluid, since the stress is always isotropic, all three viscometric functions are zero:

$$\tau(\kappa) = \sigma_1(\kappa) = \sigma_2(\kappa) = 0 \tag{71}$$

For the Newtonian fluid the two normal stress functions are zero and the viscosity function reduces to a constant:

$$S = \tau(\kappa) = \eta_0 \kappa, \qquid \kappa = \lambda(S) = S/\eta_0, \qquad \eta_0 = \text{constant} \tag{72}$$

$$\sigma_1(\kappa) = \sigma_2(\kappa) = 0 \tag{73}$$

For the viscoinelastic fluid there is no special restriction on $\tau(\kappa)$. However, it can be shown that the normal stress functions are equal for all visco-inelastic fluids:

$$\sigma_1(\kappa) = \sigma_2(\kappa) \tag{74}$$

For fluids of the differential type there are no special restrictions on any of the viscometric functions.

The second-order fluid has very simple forms for the viscometric functions:[34]

$$\tau(\kappa) = \eta_0 \kappa, \qquad \eta(\kappa) = \eta_0, \qquad \lambda(S) = S/\eta_0 \tag{75}$$

$$\sigma_1(\kappa) = (\beta + 2\gamma)\kappa^2 \tag{76}$$

$$\sigma_2(\kappa) = \beta\kappa^2 \tag{77}$$

That is, the viscosity function reduces to a constant, as it does for the Newtonian fluid, but the normal stress functions are nonzero and therefore, as we shall see, the second-order fluid does exhibit normal stress effects. The

[42] B. D. Coleman and W. Noll, *J. Appl. Phys.* **30**, 1508 (1959).
[43] B. D. Coleman, *Arch. Rational Mech. Anal.* **9**, 273 (1962).

dependence on κ^2 is to be expected, if we remember that this fluid approximates the behavior of the general memory fluid in slow flows. If σ_1 and σ_2, which by equations (55) and (56) are even functions of κ that go to zero with κ, are smooth functions of κ, the κ^2 dependence that appears here may be considered the first nonzero term of a Maclaurin expansion.

We also mention here a relation proposed by Weissenberg[44,45] although, as far as we know, it did not stem from any general constitutive relation. In our notation, it states that

$$t_{\langle 11 \rangle} - t_{\langle 33 \rangle} = \sigma_1(\kappa) = 0 \tag{78}$$

Weissenberg based this proposal on analogies that he believed should exist between the behavior of elastic solids and viscoelastic fluids.

It has been pointed out[46] that in some of the special theories to which allusion was made in Section II4b the two normal stress functions are simply related. Their ratio is independent of κ:

$$\sigma_1(\kappa)/\sigma_2(\kappa) = \text{constant} \tag{79}$$

For example, the constant is unity for the viscoinelastic fluid, zero for the Weissenberg fluid, and $(\beta + 2\gamma)/\beta$ for the second-order fluid.

III. Specific Viscometric Flows: Theory and Experiment

The basic theory of viscometric flows has been outlined in the previous section. Here we show how this general theory applies to specific examples of this class of motions. We include in this discussion methods of calculating the viscometric functions from experimental data and, where it is of interest, the corresponding results for special fluids. Following the theory, we deal with the experimental investigation of each flow, including laboratory techniques and examples of the data that have been obtained

We shall not devote much space to the experimental methods of determining the viscosity function, or its equivalents $\lambda(S)$ and $\tau(\kappa)$; the literature on this subject is already large. The older work, which contains much that is still of great utility today, is very well covered in the book by Philippoff.[47] A more recent volume[48] describes in considerable detail a number of commercially available viscometers. In this treatise this problem has been dis-

[44] K. Weissenberg, *Nature* **159**, 310 (1947).

[45] P. U. A. Grossman, *Kolloid-Z.* **174**, 97 (1961).

[46] H. Markovitz, *Trans. Soc. Rheol.* **1**, 37 (1957).

[47] W. Philippoff, "Viskosität der Kolloide." Steinkopf, Dresden, 1942; reprinted by Edward Brothers, Ann Arbor, 1944.

[48] J. R. Van Wazer, J. W. Lyons, K. Y. Kim, and R. E. Colwell, "Viscosity and Flow Measurement; a Laboratory Handbook of Rheology." Wiley (Interscience), New York, 1963.

cussed in a number of chapters.[26,49-51] A recent monograph[4] discusses experimental questions with the same point of view that is adopted here and gives more references to the original literature. The results reported in this literature confirm one aspect of the theory of the memory fluid: the same $\eta(\kappa)$ is obtained from experiments in different types of flow. Thus, the viscosity function is indeed a valid and useful concept.

Our discussion of normal-stress experiments follows that given in the monograph cited above.[4] Other discussions can be found elsewhere.[13,15] Before proceeding with the specific flows we shall deal with some experimental considerations that apply generally.

1. GENERAL EXPERIMENTAL CONSIDERATIONS[4]

The first task of the experimenter is to achieve in his apparatus flow fields that correspond to those envisaged in the theory. Clearly, if he does not do this, the readings of dials on the most precise electronic attachment have little significance. As a beginning this implies that the "geometry" of the region where the flow occurs follows the description on which the theory is based: for example, the tubes have bores that are good cylinders, and the cylinders are coaxial and rotate about a fixed axis. This prerequisite can usually be satisfied by careful attention to design and details of construction.

A less tractable difficulty in achieving the desired flow is that of "end effects" in some experiments. The theory assumes that every particle of the fluid has always experienced the represented motion. This will not be true in many cases near the edge of the apparatus because of, for example, the presence of other rigid surfaces or the entry of new material from another flow condition. Unfortunately, some experimental observations are unavoidably affected, at least in part, by portions of the fluid not partaking of the "idealized" flow. Such effects are sometimes more serious for the memory fluid, because the past history of the deformation witnessed by the material particles influences its current behavior. Methods of correcting for these end effects have been mentioned in previous chapters of this treatise[7,49] and will be discussed again below.

The theory of various types of flow that will be presented indicates how the viscometric functions can be determined from measurements of velocities, rates of flow, angular speeds, torques, forces, thrusts, etc. Nowadays the

[49] B. A. Toms, in "Rheology: Theory and Applications" (F. R. Eirich, ed.), Vol. 2, Chapt. 12. Academic Press, New York, 1958.

[50] S. H. Maron and I. M. Krieger, in "Rheology: Theory and Applications" (F. R. Eirich, ed.), Vol. 3, Chapt. 4. Academic Press, New York, 1960.

[51] R. N. Weltman, in "Rheology: Theory and Applications" (F. R. Eirich, ed.), Vol. 3, Chapt. 6. Academic Press, New York, 1960.

experimental rheologist has available, in addition to the classical mechanical techniques of measuring these quantities, a host of electrical, magnetic electronic, and optical tools that enable him to increase the precision of his results and the range of variables that he can cover. The particular combination of components that he chooses and the accuracy with which his instrument is constructed depend on a number of factors, among which are his ingenuity and the availability of time and money. The most important consideration, however, should be the purpose of his investigation. This decides the level of accuracy to be sought, the range of forces and speeds to be encountered, and the chemical and physical natures of the specimens the apparatus must tolerate. The engineer designing an extruder for a molten plastic will require rheological instruments that differ from those suited to the physical chemist interested in the properties of dilute solutions; the specifications will be more demanding in some respects and less demanding in others.

In the theory it is necessary to include statements of boundary conditions. The assumption is made that the fluid layer next to a rigid surface moves with that surface. This is known as the *adherence* assumption. The validity of this hypothesis in the case of many fluids has been demonstrated in a variety of careful experiments over the past century, even in such cases as that of mercury on glass, in which the sample does not wet the surface.[52-55] Direct observation of flow lines and compliance with the theory in work with many types of fluids give confidence in the wide range of applicability of the adherence assumption.

Deviations from the adherence condition have been demonstrated, however, in two types of fluid of extremely different character: rarefied gases, constituting a classical case widely discussed in treatises on gas dynamics[56] and, of more interest to rheologists, fluids with considerable elastic character. The observed slippage on metal of an unvulcanized rubber[57] subjected to shearing tractions is not so surprising with a material which, though a fluid, has so much "elastic character" that its fluid nature is hardly evident. Slippage has also been reported with concentrated polymer solutions[49] and molten polymers.[58,59] If slippage does occur, then the theory cannot

[52] M. Brillouin, "Leçons sur la Viscosité des Liquides et des Gaz." Gauthier-Villars, Paris, 1907.

[53] E. C. Bingham, "Fluidity and Plasticity." McGraw-Hill, New York, 1922.

[54] S. Goldstein (ed.), "Modern Developments in Fluid Mechanics." Oxford Univ. Press (Clarendon), London and New York, 1938.

[55] H. L. Dryden, F. D. Murnaghan, and H. Bateman, "Hydrodynamics." Dover, New York, 1956.

[56] See, for example, Refs. 54 and 55.

[57] M. Mooney, *in* "Rheology: Theory and Applications" (F. R. Eirich, ed.), Vol. 2, Chapt. 5, Academic Press, New York, 1958.

[58] J. J. Benbow and P. Lamb, *SPE* (*Soc. Plastics Engrs.*) *Trans.* **3**, 16 (1963).

[59] B. Maxwell and J. C. Galt, *J. Polymer Sci.* **62**, 850 (1962).

apply in its entirety. In some cases those results which are not based on the boundary conditions may still be valid. A number of proposals have been made for boundary conditions to be used when the adherence condition is not met.[6,55,57] We shall not deal further with slipping fluids here.

Another assumption of the theory is that the same material functions can be used throughout the fluid, that is, that the fluid is homogeneous. In the flow of some suspensions this condition is not met; the concentration of the suspended particles becomes nonuniform. We cannot expect the theory to be applicable under such circumstances.

2. COUETTE FLOW

a. Theory[37,38]

Couette flow is that viscometric flow [equation (62)] with a velocity field whose contravariant components in cylindrical coordinates are

$$\dot{x}^1 = 0, \qquad \dot{x}^2 = \omega(r), \qquad \dot{x}^3 = 0 \tag{80}$$

where we identify x^1 with r, x^2 with θ, and x^3 with z. Thus, the only nonzero component of the velocity is the one in the θ direction; its physical component, the actual velocity, is $r\omega(r)$, where $\omega(r)$ is the angular velocity. By equations (69), (A5) of the Appendix, and (80) the rate of shear is

$$\kappa = r \frac{d\omega}{dr} \tag{81}$$

The components of the stress are then given, according to equations (65) to (68), in terms of the viscometric functions by the following equations:

$$t_{\langle rz \rangle} = t_{\langle \theta z \rangle} = s_{\langle rz \rangle} = s_{\langle \theta z \rangle} = 0 \tag{82}$$

$$t_{\langle r\theta \rangle} = s_{\langle r\theta \rangle} = \tau(\kappa) = S \tag{83}$$

$$t_{\langle rr \rangle} - t_{\langle zz \rangle} = s_{\langle rr \rangle} - s_{\langle zz \rangle} = \sigma_1(\kappa) = \hat{\sigma}_1(S) \tag{84}$$

$$t_{\langle \theta\theta \rangle} - t_{\langle zz \rangle} = s_{\langle \theta\theta \rangle} - s_{\langle zz \rangle} = \sigma_2(\kappa) = \hat{\sigma}_2(S) \tag{85}$$

Since, by equation (81), κ is a function of r alone, then so will be $t_{\langle r\theta \rangle}$, $t_{\langle rr \rangle} - t_{\langle zz \rangle}$, and $t_{\langle \theta\theta \rangle} - t_{\langle zz \rangle}$. From the normalization condition, equation (34), it can be seen that, in fact, all the components of the stress deviator $t_{\langle \ \rangle}$ are functions of r only.

We now investigate whether flow between rotating coaxial cylinders is compatible with the steady Couette flow, equation (80). This means that the boundary conditions and balance of momentum equation, equation (13), must be satisfied also.

We assume that flow occurs between two infinite coaxial cylinders with radii R_1 and R_2 ($R_1 < R_2$). The outer cylinder rotates with angular speed Ω while the inner is stationary. We employ a cylindrical coordinate system whose z axis coincides with the axis of the cylinders.

It can be shown that equation (13) reduces, in cylindrical coordinates, to the three equations

$$\frac{\partial t_{\langle rr \rangle}}{\partial r} + \frac{1}{r}(t_{\langle rr \rangle} - t_{\langle \theta\theta \rangle}) - \frac{\partial(\rho\psi + p)}{\partial r} = -\rho r\omega^2 \qquad (86)$$

$$\frac{\partial t_{\langle r\theta \rangle}}{\partial r} + \frac{2}{r}t_{\langle r\theta \rangle} - \frac{\partial(\rho\psi + p)}{\partial\theta} = 0 \qquad (87)$$

$$-\frac{\partial(\rho\psi + p)}{\partial z} = 0 \qquad (88)$$

for the flow described by equation (80) when advantage is taken of equation (82) and the fact that all of the components of the stress deviator are independent of θ and z.

From equation (88) we see that $\rho\psi + p$ cannot be a function of z. Furthermore, since ω and all the $t_{\langle \ \rangle}$ are functions of r alone, equations (86) and (87) show that $\partial(\rho\psi + p)/\partial r$ and $\partial(\rho\psi + p)/\partial\theta$ must also be independent of z and θ. Therefore, $\rho\psi + p$ is of the form

$$\rho\psi + p = f(r) + o\theta \qquad (89)$$

where o is a constant. As a matter of fact, o must be zero in our physical situation, since we wish $\rho\psi + p$ to be single-valued. Thus, equation (89) becomes

$$\rho\psi + p = f(r) \qquad (90)$$

and equation (87) simplifies to

$$\frac{\partial t_{\langle r\theta \rangle}}{\partial r} + \frac{2}{r}t_{\langle r\theta \rangle} = 0 \qquad (91)$$

whose solution is

$$t_{\langle r\theta \rangle} = \mu/r^2 \qquad (92)$$

where μ is a constant. The physical significance of μ becomes clear if we calculate M, the moment per unit height exerted on the fluid inside the cylindrical surface, with $r = $ constant:

$$M = t_{\langle r\theta \rangle}(2\pi r)r \qquad (93)$$

By equation (92), $\mu = M/2\pi$ and, therefore, by equation (83),

$$t_{\langle r\theta\rangle} = \frac{M}{2\pi r^2} = \tau \left(r\frac{d\omega}{dr}\right) = S(r) \tag{94}$$

We thus have arrived at the equation that has been used by rheologists in their study of shear-dependent viscosity in steady Couette flow.[60] From this equation the differential equation for ω can be written in terms of the rate of shear function λ, defined in equation (59):

$$\frac{d\omega}{dr} = \frac{1}{r}\lambda\left(\frac{M}{2\pi r^2}\right) \tag{95}$$

On the assumption that the liquid adheres to the cylindrical walls, the boundary conditions are

$$\omega(R_1) = \Omega, \qquad \omega(R_2) = 0 \tag{96}$$

Thus the angular velocity of a material particle at a distance r from the axis is

$$\omega(r) = \frac{1}{2}\int_{S_2}^{S} \frac{\lambda(S)}{S}\,dS = W(S) - W(S_2) \tag{97}$$

and, in particular,

$$\Omega = \omega(R_2) = \frac{1}{2}\int_{S_2}^{S_1} \frac{\lambda(S)}{S}\,dS = W(S_1) - W(S_2) \tag{98}$$

where we write S_1 for $S(R_1)$ and S_2 for $S(R_2)$, and where

$$W(S) = \int_0^S \frac{\lambda(S)}{S}\,dS \tag{99}$$

Methods of solving this integral equation for $\lambda(S)$, which enable us to determine the material function $\lambda(S)$ from experimental data, have been discussed elsewhere in this treatise.[7] We merely cite the results here. If the gap is small, (that is, if $R_2 - R_1 \ll R_1$), the approximation

$$\lambda(S_1) \approx \frac{R_1\Omega}{R_2 - R_1} \tag{100}$$

---

[60] See, for example, Section III of Ref. 7.

may be used. For larger gaps the quantity

$$\Upsilon(S_1) = 2M \frac{d\Omega}{dM} \tag{101}$$

is computed from the data, and then $\lambda(S_1)$ can be calculated from

$$\lambda(S_1) = \sum_{n=0}^{\infty} \Upsilon(\beta^n S_1) \tag{102}$$

where

$$\beta = R_1{}^2/R_2{}^2 \tag{103}$$

The series in equation (102) converges faster, the larger the gap.

Thus the "traditional" rheological approach is able properly to account for the velocity distribution and for the relation between applied torque and speed of rotation in Couette viscometers. This result comes essentially from one of the three equations of Cauchy's first law of motion, equation (13) [or equation (87)]. This one-dimensional approach, however, ignores the other two equations and thus fails to account for another phenomenon which occurs at the same time: the normal stress effect.

From equations (86) and (90), from the definition given by equation (32), and from the assumption that ψ is a constant,[61] the differential equation for the normal stress component $s_{\langle rr \rangle}$ is seen to be

$$\frac{ds_{\langle rr \rangle}}{dr} = \frac{1}{r}(t_{\langle rr \rangle} - t_{\langle \theta\theta \rangle}) - \rho r \omega^2 \tag{104}$$

or, in terms of the normal stress functions, according to equations (84) and (85),

$$\frac{ds_{\langle rr \rangle}}{dr} = \frac{1}{r}[\hat{\sigma}_1(S) - \hat{\sigma}_2(S)] - \rho r \omega^2 \tag{105}$$

From the integration of this equation,

$$s_{\langle rr \rangle}(r) = \int_{R_1}^{r} \frac{1}{r}[\hat{\sigma}_1(S) - \hat{\sigma}_2(S)] \, dr - \int_{R_1}^{r} \rho r \omega^2 \, dr + s_{\langle rr \rangle}(R_1) \tag{106}$$

we see that $s_{\langle rr \rangle}$ is a function only of r, since S and ω are. The other two normal

---

[61] If the only body force is due to gravity, the correction for the inaccuracy of this assumption is readily made: to every normal stress component add $-\rho g h$, where ρ is the density of the fluid, g is the acceleration due to gravity, and h is the vertical height above some reference level.

stress components, $s_{\langle\theta\theta\rangle}$ and $s_{\langle zz\rangle}$ can be calculated from equations (106), (84), and (85).

The difference $\Delta s_{\langle rr\rangle}$ between the thrust[62] exerted by the fluid on the cylindrical wall at R_1 and that on the wall at R_2 can be determined by substituting $r = R_2$ in equation (106):

$$\Delta s_{\langle rr\rangle} = s_{\langle rr\rangle}(R_2) - s_{\langle rr\rangle}(R_1)$$

$$= \int_{R_1}^{R_2} \frac{1}{r}[\hat{\sigma}_2(S) - \hat{\sigma}_1(S)]\,dr - \mathscr{I} \tag{107}$$

where

$$-\mathscr{I} = -\int_{R_1}^{R_2} \rho r[\omega(r)]^2\,dr < 0 \tag{108}$$

is seen to be the part of the thrust difference due to the centrifugal force. The inequality holds, because the integrand is positive. If we define a *corrected thrust difference* as the part of the thrust difference due to normal stress effects,

$$\Delta s^c = \Delta s_{\langle rr\rangle} + \mathscr{I} \tag{109}$$

then equation (107) may be written, with the help of equation (94), as

$$\Delta s^c = \frac{1}{2}\int_{S_2}^{S_1} \frac{1}{S}[\hat{\sigma}_2(S) - \hat{\sigma}_1(S)]\,dS \tag{110}$$

This equation has precisely the same form as equation (98) and thus can be inverted in the same way:

$$\hat{\sigma}_2(S_1) - \hat{\sigma}_1(S_1) = \sum_{n=0}^{\infty} \Psi(\beta^n S_1) \tag{111}$$

where β has been defined in equation (103) and

$$\Psi(S_1) = 2M\frac{d\Delta s^c}{dM} = 2S_1\frac{d\Delta s^c}{dS_1} \tag{112}$$

Compare equations (101) and (102). The second equality of equation (112), useful for the calculation of Ψ from experimental data, follows from equation (94). These equations indicate how the difference $\hat{\sigma}_2 - \hat{\sigma}_1$ can be computed

---

[62] By *thrust* we mean normal force per unit area exerted by a fluid on a surface. By the sign convention used here it has the opposite sign from the corresponding stress component.

from measurements of the thrust difference as a function of the speed of rotation. If the relative gap $(R_2 - R_1)/R_1$ is small, equation (107) can be approximated by

$$\Delta s_{\langle rr \rangle} \approx \frac{R_2 - R_1}{R_1}[\hat{\sigma}_2(S_1) - \hat{\sigma}_1(S_1)] \tag{113}$$

Thus it is not surprising that thrust differences are seldom large enough to be measured when $(R_2 - R_1)/R_1$ is small. Another approximation, sometimes useful when the relative gap is not small, is also available.[46]

b. Special Fluids

For the *Newtonian fluid* the integration in equation (97) can be carried out after the expression for $\lambda(S)$ from equation (72) is inserted:

$$\omega_N(r) = \Omega \frac{R_1{}^2}{R_2{}^2 - R_1{}^2} \cdot \frac{R_2{}^2 - r^2}{r^2} \tag{114}$$

where the subscript N indicates that the result is valid for the Newtonian fluid. According to equation (71), $\sigma_1(\kappa) = \sigma_2(\kappa) = 0$, and thus by equation (84) and (85) the three normal stresses $t_{\langle rr \rangle}$, $t_{\langle \theta\theta \rangle}$, $t_{\langle zz \rangle}$ are equal. The thrust difference $\Delta s_{\langle rr \rangle}$ is then given by the centrifugal-force term in equation (107):

$$\Delta s_{\langle rr \rangle} = -\mathscr{I}_N = -\int_{R_1}^{R_2} \rho r [\omega(r)]^2 \, dr < 0 \tag{115}$$

In fact, the integration here can be readily performed if equation (114) is used for $\omega(r)$. The thrust difference is then

$$\Delta s_{\langle rr \rangle} = -\frac{\rho R_1{}^2 R_2{}^4 \Omega^2}{2(R_2{}^2 - R_1{}^2)^2}\left[1 - \left|\frac{R_1}{R_2}\right|^4 - 4\left|\frac{R_1}{R_2}\right|^2 \ln \frac{R_2}{R_1}\right] = -\mathscr{I}_N \tag{116}$$

For the *viscoinelastic fluid*, since $\lambda(S)$ is quite arbitrary, an explicit integration of equation (97) to obtain the distribution of angular velocities $\omega(r)$ is not possible. From equations (74), (84), and (85) two of the normal stresses are seen to be equal:

$$s_{\langle rr \rangle} = s_{\langle \theta\theta \rangle} \tag{117}$$

A further deduction from equations (74) and (110) is that the corrected thrust difference is zero. Thus the relation

$$\Delta s_{\langle rr \rangle} = -\int_{R_1}^{R_2} \rho r [\omega(r)]^2 \, dr < 0 \tag{118}$$

holds for the viscoinelastic fluid as well as for the Newtonian fluid, but it is to be noted that equation (116) will not hold, because equation (114) doesn't. This result furnishes the basis for a qualitative test, which may be useful in deciding whether the viscoinelastic fluid is an adequate model for representing the properties of a given material: if the thrust difference is found to be greater than zero, this model can be rejected. Of course, the opposite does not hold.

Since the rate of shear function λ has the same form for the *second-order fluid* and for the Newtonian fluid, as can be seen from equations (72) and (75), the distribution of angular velocities (114) is also the same. As a result, the integrations indicated in equations (107) can be performed explicitly after the values for σ_1 and σ_2 given by equation (76) and (77) are substituted:[34]

$$\Delta s_{\langle rr \rangle} = -\mathscr{I}_N - 2\gamma\Omega^2 \left| \frac{R_2{}^2 + R_1{}^2}{R_2{}^2 - R_1{}^2} \right| \tag{119}$$

From this and equation (116) the thrust difference $\Delta s_{\langle rr \rangle}$ is seen to be proportional to Ω^2. Furthermore, because of the inequality stated in (43), $\gamma < 0$, this equation leads to the inequality

$$\Delta s^c = \Delta s_{\langle rr \rangle} + \mathscr{I}_N > 0 \tag{120}$$

That is, the effect of the γ term is to tend to make the thrust on the inner wall greater than that on the outer, an effect opposite to that expected from centrifugal force. Because of the status of the second-order fluid as an approximation to the memory fluid in slow flows we must conclude that equation (120) is true for all memory fluids in Couette flow, at least at low speeds of rotation.

c. Experiment

The basic measurements needed to determine $\sigma_2 - \sigma_1$ are those of the dimensions R_1 and R_2, the speed of rotation Ω, and the thrust difference $\Delta s_{\langle rr \rangle}$. It is also necessary to know the rate of shear function $\lambda(S)$ from some type of viscosity measurement. The $W(S)$ function defined in equation (99) can then be calculated by numerical integration. A graph of $\Omega(S_1)$, computed according to equation (98), can then be used to read off values of S_1 corresponding to the speed of rotation. The integral in equation (108), giving the centrifugal correction $\mathscr{I}$, can be evaluated numerically, after the velocity distribution $\omega(r)$ is calculated on the basis of equation (97). The corrected thrust difference, which can then be calculated from equation (109), is plotted against S_1, and $\Psi(S_1)$ is determined by using equation (112). The summation indicated in equation (111) then permits the calculation of $\sigma_2 - \sigma_1$.

A suitable apparatus for measuring thrust differences in Couette flow is shown in Fig. 2.[4] The flow field is established in the annulus between two coaxial glass tubes, each of which has a small, carefully drilled hole through its wall. The stationary outer tube is made of precision bore tubing, and the outer surface of the rotating inner tube is ground to form a good cylinder.

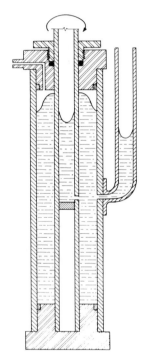

FIG. 2. Schematic diagram of an apparatus used for normal-stress measurements in Couette flow.

The thrust exerted on the outer wall of the annulus is indicated by the level of the fluid in a side tube placed over the hole in the outer tube. The level of the fluid inside the inner tube is determined by the thrust of the fluid on the inner wall of the annulus and by the pressure distribution that arises from the centrifugal force in the fluid inside the tube owing to its rigid rotation. To make it easier to see the level of the fluid in the inner tube the level of the fluid in the annulus may be raised by applying a suction through a hole in the top metal fitting. It is assumed that the tube is long enough to eliminate end effects, if the thrust difference does not change as one alters the distance of the holes from the surface of the fluid and from the bottom of the tubes.

In this instrument as well as in those used for measuring thrust differences in other flows it is found that extreme care must be exercised in the construction to avoid spurious results. Even small misalignments cause changes in thrust differences, when the direction of rotation is reversed, for example. Luckily it is found that, if an average is made of the $\Delta s_{\langle rr \rangle}$ in the two directions, the correct result is obtained if the misalignments are small.

Figure 3 is a photograph[63] of a similar apparatus, in which the fluid is a concentrated solution of polyisobutylene. A glance at this picture is sufficient

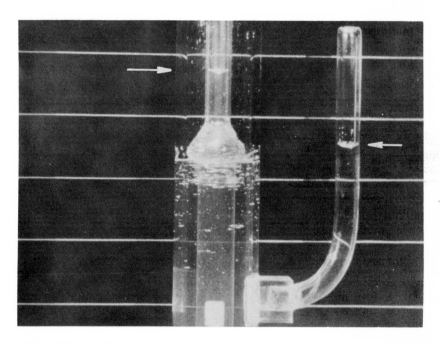

FIG. 3. Photograph of Couette-flow normal-stress apparatus filled with concentrated solution of polyisobutylene in decalin.[63]

to allow us to draw an interesting conclusion. Since the thrust on the inner cylinder is greater than that on the outer, that is, $\Delta s_{\langle rr \rangle} > 0$, inequality (115) tells us that this fluid is not Newtonian. Furthermore, the solution cannot be a non-Newtonian fluid of the viscoinelastic type because by inequality (118), the thrust difference is negative for such a fluid. Equation (107), in fact, shows that the inequality

$$\sigma_2(\kappa) > \sigma_1(\kappa)$$

[63] H. Markovitz, Rheological Behavior of Fluids. A film produced by Educational Services, Inc., Watertown, Massachusetts, for the National Committee for Fluid Mechanics Films under a grant from the National Science Foundation.

must hold, at least over part of the gap.[64] We now turn to quantitative measurements.

Data obtained on a 5.4 % solution of polyisobutylene in cetane are shown in Fig. 4, in which $\rho g(\Delta \bar{h})$ is plotted against the speed of rotation.[65,66] Here

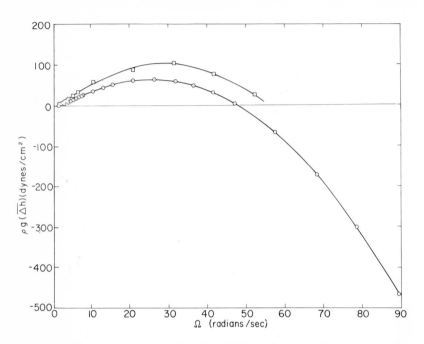

FIG. 4. Normal-stress data in Couette flow for a 5.4 % solution of polyisobutylene in cetane. For data indicated by ⊙, $R_1 = 0.500$ cm, $R_2 = 1.270$ cm; for those indicated by □, $R_1 = 0.500$ cm, $R_2 = 0.743$ cm.[66]

ρ is the density of the fluid, g is the acceleration of gravity, and $\Delta \bar{h}$ is the height of fluid in the center tube above that in the side-arm tube (averaged over the values obtained in the two senses of rotation). Data were obtained from two sets of tubes that differed in the diameter of the outer tube.

Following the procedure outlined in the beginning of this subsection,

---

[64] The photograph also shows the more commonly observed phenomenon of "climbing" around the rotating tube in the annulus. Because of the complicated flow pattern that exists in the neighborhood of the top surface of the fluid it would be difficult to formulate an exact theory of this effect. Semiquantitative discussions have been published.[2,4]

[65] Were it not for the centrifugal effect arising from the rigid rotation of the fluid inside the inner tube, $\rho g(\Delta h)$ would be equal to $\Delta s_{\langle rr \rangle}$.

[66] H. Markovitz and D. R. Brown, unpublished data.

Δs^c can be calculated[67] after $\lambda(S)$ was evaluated for this solution in a viscometer employing Poiseuille flow. From the logarithmic plot in Fig. 5 it is seen that the graph of Δs^c has nearly a slope of 2 at low values of S_1 and thus is approaching the behavior expected, according to equation (119), for a second-order fluid (which is the slow flow approximation to the memory fluid). The positive sign of Δs^c here is expected on the basis of equation (120).

Figure 5 also contains a graph of the auxiliary function Ψ calculated according to its defining equation, equation (112), and of $\hat{\sigma}_2 - \hat{\sigma}_1$ computed

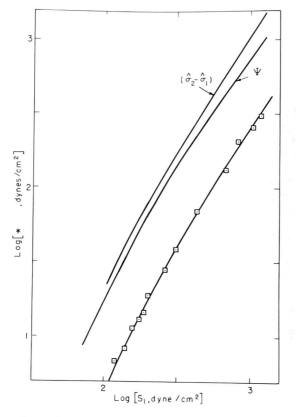

FIG. 5. Normal stress data in Couette flow for a 5.4 % solution of polyisobutylene in cetane; $R_1 = 0.500$ cm, $R_2 = 0.743$ cm.[66] The ordinate is the logarithm of the value of the function indicated next to each curve. The points indicate values of Δs^c.

[67] The centrifugal effect arising from the rigid rotation of the fluid inside the inner tube is independent of the nature of the material. Correction for this effect is based on the measurements of the observed thrust difference for a Newtonian fluid.

from Ψ by means of equation (111). In Fig. 6 it is shown that good agreement is found between values of $\hat{\sigma}_2 - \hat{\sigma}_1$ obtained with the two sets of tubes.[68]

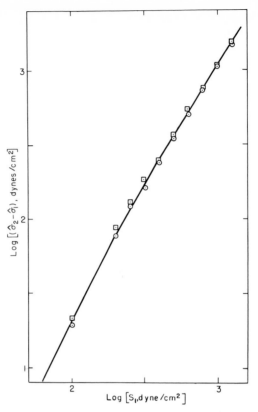

FIG. 6. Plot of $\hat{\sigma}_2 - \hat{\sigma}_1$ for a 5.4% solution of polyisobutylene in cetane. Points indicated by $\odot$ were calculated from data obtained in a Couette instrument with $R_1 = 0.500$ cm and $R_2 = 1.270$ cm; for those indicated by $\square$, $R_1 = 0.500$ cm and $R_2 = 0.743$ cm.[68]

3. AXIAL FLOW IN CYLINDERS

a. Theory[38]

In this subsection we deal with two types of axial flow: flow in a cylinder parallel to the axis (Poiseuille flow) and flow between two fixed coaxial cylinders parallel to the axis. In both of these flows the velocity field has the form

$$\dot{x}^1 = 0, \qquad \dot{x}^2 = u(r), \qquad \dot{x}^3 = 0 \qquad (121)$$

[68] H. Markovitz, *J. Polymer Sci.* **B3**, 3 (1965).

in cylindrical coordinates, where $x^1 = r$, $x^2 = z$, $x^3 = \theta$, and $u(r)$ is the velocity of a material point parallel to the axis. As a result, the two flows can be discussed together up to the point at which the boundary conditions are specified. Equations (69), (121), and (A5) of the Appendix yield

$$\kappa = du/dr \tag{122}$$

and equations (65) to (68) give the following information about the stress components:

$$t_{\langle r\theta\rangle} = t_{\langle z\theta\rangle} = s_{\langle r\theta\rangle} = s_{\langle z\theta\rangle} = 0 \tag{123}$$

$$t_{\langle rz\rangle} = s_{\langle rz\rangle} = \tau(\kappa) = S \tag{124}$$

$$t_{\langle rr\rangle} - t_{\langle\theta\theta\rangle} = s_{\langle rr\rangle} - s_{\langle\theta\theta\rangle} = \sigma_1(\kappa) = \hat{\sigma}_1(S) \tag{125}$$

$$t_{\langle zz\rangle} - t_{\langle\theta\theta\rangle} = s_{\langle zz\rangle} - s_{\langle\theta\theta\rangle} = \sigma_2(\kappa) = \hat{\sigma}_2(S) \tag{126}$$

Because u is a function of r alone, so will be κ and all of the components of the stress deviator.[69] Cauchy's first law of motion (13) reduces to

$$\frac{\partial t_{\langle rr\rangle}}{\partial r} + \frac{1}{r}(t_{\langle rr\rangle} - t_{\langle\theta\theta\rangle}) - \frac{\partial(\rho\psi + p)}{\partial r} = 0 \tag{127}$$

$$\frac{\partial t_{\langle rz\rangle}}{\partial r} + \frac{1}{r}t_{\langle rz\rangle} - \frac{\partial(\rho\psi + p)}{\partial z} = 0 \tag{128}$$

$$-\frac{\partial(\rho\psi + p)}{\partial\theta} = 0 \tag{129}$$

An analysis similar to that described in Section III2 on Couette flow leads to the conclusion that:

$$t_{\langle rz\rangle} = -\frac{1}{2}fr + \frac{b}{r} \tag{130}$$

$$\rho\psi + p = -fz + h(r) \tag{131}$$

$$\frac{dh}{dr} = \frac{dt_{\langle rr\rangle}}{dr} + \frac{1}{r}(t_{\langle rr\rangle} - t_{\langle\theta\theta\rangle}) \tag{132}$$

---

[69] We shall shorten our discussion of this and other flows, because many parts of the development closely parallel those already given in considerable detail in our treatment of Couette flow. For further amplification the reader may refer to that treatment or to Refs. 4, 37 and 38.

where f and b are constants and $h(r)$ is a function of r. The physical significance of f can be seen from equation (131). Since the stress deviators are independent of z,

$$-\frac{\partial(\rho\psi + p)}{\partial z} = -\frac{\partial(\rho\psi + s_{\langle zz \rangle})}{\partial z} = f \tag{133}$$

If we consider $-s_{\langle zz \rangle}$ the driving force per unit area due to contact forces and $-\rho\psi$ that due to body forces, then f is the gradient of the total driving force per unit area,[70] which we call the *pressure head*, for short. For us to proceed further the boundary conditions must be introduced. We discuss the two types of axial flow separately.

(1) *Poiseuille flow.* The flow is assumed to occur in a tube of radius R. On the assumption that the fluid adheres to the cylindrical surface we require that

$$u(R) = 0 \tag{134}$$

The second condition is that the stress be continuous at $r = 0$. As a result, the constant b of equation (130) must be zero and

$$t_{\langle rz \rangle} = -\tfrac{1}{2}fr = \tau(\kappa) = S \tag{135}$$

where the last two equalities come from equation (124). The rate of shear can then be written

$$\kappa = \lambda(S) = \lambda(-\tfrac{1}{2}fr) = -\lambda(\tfrac{1}{2}fr) = \frac{du}{dr} \tag{136}$$

where equations (59), (135), (55), and (122) have been used. The last equality is the basis of the usual analysis of Poiseuille flow;[71] we record these results here for reference. From equation (136) the velocity field is found to be given by the expression

$$u(r) = \int_r^R \lambda(\tfrac{1}{2}fr)\, dr \tag{137}$$

and Q, the *volume rate of flow*, by the equation,

$$Q = \frac{8\pi}{f^3} \int_0^{S_R} S^2 \lambda(S)\, dS \tag{138}$$

where

$$S_R = \tfrac{1}{2}fR \tag{139}$$

---

[70] Except for the body force, correction for which is readily made if it is due to gravity,[61] f is also the gradient in the z direction of each of the normal stresses $s_{\langle zz \rangle}$, $s_{\langle rr \rangle}$, and $s_{\langle \theta\theta \rangle}$.

[71] See, for example, Section II of Ref. 7.

Equation (138) suggests that data obtained at various f and R can be reduced to a single graph, if $D = 4Q/\pi R^3$ is plotted against S_R; the inversion of this equation yields an equation useful for the determination of $\lambda(S)$ from experimental measurements:

$$\lambda(S_R) = D \left(\frac{3}{4} + \frac{1}{4} \frac{d \ln D}{d \ln S_R} \right) \tag{140}$$

The traditional rheological analysis of Poiseuille flow is thus in complete agreement with the results for the memory fluid with regard to the shear stress, velocity distribution, and rate of flow. It is silent however with regard to the normal stresses. From equations (132), (125), and (131) it can be shown that

$$s_{\langle rr \rangle} = fz + \rho \psi + \int_r^R \frac{1}{\xi} \hat{\sigma}_1[S(\xi)] \, d\xi + c \tag{141}$$

where c is a constant. If $s_{\langle rr \rangle}$ is known at any point (say at some boundary), then c is determined, and the value of $s_{\langle rr \rangle}$ can be calculated at any other point. The other normal stresses, $s_{\langle zz \rangle}$ and $s_{\langle \theta\theta \rangle}$, can then be obtained from equations (125) and (126). It does not seem possible to determine normal stress functions from the tractions on the boundary surface, the cylinder at $r = R$. According to equation (141), measurements of $s_{\langle rr \rangle}$ at various values of z will only yield the constant f.

(2) *Axial Flow between Coaxial Cylinders.* We now consider an axial flow between two stationary coaxial cylinders of radii R_1 and R_2. Again on the assumption of adherence of the fluid to the cylinders, the boundary conditions are

$$u(R_1) = u(R_2) = 0 \tag{142}$$

From equations (122), (59), and the second equality of (142) the velocity field is found to be

$$u(r) = \int_{R_2}^r \lambda(S(r)) \, dr \tag{143}$$

where

$$S(r) = -\frac{1}{2} fr + \frac{b}{r} \tag{144}$$

as may be verified from equations (130) and (124). The boundary condition at R_1 will be satisfied if $r = R_1$ is substituted in equation (143):

$$u(R_1) = 0 = \int_{R_2}^{R_1} \lambda(S(r)) \, dr \tag{145}$$

The constant b can be evaluated from this equation by assuming an empirical form[72] for $\lambda(S)$ or by using numerical methods.

The radial normal stress component $s_{\langle rr \rangle}$ is given by an expression of the form of equation (141) here also. The thrust difference on the cylindrical walls at a given value of z is then[73]

$$\Delta s_{\langle rr \rangle} = s_{\langle rr \rangle}(R_2) - s_{\langle rr \rangle}(R_1) = -\int_{R_1}^{R_2} \frac{1}{r} \hat{\sigma}_1(S(r)) \, dr \qquad (146)$$

The only special fluid we mention here is one that obeys Weissenberg's conjecture [cf. equation (78)], $\sigma_1 = 0$. For such a fluid the thrust difference is zero. Thus, such a relation cannot adequately describe the properties of a fluid that exhibits a thrust difference in this type of flow.

b. Experiment[4]

Equation (146) indicates how σ_1 determines the thrust difference on the cylindrical walls when the fluid flows axially in the annulus between two stationary cylinders. The calculation of σ_1 from the data can be performed by a numerical procedure or by assuming empirical forms for the viscometric functions.[74,75] Besides the thrust difference $\Delta s_{\langle rr \rangle}$ and the dimensions of the apparatus the viscosity function for the fluid must be known.

In the apparatus[74] shown in Fig. 7 the thrust difference across the annulus is measured by a pressure transducer, in which fluid is introduced to both sides of the diaphragm.[76] Provision is made for reversal of flow, so that effects of misalignment may be averaged out. Of course, care must be taken that the points of measurements in the two tubes correspond to the same values of z; otherwise, the pressure head will contribute to the thrust difference. The flow is maintained by applying a constant pressure to the reservoir of the test fluid. Provision is also made for observing the volume rate of flow and the pressure drop along the annulus and along capillaries inserted in the system.

In a more recent apparatus[75] the thrust difference is measured with a micromanometer, and the flow is maintained by a pump.

Data obtained on a 0.9 % solution of hydroxyethyl cellulose (Natrosol) in water are shown in Fig. 8, where the measured thrust difference is plotted

[72] A. G. Fredrickson and R. B. Bird, *Ind. Eng. Chem.* **50,** 347 (1958).

[73] Here we also assume that ψ depends only on z. This will be the case, if gravity supplies the only body force and the cylinders are vertical. If the latter is not the case, a correction may be made.[61]

[74] J. W. Hayes and R. I. Tanner, *Proc. 4th Intern. Congr. Rheol., Providence, 1963* Vol. **3,** p. 389 (1965). Wiley, New York.

[75] J. D. Huppler, *Trans. Soc. Rheol.* **9**:**2,** 173 (1965).

[76] Measurements can be made much more quickly with the transducer than by the older technique, in which the test substance is its own manometric fluid.

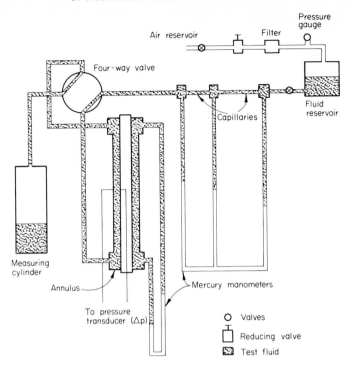

FIG. 7. Schematic diagram of an apparatus used for normal-stress measurements in axial flow between coaxial cylinders.[74]

against the pressure head.[75] Here, too, a qualitative observation is informative. The fact that a thrust difference is observed shows, according to equation (146), that $\sigma_1 \neq 0$ for this solution. Thus, not only is the fluid not Newtonian, but also the Weissenberg relation (78) does not hold. The normal stress function $\sigma_1(\kappa)$ calculated from these data is shown in Fig. 9.[75]

4. CONE-PLATE FLOW

a. Theory [4]

Here we assume a simple viscometric flow with a velocity field having the contravariant components

$$\dot{x}^1 = 0, \qquad \dot{x}^2 = \omega(\theta), \qquad \dot{x}^3 = 0 \qquad (147)$$

in a spherical coordinate system with $x^1 = \theta$, $x^2 = \phi$, and $x^3 = r$. Each material particle travels in a circle of radius $r \sin \theta$ with an angular velocity $\omega(\theta)$ and thus has a linear speed $r\omega(\theta) \sin \theta$. All points on the conical surfaces

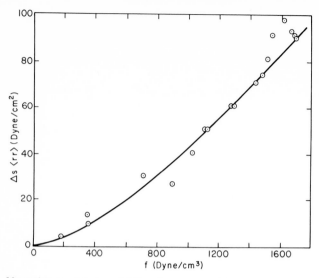

FIG. 8. Normal-stress data in axial flow between coaxial cylinders for a 0.9% solution of hydroxyethyl cellulose in water.[75]

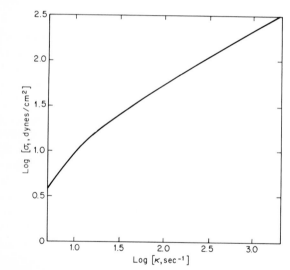

FIG. 9. The normal stress function σ_1 calculated for a 0.9% solution of hydroxyethyl cellulose in water from the data in Fig. 8.[75]

θ = constant have the same angular velocity. From equations (69), (A6) of the Appendix, and (147) the rate of shear is seen to be

$$\kappa = \sin \theta \, \frac{d\omega}{d\theta} \qquad (148)$$

Following the methods outlined in the discussion of Couette flow it can be shown that Cauchy's first law (13), expressed in spherical coordinates, reduces here to the following equations:

$$\frac{\partial s_{\langle\theta\theta\rangle}}{\partial r} - \frac{1}{r}[\hat{\sigma}_1(S) + \hat{\sigma}_2(S)] = -\rho r^2 \sin^2\theta \,[\omega(\theta)]^2 \tag{149}$$

$$\frac{\partial s_{\langle\theta\theta\rangle}}{\partial\theta} + \cot\theta\,[\hat{\sigma}_1(S) - \hat{\sigma}_2(S)] = -\rho r^2 \sin\theta\cos\theta\,[\omega(\theta)]^2 \tag{150}$$

$$S = \frac{c}{\sin^2\theta} \tag{151}$$

where c is a constant.

Now a difficulty arises. Equations (149) and (150) are found to be incompatible. This can be verified by calculating $\partial[\partial s_{\langle\theta\theta\rangle}/\partial r]/\partial\theta$ from equation (149) and $\partial[\partial s_{\langle\theta\theta\rangle}/\partial\theta]/\partial r$ from equation (150). If these derivatives are to be equal (as they should be, seeing that they differ only by a change in the order of taking the derivatives), then

$$\frac{d}{dS}[\hat{\sigma}_1(S) + \hat{\sigma}_2(S)]c\cos\theta = -\rho r[\omega(\theta)]^2 \sin^5\theta\,\frac{d\omega}{d\theta} \tag{152}$$

The left side here is a function of θ alone, but the right side is also a function of r, and so the equation cannot be satisfied. The implication here is that a flow of the type assumed in equation (147), where all particles travel in circles about the axis of the cone, is not possible,[77] because Cauchy's equation cannot be satisfied.

We now discuss conditions under which equation (152) might be approximately satisfied, and thus a flow can exist which can be approximately described by equation (147). This cannot be accomplished by restricting ourselves to slow speeds, at which the inertia effects are negligible[78] and the right side is approximately zero. The left side will be precisely zero, only if the fluid happens to have the property that $d[\hat{\sigma}_1(S) + \hat{\sigma}_2(S)]/dS = 0$ over the range of S occurring in the experiment.[79] In general, non-Newtonian fluids do not have this property. There are circumstances in which the left side of equation (152) is approximately zero even for the general memory fluid. Let us assume that the flow takes place in a fluid between a cone rotating about its axis with an angular speed Ω and a rigid stationary disc. The axis of the cone is taken to be the polar axis of the spherical coordinate system. The plane

---

[77] This result is true even for the Newtonian fluid, where $\hat{\sigma}_1(S) = \hat{\sigma}_2(S) = 0$. Here the left side of equation (152) is zero, but the right side isn't.

[78] Such a limitation is found to be sufficient for torsional flow, which is discussed below.

[79] For equation (152) to be satisfied for all S the condition $\hat{\sigma}_1(S) + \hat{\sigma}_2(S) = 0$ must be met, because $\hat{\sigma}_1(0) = \hat{\sigma}_2(0) = 0$ [cf. equation (56)].

$\theta = \frac{1}{2}\pi$ is taken to be the plane of the disc. The cone is assumed to coincide with the conical surface $\theta = \frac{1}{2}\pi - \alpha$. The equation will hold approximately if inertia is negligible and if α is so small that

$$\cos \theta \approx 0 \quad \text{and} \quad \sin \theta \approx 1$$

throughout the fluid;[80] both sides of equation (152) are then approximately zero. With these approximations, equations (148) and (151) allow us to conclude that

$$S = s_{\langle \theta \phi \rangle} = c \tag{153}$$

$$\kappa = \frac{d\omega}{d\theta} = \lambda(c) = \text{constant} \tag{154}$$

Equations (149) and (150) then simplify to

$$\frac{\partial s_{\langle \theta \theta \rangle}}{\partial \theta} = 0 \tag{155}$$

$$r\frac{\partial s_{\langle \theta \theta \rangle}}{\partial r} = \frac{\partial s_{\langle \theta \theta \rangle}}{\partial \ln r} = \sigma_1(\kappa) + \sigma_2(\kappa) \tag{156}$$

To the approximation that $\kappa = \text{constant}$ [cf. equation (154)] $\sigma_1(\kappa)$ and $\sigma_2(\kappa)$ and, in fact, all the components of the stress deviator $t\langle\ \rangle$ will have values that are independent of position throughout the fluid.

From equations (153) and (154) and the boundary conditions expressing adherence,

$$\omega(\tfrac{1}{2}\pi) = 0 \tag{157}$$

$$\omega(\tfrac{1}{2}\pi - \alpha) = \Omega \tag{158}$$

the equations usually used[81] in cone–plate viscometry can be obtained:

$$M = \tfrac{2}{3}\pi R^3 \tau(\kappa) \tag{159}$$

where

$$\kappa = \Omega/\alpha \tag{160}$$

and M is the torque necessary to rotate the cone when the fluid fills the gap out to the radius R.

Since κ is independent of r, the distribution of normal stresses is given by the indefinite integral of equation (156),

$$s_{\langle \theta \theta \rangle}(r) = [\sigma_1(\kappa) + \sigma_2(\kappa)] \ln r + \text{constant} \tag{161}$$

---

[80] This equation will not hold in those cone–cone instruments in which θ is very far removed from $\frac{1}{2}\pi$ regardless of how small α is.

[81] Cf. Ref. 7, Section VII.

Thus, for all memory fluids, if the thrust exerted by the fluid on the disc, $-s_{\langle\theta\theta\rangle}$, is plotted against $\ln r$, the natural logarithm of the distance from the axis of rotation, a straight line should result. Furthermore, the slope of that line is $-[\sigma_1(\kappa) + \sigma_2(\kappa)]$.

The measurement of the total normal force F acting on the plate has also been proposed as a method of determining normal stress functions. If the velocity field (147) extends to $r = R$, and if the thrust $-s_{\langle\theta\theta\rangle}(R)$ is known to have the value p_θ at $r = R$, the integration constant of equation (161) can be determined, and we can write

$$s_{\langle\theta\theta\rangle}(r) = -p_\theta + [\sigma_1(\kappa) + \sigma_2(\kappa)] \ln (r/R) \tag{162}$$

The total force can be calculated by integration:

$$F = -2\pi \int_0^R s_{\langle\theta\theta\rangle} r\, dr = \pi R^2 p_\theta + \tfrac{1}{2}\pi R^2 [\sigma_1(\kappa) + \sigma_2(\kappa)] \tag{163}$$

If the free surface of the fluid is indeed the spherical surface $r = R$, and if the velocity field (147) extends to that interface, then the radial component of the thrust, $-s_{\langle rr\rangle}(R)$, will be equal to the ambient pressure p_0:

$$s_{\langle rr\rangle}(R) = -p_0 \tag{164}$$

The thrust p_θ can then be written

$$p_\theta = -s_{\langle\theta\theta\rangle}(R) = -s_{\langle rr\rangle}(R) + [s_{\langle rr\rangle}(R) - s_{\langle\theta\theta\rangle}(R)] = p_0 - \sigma_1(\kappa) \tag{165}$$

where equation (67) and the identification $x^1 = \theta$ and $x^3 = r$ were used to obtain the last equality. Substitution of this equation into equation (163) then leads to the result

$$F^* = F - \pi R^2 p_0 = \tfrac{1}{2}\pi R^2 [\sigma_2(\kappa) - \sigma_1(\kappa)] \tag{166}$$

where F^* is the normal force in excess of that due to the ambient pressure.

b. Experiment[4]

Since the theory indicates that the velocity field cannot be precisely of the form equation (147), it is of special interest to map experimentally the flow pattern that does exist. By following the motion of small suspended particles Roberts [82–84] concluded that, except near the edge of the cone, equation (147) is indeed a good approximation to the velocity field, if $\alpha \leqslant 4°$.

[82] J. E. Roberts, ADE Rep. 13/52, Armament Design Establishment, Knockholt, Kent, 1952.

[83] J. E. Roberts, Proc. 2nd Intern. Congr. Rheol., Oxford, 1953, p. 350. Butterworth, London and Washington, D.C., 1954.

[84] A. Jobling and J. E. Roberts, J. Polymer Sci. 36, 421 (1959).

With a velocity field described by equation (147) the distribution of normal traction is given by equation (161).

Figure 10 is a schematic diagram of a type of apparatus used for normal-stress measurements. The cone rotates. The thrust is determined at various points on the disc by measuring the height to which the test fluid rises in glass tubes which are placed over small holes in the stationary disc. Cylindrical walls being attached to both the cone and the disc, the absolute value of the recorded thrust can have no significance, because of the complicated flow which occurs near the edges; it is only the relative values sufficiently far from the edge that are of physical importance.

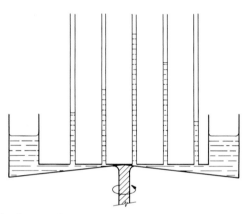

Fɪɢ. 10. Schematic diagram of an apparatus used for normal-stress measurements in cone–plate flow.

In other instruments for this purpose the thrust is measured by using a pressure transducer in a small chamber connected to each hole in the stationary plate.[85] Sometimes no cylindrical wall is used above the cone and disc.[82,85] Because of the small angle α the gap at the edge is small, and the fluid is held in by surface tension.

Even slight deviations from trueness of the surface or from coaxial alignment can have a large effect on the thrust distribution in cone–plate flow.[85] Fortunately, however, a correction for slight misalignments can be obtained by using the average, $h_a(r)$, of the two heights of rise of fluid observed for the two senses of rotation at a given distance r from the axis of rotation.

The approximate theory that we are using assumes negligible inertia effects and thus is expected to be useful at low angular speeds. It is found empirically that it is possible to extend the range of speeds for which the

[85] N. Adams and A. S. Lodge, *Phil. Trans. Roy. Soc.* **A256,** 149 (1964). This paper contains an excellent discussion of sources of error in cone–plate normal-stress measurements.

theory is applicable if a "centrifugal correction" $h_c(r)$ is made.[85,86] This correction is equal to the value $h_a(r)$ for a Newtonian fluid in the same instrument at the same speed of rotation. From the corrected value $\bar{h}(r) = h_a(r) - h_c(r)$ the semilogarithmic slope $\partial s_{\langle\theta\theta\rangle}/\partial \ln r$ is calculated as $-\rho g \, d\bar{h}/d \ln r$.

A photograph showing the distribution of tractions has already appeared in this treatise.[87]

The quantitative data in Fig. 11 for a solution of polyisobutylene show that the straight lines expected from the semilogarithmic plots on the basis

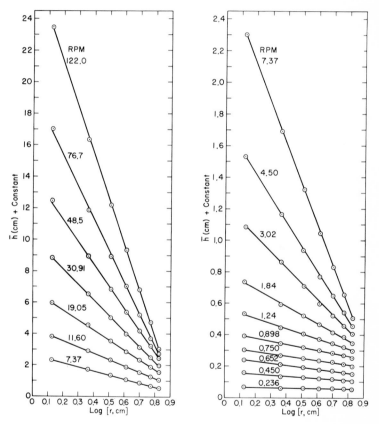

FIG. 11. Normal-stress data in cone–plate flow for a 6.9% solution of polyisobutylene in cetane.[86] The curves are labeled with values of Ω in revolutions per minute. In this instrument α is 2°. The curves have been shifted vertically by various amounts to avoid crowding.

[86] H. Markovitz and D. R. Brown, *Proc. Intern. Symp. Second-Order Effects Elasticity, Plasticity and Fluid, Haifa, 1962* p. 585. MacMillan, New York, 1964.
[87] Plate 1 of Ref. 15.

ᶠ ᵊquation (161) are indeed obtained over a wide range of speeds of rotation.[86] Figure 12 shows that the same distribution of tractions is obtained for a given κ at two different values of the angle α.[85] These results and similar ones obtained in other laboratories,[82] together with the direct observation of the flow lines, combine to give confidence in the applicability of the approximate theory. Figure 13 shows values of $[\sigma_1(\kappa) + \sigma_2(\kappa)]$ calculated from the slopes of the lines in Fig. 11.

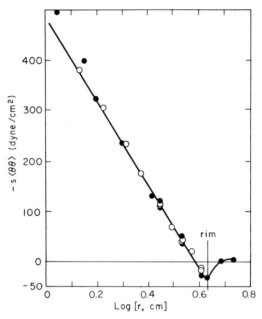

FIG. 12. Normal-stress data in cone–plate flow for a 0.058 g/ml solution of polyisobutylene in decalin. Here $\kappa = 22.8 \text{ sec}^{-1}$, $\alpha = 5.17°$ for points $\bigcirc$, and $\alpha = 3.23°$ for points $\bullet$. The ordinate represents the normal thrust (averaged over the two senses of rotation and corrected for centrifugal force) exerted by the fluid above the ambient pressure.[85]

In a number of investigations, conclusions have been reached on the basis of experimental results that depend, at least in part, on the *flow at the boundary*. These are of two types. In one the value of the thrust at the bounding surface, p_θ, is of direct interest; in the other F^*, the total normal force in excess of that due to ambient pressure, is measured. Of course, disturbance to the flow is minimized by allowing a free boundary.

For the first case conflicting results have been reported. In one series of investigations[15,82,83] it was concluded that p_θ is equal to the ambient pressure p_0 for a number of polymers. By equation (165) the conclusion was then reached that the Weissenberg relation $\sigma_1(\kappa) = 0$ held. In a more recent extensive investigation[85] the tractions were measured at many points in the

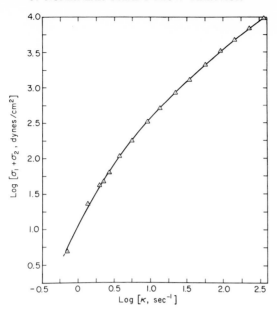

FIG. 13. Graph of $\sigma_1 + \sigma_2$ for a 5.4% solution of polyisobutylene in cetane, calculated from the data of Fig. 11.

neighborhood of the boundary. As seen in Fig. 12, the thrust at the rim, p_θ, is found to be less than the ambient pressure. Even this careful investigation left some unanswered questions, however. It was found that p_θ was rather insensitive to the shape of the bounding surface; for example, it made little difference whether the surface was spherical of radius R or was distorted by having an excess of fluid. It had been expected that p_θ would be sensitive to the orientation of that surface and the possible distortion of flow due to neighboring fluid outside the gap. Further, the value of $\sigma_1(\kappa)$, calculated as $p_0 - p_\theta$ by means of equation (165), was found to differ considerably from that calculated from traction distributions in cone–plate and torsional-flow methods.[88] Moreover, the conclusion that σ_1 is equal to 0, based on the equality of p_θ and p_0, is in conflict with a considerable body of evidence at this time.[46,89]

Because of this uncertainty of knowledge concerning flow conditions at the boundary we shall not quote here further results based on these methods; the interested reader may find reports of such investigations elsewhere.[15,90–92]

[88] See Equation (178).

[89] See Sections III3b and IV1 and Figs. 9 and 19.

[90] W. Philippoff, *J. Appl. Phys.* **27**, 984 (1956).

[91] R. F. Ginn and A. B. Metzner, *Proc. 4th Intern. Congr. Rheol., Providence, 1963* Vol. **2**, p. 583. Wiley, New York, 1965.

[92] M. Reiner, *Proc. 4th Intern. Congr. Rheol., Providence, 1963* Vol. 1, p. 267. Wiley, New York, 1965.

5. TORSIONAL FLOW

a. Theory

For torsional flow we envisage a simple viscometric flow in which the material particle travels in a circular path of radius r with an angular velocity ω, which depends on the distance z from a fixed plane; thus, the velocity is $r\omega(z)$. The contravariant components of the velocity field are then

$$\dot{x}^1 = 0, \qquad \dot{x}^2 = \omega(z), \qquad \dot{x}^3 = 0 \qquad (167)$$

in a suitably oriented cylindrical coordinate system, where we have identified x^1 with z, x^2 with θ, and x^3 with r.

Following the steps that were used in the discussion of other flows, we find that

$$\kappa = r \frac{d\omega}{dz} \qquad (168)$$

$$t_{\langle rz \rangle} = t_{\langle r\theta \rangle} = s_{\langle rz \rangle} = s_{\langle r\theta \rangle} = 0 \qquad (169)$$

$$t_{\langle z\theta \rangle} = s_{\langle z\theta \rangle} = \tau(\kappa) \qquad (170)$$

$$t_{\langle zz \rangle} - t_{\langle rr \rangle} = s_{\langle zz \rangle} - s_{\langle rr \rangle} = \sigma_1(\kappa) \qquad (171)$$

$$t_{\langle \theta\theta \rangle} - t_{\langle rr \rangle} = s_{\langle \theta\theta \rangle} - s_{\langle rr \rangle} = \sigma_2(\kappa) \qquad (172)$$

We note that κ and all the physical components of the stress deviator $t_{\langle\ \rangle}$ depend only on z and r.

In this flow, as in cone–plate flow, the assumed motion, equation (167), is incompatible with the dynamical equations (13). Here it is found that the difficulty can be removed if the approximation of negligible inertia is made, an approximation that is expected to be useful at slow flows.

We now assume that the torsional flow occurs between two parallel discs of radius R, which lie in the planes $z = 0$ and $z = l$ with their centers on the z axis. The disc at $z = l$ rotates with angular speed Ω, while the other one is stationary. The boundary conditions and the dynamical equations (with $\rho = 0$) are satisfied if

$$\omega(z) = \frac{\Omega}{l} z, \qquad \kappa = \frac{\Omega}{l} r \qquad (173)$$

and

$$\frac{ds_{\langle zz \rangle}}{d\ln r} = \frac{d\sigma_1(\kappa)}{d\ln \kappa} + \sigma_2(\kappa) \qquad (174)$$

To maintain the flow a torque M, given by

$$M = \frac{2\pi R^3}{\kappa_R{}^3} \int_0^{\kappa_R} \kappa^2 \tau(\kappa)\, d\kappa, \qquad \kappa_R = \frac{R\Omega}{l} \qquad (175)$$

must be applied. To determine the material function τ from measurements of M as a function of Ω it is useful to invert equation (175). This can be accomplished by a method similar to that employed in obtaining equation (140), with the result:

$$\tau(\kappa_R) = E(\kappa_R)\left[\frac{3}{4} + \frac{1}{4}\frac{d \ln E(\kappa_R)}{d \ln \kappa_R}\right], \qquad E(\kappa_R) = \frac{2M}{\pi R^3} \qquad (176)$$

The distribution of thrusts, $-s_{\langle zz\rangle}(r)$, on the discs can be obtained by integration of equation (174):

$$s_{\langle zz\rangle}(r) - s_{\langle zz\rangle}(0) = \sigma_1(\kappa) + \int_0^\kappa \frac{1}{\zeta}\sigma_2(\zeta)\,d\zeta \qquad (177)$$

Equations (174) and (177) suggest a method of plotting data that will give a single curve from measurements of the thrust distribution, $s_{\langle zz\rangle}(r)$, at various speeds of rotation, Ω, and gaps l. Either $ds_{\langle zz\rangle}/d \ln r$ or

$$s_{\langle zz\rangle}(r) - s_{\langle zz\rangle}(0)$$

can be graphed as a function of κ.

Before closing this discussion we wish to remind the reader of the difference between the theory of torsional and cone–plate flow and that applicable to the cylindrical flows in the preceding subsections. In the former cases a solution was not possible without making approximations (negligible inertia effects and also $\cos \theta \approx 1$ for cone–plate flow). Such approximations were not needed in Couette or axial flows, for which the exact theory includes the effects of inertia.[93] Actually, there is no mathematical proof that the laminar flows, equation (147) and (167), will occur, even if the experimental conditions corresponding to the approximate theory are met. However, experimental observation of flow lines and other checks with predictions of the theory lend confidence in the applicability of these approximations in these limiting situations.

b. Experiment[4]

One form of apparatus for making normal-stress measurements in torsional flow between parallel plates is shown in Fig. 14. The plates are the bases of two cylindrical cups, the outer one rotating while the inner one is stationary. Construction and details of measurement[85,86,94] are similar to those already

---

[93] However, this does not imply that the simple motions envisaged in the discussions of those flows will always occur. In fact, it is known that at sufficiently high speeds more complicated motions (Taylor instabilities and turbulence, for example) take place, even for a Newtonian fluid. The explanation of these phenomena lies in the consideration of questions of stability and uniqueness, questions that we shall not discuss here.

[94] H. W. Greensmith and R. S. Rivlin, *Phil. Trans. Roy. Soc.* **A245**, 399 (1953).

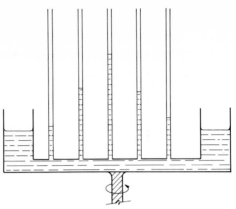

FIG. 14. Schematic diagram of an apparatus used for normal-stress measurements in torsional flow.

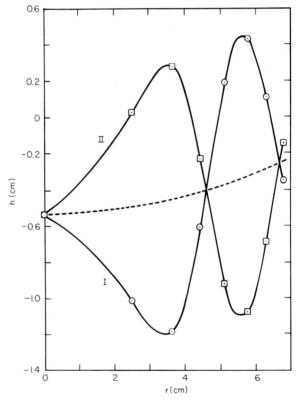

FIG. 15. Normal stress data in torsional flow for a lubricating oil showing the effect of non-parallelism of discs.[94] Curve I, clockwise rotation; curve II, counterclockwise rotation. Broken line is $h_a(r)$.

described in the section on cone–plate flow. Here, too, slight misalignments can greatly affect the distribution of thrusts. This is seen in Fig. 15, which shows $h(r)$, the height of rise of fluid in the glass tube, for the two senses of rotation when the plates deviate by about 0.3° from being parallel.[94] The extent of the discrepancy increases, the more viscous the test fluid. Fortunately, when the misalignment is small, the average, $h_a(r)$, of the values of $h(r)$ for the two directions of rotation provides a correction. A centrifugal correction[94] $\bar{h}_c(r)$, defined in our discussion of cone–plate flow, is found to be useful in obtaining a corrected average height of rise, $\bar{h}(r) = h_a(r) - \bar{h}_c(r)$.

Figure 16 shows $\bar{h}(r)$ data[94] obtained from a solution of polyisobutylene in tetralin in torsional flow for four different values of the gap l. All of these data fall together on a single curve when plotted as $\bar{h}(0) - \bar{h}(r)$ versus $\kappa = \Omega r/l$. See Fig. 17.[94] This result is expected on the basis of equation (177) and thus lends confidence to the applicability of the approximate theory (neglect of inertia), on which it is based.

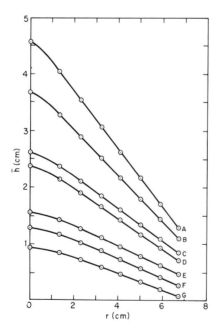

FIG. 16. Normal-stress data in torsional flow for a 6% solution of polyisobutylene in tetralin.[94] The speeds of rotation, Ω, in revolutions per minute, and the gap l, in centimeters, respectively, for the various curves are: A, 28.54, 0.185; B, 41.24, 0.343; C, 15.92, 0.185; D, 30.81, 0.503; E, 27.28, 0.503; F, 41.22, 0.805; G, 35.49, 0.805. The curves have been shifted vertically to increase clarity.

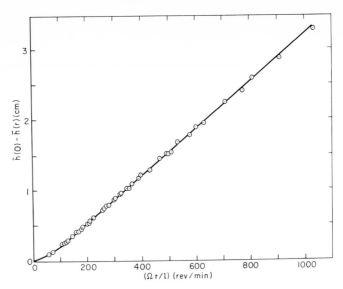

FIG. 17. Normal-stress data in torsional flow for a 6% solution of polyisobutylene in tetralin.[94] The data include all those shown in Fig. 16.

IV. Normal Stress Functions

1. EXPERIMENTAL RESULTS

Two different types of measurement are required to determine both the normal stress functions, σ_1 and σ_2. With three types it is possible to perform a test of consistency. Up to the time of this writing very few experimental results of this kind have been reported; by the time of publication more will probably be available, because several groups are actively engaged in research programs in this field.

One possible combination of flows that can be used for determining the normal stress functions is the combination of Couette flow, which according to equation (111) yields $\sigma_2 - \sigma_1$, and cone–plate flow, which according to equation (161) gives $\sigma_2 + \sigma_1$. From these the individual functions σ_1 and σ_2 are not difficult to find. Results of such a calculation are plotted in Fig. 18 for a 5.4% solution of polyisobutylene in cetane.[66] The third viscometric function $\tau(\kappa)$ is shown in the same figure. For this material σ_1 is of the order of one-half to one-third of σ_2. At low rates of shear the shear stress function τ is much greater than the normal stress functions, but at the high rates of shear $\sigma_2 > \tau$ and σ_1 approaches equality with τ.

For a 0.9% solution of hydroxyethyl cellulose in water values of $\sigma_1(\kappa)$ were calculated from thrust difference measurements in axial flow between

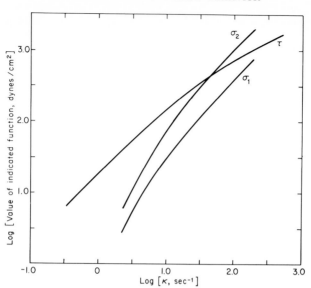

FIG. 18. The viscometric functions for a 5.4% solution of polyisobutylene in cetane.[66]

coaxial cylinders according to equation (146) and values of $(\sigma_1 + \sigma_2)$ from cone–plate flow according to equation (161). Figure 19 shows these viscometric functions together with τ.[75] For this solution σ_1 has values of the order of one-tenth those of σ_2. Throughout the range of measurements, which do not extend to low enough rates of shear to approach the region of second-order fluid behavior, $\sigma_2 > \tau$.

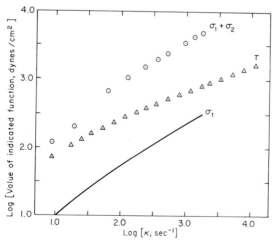

FIG. 19. The viscometric functions for a 0.9% solution of hydroxyethyl cellulose in water.[75]

In principle, σ_1 can be calculated from a combination of cone–plate and torsional flow data. If equation (156) is subtracted from equation (174) and the resulting expression integrated, it is found that

$$\sigma_1(\kappa) = \kappa \int_0^\kappa \frac{1}{\kappa'^2}\left(\frac{\partial s_{\langle zz \rangle}}{\partial \ln r} - \frac{\partial s_{\langle \theta\theta \rangle}}{\partial \ln r}\right) d\kappa' \tag{178}$$

where it is understood that both derivatives in the integrand are evaluated at the rate of shear, κ'. However, for those fluids for which data have been reported[46,85,86,95] the difference that appears in the integrand is so small that the values calculated for $\sigma_1(\kappa)$ are of low accuracy.

A test of consistency has been performed with data obtained from a solution of polyisobutylene. By using the values of σ_1 and σ_2 graphed in Fig. 18 the results of normal stress exponents in torsional flow were calculated according to equation (177) and compared with the experimental data. The result is shown in Fig. 20.[96] Rather good agreement is found throughout.

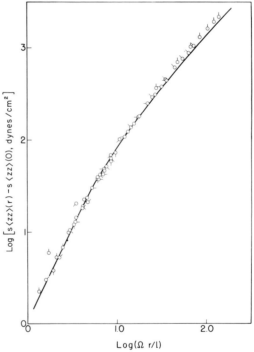

FIG. 20. Normal-stress data in torsional flow, for a 5.4% solution of polyisobutylene in cetane.[96] The points are experimental values; the line is calculated.

[95] H. Markovitz and D. R. Brown, *Trans. Soc. Rheol.* **7**, 137 (1963).

[96] H. Markovitz, *Phys. Fluids* **8**, 200 (1965).

The small discrepancy at the highest rates of shear may be due in part to cumulative errors arising from the numerical integration that had to be performed and in part to the partial breakdown of the approximate theory for torsional and cone–plate flow at the higher speeds of rotation. It must be admitted that this test of consistency, for the same reasons that equation (178) does not give a good method of calculating $\sigma_1(\kappa)$, is not as demanding as would be a comparison of the calculated values of $\sigma_1(\kappa)$ with those obtained from axial flow measurements. Such a comparison at the time of this writing has not been carried out.

The concept of *reduced variables* has been a very useful one in linear viscoelasticity.[35] Data obtained at various temperatures (and, somewhat less successfully, various concentrations) can be reduced to a single curve by appropriate shifting of logarithmic plots. Similar reductions have been proposed for the viscometric functions.[27] For fluids the reduced variables can be defined as[97]

$$\kappa_r = \kappa \eta_0 J_e / \eta_0^* J_e^* \tag{179}$$

$$(\sigma_i)_r = \sigma_i J_e / J_e^*, \qquad i = 1, 2 \tag{180}$$

$$\eta_r = \eta / \eta_0 \tag{181}$$

where η_0^* and J_e^* are, respectively, the values of the zero-shear viscosity and steady-state compliance in the standard state. For temperature reduction it is usually assumed[35] that

$$J_e / J_e^* = T^* \rho^* / T \rho \tag{182}$$

and therefore that

$$\kappa_r = \kappa \eta_0 T^* \rho^* / \eta_0^* T \rho \tag{183}$$

$$(\sigma_i)_r = \sigma_i T^* \rho^* / T \rho \tag{184}$$

where ρ^* is the density at the standard temperature T^*. From the theory of viscometric flows it can be seen that the shear stress τ and various thrust gradients can also be reduced by formulas similar to equation (180). For example, from equations (174) and (184) we see that we can define a reduced normal-stress gradient in torsional flow,

$$\left(\frac{ds_{\langle zz \rangle}}{d \ln r} \right)_r = \frac{T^* \rho^*}{T \rho} \left| \frac{ds_{\langle zz \rangle}}{d \ln r} \right| \tag{185}$$

and from equations (181) and (183) a reduced shear stress,

$$\tau_r = \tau T^* \rho^* / T \rho \tag{186}$$

[97] H. Markovitz, *J. Phys. Chem.* **69**, 671 (1965).

A temperature superposition of this kind for data on a 15.2% solution of polystyrene in toluene is shown in Fig. 21.[98] Both shear-stress and torsional normal-stress data obtained on this solution at four temperatures between 20 and 50°C. are shown in the top figure. The graphs containing the reduced variables defined in equations (183), (185), and (186) are plotted in the bottom figure. The shear-stress data appear to superpose better than the normal-stress data. As in linear viscoelasticity, a corresponding concentration superposition does not appear to be as successful.[95,99]

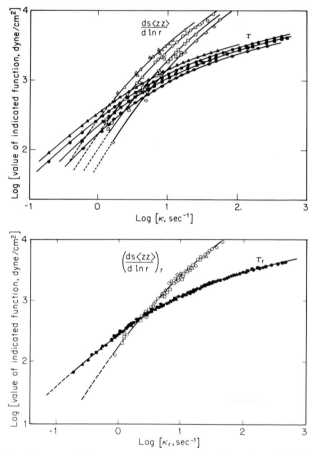

FIG. 21. Normal-stress and shear-stress data in torsional flow for a 15.2% solution of polystyrene in toluene;[98] the data were obtained at temperatures of (△) 20, (○) 30, (□) 40 and (◇) 50°C. The symbols with dots indicate normal-stress data; the blacked-in symbols, shear-stress data. Plot of the reduced variables are given in the bottom figure.

[98] T. Kotaka, M. Kurata, and M. Tamura, *Rheol. Acta* **2**, 179 (1962).
[99] M. C. Williams, *A.I.Ch.E. J.* **11**, 467 (1965).

2. MOLECULAR THEORIES

Almost all published normal-stress data have been obtained with concentrated solutions of high polymers. It is understandable that no fully successful molecular theory for the mechanical properties of such complex materials has yet been developed. One proposed model has been that of a polymer network with temporary junctions,[100,101] which could arise from entanglements or associations due to secondary forces. It is assumed that some junctions break and others form during the flow of the solution. After simplifying assumptions that enable the calculation to proceed are made, it is found that the most simple model corresponds to a second-order fluid with $\beta = -2\gamma$. From equations (76) and (78) it is seen that the Weissenberg relation holds here. However, when the model is modified to make it more realistic,[101] shear-dependent viscosity appears, and $\sigma_1 \neq 0$.

Normal stress effects also can be calculated on the basis of molecular theories of dilute solutions of various kinds of particles. These theories are of the type discussed in other chapters of this treatise[102-104] dealing with the viscosity of macromolecules.[105] When the calculations are extended to include all stress components, it is possible to calculate the normal stress functions corresponding to these models. In the case of dilute suspensions of dumbbells,[106,107] rigid rods,[108] and "pearl necklaces",[109] the Weissenberg relation (78) and a constant viscosity function are derived in the most simple form of the theory. If hydrodynamic interaction is included,[106,108] or if the spheres of the dumbbell are replaced with ellipsoids, then $\sigma_1 \neq 0$, and the simplicity of the Weissenberg relation no longer obtains.[106]

The complexity of large molecules is not a necessary prerequisite for

[100] A. S. Lodge, *Trans. Faraday Soc.* **52**, 120 (1956).

[101] M. Yamamoto, *J. Phys. Soc. Japan* **11**, 413 (1956); **12**, 1148 (1957); **13**, 1200 (1958); *Rept. Progr. Poly. Phys. Japan* **7**, 125 (1964); *J. Phys. Soc. Japan* **19**, 739 (1964).

[102] J. Riseman and J. G. Kirkwood, *in* "Rheology: Theory and Applications" (F. R. Eirich, ed.), Vol. 1, Chapt. 13. Academic Press, New York, 1956.

[103] H. L. Frisch and R. Simha, *in* "Rheology: Theory and Applications" (F. R. Eirich, ed.), Vol. 1, Chapt. 14. Academic Press, New York, 1956.

[104] B. H. Zimm, *in* "Rheology: Theory and Applications" (F. R. Eirich, ed.), Vol. 3, Chapt. 1. Academic Press, New York, 1960.

[105] Apparently, the distinguished originators of some of these molecular theories did not perceive that normal stress effects were contained within the framework of their models. This illustrates again that one may be expecting too much in hoping that molecular theories will precede the development of the phenomenological background.

[106] H. Giesekus, *Proc. Intern. Symp. Second-Order Effects Elasticity Plasticity and Fluid Dynamics, Haifa, 1962* p. 553. MacMillan, New York, 1964. This paper summarizes this author's extensive research in this area.

[107] S. Prager, *Trans. Soc. Rheol.* **1**, 53 (1957).

[108] T. Kotaka, *J. Chem. Phys.* **30**, 1566 (1959).

[109] M. C. Williams, *J. Chem. Phys.* **42**, 2988 (1965).

a molecular theory to lead to normal stress effects. On the basis of a rigorously developed kinetic theory, it has been shown that a rarefied Maxwellian[110] gas exhibits normal stress effects with the Weissenberg relation holding.[111]

V. Nonviscometric Flows

The class of viscometric flows is a very limited class of motions that are rarely achieved in practice even approximately, except under very strictly controlled laboratory experiments. The three functions that determine the behavior of the memory fluid in viscometric experiments represent only a small facet of the total behavior expressed in the functional $\mathfrak{G}_{ij}$. Another facet shows itself in the shear relaxation modulus (or other equivalent linear viscoelastic function), which governs experiments of the type usually performed in linear viscoelasticity. Other aspects of $\mathfrak{G}_{ij}$ reveal themselves in other experiments. One example of a nonviscometric flow, for which a theory exists for the general memory fluid, is steady extension, which is discussed briefly below. The stresses here are not in general determined by the viscometric functions.

Theories of more complicated flows have not been worked out for the general memory fluid. This is not very surprising when one considers that even for the Newtonian fluid the number of exact solutions is rather limited. Some of these flows have been investigated for special constitutive relations. Steady flow through tubes of noncircular cross section is an example of such a motion. As described below, it is found that the flow pattern differs qualitatively from that which occurs in a Newtonian fluid.

There are many other flow situations, some of great practical importance, in which it is experimentally found that non-Newtonian fluids behave qualitatively differently from Newtonian fluids. We only mention a few here. One of the most widely studied examples, of special significance in extrusion processes, is the increase in diameter of a stream of non-Newtonian fluid on emerging from a tube. This contrasts with the decrease found in the case of the Newtonian fluid. Another case, that has a special bearing on mixing studies, has been demonstrated in an artistically beautiful and scientifically important movie film, in which Giesekus[112] shows the flow of a concentrated polymer solution around a rotating sphere, around a cone, and around stirrers. With the rotating sphere, for example, the fluid spirals in at the equator and outward at the poles, which is again opposite to the behavior expected of a Newtonian fluid.

[110] In a Maxwellian gas molecules repel each other according to the inverse fifth power of their separation.

[111] C. Truesdell, *Arch. Rational Mech. Anal.* **5,** 55 (1956).

[112] H. Giesekus, *Proc. 4th Intern. Congr. Rheol., Providence, 1963* Vol. 1, p. 249. Wiley, New York, 1965.

1. STEADY EXTENSION[112a]

Viscometric flows do not exhaust the class of motions of the incompressible memory fluid for which exact solutions can be obtained. *Steady extension*[2,113] is a motion where the fluid is lengthened (or shortened) in such a way that the fractional change of length per unit time is constant. For example, if the material is in the form of a circular cylinder of length $L(t)$ and radius $R(t)$, the motion is taken to be such that

$$\frac{1}{L}\frac{dL}{dt} = -\frac{2}{R}\frac{dR}{dt} = a \tag{187}$$

where a is a constant. This is not a viscometric flow. The stresses are determined, not by the viscometric functions, but by other functions which are determined by the memory functional $\mathfrak{G}_{ij}$.

Steady extension and the viscometric flows are examples of a class of motions in which an observer moving with the particle can always orient himself such that the history of strain does not change with time. These motions, called *flows with constant stretch history*,[114] include all those for which exact solutions are known for the general memory fluid.

2. STEADY FLOW THROUGH TUBES

We consider the flow through a stationary tube of arbitrary cross section. It is assumed that each particle moves in a straight line parallel to the wall with constant speed. An analysis[2,116] shows that such a flow is possible only under two very special circumstances. One is that, if the surfaces of constant velocity are coaxial circular cylinders (as they are in the cases of Poiseuille flow and axial flow between stationary cylinders discussed above) or parallel planes (as they are in the viscometric flow which takes place between stationary parallel planes), then such a flow is possible for any memory fluid. The other is that, if the cross section of the tube is arbitrary, then the flow is possible only if the viscometric functions of the fluid are interrelated by

$$\sigma_1(\kappa) = c\kappa^2\eta(\kappa) \tag{188}$$

where c is a constant. Under other circumstances it is expected that the motion will be more complicated; the particles will not move in straight lines. It is not possible to calculate the path for the general memory fluid.

[112a] See Chapter 7 of this volume.

[113] B. D. Coleman and W. Noll, *Phys. Fluids* **5**, 840 (1962).

[114] They were originally called *substantially stagnant motions*.[43,115]

[115] B. D. Coleman, *Trans. Soc. Rheol.* **6**, 293 (1962).

[116] W. O. Criminale, Jr., J. L. Ericksen, and G. L. Filbey, Jr., *Arch. Rational Mech. Anal.* **1**, 410 (1958).

There are several *special fluids*, for which equation (188) holds. The constant c assumes the value zero not only for the Newtonian fluid (73) but also for the Weissenberg fluid (78). For the second-order fluid equation (188) is satisfied with $c = (\beta + 2\gamma)/\eta_0$, as can be seen from equations (75) and (76). For these fluids, therefore, rectilinear flow may occur in tubes of any cross section.

The calculation of the flow lines has been carried out in a particular case in which straight-line flow is not possible. In the flow of a fourth-order fluid[31] (cf. Section II4b) through a pipe of elliptical cross section a flow perpendicular to the axis is superimposed on the flow parallel to it. The streamlines of this transverse flow are given in Fig. 22. Here each particle performs a distorted helical motion about an axis in its quadrant.[117]

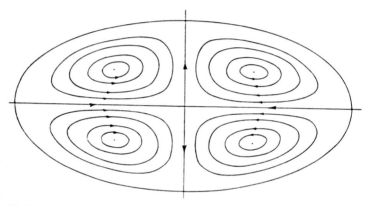

FIG. 22. Streamlines of the transverse flow component in the flow of a fourth-order fluid through a pipe of elliptical cross section.[30]

The second-order fluid (and, a fortiori, higher-order fluids) exhibits another behavior different from that of the Newtonian in this type of flow. The stress components are not constant over a cross section in tubes of noncircular cross section. Explicit theoretical calculations of the stress distribution in tubes of elliptical, rectangular, and equilateral triangular cross sections have been carried out.[118] The existence of such an effect has been demonstrated by Kearsley[119] in his experiments on the flow of polymer solution through tubes of rectangular cross section. He finds unequal normal thrusts exerted by the fluid at the mid-points of neighboring sides at a given level.

[117] Such motions have been observed by E. Kearsley at the U.S. National Bureau of Standards and by H. Giesekus, *Rheol. Acta* **4**, 85 (1965).

[118] A. C. Pipkin and R. S. Rivlin, *Z. Angew Math. Phys.* **14**, 738 (1963).

[119] E. Kearsley, private communication.

VI. Appendix

In this chapter we use the notations and conventions commonly used in tensor analysis, which have also been used by other authors in this treatise. Since expressions in terms of the components of the tensor are more likely to be familiar to the reader, we use this notation rather than the direct notation used in some works on continuum mechanics. We limit our discussion to three-dimensional systems and the use of orthogonal coordinates.

The symbols x^1, x^2, x^3 represent the coordinates of points in space with respect to some specified orthogonal coordinate system. In the symbol x^i the i (or any other letter that may appear as an index) may take on the value 1, 2, or 3. Thus, the equation

$$\xi^i = \xi^i(x^k) \tag{A1}$$

stands for the three equations

$$\xi^1 = \xi^1(x^1, x^2, x^3) \tag{A2}$$

$$\xi^2 = \xi^2(x^1, x^2, x^3) \tag{A3}$$

$$\xi^3 = \xi^3(x^1, x^2, x^3) \tag{A4}$$

We refer to equation (A1) variously as "an equation" or "equations."

For the contravariant components of a vector we use a similar notation, such as b^i; for the corresponding covariant components we write b_i. The covariant, mixed, and contravariant components of a tensor are indicated by s_{ij}, $s^i{}_j$, and s^{ij}, for example. We sometimes write "the tensor s_{ij}" as an abbreviation for "the tensor whose covariant components are s_{ij}." In Cartesian coordinates there is no distinction among the various types of components; we lower all indices in this case.

All indices may take on the values 1, 2, or 3 except when otherwise indicated. In summations $\sum_k$ means $\sum_{k=1}^{3}$.

The values of the nonzero components g_{ij} of the metric tensor in the cylindrical coordinates $x^1 = r$, $x^2 = \theta$, and $x^3 = z$ are given by

$$g_{11} = g_{33} = 1, \qquad g_{22} = r^2 \tag{A5}$$

and in the spherical coordinates $x^1 = \theta$, $x^2 = \phi$, and $x^3 = r$ by

$$g_{11} = r^2, \qquad g_{22} = r^2 \sin^2 \theta, \qquad g_{33} = 1 \tag{A6}$$

The physical components of a vector whose contravariant components are v^i are $\sqrt{g_{ii}}\, v^i$. For tensors physical components are indicated by placing the indices (or the corresponding coordinate) in angular brackets as, for example, $t_{\langle ij \rangle}$ or $t_{\langle rr \rangle}$.

ACKNOWLEDGMENT

Most of this review was written while the author held a Fulbright–Hayes Lectureship (1964–1965) at the Weizmann Institute of Science, Rehovoth, Israel.

Nomenclature

(The number in parentheses gives the equation in which this symbol is defined mathematically.)

$A_j^{(n)i}$	nth Rivlin–Ericksen tensor (26), (27)	$t_{\langle ij \rangle}$	Stress deviator tensor (physical components)
b_i	Body force density per unit mass		
D_{ij}	Rate of deformation tensor (8)	T	Absolute temperature
E_{ij}	Infinitesimal strain tensor	u	Velocity
f	Gradient of driving force (pressure head) (131)	W_{ij}	Vorticity tensor (9)
		x^i	Coordinates
F	Total normal force	x_i	Coordinates (Cartesian)
F_{ij}	Relative deformation gradient (2)	$\dot{x}^i$	Velocity
g	Acceleration due to gravity	$\ddot{x}^i$	Acceleration
g_{ij}	Metric tensor	β, γ	Normal stress coefficient for second-order fluid (37)
$G(t)$	Stress-relaxation modulus		
h	Height of fluid	δ_{ij}	Kronecker delta
$\mathscr{I}$	Centrifugal-force integral (108)	η	Viscosity function (57)
J_e	Steady-state compliance	κ	Rate of shear
J_{ij}	Finite strain tensor (4)	λ	Rate of shear function (59)
Q	Volume rate of flow	ξ_i	Coordinates
r	Radius	ρ	Density
R	Radius	σ_1, σ_2	Normal stress functions (κ as argument) (53)
s_{ij}	Stress tensor (covariant components)	$\hat{\sigma}_1, \hat{\sigma}_2$	Normal stress functions (S as argument) (60)
$s_{\langle ij \rangle}$	Stress tensor (physical components)		
S	Shear stress (58)	τ	Shear stress function (53)
t	Time	ψ	Potential of b_i (10)
t_{ij}	Stress deviator tensor (covariant components)	ω	Angular velocity
		Ω	Angular velocity

CHAPTER 7

SOME QUANTITATIVE CONSIDERATIONS ABOUT SPINNING

H. F. Mark

I. Introduction

The conversion of a fluid polymeric system, such as a melt, solution, emulsion, suspension, or monomer mixture, into a filament with significant mechanical and thermal properties involves *much more than giving form.* When the fluid system emerges from the spinnerette it has, in general, a somewhat non-Newtonian character with a viscosity between 100 and 2000 poises; when the filament is wound up, it is an essentially solid material with softening ranges above 150°C. and with rigidities (elastic modul) 20 g./den. While the thread solidifies, many important processes occur, which are essential for the structure and properties of the resulting fibers.

In the case of *melt* spinning,[1] *chain folding, orientation,* and *crystallization* of the individual macromolecules take place in mutual inerrelation and determine the character and behavior of the resulting solid-state system. In *solution* spinning of cellulose acetate, polyvinylalcohol, polyacrylonitrile, or other polymers the macromolecules are desolvated and precipitated; the sol is converted to a gel which, in turn, is oriented and, being progressively desolvated, may or may not produce laterally ordered and virtually crystalline domains. The spinning of cellulose xanthate solutions is controlled by additional *chemical processes,* such as the decomposition of the xanthate groups and the interaction of the decomposition products with the other ingredients of the spinning solution and the spinning bath which, in turn, affect the colloidal and physicochemical events of coagulation, syneresis, orientation, cross-linking, and eventually crystallization.[2] *Emulsion* and

---

[1] A. Ziabicki *et al., J. Appl. Polymer Sci.* **2,** 19, 24 (1959); *Kolloid-Z.* **171,** 51, 111 (1960); **175,** 14 (1960).

[2] V. Elsaesser, *Kolloid-Z.* **111,** 174 (1948); **112,** 120 (1949); **113,** 37 (1949); A. Pupke, *ibid.* **125,** 171 (1952).

suspension spinning processes are less developed and less analyzed in respect to contributing factors, which obviously are the precipitation and coalescence of the individual particles embedded in a solidifying matrix and their stream-lining into a homogeneous, coherent filament. Finally, the formation of a continuous thread from a monomeric mixture or prepolymer through polymerization and copolymerization of its components is still in the state of early approaches and requires a coordination of polymerization kinetics—initiation, propagation, chain transfer, cross-linking and termination—with the orientation and lateral ordering of the resulting polymeric system.

It is obvious that melt spinning is the least complicated procedure and, as a consequence, has been analyzed by several workers in a somewhat rational and quantitative fashion;[1] it may, therefore, be justified to focus this comprehensive report primarily on a presentation of the mechanism of melt spinning.

II. Melt Spinning

The tempering of cylindrical wires, tubes, and pipes that are passed through cooling liquids has been frequently treated in the past with the aid of adequate partial differential equations down to textbook-level presentations. Those tests provide a convenient starting point for the following considerations.

Suppose a molten polymer is extruded from a cylindrical orifice at point $x = x_0$ at time $t = t_0$ with radius $r = r_0$ and jet velocity $v = v_0$ in the x direction. Then the volume of the melt that is delivered from the spinneret during dt is given by

$$dV_0 = (v_0 r_0^2 \pi) \, dt \tag{1}$$

and its weight is

$$dW_0 = (v_0 r_0^2 \pi d) \, dt \tag{2}$$

where d is the specific gravity of the melt. Its temperature at the orifice is T_0 and its thermal characteristics are:

Isobaric heat capacity, c_p
Coefficient of heat transfer at the filament surface, h
Heat conductivity of melt, k_p

ɔled at a distance x from the orifice by radiation ʃeam of temperature T_a, the amount of heat lost ring dt is given by

$$T_a) + 2\pi r e \tau (T^4 - T_a^4)] \, dx \, dt \tag{3}$$

ent at distance x from the orifice, e is the specific

emissivity of the filament surface, and τ is the Stefan–Boltzmann radiation constant.

Through this heat loss the temperature of the segment will be decreased by dT, which is given by

$$r^2\pi dc_p\, dx\, dT = 2r\pi[h(T - T_a) + e\tau(T^4 - T_a^4)]\, dx\, dt \qquad (4)$$

where d is the specific gravity of the segment dx. Since the temperature change with distance from the orifice is

$$\frac{dT}{dt} = \frac{\partial T}{\partial t} + \frac{\partial T}{\partial x}\frac{\partial x}{\partial t}$$

we get

$$rc_p d\left(\frac{\partial T}{\partial t} + v\frac{\partial T}{\partial x}\right) = 2h[(T - T_a) + 2e\tau(T^4 - T_a^4)] \qquad (5)$$

In this equation the dependent variables r and T are functions of x and t, whereas c_p, d, h, e, τ, and T_a and $dx/dt = v$ are parameters and variables that establish the physical character of the system.

The first term on the right side represents the cooling through convection; the second, that through radiation. Whenever, and as long as, the temperature T is below 500°C., radiation cooling amounts to only a few percent of convection cooling and can be safely neglected.

Thus the *heat balance relation* of the moving thread leads to

$$rc_p d\left(\frac{\partial T}{\partial t} + v\frac{\partial T}{\partial x}\right) - 2h(T - T_a) = 0 \qquad (6)$$

The segment dx at point x moves downward under the influence of *two* forces:

(a) The *force* that is caused by its own weight. It is given by

$$\int_x^{x_w} \pi r^2 d\, g\, dx$$

where g is the constant of gravity and x_w is the distance between the spinneret and the wind up roll.

(b) The *force* F_w, which is provided by the rotation of the windup roll. This force produces a *tension* $F_w/\pi r^2$.

The sum of the forces (a) and (b) is balanced by the acceleration of the moving thread, which is given, per second, by

$$\pi r^2 d(v_w - v)$$

where v_w is the filament velocity at the windup roll. Thus, the balance of forces may be expressed by

$$F_w + \int_x^{x_w} \pi r^2 d\, g\, dx - \pi r^2\, d(v_w - v) = 0 \tag{7}$$

Air friction has been neglected because of its small value; this is justified in melt and solution dry spinning, but not when the thread is spun in a bath of any kind.

In real spinning the three terms of equation (7) do not add up to zero because of internal and external friction losses, but to a certain force F, which produces the spinning tension $F/r^2\pi$. This gives

$$F = F_w + \int_x^{x_w} \pi r^2 d\, g\, dx - \pi r^2\, d(v_w - v)$$

As the filament is exposed to this spinning tension $S = F/r^2\pi$, its cross section $Q = r^2\pi$ decreases, and one can define a *spinning viscosity* η as the ratio of the tension and the relative increase in length or decrease of the cross section $dQ = -(2\, dr/r)Q$:

$$\eta_x = -F\frac{dx}{dQ}\frac{dt}{dx}; \qquad \eta_r = -\frac{F}{Q}\cdot\frac{Q\, dt}{dQ} \tag{8}$$

This gives

$$F = -\eta\frac{\partial Q}{\partial t} - \eta v\frac{\partial Q}{\partial x} \tag{9}$$

Equation (9) represents the relation of *force balance*.

Finally, we have the *equation of continuity*, which indicates that no material can be gained or lost during the spinning process. Into a cylinder of length Δx and radius r at position x flows a certain mass M of polymer per unit time, and it is evident that the same mass M must emerge per unit time at position $x + dx$ with a radius $r - dr$ or a cross section $Q - dQ$. This gives, assuming incompressibility,

$$\frac{\partial M}{\partial x}\Delta x = -\frac{\partial(Qd)}{\partial t}\Delta x \tag{10}$$

Since $M/d = Qv$, this equation can be written

$$v\frac{\partial Q}{\partial x} + \frac{\partial Q}{\partial t} = -Q\frac{\partial v}{\partial x} \tag{11}$$

If equations (9) and (11) are combined, there results

$$\frac{F}{Q} = \eta\frac{\partial v}{\partial x} = S \tag{12}$$

Thus the spinning viscosity η is given by the ratio between the spinning tension and the velocity increase:

$$\eta = \frac{S}{\partial v/\partial x}$$

In summary, at high take-up speeds, negligible radiation heat losses, and incompressible melts the three simultaneous partial differential equations that describe spinning are

$$v \frac{\partial T}{\partial x} + \frac{\partial T}{\partial t} = \frac{2h}{c_p rd}(T - T_a) \tag{6}$$

$$v \frac{\partial Q}{\partial x} + \frac{\partial Q}{\partial t} = -Q \frac{\partial v}{\partial x} \tag{11}$$

$$S = \eta \frac{\partial v}{\partial x} \tag{12}$$

In the steady state of melt spinning all partial derivatives concerning respect to time vanish, and we arrive at the two equations

$$\frac{dT}{dx} = \frac{2h}{vc_p rd}(T - T_a) \tag{13}$$

$$v \frac{dQ}{dx} = -Q \frac{S}{\eta} \tag{14}$$

Both equations may immediately be interpreted with respect to their physical meaning. If we consider that at point x a cylindrical volume element of the moving thread of length Δx its surface area A is given by $2r\pi\Delta x$ and its total heat content H is given by $r^2\pi\Delta x\, dc_p$, it is obvious that the decrease in temperature will be given by the ratio A/H times the heat transfer coefficient h and the temperature gradient $(T_a - T)$. This is precisely what equation (13) expresses. Further, the relative decrease of the cross section Q at point x; $1/Q(dQ/dx)$ will be proportional to the spinning tension S and inversely proportional to the spinning viscosity η.

Equations (13) and (14) can be solved if the heat transfer coefficient h and the spinning viscosity η are given as functions of x and t or x and v; the spinning tension S is an unknown parameter, which must be determined if one of the boundary conditions is to be satisfied, namely that of the *take-up speed* v_w at the windup roll.

The solutions of equations (13) and (14) give T and Q of the filament as functions of x and t or of x and v and permit the prediction of the temperature and the cross section (denier) of the moving thread at any point between the spinneret and the take-up roll.

Measurements of the cooling rate of monofilaments of polyethylene, polypropylene, Saran, nylon, and Terylene have shown that the *heat transfer coefficient h* of organic polymers moving with velocities of 10^2 to 10^4 cm./sec. and being cooled by a stream of air that flows perpendicularly to the fiber at velocities of 10^2 to 10^3 cm./sec. can be expressed in terms of

$$h \sim \frac{\text{cal.}}{\text{cm.}^2 \times \text{sec.} \times \text{deg.}}$$

with numerical values ranging from 0.05 to 0.50.

The *spinning viscosity* η is a characteristic function of temperature which must be determined experimentally for every individual polymer but is bound to have a similar algebraic character for all polymeric melts and can be expected to be of the same form as the temperature coefficient of the melt viscosity measured by the flow in a capillary. Its general form is of the type

$$\log \eta = \log A + E/RT$$

where E is the activation energy of the segment movement necessary to produce viscous flow, and A is connected with the size of the moving volume.[2a]

The isobaric *heat capacity* at constant pressure c_p has been measured for several polymers; its value changes somewhat with the chemical character of the material but always remains within the limits 0.50 and 1.50 cal./g./deg.

Let us first consider equation (14), which gives for constant values of the tensile force S and the spinning viscosity η an exponential decrease of the cross section down to zero:

$$\frac{Q}{Q_0} = K \exp\left[-\frac{S}{\eta}\frac{x}{v}\right] \tag{15}$$

This indicates that the diameter of the thread decreases continuously until it breaks; no spinning is possible under such conditions. To avoid this "catastrophe" we must either see to it that S decreases with x or that η increases. The spinning tension S is determined by two boundary conditions:

(a) At the spinneret, where it is zero and where the delivered mass per second is $v_0 Q_0 d$.
(b) At the windup, where it is F_w/Q_w and where the mass taken up is $v_w Q_w d$.

The spinning tension *increases* along the spinning path while Q decreases. Hence the *stabilization* of the spinning must be due to a *strong increase* of

---

[2a] See, e.g., Chapter 1 of this volume.

the spinning viscosity η. This is, of course, the case in all forms of spinning, including the special case of melt spinning by the cooling of the thread and its resulting solidification. Superimposed on this effect can also be progressive orientation and even crystallization of the streaming linear macromolecules of the melt.

Expanding the exponential function in equation (15) to its first two terms, one obtains for the decrease of the cross section Q with increasing distance x from the spinneret

$$Q = \frac{Q_0}{1 + (S/\eta)x/v} \tag{16}$$

Since the cross section Q must be Q_0 at $x = 0$ and Q_w at $x = x_w$, this equation can also be rewritten

$$Q = \frac{Q_0}{1 + x(Q_0 - Q_w)/(Q_w x_w)} \tag{17}$$

This relation contains only directly measurable quantities, which refer to the spinning operation, namely:

Q_0 the cross section at the spinneret
Q_w = the cross section at the windup
x_w = the distance between the face of the spinnerette and the windup roll

It represents a hyperbolic relation between cross section and stream profile and is, in general, well supported by experiment.

The replacement of the exponential in equation (15) with the first term $1 + (S/\eta)x/v$ amounts to the linear equation

$$\eta = S(\eta_0/S_0 + ax) \tag{18}$$

which indicates a linear increase of the spinning viscosity with the distance of the spinneret at constant spinning tension S. Solution of equation (15) at better approximations leads to more pronounced dependance of η on x, which can be empirically expressed by

$$\eta = S(\eta_0/S_0 + ax^n) \tag{19}$$

where n is between 2 and 3.

Usually one expresses the final cross section Q_w in percent P of Q_0:

$$Q_w = \frac{Q_0}{100}P$$

and can rewrite equation (17) in the form

$$Q = \frac{Q_0}{1 + [(100 - P)/PL]x} \tag{20}$$

where $L = x_w$ is the length of the spinning path.

III. Comparison with Experiment

Numerous experimental data on the stream profile of melt spun polymers exist. Let us first consider those obtained with 6 nylon.

All profiles exhibited *clearly hyperbolic character*, even though the values of Q_0 were varied from 0.031 to 0.0020 cm.^{2}, the takeup velocities covered a range of 150 to 1900 m./min., and the temperature was varied from 255 to 285°C. The values for the spinning length L covered the range of 65 to 200 cm. A steep slope always exists at the beginning of the profile ($x = 0$), which agrees within the limits of error with

$$\frac{dQ}{dx} = 0 = -\frac{Q_0(100 - P)}{PL}$$

as required by equation (20).

Tables I, II, and III contain experimental data on polycaprolactam (6 nylon), in which the parameters Q_0, P, L, v, and T were varied over wide ranges without substantially affecting the general shape of the profile indicated by equation (22).

TABLE I

STREAM PROFILE IN SPINNING OF 6 NYLON*

	Distance x, cm.								
	0	25	50	75	100	125	150	175	200
Q/Q_0 observed	1.00	0.65	0.46	0.37	0.31	0.27	0.24	0.21	0.20
Q/Q_0 theoretical	1.00	0.67	0.50	0.40	0.33	0.28	0.25	0.22	0.20

* Parameters $Q_0 = 31 \times 10^{-3}$ cm.^{2}, $P = 20\%$, $L = 200$ cm., $T = 265$°C., $v = 400$ m./min.

TABLE II

STREAM PROFILE IN SPINNING OF 6 NYLON*

	Distance x, cm.					
	0	25	50	75	100	125	150
Q/Q_0 observed	1.00	0.42	0.26	0.16	0.15	0.11	0.10
Q/Q_0 theoretical	1.00	0.40	0.25	0.18	0.14	0.11$_7$	0.10

* Parameters $Q_0 = 124 \times 10^{-3}$ cm.^{2}, $P = 10\%$, $L = 150$ cm., $T = 265$°C., $v = 300$ m./min.

TABLE III

STREAM PROFILE IN SPINNING OF 6 NYLON*

| | | | | | Distance x, cm. | | | | | |
	0	10	20	30	40	50	60	70	80	90	100
Q/Q_0 observed	1.00	0.45	0.30	0.20	0.15	0.12	0.11	0.09	0.08	0.08	0.08
Q/Q_0 theoretical	1.00	0.48	0.32	0.24	0.19	0.16	0.14	0.12	0.10	0.09	0.08

* Parameters $Q_0 = 75 \times 10^{-4}$ cm.^{2}, $P = 8\%$, $L = 100$ cm., $T = 275°C.$, $v = 500$ m./min.

Corresponding figures are given for the melt spinning of polyethylene-terephthalate in Table IV and for the melt spinning of isotactic polypropylene in Table V.

TABLE IV

STREAM PROFILE IN SPINNING OF POLYETHYLENETEREPHTHALATE*

INTRINSIC VISCOSITY OF THE POLYMER = 0.75

| | | | Distance x, cm. | | | |
	25	50	75	100	125	150
Q/Q_0 observed	1.00	0.30	0.20	0.16	0.12	0.12
Q/Q_0 theoretical	1.00	0.29	0.27	0.17	0.14	0.12

* Parameters $Q_0 = 31 \times 10^{-3}$ cm.^{2}, $P = 12\%$, $L = 150$ cm., $T = 295°C.$, $v = 400$ m./min.

TABLE V

STREAM PROFILE IN SPINNING OF POLYPROPYLENE*

| | | | | Distance x, cm. | | | | | |
	0	25	50	75	100	125	150	175	200
Q/Q_0 observed	1.00	0.37	0.20	0.14	0.10	0.08	0.07	0.07	0.07
Q/Q_0 theoretical	1.00	0.38	0.23	0.16	0.13	0.11	0.09	0.08	0.07

* Melt index of polymer, 2.5; parameters $Q_0 = 124 \times 10^{-3}$ cm.^{2}, $P = 7\%$, $L = 200$ cm., $T = 262°C.$, $v = 350$ m./min.

IV. Conclusion

All experimental data indicate a very sharp reduction of Q/Q_0 in the early stages of the filament formation, which is in conformity with the hyperbolic character of equation (17). The theoretical analysis shows that this is caused by the fact that S/η is virtually constant along the spinning profiles. The undesirable consequence is that practically all dimensional and structural changes in the fiber, as it forms, occur on the first 20 to 25% of the spinning path, and all the rest of the profile is wasted with respect to incorporating into the filament certain advantageous structural features such as orientation, crystallization, and chain unfolding.

It is well known from the spinning of viscose rayon that the slowing down of the fiber-forming process and the lengthening of the distance over which the coagulation and regeneration of the filament occurs have made possible very substantial improvements in fiber properties, which have led to the modern modified rayons and polymeric fibers.

This might provoke the thought that a similar lengthening of the distance over which, in melt spinning, the filament cross section goes from its original to its final value might influence the details of the solid-state characteristics of the spun yarn, such as the proportion of completely straightened-out chains, the degree of their lateral order, the percentage of regularly folded chains, the amount of substantially disordered domains, and the nature of the spherulitic structure. Such a lengthening could be achieved if the spinning viscosity η would increase with x more rapidly than it does according to equation (19). It might be achieved in a polyblend if one component precipitated during cooling, or in a polymer filled with a finely divided material that is liquid at the spinning temperature but solidifies at somewhat lower temperatures. It is also possible that polymers of a higher D.P. would automatically effect it, because their viscosity increase on cooling is larger than that of species of lower D.P. It is also possible that a rapid cross-linking reaction, which occurs during the very short time of the filament-forming (10^{-2} to 10^{-3} sec.), would produce a substantial increase in viscosity or consistency.

It appears that, in general terms, a combined hydrodynamic, rheological, and thermal analysis of melt spinning not only leads to a reasonable agreement with the observed data but also provides certain information on desirable and perhaps even possible improvements.

If one wanted to set up a corresponding system of differential equations for solution dry spinning as it is carried out with polyacrylics and with cellulose acetate, it would be necessary to introduce additional terms that would take care of the consequences of solvent evaporation: the cooling through the heat of vaporization, the reduction of the cross section due to the gradual removal of the solvent, and the increase in spinning viscosity as a consequence of increasing polymer concentration in the forming filament. This can be done without very much difficulty, but it would not be worth while to reproduce here the resulting expressions because there exist at present not enough reliable data to permit a conclusive test of the expressions.

Spinning of a solution in a bath, such as viscose rayon or polyurethane spinning, adds certain other features, such as bath drag, diffusion of reagent into the filament, coagulating of the sol to a gel, and chemical decomposition of the xanthate gel to cellulose. Again, a more detailed analysis would not be of great value at this time because of lack of experimental data that could be used to either confirm or disprove the validity of any theoretical expression.

Nomenclature

a	Constant	F	Force
d	Density, also differential	L	Length of spinning path
g	Gravitational acceleration	M	Mass
h	Coefficient of surface heat transfer	P	Percent of original cross section Q_0
r	Cylindrical coordinate	Q	Cross section of filament
t	Time	S	Stress, tension
v	Velocity	T	Temperature
x	Coordinate of spinning direction	V	Volume
		W	Weight
A, E, K	Constants	∂	Partial differential
D.P.	Degree of polarization	η	Coeficient of viscosity

Chapter 8

THIXOTROPY AND DILATANCY

Walter H. Bauer

and

Edward A. Collins

I. Introduction

1. THIXOTROPY

When a reduction in magnitude of rheological properties of a system, such as elastic modulus, yield stress, and viscosity, for example, occurs reversibly and isothermally with a distinct time dependence on application of shear strain, the system is described as thixotropic. The phenomenon is called thixotropy, according to current general usage. Originally conceived as an unusual property of very special materials, sol–gel systems such as aqueous iron oxide dispersions, thixotropy in the sense described above has been found to be exhibited by a great many and a large variety of systems. The original meaning has been extended far beyond that intended when the term was first proposed. The concept of thixotropy arose in the study of certain colloidal dispersions. These were capable of stable existence as solid materials with a low modulus of elasticity when under a low shearing stress such as a few inches of gravitational head. They had sufficient rigidity to sustain their own weight in an inverted test tube, exhibiting a "yield value" or upper limit of stress under which no permanent deformation took place in the time the observer chose to wait. When vigorously mechanically agitated in a container, the dispersions transferred to a freely flowing liquid state, with viscosities of the order of the dispersing liquid. After a sufficient period of rest time, the systems returned to their former state, developing the original rigidity and yield stress. This reversible gel-sol-gel transformation occurred isothermally, in contrast to the more familiar temperature controlled gel-sol changes. Thixotropy was thus originally associated with a type of structure in a disperse system. The characteristic rheological behavior was only one of several properties, such as conductivity and dielectric constant, which depended measurably on the breakdown and buildup of structure in colloidal systems undergoing mechanical shearing or subsequent rest, under isothermal conditions. Through the period since the introduction of the concept, thixotropy has assumed the broader and rheologically oriented significance of present general usage.

The phenomenon which led to the concept of thixotropy was reported in 1923 by Schalek and Szegvari[1] who were studying iron oxide aqueous dispersions. They noted, "These gels have the remarkable property of becoming completely liquid through gentle shaking alone, to such an extent that the liquified gel is hardly distinguishable from the original sol. These sols were liquified by shaking, solidified again after a period of time ... the change of state process could be repeated a number of times without any visible change in the system." They quoted from a communication from Peterfi, who coined the term thixotropy,[2,3] "Peterfi found ... that the originally quite stiff living cell content assumed a fluid consistency when agitated ... , after a passage of a certain time the cell content returned to its original state." It is interesting to note this early observation of thixotropy in the contents of living cells. An attempt to find a quantitative measure of the property was made by Freundlich and Rosenthal[4] in work on the system described by Schalek and Szegvari. They noted that the time required for resolidification of the liquefied Fe_2O_3 sols was reproducible, and they recommended the "solidification speed," the reciprocal of the resolidification time, as a quantitative characteristic of the system. The term "thixotropy" was introduced in the literature by Freundlich and Bircumshaw[5] in the title of a report on the behavior of aluminum hydroxide gels. A definition of the property was given by Freundlich and Rawitzer,[6] "By thixotropy is

[1] E. Schalek and A. Szegvari, *Kolloid-Z.* **32**, 318 (1923); **33**, 326 (1923).

[2] T. Peterfi, *Arch. Entwicklungsmech. Organ.* **112**, 680 (1927).

[3] T. Peterfi, *Verhanitlungen 3 Intern. Zellforschung-Kongr., Arch. exp. Zellf* **15**, 373 (1934).

[4] H. Freundlich and A. Rosenthal, *Kolloid-Z.* **37**, 9 (1925).

[5] H. Freundlich and Bircumshaw, *Kolloid-Z.* **40**, 19 (1926).

[6] H. Freundlich and W. Rawitzer, *Kolloid-Z.* **41**, 102 (1927).

meant the phenomenon of concentrated sols ... which solidify to gels which may again be liquified to sols. The resolidification occurs repeatedly, at constant temperature with a constant speed." Within a short time, what was originally thought to be a singular characteristic of the iron oxide sol studied by Schalek and Szegvari was found to exist for many systems. A quantitative measure was needed in order to provide comparison of effects of such variables as concentration and additive electrolytes, as well as the nature of suspended material and suspension medium. In 1928 Freundlich and Soellner[7] described the measurement of the solidification time for iron oxide sols, "4 cc of the ... sols were placed in a tube of jena glass which was ... stoppered, were completely liquified by shaking, and then the time was measured after which the sample no longer had a tendency to flow after the test tube was inverted." An effort was made to use a Couette rotary cylinder apparatus to measure elastic properties of iron oxide gels[8] with the qualitative conclusion that the resistance to shear increased during the process of solidification in the sol-gel transformation, and that the degree of liquefication depended on the applied shear.

It was soon found that aluminum hydroxide sols,[5,9] vanadium pentoxide sols,[10] starch pastes,[11] and the aqueous gels of gelatin,[12] pectin and sugar, and certain quinine derivatives exhibited thixotropic properties, as described by Ostwald[13] in a discussion of gels and jellies. Freundlich suggested that many gels, if not all, could be made to undergo a reversible, isothermal gel-sol transition through mechanical action such as shaking or stirring, in a general treatment of thixotropy.[14] In this discussion, Freundlich ascribed the origin of the word thixotropy to a combination of "thixis," stirring or shaking, and "trepo," turning or changing, based upon the suggestion of Peterfi.[3] He was convinced that the occurrence of thixotropy was widespread in nature, and that knowledge concerning the subject would contribute greatly to the understanding of biological phenomena.

A primary interest of early investigators of colloidal dispersion was the micro-organization of matter that led to the existence of internal structures allowing reversible gel-sol-gel transformations to take place on addition and withdrawal of heat at critical temperature ranges, with time dependence. The discovery that these changes in elasticity and flow consistency could be induced by mechanical agitation, reversibly, and with time dependence,

[7] H. Freundlich and K. Soellner, *Kolloid-Z.* **44**, 309 (1928).
[8] H. Freundlich and W. Rawitzer, *Kolloidchem. Beih.* **25**, 231 (1927).
[9] M. Aschenbrenner, *Z. Physik. Chem.* **127**, 415 (1927).
[10] H. Zocher and H. Abu, *Kolloid-Z.* **46**, 27 (1928).
[11] H. Freundlich and H. Nitze, *Kolloid-Z.* **41**, 206 (1927).
[12] H. Freundlich and H. Abramson, *Z. Physik. Chem.* **131**, 278 (1928).
[13] W. Ostwald, *Kolloid-Z.* **46**, 248 (1928).
[14] H. Freundlich, *Kolloid-Z.* **46**, 289 (1928).

opened up a wide field of study of the many factors affecting the thixotropy of dispersions, such as changes in dispersion degree and shape, of electrolyte concentration, and of solvation. The resolidification time was initially the generally used flow property in characterizing the thixotropic nature of the systems studied, though it was soon observed by Freundlich that this time measurement was very dependent on sample size and container dimensions.[8,15]

Once introduced, the term "thixotropy" showed itself to be viable, being taken up by many investigators, and remaining in continuous use to the present time. As the variety of systems which were shown to have shear sensitivity in flow resistance increased rapidly in number and variety, the original definition of Freundlich was no longer uniformly adhered to. His definition implied an understanding of what was meant by "solidification," "liquefaction," the "gel" and "sol" states, and "time of resolidification," terms not lending themselves readily to exact specification. These descriptions were most applicable to the systems first studied, which were low concentration dispersions of highly insoluble materials such as iron oxide and vanadium pentoxide in an aqueous medium of relatively low viscosity. With the discovery that such systems as rubber latex, zinc oxide in paraffin, and paint pigment dispersions in linseed oil also could undergo reversible time-dependent change in flow resistance with mechanical agitation, investigators showed an increasing tendency to define thixotropy in terms of rheological concepts, independently of the "gel-sol-gel" description. To all the difficulties encountered when attempts were made to make clearly defined[16] and uniformly interpreted measurements of flow properties were added the complications arising from time dependency of the various systems' properties.

The statement of Pryce-Jones[17] concerning "... the true meaning of thixotropy, namely an increase of viscosity in a state of rest and a decrease of viscosity when submitted to a constant shearing stress," considered carefully, shows that two difficult operations were inferred by his definition. First, to measure the viscosity, the ratio of shearing stress to rate of shear, requires flow to take place. It could not be measured "in a sample at rest," as required. Second, no apparatus was available in which his flow samples could be subjected to a constant shearing stress in a uniform manner. A constant driving force for test instruments does not necessarily result in the application of a uniform constant shearing stress throughout the sample. Difficulties such as those posed by Pryce-Jones' definition of thixotropy were soon apparent to him and other competent observers. In spite of their best efforts, however, the information they reported was almost always highly dependent on the dimensions of the viscometric apparatus used, on

[15] H. Freundlich, *Protoplasma* **2,** 278 (1927).
[16] E. C. Bingham, *J. Rheol.* **1,** 507 (1930).
[17] J. Pryce-Jones, *J. Oil & Color Chemists' Assoc.* **17,** 305 (1934).

the nonuniformity of distribution of shearing stresses and rates of shear throughout the test specimen, on the magnitude and the rate of increase or decrease of applied shear forces and flow rates, and on the period of time of application of the shearing forces and shearing flows. Despite the early lack of success at defining quantitative measurements of thixotropy, the investigators obtained qualitative information which showed more and more systems and substances to have the property of isothermally reversibly changing flow resistance in response to changes of shearing stress or of shearing rate. In fact, it became clear that materials exhibiting Newtonian flow over a wide range of shearing rates were the exception, and that substances generally exhibited much greater rheological complexity. As it became evident that terms such as *yield value* were dependent on specification of magnitude, rate of application, and time of application of the forces involved, or were an artifact of the scale of plotting, the requirement of a "solid" state for a thixotropic substance was neglected. Large classes of materials such as paints and nonaqueous dispersions with no readily evident solidity or "yield value" were referred to as thixotropic. Despite some reluctance of investigators to consider properties which are evidenced in time intervals too short for ready measurement or in time intervals too long for convenient waiting, the number of systems shown to exhibit measurable time dependency of flow properties such as viscosity, elastic modulus, relaxation of stress and strain, and shear thinning after mechanical agitation has steadily increased. It is most likely that the term thixotropy has survived because of this common thread of existence of time-dependent rheological properties in spite of the varying definitions given it by a host of investigators since the coining of the word.

2. DILATANCY

Anyone who has walked on moist sand on the beach has observed that it becomes firm and dry under the pressure of a foot, but becomes moist again when the pressure is removed. Similarly, concrete becomes hard under the pressure of a trowel but becomes fluid when the trowel is removed. Sand grains normally settle in a state of close packing. When they are subjected to a shear force, a certain amount of separation must take place before they can slide over one another. To this property of volume expansion with shear, Osborne Reynolds[18] gave the name dilatancy. Similar behavior has been observed with mixtures of fine powders and suitable liquids.[19,20] It is

---

[18] O. Reynolds, *Phil. Mag.* [5] **20,** 469 (1885); *Nature* **30,** 429 (1886).

[19] R. V. Williamson and W. W. Heckert, *Ind. Eng. Chem.* **23,** 667 (1931).

[20] R. V. Williamson, *J. Phys. Chem.* **35,** 354 (1931).

of interest to note that Sir William Thomson (Lord Kelvin) predicted on purely theoretical grounds in 1875 (Encyclopaedia Britannica) that a shearing stress may produce, in a truly isotropic solid, condensation or dilation. This phenomena of dilatancy has been investigated experimentally by Jenkins,[21] Brown and Hawksley,[22] Andrade and Fox,[23] Metzner and Whitlock,[24] and others.[25-28]

The original meaning of dilatancy as defined by Reynolds was extended by Freundlich and co-workers[23,29-31] to include any system which increases in viscosity with increasing *rate* of shear whether the system was composed of granular particles or not, since Reynolds had also noted increases in rigidity upon distortion of his systems. The term was broadened further to include systems which increase in viscosity with increasing *amount* of shear.[32,33]

Dilatancy has also been considered to be the opposite to thixotropy[29,30 34-38] but was distinguished from rheopexy.[39,40] Rheopexy was the term introduced by Freundlich[39,40] to describe the accelerated setting of a thixotropic gel under gentle rhythmic shearing, although some authors[41] have defined rheopexy as opposite to thixotropy, or antithixotropy.[42] The distinction between dilatant and rheopectic properties is quite clear. Although both systems exhibit an increase in viscosity with increasing rate of shear, in a rheopectic fluid the viscosity increases with time of shear at a constant rate of shear. A dilatant system, after being sheared, liquifies almost instantane-

[21] C. F. Jenkins, *Proc. Roy. Soc.* **A131**, 53 (1931).

[22] R. L. Brown and P. G. W. Hawksley, *Fuel* **26**, 159 (1947).

[23] E. N. da C. Andrade and J. W. Fox, *Proc. Phys. Soc.* (*London*) **B62**, 483 (1949).

[24] A. B. Metzner and M. Whitlock, *Trans. Soc. Rheol.* **2**, 239 (1958).

[25] F. K. Daniel, *India Rubber World* **101**, No. 4, 33 (1940).

[26] A. Jobling and J. E. Roberts, *in* "Rheology of Disperse Systems" (C. C. Mill, ed.), p. 127. Pergamon Press, Oxford, 1959.

[27] A. Jobling and J. E. Roberts, *Brit. J. Appl. Phys.* **9**, 235 (1958).

[28] A. S. Roberts, *J. Chem. Eng. Data* **8**, 440 (1963).

[29] H. Freundlich and A. D. Jones, *J. Phys. Chem.* **40**, 1217 (1936).

[30] H. Freundlich and F. Juliusberger, *Trans. Faraday Soc.* **30**, 333 (1934).

[31] H. Freundlich and H. L. Roder, *Trans. Faraday Soc.* **34**, 308 (1938).

[32] P. S. Roller, *J. Phys. Chem.* **43**, 457 (1939).

[33] P. A. Briscoe, W. J. Dunning, and R. C. Seymor, *Brit. J. Appl. Phys.* **3**, 193 (1952).

[34] S. LeSota, *Paint Varnish Prod.* **47**, 60 (1957).

[35] H. Green, "Industrial Rheology and Rheological Structures," Wiley, New York, 1949.

[36] E. A. Hauser, "Colloidal Phenomena," p. 225. McGraw-Hill, New York, 1939.

[37] G. W. Scott-Blair, "An Introduction to Industrial Rheology," McGraw-Hill (Blakiston), New York, 1938.

[38] G. Mozes and E. Vamos, *Intern. Chem. Eng.* **6**, 150 (1966).

[39] H. Freundlich and F. Juliusberger, *Trans. Faraday Soc.* **31**, 920 (1935).

[40] F. Juliusberger and A. Pirquet, *Trans. Faraday Soc.* **32**, 445 (1936).

[41] D. W. Dodge, *Ind. Eng. Chem.* **51**, No. 7, 839 (1959).

[42] A. I. Lenov and G. V. Vinogradov, *Dokl. Akad. Nauk SSSR* **155**, 406 (1964).

ously, whereas a finite time is required before a rheopectic system will return to the liquid state. In addition, a dilatant system unlike a rheopectic system can be accompanied by a volume change.

Much confusion has appeared in the literature as a result of disagreement in the terminology.[43] For example, the polymethyl methacrylate solutions in diphenyl investigated by Peterlin and Turner[44,45] clearly do not conform to the original definition of rheopexy as postulated by Freundlich but more appropriately describe dilatant, or more simply, shear rate thickening behavior. On the other hand, abnormal dilatancy as used by Sato,[46] negative thixotropy[47,48] or antithixotropy[49,50] describe more closely the original intended meaning of rheopexy, although an attempt is made to separate behavior according to the speed of response to shear rate. Rheopexy has been used in its original meaning by several authors.[51-60]

In general, the term dilatancy has been used to describe both volumetric dilation under shear and the increase in viscosity with increasing rate of shear. Although the latter definition, which is the opposite of pseudo-plasticity or more appropriately, of shear rate thinning, has been the most common in modern usage[28,41,61-69] it should be emphasized that the two

[43] W. F. Bon, *Chem. Weekblad* **33,** 45 (1936).

[44] A. Peterlin and D. T. Turner, *Nature* **197,** 488 (1963); *Polymer Letters* **3,** 517 (1965).

[45] A. Peterlin, C. Quan, and D. T. Turner, *Polymer Letters* **3,** 521 (1965).

[46] K. Sato, *Science (Tokyo)* **13,** 165 (1943).

[47] G. S. Hartley, *Nature* **142,** Suppl., 161 (1938).

[48] J. Eliassaf, A. Silberberg, and A. Katchalsky, *Nature* **176,** 1119 (1955).

[49] J. Crane and D. Schiffer, *J. Polymer Sci.* **23,** 93 (1957).

[50] R. L'Hermite, *Ann. Inst. Tech. Batiment. Trav. Publ.* No. 5, 92 (1949).

[51] J. Pryce-Jones, *Kolloid-Z.* **129,** 96 (1952).

[52] J. B. Yannas and R. N. Gonzales, *Nature* **191,** 1384 (1961).

[53] C. G. Albert, *Tappi* **34,** 453 (1951).

[54] R. N. Weltman, *Rheol. Theory Appl.*, **3,** 189 (1960).

[55] G. W. Scott-Blair, "A Survey of General and Applied Rheology," p. 58, Pitman, New York, 1945.

[56] G. E. Alves, D. F. Boucher, and R. L. Pigford, *Chem. Eng. Prog.* **48,** 385 (1952).

[57] W. A. Weyl and W. C. Ormsby, *Rheol. Theory Appl.* **3,** 278 (1960).

[58] M. Reiner, "Deformation Strain and Flow," Wiley (Interscience), New York, 1960.

[59] S. Peter, *Rheol. Acta* **3,** 178 (1964).

[60] I. Steg and D. Katz, *J. Appl. Polymer Sci.* **9,** 3177 (1965).

[61] E. L. McMillen, *J. Rheol.* **3,** 75 (1932).

[62] A. B. Metzner, *Advan. Chem. Eng.* **1,** 77 (1956).

[63] L. Dintenfass, *J. Oil & Color Chemists' Assoc.* **40,** 761 (1957).

[64] L. Dintenfass, *J. Appl. Chem.* **8,** 349 (1958).

[65] E. T. Severs, "Rheology of Polymers," Reinhold, New York, 1962.

[66] R. McKennell, *Chem. Products* **18,** 267 (1955).

[67] E. Varley, *Quart. Appl. Math.* **19,** 331 (1962).

[68] D. Fensom and J. H. Greenblatt, *Can. J. Res.* **26B,** 215 (1948).

[69] W. Gallay and I. E. Puddington, *Can. J. Res.* **22B,** 161 (1944).

effects can occur separately[70] or under different conditions[24,71] in the same system.

Although there has been no final official stand taken on the subject of definitions, there are many who advocate the discontinuance of usage of terms such as dilatant and pseudoplastic and propose to use instead shear rate thickening and shear rate thinning, respectively. A definition for a dilatant system which agrees with much of the current usage is isothermal reversible shear rate thickening with no measurable time dependence. According to the nomenclature proposed by the Joint Committee on Rheology at the international Congress on Rheology,[72] by shear rate thickening is meant the increase of viscosity with shear rate. It would follow that shear thickening behavior with measurable time dependence would be called rheopectic.*

II. Measurements of Thixotropy and Dilatancy

1. THIXOTROPIC INVESTIGATIONS

In 1932 McMillen[61,73,74] discussed the quantitative measurement of thixotropy, suggesting that it was appropriate to measure the properties of thixotropic systems at various stages of transformation in terms of fundamental units commonly used to describe the physical nature of solids and liquids. He proposed four measurements, the minimum viscosity (or fluidity) attained by violent agitation, the rate of change of viscosity with time after agitation under a specified small constant shearing stress, the maximum elastic limit or yield value attained by the system, and the rate at which the elastic limit or yield value increases after agitation ceases. He noted that the modulus of elasticity also may change with time, but did not suggest its use as a possible method of measuring thixotropic changes. He recommended the measurement of the fluidity at frequent intervals after agitation by maintaining a constant small shearing stress and measuring the corresponding rate of shear, in a rotating cylinder viscometer. Defects in this method were the difficulty of defining the agitated reference state and the effect of the continuous shearing on the viscosity. In addition, he was unable to avoid the dependence on the dimensions of the viscometric apparatus, which permitted the measurement only of the gross rotational speed and the torque applied upon the thin walled cylinder rotating in the test fluid. The assumption could

* Cf. also Chapter 9 of this volume.

[70] J. J. Hermans, "Flow Properties of Disperse Systems," p. 10. Wiley (Interscience), New York, 1953.
[71] J. Pryce-Jones, *Proc. Univ. Durham Phil. Soc.* **10**, 427 (1948).
[72] J. M. Burgers and G. W. Scott-Blair, *Proc. 1st Internat. Congr. Rheol., Schevaningen, Holland 1948* p. VI. North-Holland Publ., Amsterdam, 1949.
[73] E. L. McMillen, *J. Rheol.* **3**, 164 (1932).
[74] E. L. McMillen, *J. Rheol.* **3**, 179 (1932).

not be made that a calibration of the apparatus carried out with a Newtonian standard liquid was valid for all liquids and that the shearing stress-rate of shear ratio for the thixotropic materials varied only with time of shearing and not with shear rate. McMillen found that the maximum fluidity after agitation, the maximum elastic limit reached on rest, and the rate of increase of the elastic limit with time were not quantitatively measurable in the rotating cup apparatus. He suggested that all materials having a yield value or elastic limit and deviating from a straight-line flow curve of shear stress versus shear rate would be thixotropic. The important observation was made by McMillen that for thixotropic materials rate of shear–shearing stress curves determined at successively increasing shear stress would not coincide with similar curves obtained at successively decreasing shear stress. He thus called attention to the importance of the shear history of thixotropic materials.

He recommended very strongly that capillary instruments not be used for study of thixotropic material because of the varying intensity of shear rate across the diameter of the capillary in flow and because fresh undisturbed material entered the capillary[75] and was subjected to an unmeasured degree of agitation. McMillen made an observation which has considerable validity even today: "In the agitation of a thixotropic material, the energy put into the system is used in two ways: (1) To overcome purely viscous resistance; (2) to break down structure and give a random distribution of particles of the dispersed phase in place of a more or less orderly arrangement; this latter energy may be considered as a sort of potential energy, being regained in some other form as the material sets and the structure rebuilds." In this statement McMillen foreshadows the current emphasis on the building up of normal stress in flow.

Through a long period, Pryce-Jones[17,51,76–78] maintained a consistent but restricted view of the measurement of thixotropy based on the conviction that the shear stress-rate of shear ratio cannot be quantitatively defined for flow of thixotropic shear sensitive materials in available viscometric apparatus. He proved the existence of thixotropic properties in a wide variety of materials and demonstrated qualitative differences in specially designed apparatus. He proposes holding strictly to the definition of thixotropy originally given by Freundlich, "the reversible isothermal gel-sol-gel transformation induced by mechanical stirring and subsequent rest." The requirement of Pryce-Jones for thixotropic sols is very strict, "their viscosities in the sol state are independent of the shear rate or the time of shearing." He thus restricts thixotropy to a class of dispersions which proceed from the

[75] R. Bulkley and F. G. Bitner, J. Rheol. 1, 269 (1930).
[76] J. Pryce-Jones, J. Oil & Color Chemists' Assoc. 19, 295 (1936).
[77] J. Pryce-Jones, J. Sci. Instr. 18, 39 (1941).
[78] J. Pryce-Jones, J. Oil & Color Chemists' Assoc. 26, 3 (1943).

gel state directly to a final sol condition with unique rheological properties on shearing, a very small class of materials. His definition of thixotropy excludes most of the systems studied by Freundlich and co-workers, for which the viscosity of the material was a function of the rate of shear applied as well as the time of shear or of rest.

In apparatus designed by Pryce-Jones for the study of thixotropy, he was able to demonstrate many striking differences in materials, on which he based a complex classification of materials. His measurements greatly extended the list of substances exhibiting time-dependent rheological properties, even if his severely limited definition of thixotropy has not been generally accepted. In the rheological apparatus devised by Pryce-Jones for studying thixotropy, provision was made for the measurement only in arbitrary units. Few systems studied met his ideal requirement of reversible isothermal transition from a rigid state on application of mechanical stress to a liquid state with a constant viscosity, to which systems Pryce-Jones restricted the term thixotropy.

For a large number of materials including paints, he found that the yield value and the shear stress–shear rate ratio in flow were functions of the rate of application of stress, of the shear rate, and of the time of shear or rest, and that elastic recoil was exhibited. Such dispersions Pryce-Jones called false body dispersions. While his work did not give quantitative measurements of thixotropy, Pryce-Jones' investigations were very valuable in demonstrating the widespread existence of time-dependent rheological properties in cream, gelatine, starch pastes, adhesives, petroleum jelly, polishes, emulsions, paints, and enamels. He showed the advantages of continuous time records of flow variables concerned, often termed hysteresis diagrams. In contrast to the practice recommended by McMillen for Couette-type viscometers, i.e., the use of a constant stress to rotate the inner cylinder and the recording of shear rate, Pryce-Jones preferred rotating the outer cylinder at constant speed while observing the change in shear stress. Since thixotropic materials may have properties that depend on the stress history or on the shear rate history, or on both, there appears to be no distinct merit in either choice of controlled variable. The most information could be obtained if both methods of operation were applied in studies of thixotropy. The structural factors in a disperse system which give it thixotropic characteristics may be expected to lead to an equilibrium state in continued shear at constant stress, or to another equilibrium state in continued shear at a constant shear rate. The two types of measurement give independent information, and no a priori virtue can be ascribed to either. Both Pryce-Jones and McMillen agree that it is more possible to reach a defined flow state throughout the sample in the repeated path of Couette or other rotary types of instruments, in which the same sample of test material is sheared between rotating elements, than is possible when a capillary instrument is used and fresh sample is continually injected. This conclusion is not necessarily correct, as may be seen from an investigation of shear- and time-sensitive greases, in which both rotary cone plate and pressure capillary viscometric systems were used.[79] While limiting states were reached most expeditiously in the rotary instrument used, it was found that significant information as to changes in grease structure could also be obtained from capillary measurements. It is apparent that when time- and stress-dependent systems are studied in viscometric apparatus, no apparatus is uniquely applicable. In fact, the structure of the system may best be understood from comparisons among the results obtained in viscometers of widely differing flow patterns.

The importance of time of shearing as well as of the rate of shear as factors affecting the viscous resistance in flow exhibited by thixotropic materials was clearly demonstrated by Pryce-Jones in his continuous recording of stress relaxation in samples previously subjected to shearing. Both the rate of shear and the time of shearing were controlled, and the relaxation of stress in a torsion pendulum attached to a bob immersed in the test sample was recorded. McMillen had attempted to follow changes in viscosity with time after mechanical agitation of a thixotropic material in a rotational cone type of viscometer by recording the shear rate at intervals of time while the system was moving at a very low state of stress. During the relaxation of the torsion in the instrument of Pryce-Jones there was also continuous low rate of shear flow, and thus the methods were not basically different in that respect. In work of Pryce-Jones, however, the test sample was mechanically agitated in a controlled manner at the beginning of a series of rest periods, and allowed to rest for a specified time before the flow stress relaxation with time was recorded. Both observers' methods were subject to the same limitations, in that the process of measurement of viscosity itself disturbed the rate of restoration of viscosity in a system which had previously been placed in a poorly defined reference state by arbitrary mechanical agitation.

In presenting a general theory of thixotropy and viscosity, Goodeve[80,81] took an even broader view of thixotropy than that of Freundlich,[82] in contrast to Pryce-Jones. Goodeve defined thixotropy as "any isothermal reversible decrease of viscosity with increase of rate of shear," omitting the requirement of gel-sol-gel transformation. Though his views emphasized

[79] W. H. Bauer, D. O. Shuster, and S. E. Wiberley, *Trans. Soc. Rheol.* **4**, 315 (1960).

[80] C. F. Goodeve, *Trans. Faraday Soc.* **35**, 342 (1939); *Proc. 1st Intern. Congr. Rheol, Scheveningen, Holland, 1948* p. II5. North-Holland Publ., Amsterdam, 1949.

[81] C. F. Goodeve and G. W. Whitfield, *Trans. Faraday Soc.* **34**, 511 (1938).

[82] H. Freundlich, "Thixotropy." Hermann et Cie, Paris, 1935.

only the shear stress–shear rate relationship, being characterized by Scott-Blair[37] as "only applicable for comparatively simple systems," Goodeve identified some of the structural elements leading to thixotropy. He drew attention to the way in which various structures could be expected to be affected by shear and thermal conditions. In his theory, a steady state is set up in a fluid undergoing continuous shear, as a consequence of the shear breaking and thermal breaking rates of links in the sheared system and of the shear making and thermal making of links. He was thus able to take into account the possible effects of such structural factors as Brownian motion, orientation, energy of activation of repulsive forces, collision rates, particle shapes and other dispersion micro-properties in his "impulse theory" of thixotropy. Considering these effects, Goodeve proposed a quantitative theory of viscosity in which a "coefficient of thixotropy" was defined as "the limiting slope of the viscosity–reciprocal shear curve as the shear approaches a high value." Such a definition of a quantitative measure of thixotropy is restricted in usefulness to the limited number of materials for which such a limit exists. In addition, the time dependence of rheological properties of thixotropic materials, of great practical importance, is ignored.

A rotary viscometer was described by Goodeve and Whitfield[81] for making measurements of the shearing stresses developed in a test system at continuously variable flow rates, in a modified Couette apparatus. In this viscometer, cylinders were replaced by coaxial cones in which a layer of the thixotropic liquid was sheared between a moving conical surface and a stationary conical plug which could be adjusted in a direction parallel to the axis of rotation of the outer cone. Thus, the effective shear rate could be changed continuously by change in the gap distance. It was hoped that this would provide a more satisfactory means of obtaining variable shear than that afforded by changing the speed of rotation in a concentric cylinder apparatus. An obvious defect, however, is the changing flow pattern in the sheared medium as the gap separation is varied and the consequent dependence on the dimensions of the moving system for materials exhibiting non-Newtonian behavior. The apparatus devised by Goodeve-Whitfield could thus not be successfully applied to the direct measurement of the rates of destruction or recovery of flow consistency in thixotropic materials under shear or at rest after shear, highly important features of their theory of thixotropic systems. However, the method of Goodeve and Whitfield for continuous variation of shear by change of the gap in a coaxial conical arrangement has sufficient advantages to have been employed in important investigations of thixotropy by Rehbinder.[83]

[83] P. A. Rehbinder, *Discussions Fara   · Soc.* **18**, 151 (1954).

Green and Weltman[35,54,84] recommended employment of shear stress–shear rate flow curves obtained in a rotary Couette-type viscometer in cycles of increasing and decreasing rates of controlled shear rate in order to characterize thixotropic behavior for various disperse systems. The flow curves thus obtained with automatic recording viscometers furnished striking demonstrations of the behavior of thixotropic systems under dynamic flow conditions, of great use in the comparison of systems. Since the change of consistency of thixotropic materials in mechanical agitation depends on both the magnitude of the shear rate applied and the time of shearing at a given shear rate, the viscosity changes from these two sources are not readily identifiable in the flow curves obtained in cycles of increasing and decreasing shear rate. As illustrated in the idealized diagram of Fig. 1(a), a thixotropic substance, with a history such that at time t_1 and shear rate D_1 it has the

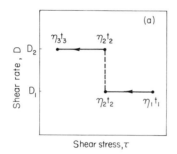

FIG. 1a. Shear stress developed in time for an ideal thixotropic substance subjected successively to the constant shear rates D_1 and D_2, where $D_2 > D_1$. η_1, viscosity of substance at arbitrary time t_1; η_2, limiting viscosity reached at time t_2 at shear rate D_1; η_3, limiting viscosity reached at time t_3 after sudden increase of shear rate to constant D_2.

viscosity $\eta_1 = \tau/D_1$, will decrease in viscosity when continuously sheared at constant D_1 until some limiting value of the viscosity, η_2, is approached after an elapsed time of $t_2 - t_1$. If at t_2 the applied shear rate is suddenly changed to D_2, and shear is continued at the constant shear rate D_2, the thixotropic material decreases in viscosity from the value of η_2 at t_2 to a lower viscosity, η_3, at a time t_3. At this time, destructive and restorative processes are again in equilibrium at the shear rate D_2, and no further significant change in the viscosity takes place. In Fig. 1(b), the corresponding rates of change of viscosity are shown for the two periods of viscosity change portrayed in Fig. 1(a). From an initial rate of change of viscosity, $(d\eta/dt)_1$ at time t_1 and viscosity η_1, the rate of change of viscosity approaches zero by the time t_2, and the viscosity approaches the limit η_2, all occurring at the

[84] H. Green and R. N. Weltman, Colloid Chem. **6**, 328 (1946).

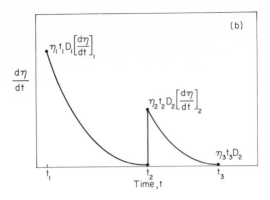

FIG. 1b. Path of rate of change of viscosity, $d\eta/dt$, with time for process described in (a). Thixotropic substance in arbitrary state of viscosity η_1, and rate of change of viscosity $(d\eta/dt)_1$, at initial time t_1, subjected to constant shear rate D_1 until $d\eta/dt$ falls to zero at time t_2, and a limiting viscosity η_2 is reached. When the constant shear rate D_2, where $D_2 > D_1$, is applied suddenly at t_2, an initial rate of change of viscosity $(d\eta/dt)_2$ is established in the ideal thixotropic substance, falling to zero at t_3, when the limiting viscosity η_3 corresponding to shear rate D_2 is reached. Noted values of η, t, and D are identical (a) and (b).

fixed shear rate D_1. Now when the shear rate is suddenly raised to D_2, change in viscosity again sets in, at an initial rate of $(d\eta/dt)_2$. The viscosity continues to decrease while the thixotropic material is sheared at D_2 during the time $t_3 - t_2$, until at time t_3 the rate of change of viscosity has fallen effectively to zero, and the viscosity has approached the new equilibrium value at D_2 of η_3.

From the foregoing analysis, the flow curve for a thixotropic substance in the portion of a cycle with decreasing shear rate will lie to the low-stress side of the portion of the flow curve obtained in the period of increasing shear rate, as illustrated in Fig. 3. A characteristic "hysteresis" flow curve will be obtained for a complete cycle of shear, two of which are shown in Fig. 2(a), for an ideal thixotropic substance whose rate of viscosity decrease under shear and rate of recovery under reduced shear gradient or at rest are both appreciable in the time scale of the measurements. In the cycle ABA, material with the initial viscosity η_1 at time t_0 is subjected to shear at increasing shear rate till the shear rate D_1 is reached at time t_1, when the viscosity has been reduced to η_2. Under the reduced shear rates in the BA portion of the cycle of decreasing shear rate, partial recovery of the viscous structure has taken place, and, when the cycle is ended, the viscosity of the material has been restored to the value η_y at the time t_y. As indicated in Fig. 2(b), the thixotropic material, after a sufficient time of rest, continues recovery until the original viscosity η_1 is reached. If the cycle ABCA is considered,

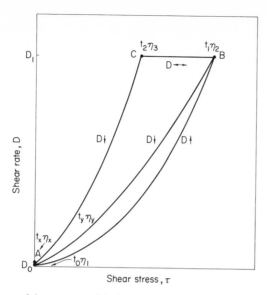

FIG. 2a. Change of shear stress τ with change in shear rate D for an ideal thixotropic material subjected to cycles of increasing and decreasing rate of shear D. The material is brought from arbitrary initial state A at rest at time t_0 and viscosity η_1 to viscosity η_2 in state B at time t_1, subjected to shear at continually increasing shear rate. In cycle ABA, the material is subsequently restored to the original state A with viscosity η_1 after the shear is reduced to zero at time of rest. In cycle ABCA, the thixotropic material after reaching state B is subjected to the constant shear rate initially reached at time t_1 until the viscosity falls to η_3 at time t_2. The shear rate is then continually decreased to zero at time t_x when the viscosity is η_x. Finally, the system is allowed to rest until the original state A with viscosity η_1 is regained, after sufficient time.

it is seen that at the constant shear rate D_1 the material under shear reduces in viscosity until the limiting equilibrium viscosity η_3 is reached at the time t_2. During the remainder of the cycle, the shear rate is reduced to zero from D_1, and the viscosity is partially restored during this portion of the cycle to the value η_x at the time t_x, eventually recovering its original viscosity of η_1 after rest. The course of the viscosity change during the cycle ABCA is analyzed in Fig. 2(b), for an ideal thixotropic substance.

The area or size and the shape of "hysteresis" loops thus depend on the rate of increase and the rate of decrease of shear rate in the cycles. If the rate of change of shear rate in the parts of a cycle were so extremely slow that equilibrium continually was reached between destructive and constructive processes, a limiting equilibrium flow curve would be approached, with a vanishing loop area. The area enclosed in a loop was interpreted by Green and Weltman[84] as a measure of the extent of "thixotropic" breakdown in a cycle. Since the speed of restoration of structure must be a major factor in

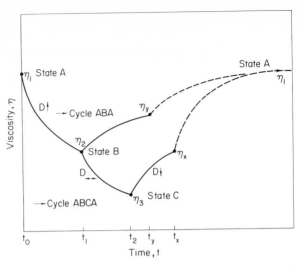

FIG. 2b. Change of viscosity η with time t in ideal thixotropic material subjected to cycles of increasing and decreasing rate of shear D. Material initially in arbitrary state A with viscosity η_1 is sheared at progressively increasing shear rate until viscosity η_2 in state B is reached at time t_1. In cycle ABA, the material is sheared at progressively decreasing shear rate until the viscosity η_y is reached at zero shear rate at time t_y, and is allowed to rest a sufficient time until the original state A with viscosity η_1 is again reached. In cycle ABCA, the material is sheared at constant shear rate after reaching state B, until time t_2, when the viscosity is η_3. The shear rate is then reduced to zero at time t_x when the viscosity is η_x, and finally the system at rest reaches the original viscosity η_1, after sufficient time. Noted values of η and t are identical in (a) and (b).

determining the size of the area enclosed in the flow curve loop, a material showing the largest area in cycles of similar controlled shear rate variation applied to a group of thixotropic substances may be said to be the most thixotropic, or to exhibit the highest degree of thixotropy.

Interpreted in this way, the flow curve loops give information closely analogous to that obtained from the measurement of "setting times," and similarly, materials with the longest "setting times" were said to be most thixotropic. Though very useful in qualitative demonstrations of thixotropy, the flow curves obtained in cycles of shear rates are not completely interpretable for quantitative measurements of thixotropy. The extent of approach to equilibrium is usually not known. Thus, unless flow continues to equilibrium, the effects of elapsed time of shear at a specified stress are undetermined so that interpretability of the results is usually limited. In spite of the problems of interpretation, the flow curve loops of the type introduced by Green and Weltman are extremely valuable in industrial practice,[85] as they furnish a

[85] P. Pierce and V. M. Donegan, *J. Paint Technol.* **38**, 1 (1966).

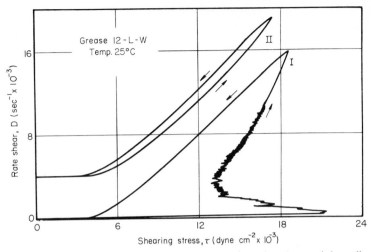

FIG. 3. Flow curves for lithium stearate-oil dispersion, showing "hysteresis loops," uniformly increasing and decreasing shear rate in shear flow. Cycle time, 300 sec. Cycle II follows cycle I after 600 sec rest.

sensitive means of noting changes in thixotropic systems caused by changes in formulation, processing, and control. In addition, there is the important practical advantage that modern automatically controlled and recording Couette or cone-plate types of viscometric apparatus with rotary elements will provide the flow curve information with great rapidity.

It is generally expected that all thixotropic materials will show a hysteresis loop in the flow curve obtained in cycles of increasing and decreasing shear rate, since, for thixotropy to exist, the rate of destruction of viscous properties when shear is initially applied must be greater than the rate of recovery. As is the case in all rheological investigations, however, the scale of forces measurable and the rapidity of measurements characteristic of the experimental apparatus determine the limits of usefulness of a criterion for the existence of thixotropy, such as the appearance of hysteresis loops. For example, aqueous dispersions of carboxy polymethylene, "Carbopol," exhibit pronounced reversibly recoverable yield stresses and become extremely shear sensitive viscous fluids, isothermally and reversibly, under sufficient large mechanical agitation. These are basic characteristics of thixotropic dispersions. However, the speed of recovery of the Carbopol dispersions is so great that, in an investigation carried out with a modern automatically controlled and recording viscometer,[86] the flow curves obtained in cycles of increasing and decreasing rates of shear showed no

[86] W. H. Fischer, W. H. Bauer, and S. E. Wiberley, *Trans. Soc. Rheol.* **5,** 221 (1961).

hysteresis loops within the experimental limitations of the speed of measurement. The Carbopol dispersions should thus be classified simply as shear thinning rather than as thixotropic, according to most definitions of thixotropy.

Limitations on the use of flow curve hysteresis loops as a criterion for thixotropy also arise when the rate of recovery of viscous properties in rest after destructive shearing is extremely slow instead of extremely fast. Greases formed by dispersions of soaps in oils are frequently described as thixotropic,[87] and they certainly exhibit limiting yield stresses, time-dependent shear rate thinning in flow, and give pronounced hysteresis loops in cycles of varying shear rate. However, as shown in a study of lithium soap-oil greases[88] (as illustrated in Fig. 3) it was found that the destruction of viscous structure in the greases in periods of increasing shear rate was not reversibly and isothermally recovered in the period of time of decreasing shear rate, nor in long subsequent rest periods. Since the recovery of structure must involve favorable reorientation, reaggregation or regrowth of lithium stearate crystallites in a very viscous medium in which the rate of diffusion, desorption of oil from crystal surfaces, and migration of soap micro-particles are very slow, it is not surprising that recovery does not occur isothermally during the practical times of rest. However, the seemingly permanent loss of the viscous structure is reversibly restored when the temperature is temporarily raised to produce conditions of higher mobility and later restored to the original test temperature, as demonstrated by Weltman and Kuhns[89] and other investigators. Since the structure breakdown of the greases was not recovered isothermally in practical times of rest, Weltman[54] preferred that the flow curve hysteresis loops exhibited should not be considered as caused by thixotropic behavior.

In his quantitative studies of thixotropic materials, Dintenfass emphasized the importance of establishing flow equilibrium states as reference points for making comparison of changes in rheological properties with time. In effect, this provides reference conditions which have a controlled history, a basic requirement for the investigation of thixotropic materials whose rheological state is dependent not only on stress and strain variables, but also upon history.[90,91] In investigations of thixotropy,[92,93] as a quantitative measure, Dintenfass has employed the "recovery time" required for a

[87] S. J. Hahn, T. Ree, and H. Eyring, *Ind. Eng. Chem.* **51**, 856 (1959).

[88] W. H. Bauer, A. P. Finkelstein, and S. E. Wiberley, *ASLE Trans.* **3**, 215 (1960).

[89] R. N. Weltman and P. W. Kuhns, *Lubrication Eng.* **13**, 43 (1957).

[90] T. Alfrey, Jr. and C. E. Rodewald, *J. Colloid Sci.* **4**, 283 (1949).

[91] S. Thornton, *Proc. Phys. Soc.* (*London*) **B66**, 115 (1953).

[92] L. Dintenfass, *Kolloid-Z.* **163**, 48 (1959).

[93] L. Dintenfass, *Rheol. Acta* **2**, 187 (1962).

sheared material to arrive at a new constant viscosity when equilibrium is again reached after sudden change to a lower shear rate from a higher applied shear rate at which previously an equilibrium state of constant viscosity had been obtained. The "thixotropic recovery time" thus measured has the advantage that it is a property measured after the thixotropic system is first brought to state of defined rate of shear at which it is retained until the previous history is destroyed. The recovery time subsequently measured is referred to a final state of defined rheological condition in which recovery processes are again in equilibrium at a specified shear rate. For the special systems studied by Dintenfass, such as pigment dispersions in organic liquids, he found that an upper value of shear could be determined beyond which continued shearing had no effect on the viscosity; this he called the "critical" shear rate. The logarithm of viscosities measured when equilibrium had been reached in shear plotted against the logarithm of shear rate gave a straight line, whose "thixotropic slope" Dintenfass proposed as another quantitative measure of thixotropy. The "thixotropic slope" and the "critical shear rate" thus defined are analogous to the "coefficient of thixotropy" which Goodeve and Whitfield defined as the limiting slope of viscosity reciprocal shear rate curves, quantities most applicable to comparatively simple systems, It is not likely that they can be determined unambiguously for most complex thixotropic systems. The "recovery time" for reaching shear equilibrium between two reference states of defined rheological state would appear to be likely to have more general application and to be more clearly defined.

2. DEFINITIONS OF THIXOTROPY AND DILATANCY

The work of the various investigators cited in the foregoing discussions, though it encompasses only a small selection from the studies conducted, gives a representative view of attempts to find a quantitative measurement of thixotropy. It is clearly evident that there has been both a diversity in definition of the concept of thixotropy and also a lack of agreement on the experimental arrangements required to produce precisely measured and unambiguous assessments of viscous properties in rheologically defined systems. A definition of thixotropy, to be valuable, must take into account the usage that actually and currently exists. Usage, nevertheless, has such a range that a definition reflecting all views may have the effect of confusion rather than clarification. Limitation of a concept may make it more practically useful than if it were extended to have a very wide application. A practical mean is most acceptable. Further, a definition of thixotropy should be made in defined rheological terms. In this respect, one only has to consider the seventy page report of the Joint Committee on Rheology of the International

Council of Scientific Unions on the "Principles of Rheological Nomenclature"[72] to realize the complexity of the subject of definition of rheological terms. Since the report, though no longer new, is a model of clarity, emphasizing the basic quantities of force, mass, and time in definitions, no attempt will be made to improve the nomenclature therein with respect to variables based on flow stress and strain, relating to thixotropy.

The area of agreement among the investigators whose work was reviewed points to the following definition:

> A thixotropic system exhibits a time dependent, reversible, and isothermal decrease of viscosity with shear in flow.

This definition omits the requirement, for instance, of a clear gel-sol-gel transformation. On the other hand, it includes the restriction of time dependency, omitted in some current definitions. The term "time dependent" is itself limited by the speed of measurement possible, which gives the definition a certain ambiguity. In this respect, Jobling and Roberts[26] comment, "... the time lag required before the original rest structure is regained may be very short indeed and it then becomes difficult to distinguish between a thixotropic material with a very short recovery time and a material whose viscosity falls with increasing rate of shear and depends for all practical purposes only on the instantaneous rate of shear." Not only has there been diversity of definition among the investigators cited, but there has been no clear agreement on the quantitative measure of thixotropy.

Although the early usage of the term dilatancy referred to a volume expansion with the application of a shear stress, a modern and widely accepted definition is as follows:

> Dilatancy is the isothermal reversible increase of viscosity with increasing shear rate with no measurable time dependence. The process may or may not be accompanied by a detectable volume change.

Because of this dual meaning, it should always be made clear, when the term dilatancy is used, whether shear rate thickening or volume increase, or both, are meant. The simplest solution to the problem is to avoid the term and simply describe the phenomena. The term dilatancy may have different meanings to different people, but it is difficult to misunderstand what is meant by "shear rate thickening."

3. PRINCIPLES OF QUANTITATIVE MEASUREMENT OF THIXOTROPY

Limitations with respect to quantitative measurements of thixotropy are not those of thixotropy in itself. They are the same limitations that apply to the measurement of the basic rheological variables of stress and strain with time in any material system while it is deformed or while it is undergoing continuous deformation. Some general principles have become apparent. The important variables involved in the rheological state of a thixotropic material are shear stress, shear strain, rate of shear, rate of change of shear stress, rate of change of shear rate, time, and temperature. These variables should be measurable and controlled, and ideally they should be uniform throughout the material under test, since average values may have multiple origins. In addition, the flow variables for the same unit of material should be observable in time during continued shear flow. These requirements are best met in instruments with a narrow flow path, such as the concentric cylinder, concentric cone in cone, parallel plate, or cone and plate arrangements. Such instruments permit a state of flow to be set up in the sheared sample held between two members providing a controlled difference in speed of rotation, with one or both of the concentric or parallel elements in motion. Temperature also is most easily controlled when the sheared material is in a thin layer. The unit of material sheared in such rotary viscometers may be subjected to continuous observation instead of being continually renewed and removed from observation as occurs in capillary extrusion apparatus. It has been generally agreed that the requirements for rheological measurements on thixotropic systems are best met in rotary instruments, rather than in extrusion apparatus. It should be remembered, however, that one way to study how thixotropy may affect extrusion is to use an extrusion apparatus. Capillary extrusion should not be arbitrarily rejected in experiments on thixotropy.

While observers have agreed upon the general advantage of rotary, continuous path, instruments for thixotropic substances, such instruments show serious limitation with respect to substances such as aluminum soap gels in toluene or silicone oils that develop high normal stress differences in flow. These materials will move from a region of high to a lower shear and will spontaneously remove themselves from the shear region at sufficiently high shear rates. With this exception, the rotary viscometers which shear a thin layer of material in a repeated flow path and which provide measurement in time of the shear stress and the shear rate in the unit of material subjected to controlled chosen flow conditions have been most suitable for study of change in rheological properties with time of shear and time of rest. Practically, the attainment of uniform flow conditions throughout a material in shear has not yet been successfully accomplished. However, close approach has been shown to be possible when the material studied is in a thin layer so that the region of observation is well defined. Ambiguity arises as to what part of the material studied is affected by the mechanical agitation when measurements are made in a rotary-type viscometer having a wide gap between the moving and stationary walls. It is very advantageous if the viscometric apparatus permits continuous controlled variation of shear stress or of shear rate, and if automatic recording of pertinent variables measured in time is available.

The many descriptive names which have been applied to measuring instruments used in thixotropic study such as "thixotrometer" or "thixo-viscometer" are misleading in view of the complexity of thixotropy, since all useful instruments must measure the fundamental rheological variables. Study of thixotropy simply adds the requirement that these variables be measured in time; that is, that the rheological history be established. A guiding principle in a significant investigation of thixotropy is the measurement of the rheological state of a material at selected intervals in a period of time in which the sample studied has at all times a known shear history.

A variety of programs of measurement have been employed in rheological measurements by various investigators for providing information about thixotropic material. In any program of study, the observations of Thornton[91,94] should be taken into account. He points out that for a normal thixotropic material the viscosity is dependent on the previous shear history as well as on the currently applied shear stress or on the shear rate maintained in the sample. In a quantitative study of the time dependence of viscosity, the thixotropic material must first be brought to a defined state of shear. This requirement is met if the effect of previous shear history is removed when a thixotropic sample is subjected to a constant shearing stress or shear rate, maintained for a sufficient time that a constant equilibrium viscosity is reached. Quantitative study of viscosity as a function of time may be carried out for this type of system by procedures such as the following. If shear stress is to be the controlled shear variable, a selected constant shear stress τ_0 may be applied for time sufficiently long for constant viscosity η_0 to be reached. Subsequently a greater or a smaller shear stress τ_x or τ_y may be applied, and the shear rate D may be measured so that viscosity is obtained as a time function of the chosen shear stress, $\eta = F(\tau_0, \tau_x, t)$ where the viscosity $\eta = \tau/D$. At some time t_y, a limiting viscosity will be reached corresponding to τ_x, as illustrated for an ideal case in Fig. 4(a).

Alternatively, after initial viscosity equilibrium at τ_0 is reached, the applied shear stress may be continuously varied, and the shear rate measured. In this case the viscosity is a function of both stress and time, $\eta = F(\tau_0, \tau, t)$. An infinite number of patterns is possible according to the rate of change of stress chosen. If the stress is returned to the original value τ_0, a "hysteresis" loop will be obtained, Fig. 4(b).

If shear rate is the variable to be controlled, as is frequently most convenient, a constant shear rate D_0 may be maintained in the sample until an equilib-

[94] S. Thornton and D. Rae, *Proc. Roy. Soc.* **B66,** 120 (1953).

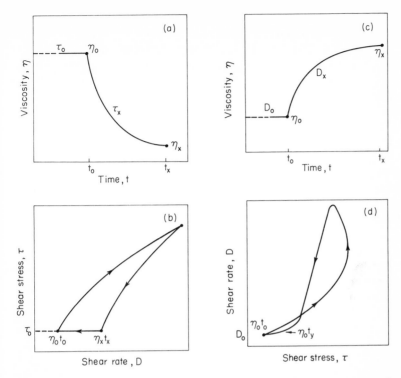

FIG. 4. (a) Viscosity as a function of time for an ideal thixotropic material, initially in shear equilibrium at time t_0, viscosity η_0, and stress τ_0, subjected to flow at a constant shear stress τ_x until equilibrium is reached at time t_x and viscosity η_x. (b) Shear rate as a function of varying stress τ and time t for an ideal thixotropic material initially in equilibrium at time t_0, viscosity η_0 and stress τ_0 subjected to cycle of flow in increasing and decreasing shear stress, and returned to equilibrium at η_0 in sufficient time at constant shear stress τ_0 after reaching viscosity η_x at time t_x. (c) Viscosity as a function of time for an ideal thixotropic material, initially in shear equilibrium at time t_0, viscosity η_0, and constant shear rate D_0, subjected to flow at a constant shear rate D_x until equilibrium is reached at time t_x and viscosity η_x. (d) Shear stress as a function of varying shear rate D and time t for an ideal thixotropic material initially in equilibrium at time t_0 and viscosity η_0, subjected to a cycle of increasing and decreasing shear rate until, after sufficient time $(t_y - t_0)$ at the original shear rate D_0, the original viscosity η_0 is restored.

rium viscosity η_0 is reached. The dependence of viscosity on particular values of shear rate, greater or less than D_0, may be measured in time, so that $\eta = G(D_0, D_x, t)$. The viscosity will approach a limiting viscosity η_x in time, as illustrated in Fig. 4(c) for the ideal case where $D_x \ll D_0$. The time interval, $t_x - t_0$ for the new equilibrium viscosity η_x to be reached is a "recovery time," defined between two equilibrium states.

Again, alternately, a volume of material may be subjected to flow at a shear rate D_0 until a state of equilibrium is reached at a constant viscosity

η_0, after which the applied shear rate may be continuously varied and the viscosity found from the corresponding shear stress developed. In this case, the viscosity becomes a function of shear rate and time,

$$\eta = G(D_0, D, t),$$

and paths followed will vary according to the rate of change of shear rate employed. If the shear rate is returned to the original value of D_0, and time allowed for the system to reach its original viscosity η_0, a "hysteresis loop" will be obtained, as illustrated in Fig. 4(d) for an ideal material. From the basic processes described in the idealized examples given, informative programs for actual thixotropic quantitative investigations may be constructed, with the elementary procedures in various combinations. For quantitative measurements to be truly significant, the thixotropic material, in initial state of known shear history, should be studied under the most uniform conditions possible of shear stress, shear rate, and temperature.

Actual programs of investigation of thixotropy may be as varied as the number of instrumental arrangements for the application and measurement of shear, stress, and time. Real thixotropic materials may exhibit time dependence in other quantitatively measurable rheological properties in addition to viscosity. Some of these are delayed stress relaxation after cessation of stress application, elastic recoil after applied shear, critical yield stress, critical solidification stress, shear moduli, and development of normal stress in flow. For instance, transitional processes in the attainment of steady flow for polyethelene and polystyrene melts occur in time, indicating the existence of thixotropy. In addition, the modulus of elastic deformation has been found to depend on the deformation rate.[95] The structural changes in thixotropic systems leading to time dependence of rheological properties may also be accompanied by changes in time of other properties such as dielectric constant[96-98] and optical birefringence.[99-101] When these properties can be correlated with the corresponding viscosity, they may furnish a means of following changes in viscosity in studies of thixotropic materials with time, without introducing shear effects arising from the test measurements.

[95] G. V. Vinogradov and I. M. Belkin, *J. Polymer Sci.* **A3**, 917 (1965).

[96] A. Voet, *J. Phys. Colloid Chem.* **51**, 1037 (1947).

[97] A. Bondi and C. J. Penther, *J. Phys. Chem.* **57**, 72 (1953).

[98] E. A. Collins, M. S. Thesis, University of Manitoba, Winnepeg, Canada (1951).

[99] R. Cerf and H. A. Scheraga, *Chem. Rev.* **51**, 185 (1952).

[100] A. V. Frisman and S. Mao, *Polymer Sci.* (*USSR*) (*English Transl.*) **6**, 37 (1964); **6**, 46 and **6**, 223 (1964).

[101] H. Wayland, *J. Polymer Sci.* **C5**, 11 (1964).

4. MEASUREMENTS OF DILATANCY

A major difficulty in early studies of dilatancy was the lack of a satisfactory technique for measuring the flow properties of a dilatant material. This problem still exists in spite of the advances that have been made in rheological instrumentation.

It is not surprising then, to find many of the early references on dilatant behavior to be of a qualitative nature, such as observing the effect of a disturbance with a spatula on a dilatant paste. For example, a paste consisting of a mixture of 5.0 g of quartz and 2.25 cc of water when left to itself appears quite fluid, but when probed with a spatula it immediately becomes dry and very resistant to the spatula. The harder one forces the spatula into the paste the greater the resistance offered. If taken into the palm of the hand and the hand closed rapidly, the material crumbles like a brittle solid. Left to themselves, the fragments unite and return to a viscous fluid. When the material becomes so dilatant that it has more characteristics of a solid than a liquid, conventional viscometric methods become quite useless.

Freundlich and Roder[31] made early measurements on dilatant pastes using a modified Searle viscometer. They attempted to simulate the movement of a spatula by pulling a small sphere through a dilatant material contained in a channel using a weight pulley system. For Newtonian materials they obtained a linear relationship between the speed of the sphere and the weight on the pulley system, as shown by curve a in Fig. 5. A dilatant material, on the other hand, resulted in curve b. At a certain speed (x) a disproportionately high resistance sets in, and any further increase of the shear stress has no effect on the speed at which the sphere moves through the system.

There is no single instrument currently available that can handle dilatant materials up to stresses where the material assumes characteristics of a solid. However, rotational and capillary instruments can be used at lower stress levels. A measure of dilatancy widely used in industry is derived from the shape of the flow curve in the region where the empirical power law relating shear stress τ and shear rate D is operative,

$$\tau = kD^n,$$

where k and n are characteristic constants.

In accordance with the specified viscosity characteristics, $n < 1$ defines the region of shear rate thinning, whereas $n > 1$ defines the region of shear rate thickening or dilatant behavior.

Recently Jobling and Roberts[26,27,102] suggested the use of a new and very general technique for mechanical testing, as proposed by Weissenberg[103] for overcoming the difficulties in handling dilatant systems in any

[102] A. Jobling and J. E. Roberts, *Rheol. Theory Appl.* **2**, 503 (1958).
[103] K. Weissenberg, *Bull. Brit. Soc. Rheol.* **43**, 6 (1955).

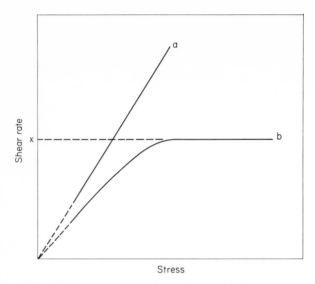

FIG 5. Flow curves for (a) Newtonian and (b) dilatant systems.

conventional rheometers. In this technique, the system to be tested is spread in a thin film on a substrate which is itself deformable. The substrate can then be made to undergo any one of a wide range of movements, the components of which in the plane of the interface are transmitted to at least the bottom layer of the sample, provided that the two materials adhere adequately. The three major advantages to this "deformable membrane technique" are summarized by Jobling and Roberts as follows:

(1) It enables mechanical tests to be made simply with any general homogeneous strain and many types of heterogeneous strains.

(2) It affords improved control over the behavior of the material in the plane of the film, as evidenced by the remarkable regularity of the patterns of "flow lines" produced if a straining action applied in this way is continued either in time or space to the point where failure occurs. In isotropic materials, the direction of this pattern can be related to the strain geometry.

(3) It extends the range of materials that can be tested in any given deformation so that, for example, actions involving tensile stresses can be applied to materials of little cohesion, such as pastes and powders.

Striking similarities of the mechanical behavior of materials of widely differing physical and chemical properties have been revealed by this technique.[27]

III. Occurrence of Thixotropy and Dilatancy

A complete listing of the different systems and the numerous publications devoted to each of these systems would be a monumental task and would detract from the purpose of this communication. However, to illustrate the variety of materials which have been considered to be thixotropic or dilatant by various investigators, some systems are listed in Appendixes I and II together with one or more pertinent references which, hopefully, would serve as a starting point for any further investigation. A very broad classification has been made according to whether the dispersion is aqueous or nonaqueous.

IV. Origin of Thixotropy and Dilatancy

1. ORIGIN OF THIXOTROPY

Most investigators of thixotropy have postulated that thixotropic materials at rest have an internal structure. This may arise from arrangements of anisometric dispersed particles having areas or points at which adhesive forces may arise, arrangements which lead to network structure of particles in a liquid dispersion medium. Such a network would provide a structure enabling the dispersion to appear to react elastically to a low stress for a limited time of application. If this time were of the order of 24 hours, for example, in a time less than that period the dispersion would be considered to be stable under the applied stress. In the interim however, under the stress, the number of effective network junctions would decrease, with the result that after about 24 hours, as postulated, continuous flow or yield would occur, establishing a yield value for the time and stress involved. Such a yield stress is, of course, a significant variable only if the time interval for application of stress to result in yield is specified. Now, when continuous flow begins to take place in the thixotropic dispersion, catastrophic breaking of the network linkages occurs as the energy of flow is added to the system and the anisometric particles assume random orientation with respect to the active regions for adhesion. If the volume of the dispersed particles is small relative to that of the dispersing liquid, and if the particle surface is generally stabilized against coagulation, the dispersion could act under continued shear as if it were a liquid of a constant viscosity determined by the nature of the dispersing medium, with viscosity nearly independent of shear rate. If addition of energy through shear were now stopped and the freely flowing dispersion were allowed to enter a state of mechanical rest, random Brownian movement would bring areas or points of adhesion of the anisometric particles into the volume of space in which linkage could again be formed. Given sufficient time, enough favorable contacts would be made so that the original density of linkages would be reached. The dispersion would have returned to the solid state, with the elastic modulus corresponding to the network rigidity. In the process of restoration of structure, the variables that affect lyophobic dispersion stability[83,104,105] (such as the viscosity of the dispersion medium and diffusion rate of the particles and factors such as ionic concentration, solvation, zeta potential, and surface active agents that affect the forces of repulsion and attraction at the surface of the particles) would determine the time required for solidification. During the period of rest before the network was restored, the growth of structure would be continuous, so that the dispersion would be considered to have a micro-structure content, even though it flowed like a liquid in the time interval of recovery.

The above reversible transformation, if carried out isothermally, is the classical gel-sol-gel thixotropic change typified by the dispersion of vanadium pentoxide needle form crystals in water.[10,75] In the case of fresh vanadium pentoxide crystals, attractive forces are greatest at the ends of rodlike particles first formed in a condensation process, so that end-to-end aggregation occurs, leading to the possibility of forming a network required to give thixotropy. It is evident that all factors influencing the stability of lyophobic colloids will affect materials in a thixotropic dispersion of the vanadium oxide type. These are, for instance, protective lyophilic layers and adsorbed ions which provide solvation and stabilizing electrical double layers for the dispersed solid. The protective factors must be strong enough to prevent complete coagulation by aggregation of the dispersed particles, yet the stabilization must not be so complete that there are no points or areas of adhesion on the dispersed particles. Such "coagulation centers"[83] are most likely to occur at corners or edges of asymmetric particles. Dispersions of spherical particles would be least favorable for providing thixotropy, since factors affecting stabilization would be likely to be uniform across the surface.[106,107]

The concentration of the solid phase must also be important for sols of the vanadium pentoxide type, since if it were too small for a significant number of end-to-end aggregations to occur, conditions leading to a network would not be present. If the concentration were too great, coagulation by growth of rods into bundles would be encouraged, and the dispersion would flocculate. The general requirements which give rise to thixotropy in such systems as the aqueous dispersions of metal oxides apply to a wide variety

[104] E. J. W. Verwey and J. T. G. Overbeek, "Theory of Stability of Lyophobic Colloids." Elsevier, Amsterdam, 1948.

[105] B. V. Derjaguin, Trans. Faraday Soc. 36, 203 and 730 (1940).

[106] G. Wieguer and C. Marshall, Z. Physik. Chem. 140, No. 1, 39 (1929).

[107] A. Pockter, Z. Physik. Chem. (Leipzig) 211, 40 (1959).

of thixotropic materials, such as pigment dispersions and clay dispersions. According to the properties of the liquid dispersion medium, its polarity, conductivity, viscosity, and wetting ability, for instance, the origin of thixotropy will require appropriate size distribution, shape, nature and area distribution of secondary forces leading to particle adhesion. Concentration of particles, sedimentation volume, and similar properties must be so adjusted that network or aggregate formation will take place, with the forces leading to structure formation just strong enough to resist destruction by the disarrangements caused by Brownian movement. These forces of aggregation must be just weak enough that complete coagulation does not occur and that shear stresses in flow will be able to disturb or destroy the structure. Under the stresses at rest or under shear stress in flow, the processes of formation and destruction of network or aggregation must be continually taking place, with the extent of structure present determined by the net result of many opposed processes. Rates of restoration of structure on reduction of shear stress clearly may be influenced by speed of orientation, leading to favored alignment and occurring by rotation or diffusion or by the elastic return of shape of deformed aggregates. When one considers the large number of thixotropic systems discovered, it is evident that the origin of thixotropy is as complex as the problem of the nature of the origin of the rheological properties of these systems.

Theoretical considerations of the fundamental processes of breakdown and recovery[42] and efforts to find constituitive equations[108] for thixotropic materials should prove of value in understanding the mechanical nature of the phenomenon of thixotropy, especially as the amount of quantitative information increases. All displacements which lead to permanent deformations of materials take place in finite times, and it is to be expected that all systems may exhibit time dependence of establishment of viscous properties under a change in shear rate or stress if these systems can develop a non-uniform micro-organization. The widespread existence of thixotropic liquids indicates the presence of significant concentrations of clusters, aggregations, and networks which, though transient, are present in amounts sufficient to give origin to thixotropy.

2. ORIGIN OF DILATANCY

Early work on dilatancy indicated that the phenomenon most frequently occurred with suspensions of solids, such as quartz powders and starch pastes at high solids concentration, and was explained simply on the basis of particle packing and particle dislocation. The extension of the term, originally intended for a volume expansion occurring with the application

[108] P. R. Paslay and A. Slibar, *Rheol. Acta* **2**, 236 (1962).

of a shear, to denote also an increasing viscosity with increasing rate of shear has broadened the scope of the definition considerably and now includes such varied systems as concentrated polymer solutions, emulsions, plastisols, polymer melts, and even metals. Thus, it would appear that dilatancy is much more widespread and the mechanism more complex than is generally believed. No single current theory accounts adequately for dilatant behavior as encompassed by the modern day usage of the term.

a Qualitative Observations

Very little quantitative work on dilatancy has been reported in the literature, the notable exceptions being the investigations of Andrade and Fox,[23] Daniel,[25] Metzner and Whitlock,[24] Jobling and Roberts,[26] and Peterlin.[109] In their studies on quartz and starch suspensions, Freundlich and Jones[29] concluded that, in order to form a dilatant paste, the particles had to be absolutely independent of one another, exactly the converse of what was proposed in an earlier paper.[39] They believed that dilatancy was reduced or destroyed if the particles adhered in the least. In this regard, the addition of electrolyte brings about coagulation and consequently reduces the dilatancy of aqueous quartz suspensions. Mutual adhesion as produced by coagulation also causes an increase in the sedimentation volume. Ehrenberg[110] and Buzagh[111] had previously established that small sedimentation volumes result from noninteracting particles. Thus, the main factors favoring a small sedimentation volume should also favor dilatancy. Brown and Hawksley[22] claim all packed systems, regular or irregular, open or closed, dilate when initially deformed, although loose packings collapse and thus the phenomenon may be masked. Pryce-Jones[51,71,78] and Dintenfass[63,112] would have a completely dispersed system behave in a Newtonian manner but, as aggregation takes place, become dilatant and, at higher aggregation, thixotropic. That is, the degree of dispersion rules the rheological behavior and, conversely, the rheological behavior could serve as a criterion of the state of the dispersion.

Ostwald and Haller[113] reported the sedimentation volume of quartz suspensions to be small in water and other polar liquids but large in carbon tetrachloride and other nonpolar liquids. Williamson and Heckert[19] observed a similar behavior for suspensions of cornstarch. These data were interpreted by Freundlich and Roder[31] to mean that the particles are

---

[109] A. Peterlin and D. T. Turner, *J. Chem. Phys.* **38,** 2315 (1963).

[110] P. Ehrenberg, *Bodenkolloide (Dresden)* p. 83 (1918).

[111] V. Buzagh, *Kolloidchem. Beih.* **32,** 114 (1930).

[112] L. Dintenfass, *J. Oil & Color Chemists' Assoc.* **43,** 46 (1960).

[113] W. Ostwald and W. Haller, *Kolloidchem. Beih.* **29,** 354 (1929); W. Haller, *Kolloid-Z.* **46,** 366 (1928).

independent in water and do not adhere to each other, whereas in nonpolar liquids they have a strong tendency for mutual adhesion. In support of these conclusions and of the effect of electrolytes, they found the zeta potential of quartz and rice starch aqueous suspensions to be -29 and -36 mv, respectively. Daniel[25] also found that one of the main factors for producing dilatancy was a high zeta potential. This is essential in preventing the attractive or adhesive forces between particles from becoming dominant. In general, the influence of electrolytes on suspensions does not result from chemical reactions with the dispersed particles, but from the effect on the zeta potential of the particles and on the solvation layer of which very little is known. Whatever the role of zeta potential may be, it is clear that it is only one of the requisites for dilatant behavior, since Freundlich and Jones[29] also reported high zeta potentials for suspensions of particles which had large sedimentation volumes and which did not exhibit dilatant behavior as, for example, graphite having a zeta potential of -52 and hematite having a zeta potential of -36 mv.

Freundlich and Roder[31] regarded dilatancy as being opposite to thixotropy and a characteristic property of close-packed systems having a small sedimentation volume. They viewed the mechanism of dilatancy as being associated with close but mobile packing of solid particles, or with their uniform distribution in a liquid medium. If there is just enough liquid present in a closely packed system to wet all the particles, any disturbance of this arrangement must produce more open spacing of the particles with the result that the medium is drawn into the widening interstices and the whole system turns "dry and hard" because of localized de-wetting. If the cause of the disturbance is eliminated, the particles return to their original dispersed or "fluid" condition. The mechanical treatment causes merely a change from one solid with near liquid internal friction to another solid with high solid internal friction, as illustrated in Fig. 6. The basic independence of the particles causes the structure formed by shear rate to break down spontaneously when the forces are removed and also allows the particles to form fairly closely packed sediments. By contrast, Freundlich regarded the

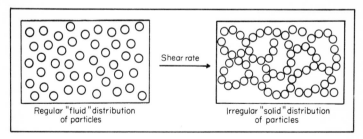

FIG. 6. Mechanism of dilatancy.

454   WALTER H. BAUER AND EDWARD A. COLLINS

particles in thixotropic systems as not independent, but with some tendency to adhere to one another. Hence the irregular distribution of piled up particles and of cavities filled with liquid corresponds to the stable solid state. On shaking, the particles are more uniformly distributed in the liquid, the system becomes fluid, but spontaneously returns to the solid coagulated state when left at rest. Gallay and Puddington[69,114] also found a correlation between dilatancy and sedimentation volume, with the former occurring in suspensions in which the latter is small.

Early investigators believed dilatancy to occur only with close-packed spherical suspensions. Freundlich and co-workers[29,31] demonstrated that irregular particles (42–44% by volume of 1- to 10-μ quartz suspension) also exhibit true dilatancy when subjected to shear, and that this also arises from an increase in the overall volume occupied by the particles. Dilatancy can then occur at lower concentrations than correspond to close packing, and is more pronounced.[25] Perhaps too much importance has been attached to the fact that starch and other materials show dilatancy at concentrations which do not correspond to close packing. So little is known about the effective (hydrated) size of the particles that it would be more advisable to calculate the mechanically effective volume of the particles from the concentration at which dilatancy is first observed.

Examination of the literature shows dilatancy is not restricted to any particle size range and has been reported for systems having particle sizes below 1 μ[115,116] in the range 1–10 μ,[117] 20–25 μ,[71] 60–90 μ,[31] 90–250 μ, 250–500 μ, and 500–700 μ[118] as well as for polymer solutions.[34,44,45,68,77] The fact that Robinson[119] did not observe dilatancy with particle sizes in the range 3 to 30 μ for concentrations up to 52% suggests that particle size is only one requisite for this phenomenon.

Although many suspensions are dilatant over the very narrow concentration range of 40–50% by volume, [24,29,31,120,121] the literature contains numerous references to the occurrence of dilatancy at much lower concentrations for a variety of systems such as 10% solution of calcium naphthenate in white spirit,[77] 25% calcium stearate in mineral oil,[69] 26% red iron oxide in aqueous sulfonated lignum solution, and 35% starch in ethylene glycol.[28] In addition, several authors have reported dilatant behavior of solutions at

[114] W. Gallay and I. E. Puddington, Can. J. Res. B21, 179 (1943).
[115] R. Forquet, Materie Plastiche 21, 371 (1955).
[116] W. D. Todd, Off. Dig. Federation Paint & Varnish Prod. Clubs 24, 98 (1952).
[117] H. Whittaker, DSc. Thesis, MIT (1937).
[118] H. de Bruijn and P. G. Meerman, Proc. 1st Intern. Congr. Rheol., 1948 p. 000. North-Holland Publ., Amsterdam, 1949.
[119] R. V. Robinson, J. Phys. & Colloid Chem. 55, 455 (1951).
[120] E. K. Fischer, "Colloidal Dispersions." Wiley, New York, 1950.
[121] E. J. W. Verwey and J. de Baer, Rec. Trav. Chim. 57, 383 (1939).

very low concentrations such as 5% nitrocellulose in *n*-butyl acetate or acetone,[68] 1% borax added to a 10% aqueous solution of polyvinyl alcohol,[34] 30% vinyl resin in cyclohexanone,[77] and 2×10^{-3} g/ml polymethylmethacrylate in chlorinated diphenyl.[44] Hartly[47] reported dilatant behavior (with time dependence) for a 0.02% solution of the copper salt of cetylphenyl-ether sulfonic acid.

Pryce-Jones[77] believed dilatancy to be an essential property of all "spinnbar" systems (spinnbarkeit refers to the property which enables certain fluids to be drawn out into long threads or fibers) although all dilatant systems do not necessarily show this property. Dintenfass[63] shared this viewpoint and was also of the opinion that dilatant polymer solutions had a tendency to climb a rotating shaft or cylinder. In other words, Dintenfass extended the term to include the "Weissenberg effect" or normal stress development. This view is also shared by Reiner.[58] Dintenfass attributed the dilatant behavior of polymer solutions to aggregation of the resin molecules into a true network. Such a network is built by two- and three-dimensional anisotropic molecules, arranged into aggregates with at least one huge aggregate extending through the total volume of the sample. At high shear rates, this network can be destroyed releasing the solvent immobilized in the aggregates, resulting in what he termed dilatant rupture.

b Quantitative Observations

Perhaps the earliest quantitative study of dilatancy encompassing both the volume dilation and the rheological connotation was made by Andrade and Fox.[23] These authors studied the effect of loading, with a rigid piston, two-dimensional hexagonal arrays of uniform cylinders. They observed increase and decrease of the overall volume in this idealized system; the descending motion of the piston was explained in terms of a slip mechanism. Metzner and Whitlock[24] made a quantitative investigation of the relationship between volumetric dilation and shear rate thickening. They observed volumetric dilatancy quite independent of dilatancy in the rheological sense. When both phenomena were observed in a single system they found that volumetric dilation, as evidenced by surface drying, always appeared at shearing stresses well below those at which the flow curve began to show shear rate thickening. The volumetric dilation is readily explained on the basis of packing as originally proposed by Osborne Reynolds, although it is questionable that it must always appear at shearing stresses below those at which the system shows shear rate thickening. A momentum transfer process between adjacent layers or streamlines was suggested to account for the increase of viscosity with increases in shear rate. When one particle moves into an adjacent layer, all of the particles in that layer experience the change in velocity. Thus, as the shear rate is increased the particles become successively more perfectly aligned into laminae, with the result that progressively more particles experience momentum transfer with adjacent laminae whenever a single particle moves laterally. This mechanism suggests that dilatancy in a given system should occur at progressively decreasing rates of shear with increasing solids concentration since the combined movement of layers should occur more readily at higher solids concentration where there are necessarily fewer voids. In the limit represented by no liquid at all between the particles, that is, completely dry powder, cohesion is lost and the particles glide freely over one another, frequently on an air cushion, thus predicting the required low resistance to shear. If completely dry powders were subjected to compressive stresses equivalent to the internal pressure, the viscosity would go to extremely high values.

There are essentially two mechanisms of dilatancy, one associated with volume expansion and the other with network formation. It is quite clear that, in a solid or a close-packed suspension, the particles must first be separated before flow can occur, that is, the Poisson ratio drops below 0.5 at the dilatancy point. In dilute suspension, however, there is no volume expansion. In this case, the effect of shear is manifested in the formation of aggregates,[43,45,99,122-124] that is, enlarged flow units which can build a network with the consequent appearance of gel behavior. It is this mechanism that accounts for observing some dilatancy prior to a steep viscosity increase.

The field of flow, however, acts in two ways. It not only builds up structure through orientation and alignment, but it also breaks intermolecular bonds that are not sufficiently strong. Whenever there is more breaking than making of structure, we have the conditions for thixotropic behavior.

The occurrence of dilatancy without volume increase in dilute (2–4%) polymer systems can be explained as follows:[48] During shear, the collision frequency between groups of different polymer molecules increases, permitting the formation of local networks and thus gel. In this case, there is sufficient free volume in the surrounding liquid, in which the gel is dispersed, to accomodate the volume increase of the gel, and thus only an increase in viscosity is observed. However, when the gel structure or elastically connected dispersed phase grows to such magnitude that its volume becomes equal to that of the liquid, dilatant behavior (volume expansion and shear rate thickening) results.

[122] H. M. Taylor, S. Chien, M. I. Gregerson, and J. L. Lundberg, Nature 207, 77 (1965).
[123] M. Joly, Biorheology 2, 75 (1964).
[124] K. M. Beazley, Trans. Brit. Ceram. Soc. 64, 531 (1965); 63, 451 (1964).

Appendix I. Thixotropic Systems

Aqueous systems		Nonaqueous systems	
System	Ref.	System	Ref.
Starch	11, 29, 31, 125	Pigments in oil	71, 92, 93, 168–170
Bentonite clays	126–130	Oils	171, 172
Attapulgite	131, 132	Asphalts	173, 174
Montmarillonite	133, 134	Paints	54, 76, 85, 169, 175–179
Drilling muds	135, 136	Alkyl paints	180–182
Soils	137	Printing inks	54, 81, 183–187
Barium sulfate	138	Greases	188–190
Metallic oxide sols	1, 5, 9, 10, 39, 40, 139–143	Aluminum soap gels	77, 88, 191–195
Latex	144–147	Polymer solutions	68, 196
Biological fluids	148, 149	Polymer melts	92, 197–203
Blood	150–154	Silver amalgams	204
Flour dough	155, 156	Plastisols	115, 116, 205, 206
Mayonnaise	51, 78	Butter	207
Emulsions	98, 145, 157	Honey	51, 78
Gelatin	12, 158	Ultramarine-castor oil	71
Gums	78, 159, 160	Barium sulfate parafin oil	71
Pigments	29, 71, 92, 161	Paint vehicles	63, 64
Ceramics	162, 163	Polyesters	60
Cellulose dispersion	164, 165	Viscose resin	208
Concrete	58, 166	Rubber	209
Egg albumin	167		

[125] A. V. Senakhov and F. I. Sadov, *Kolloidn. Zh.* **21**, 692 (1959).

[126] E. A. Hauser and C. E. Reed, *J. Phys. Chem.* **41**, 911 (1937).

[127] J. N. Mukherjee and N. C. Sen Gupta, *Indian J. Phys.* **16**, 54 (1942).

[128] J. P. Singhol and W. U. Malik, *Rheol. Acta* **3**, 127 (1964).

[129] M. E. Schishniashvili and A. L. Avsarkisova, *Kolloidn. Zh.* **21**, 364 (1959).

[130] J. L. Russell, *Proc. Roy. Soc.* **A154**, 550 (1936).

[131] A. F. Gabrysh, E. Eyring, and I. Cutler, *J. Am. Ceram. Soc.* **4**, 334 (1962).

[132] A. F. Gabrysh, T. Ree, H. Eyring, and I. Cutler, *Trans. Soc. Rheol.* **5**, 67 (1961).

[133] K. H. Hiller, *Rheol. Acta* **3**, No. 3, 132 (1964).

[134] H. G. F. Winkler, *Kolloid-Z.* **105**, 29 (1943).

[135] F. W. Jessen and C. N. Toktar, *Petrol. Eng.* **32**, B48 (1960).

[136] W. F. Rogers, "Composition and Properties of Oil Well Drilling Fluids." Gulf Publ. Co., Houston, Texas, 1953.

[137] J. K. Mitchell, *J. Soil Mech. Found. Div., Proc. Am. Soc. Civil Eng'rs.* **SM3**, 19 (1960).

[138] B. Tamamushi and Y. Sekiguchi, *Bull. Chem. Soc. Japan* **13**, 556 (1958).

[139] E. W. J. Mardles, *Trans. Faraday Soc.* **36**, 1189 (1940).

[140] W. Heller and H. L. Roder, *Trans. Faraday Soc.* **38**, 191 (1942).

[141] W. Heller, *J. Phys. Chem.* **45**, 1203 (1941).

[142] R. Guthknecht, *Bull. Soc. Chim. France* p. 55 (1946).

[143] T. Henning, *Kolloid-Z.* **72**, 73 (1935).

[144] F. H. Cotton, *Rubber Age (London)* **20**, 127 (1939).

458   WALTER H. BAUER AND EDWARD A. COLLINS

[145] E. W. J. Mardles and A. de Waele, *J. Colloid. Sci.* **6**, 42 (1951).
[146] M. Mooney, *J. Colloid. Sci.* **1**, 195 (1946).
[147] K. Edelmann and E. Horn, *Gummi Asbest* **12**, 66 (1959).
[148] L. Dintenfass, *Proc. 4th Intern. Congr. Rheol.*, *1960* Part 4, p. 489. Wiley (Interscience), New York, 1965.
[149] R. Block and L. Dintenfass, *Australian New Zealand J. Surg.* **33**, 108 (1963).
[150] L. Dintenfass, *Biorheology* **1**, 91 (1963).
[151] L. Dintenfass, *Angiology* **13**, 333 (1962).
[152] L. Dintenfass, *Circulation Res.* **11**, 233 (1962).
[153] L. Dintenfass, *Nature* **199**, 813 (1963).
[154] L. Dintenfass, *Med. J. Australia* **1**, 575 (1963).
[155] M. C. Markley, *Cereal Chem.* **14**, 434 (1937).
[156] I. Hlynka and J. A. Anderson, *in* "Rheology," *Proc. 2nd Intern. Congr. Rheol, Oxford, 1953*, (Harrison, ed.), p. 297. Academic Press, New York, 1954.
[157] L. Ya. Kremnev, *Kolloidn. Zh.* **19**, 68 (1958).
[158] E. W. Billington, *Proc. Phys. Soc. (London)* **75**, 40 (1960).
[159] E. H. deButts, J. A. Hudy, and J. H. Elliott, *Ind. Eng. Chem.* **49**, 94 (1957).
[160] K. F. Zhigach, M. Z. Finkelstein, and I. M. Timokhin, *Dokl. Akad. Nauk SSSR* **126**, 1025 (1959).
[161] F. K. Daniel, *India Rubber World* **101**, No. 3, 50 (1939).
[162] D. G. Beech and M. Francis, *Trans. Brit. Ceram. Soc.* **44**, 25 (1945).
[163] S. R. Hind, *Trans. Brit. Ceram. Soc.* **61**, 463 (1962).
[164] H. Hoffmann and N. Bruch, *Cellulosechemie* **14**, 53 (1933).
[165] R. O. Herzog, O. Kratky, and E. Peteril, *Trans. Faraday Soc.* **29**, 60 (1933).
[166] M. Reiner, *Rheol., Theory Appl.* **3**, 341 (1960).
[167] W. G. Myers and W. G. France, *J. Phys. Chem.* **44**, 1113 (1940).
[168] D. G. Osborne and S. Thornton, *Brit. J. Appl. Phys.* **10**, 214 (1959).
[169] L. Dintenfass, *J. Oil & Color Chemists' Assoc.* **41**, 125 (1958).
[170] H. J. Jeffries, *J. Oil & Color Chemists' Assoc.* **45**, 681 (1962).
[171] R. N. Weltman, *Ind. Eng. Chem., Anal. Ed.* **15**, 424 (1943).
[172] V. L. Val'dman, *Zavodsk. Lab.* **11**, 1077 (1945).
[173] C. E. Coombs and R. N. Traxler, *J. Appl. Phys.* **8**, 291 (1937).
[174] R. N. Traxler, *in* "Bituminous Materials" (A. J. Hoiberg, ed.), Vol. 1. Wiley (Interscience), New York.
[175] E. K. Fischer, *J. Colloid Sci.* **5**, 271 (1950).
[176] A. X. Schmidt and C. A. Marlies, "Principles of High Polymer Theory and Practice." McGraw-Hill, New York, 1948.
[177] J. E. Arnold, *Paint Technol.* **9**, 77 (1944).
[178] D. J. Doherty and R. Hurd, *J. Oil & Color Chemists' Assoc.* **41**, 42 (1958).
[179] W. F. Daggett, *Paint Manuf.* **27**, No. 9, 336 and 357 (1957).
[180] D. M. James, *Paint. Manuf.* **27**, No. 11, 401 and 415 (1957).
[181] W. Gotze, *Rheol. Acta* **2**, 254 (1962).
[182] J. Janusek and J. Mleziva, *Chem. Prumysl* **14**, 184 (1964).
[183] A. C. Zettlemoyer and R. R. Myers, *Rheol., Theory Appl.* **3**, 000 (1960).
[184] A. Voet, "Ink and Paper in the Printing Process." Wiley (Interscience), New York, 1952.
[185] A. A. Trapeznikov and T. G. Shalopalkina, *Kolloidn. Zh.* **19**, 232 (1958).
[186] C. C. Mill, *J. Oil & Color Chemists' Assoc.* **43**, 77 (1960).
[187] G. W. Whitfield, *Paint Manuf.* **9**, 80 (1939).
[188] S. J. Hahn, T. Ree, and H. Eyring, *NLGI Spokesman* **21**, 12 (1957).
[189] H. Utsugi, K. Kim, T. Ree, and H. Eyring, *NLGI Spokesman* **25**, 125 (1961).

Appendix II. Dilatant Systems

Aqueous systems		Nonaqueous systems	
System	Ref.	System	Ref.
Starch	31, 77, 78, 114, 120	Starch in ethylene glycol	28, 43
Zinc oxide	28	Polymer solutions	44, 51, 68, 77, 109
Pigments (CaCO$_3$, BaSO$_4$)	25, 120, 174, 175, 210	Calcium stearate in oil	69
Red FeO	28	Plastisols	54, 115, 116, 124
Smalts	71	Calcium naphthenate-white spirit	77, 78
Gum arabic	51, 77	Honey	77
Emulsion paint	78	Paints	35, 175
Concrete	58	Printing inks	35, 184
Latex	211	Ultramarine-castor oil	71
		BaSO$_4$ in polymerized oil	71
		Paint vehicles	63

[190] K. I. Klimov and B. I. Leont'ev, *Khim. i Tekhnol. Topliv i Mesel* **5**, 17 (1960).

[191] W. H. Bauer, N. Weber, and S. E. Wiberley, *J. Phys. Chem.* **62**, 106 (1958).

[192] T. G. Shalopalkina and A. A. Trapeznikov, *Kolloidn. Zh.* **22**, 735 (1960); **25**, 722 (1963).

[193] A. A. Trapeznikov, *Rheol. Acta* **1**, 617 (1961).

[194] A. A. Trapeznikov, *Kolloidn. Zh.* **23**, 626 (1961).

[195] T. I. Zatsepina and A. A. Trapeznikov, *Kolloidn. Zh.* **23**, 690 (1961).

[196] P. Rehbinder and L. I. Tschumakova, *Z. Physik. Chem.* (*Leipzig*) **209**, 1 (1958).

[197] E. Gaspar and M. G. Munns, *Brit. Plastics* **36**, 458 (1963).

[198] K. Komura, Y. Todani, and N. Nagata, *Polymer Letters* **2**, 643 (1964).

[199] H. A. Pohl and J. L. Lund, *SPE J.* **15**, 390 (1959).

[200] H. A. Pohl and C. G. Gogos, *J. Appl. Polymer Sci.* **5**, 67 (1961).

[201] R. Buchdahl, *J. Colloid Sci.* **3**, 87 (1948).

[202] J. Diennes and F. D. Dexter, *J. Colloid Sci.* **3**, 181 (1948).

[203] J. Galt and B. Maxwell, *Mod. Plastics* **42**, 115 and 189 (1964).

[204] D. R. Hudson, *Metallurgia* **29**, 207 (1944).

[205] S. K. Khanna and W. F. O. Pallett, *J. Appl. Polymer Sci.* **9**, 1767 (1965).

[206] W. D. Todd, *in* "High Polymers" (C. E. Schildknecht, ed.), Vol. X, Polymer Processes, Chap. 14, p. 551. Wiley (Interscience), New York, 1956.

[207] F. Fukada, T. Sone, and M. Fukushima, *J. Japan. Soc. Test. Mater.* p. 43 (1961).

[208] D. N. E. Cooper, *J. Textile Inst. Trans.* **53**, 560 (1962).

[209] H. Freundlich, *Trans. Inst. Rubber Ind.* **11**, 55 (1935).

[210] T. H. Daugherty, *J. Chem. Educ.* **25**, 482 (1948).

[211] C. Cawthra, G. P. Pearson, and W. R. Moore, *Trans. Plastics Inst.* (*London*) **33**, 39 (1965).

CHAPTER 9

RHEOLOGICAL TERMINOLOGY

M. Reiner and G. W. Scott Blair

> "When *I* use a word," Humpty Dumpty said in rather a scornful tone, "it means just what I choose it to mean—neither more nor less." "The question is," said Alice, "whether you can make words mean so many different things." "The question is," said Humpty Dumpty, "which is to be master,—that is all."
> —LEWIS CARROLL, *Alice Through the Looking Glass*

I. Introduction

Nothing can cause more misunderstanding and confusion in scientific communications than a lack of agreement over the use of technical terms. Although various reports and lists of rheological terms have appeared from time to time, no complete or generally accepted system has been published.

There are two ways in which scientific terms come to be defined. The first is through the attachment of a precise meaning to an ordinary word of everyday speech (one of its connotations is made more specific), whence it is arbitrarily agreed to restrict its use in technical communications to that meaning. For example, such terms as *force, energy, power, stress*, and *strain* were at one time used indiscriminately by scientists, as they still are by the general public, but now are carefully defined. The second way is through the invention of a term, usually out of Greek or Latin elements (scholars abhor combinations of the two) and, again, the defining of it precisely. In both cases it is helpful if it is possible to express the measure of the quantity which the word represents in units of mass, length, and time [M, L, T].

The rules to be followed in proposing precise definitions differ somewhat in the two cases.

In the first the meaning should not conflict with the definitions given in an ordinary standard dictionary, and it should, if possible, be one of them. From then on the question is one of usage. If a term has been used by scientists in contradictory senses, a decision must be made about which sense is the most widely understood, but it is clearly quite in order to make an apparently arbitrary decision.

In the second case the inventor's definition should be retained, unless he himself has expressed a wish to modify it, or unless it has found that the concept for which the term was proposed was in some way misconceived or meaningless. We shall later see two instances of such modification when we discuss the terms "thixotropy" and "rheopexy" (Section II).

Sometimes by chance the same term has been independently applied by different authors to quite different things (*stiffness* is a case in point). Then it seems best to drop one or both of the usages and, if necessary, introduce a new term.

Translations cause even greater difficulties. Sometimes quite subtle distinctions exist between apparently obvious synonyms, such as *Spannung* in German and *stress* in English, and *solicitation* in French and *traction* in English. Moreover, it is dangerous to make literal translations. Thus, in America *lubricity* is often used for "oiliness," that is, effectiveness as a lubricant, in England the word has this meaning but also another, and in France *lubricité* has too unpleasant a meaning in ordinary speech to be introduced in rheology.

Words describing rheological properties are, in general, of two kinds. They were once described by one of us[1] as "connotative" and "denotative," but because these terms have themselves changed in meaning in the course of time, for our present purpose we shall use the terms "measurable" and "assessable." Measurable properties can be represented by symbols and generally given dimensions [M, L, T]. But assessable properties, too, it is sometimes forgotten, may be given names, and indeed, we have a need of names for them. The rheologist would be hard put to it if he had no such terms as *body, consistency, printability,* and *covering power,* even though it is his hope to reduce their meanings eventually (as has apparently been done with *tack*)[1a] to measurable concepts.

Assessable properties are usually judged by experts handling materials, who score as "similar" those which seem to them to behave similarly. Psychorheology[2] attempts to relate such judgements to measurable physical properties with definite meanings.

In the present state of rheology it is generally a mistake to restrict a term of assessability (such as "consistency") to a specific meaning (such as variable viscosity), as was once done, unless it can be shown that the assessable property to which it commonly refers can fairly be so defined.

Below we discuss a few of the more controversial terms in detail. In compiling the Dictionary, Section III, we have attempted to follow the principles

[1] G. W. Scott Blair, *In* "Some Recent Developments in Rheology," British Rheologist's Club (V. G. W. Harrison, ed.), p. 105. United Trade Press, London, 1950.

[1a] J. D. Skewis., *Rubber Chem. and Tech.* **38,** 689 (1965).

[2] G. W. Scott Blair, "A Survey of General and Applied Rheology," 2nd ed., Pitman, New York, 1949.

outlined above. A list of proposed symbols is given in the last section, and a list of sources (not always followed faithfully) is given at the end of the chapter.

II. Discussion of Some Individual Definitions

Anelasticity. The prefixes *an* (Greek) and *in* (Latin) have the same meaning, "not." The word *elastic* is derived from Greek through Latin. Zener[3] has proposed this special use of the Greek *an*, which, although linguistically not correct, is, we feel, justified (after all, the Greek *rheo* really should not refer to elastic deformations!).

Breakage, rupture, fracture. We are advised that there are subtle distinctions of meaning between these terms, but we have not attempted here to distinguish them.

Capillary. Hemorheologists must be careful to distinguish between the rheologist's use of this term to mean a rather narrow tube of even bore, usually cylindrical, and long compared with its diameter, and the hematologist's use of it to mean a very small blood vessel.

Crowding factor, self-crowding factor. As a general rule we have included in our dictionary only terms that have been used rather widely; these two have not, but we feel that the concept of the crowding factor, named by Mooney[4] but first, in fact, proposed by Vand,[5] is likely to find increasing use in the study of the relation between viscosity and concentration in fairly concentrated polymers.

Dilatancy. See below, *Shear thinning, shear thickening.*

Dynamic viscosity. Various authorities, including a British Standards Institute Committee, authorize the use of this word to mean ordinary viscosity $[ML^{-1}T^{-1}]$ as distinct from kinematic viscosity $[L^2T^{-1}]$. We have ignored this use, however, because we feel that its more widely adopted meaning, applying to sinusoidal conditions, is the more important.

Inherent viscosity. Wagner[6] has proposed that this term means the value of reduced viscosity at an arbitrary selected finite concentration. It appears to us, however, that its use for specifying the parameter of the Arrhenius equation is the more important. See also Sources for more recent definitions.

[3] C Zener and J. H. Hollomon, *J. Appl. Phys.* **17**, 69 (1946).
[4] M. Mooney, *J. Colloid Sci.* **6**, 1962 (1951).
[5] V. Vand, *J. Phys. Colloid Chem.* **52**, 277 (1948).
[6] R. H. Wagner, *Anal. Chem.* **20**, 155 (1948).

Linear phenomena, nonlinear phenomena. One of us (G. W. S. B.) recently attended a symposium on Nonlinear Relaxation and Creep and took the opportunity to ask a number of distinguished rheologists to define "nonlinear." There was so much disagreement that we were somewhat hesitant about offering a definition—but we have done our best (cf. Dictionary).

Rate. This has many meanings, from "rate of interest" in everyday business to "rate of increase of *x* with *y*" in physics. In rheology, however, the term is so seldom used to imply any differential other than that with respect to time that we have limited it to this sense and recommended the use of "Newton's dot" to indicate it (Newton's dot is a dot over the symbol).

Rheomalaxis. Hoeppler[7] has proposed the term *Rheodestruktion* in German for the nonrecoverable breakdown of consistency of materials under shear. Because mixed Latin and Greek roots in English words offend classical scholars, we have preferred the term "rheomalaxis" ("softening in flow").

Shear thinning, shear thickening. We have derived these terms from the report of Burgers and Scott Blair (see Sources at the end of this chapter), in which "shear-*rate* thinning and thickening" are in general employed. We have shortened their terms (and omitted a few others offered in that report, which appear not to have proved useful during the years). We strongly insist that *dilatancy* be reserved for materials that are known to dilate when sheared. (See below, *Structural viscosity* and *Thixotropy, rheopexy*.)

Stiffness. This word has been given many meanings in rheology. We hope that our friends in the paper and metal-spring trades, in which it has special meanings, will excuse our preference for its use as a variable Young modulus.

Stress, strain. These are tensors, but it is hard to refute the common habit of referring to tractions and to strain components by these terms when no possibility of confusion exists.

Structural viscosity. This term requires some explanation. It has been our policy throughout the making of our dictionary to choose definitions that define phenomena without implication of any theories about their origins. When Ostwald introduced the term *Strukturviskositaet*, he certainly thought in terms of a postulated breaking down of structures within a material. Moreover, at first sight it might seem that the English equivalent, "structural viscosity," is no more than a synonym for *shear thinning*. There are occasions, however, when the latter term, although strictly correct, would be misleading such as in an experiment in which the rate of shear of, say, a damped pendulum is progressively falling and the consistency correspondingly rising. We base

---

[7] F. Hoeppler, *Kolloid Z*. **98**, 348 (1948).

our retention of "structural viscosity" on the fact that it is the viscosity, and not necessarily the material, that has a structure and would support this by quoting the Shorter Oxford Dictionary's third definition of *structural*: "Of or pertaining to the arrangement and mutual relation of the parts of any complex unity."

Thixotropy, rheopexy. These terms were invented by Freundlich and his pupils. *Thixotropy* originally referred only to "a sol-gel transformation," and it was implied that the sol was at least approximately a true fluid and the gel an elastic solid. It is used rarely in this sense, and in a personal conversation with Professor Freundlich just before his death one of us (G. W. S. B.) had his definite approval of an extension of the meaning of the term (see the Dictionary, Section III). He would *not* approve, however, its synonymity with *shear thinning*, which was proposed by Goodeve and Whitfield[8] and which is still occasionally found (e.g., in Dintenfass[9]).

The difficulty with *rheopexy* is much more serious. This term was clearly defined by Juliusburger and Pirquet[10] as "the solidification of *thixotropic* sols by gentle and regular movement" (our italics). The term should not be used for nonthixotropic systems. In recent papers—alas, too numerous to quote—it has been used with quite a number of other meanings, for all of which alternative terms are available in the Dictionary, Section III. We hope that it is not too late to correct this unfortunate confusion. (See above, *Shear thinning, shear thickening.*)

Viscoelasticity. This term is widely and wrongly used to describe the elastic behavior of a liquid, for which the correct term is *elasticoviscosity*, introduced by Jeffreys.[11] *Viscoelasticity* should be reserved for a solid exhibiting a resistance to its deformation that increases with the rate thereof.

These remarks may serve as the starting point of the exposition of a system of ideal bodies, first suggested by von Mises,[12] and later worked out in detail by one of us.[13] One cannot rigorously define real substances but must be content to define concepts that can form bases of mathematical treatment. One cannot correctly say "a rock is a solid" or "a rock is a liquid," because every material can show various forms of rheological behavior, depending on the circumstances. One may, however, say "a Hookean body is a solid" and "a Newtonian material is a liquid," where "Hookean body" and "Newtonian material" are abstract concepts defined by their respective

[8] C. F. Goodeve and G. W. Whitfield, *Trans. Faraday Soc.* **34**, 511 (1938).

[9] L. Dintenfass, *Nature* **197**, 496 (1963), and other papers.

[10] F. Juliusburger and A. Pirquet, *Trans. Faraday Soc.* **32**, 445 (1936).

[11] H. Jeffreys, "*The Earth.*" Cambridge Univ. Press, London and New York, 1924.

[12] R. von Mises, *Proc. 3rd Intern. Congr. Appl. Mech., Stockholm, 1930*.

[13] M. Reiner, *J. Sci. Instr.*, **22**, 127 (1945).

rheological equations. The equations may be represented by mechanical models, a helical spring serving for the Hookean body, a dashpot for the Newtonian material, and a slider for the St. Venant plastic. Such model units may be connected either in series (—) or in parallel (|) to form models of more complicated materials. When two bodies are connected in series, the rates of deformation are additive; when they are connected in parallel, the stresses are additive. A tree of the "rheological bodies" proposed so far is shown in Fig. 1.

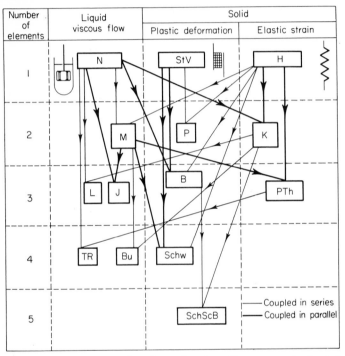

FIG. 1. Tree of rheological bodies. (For explanation of letters see under Symbols.)

III. Dictionary

The English term is printed in italics. It is followed by translations in French, German, and Russian (in that order). Translations are not given when the foreign terms are almost identical with the English. The letters in parentheses, m, f, and n, indicate the gender of the foreign noun. Dimensions and symbols are given after the definition.

abhesion   the spontaneous relinquishing of an interface

absolute dynamic modulus   the ratio of amplitude of stress and strain under sinu-soidal stress in the absence of initial forces

activation energy   the increase in the energy of a unit of mass to the level at which it will undergo a given process

active earth pressure   the total force exerted by or through a mass of earth
poussée des terres (f)
activer Erddruck (m)
Активное давление земли (n)

adhesion   the force that resists the separation of two bodies in contact
adhésion (f);   adhérence (f)
Adhäsion (f)
Адгезия (f);   сцепление (n)

adiabatic   taking place without transfer of heat

after-effect ≡ *elastic after-effect*, q.v.

after-working ≡ *after-effect*, q.v.

aggregate   a group of particles held together in clumps by weak forces or by being in a matrix

aging   a change in properties occurring gradually in time
vieillissement (m)
Altern (n)
Старение (n)

amorphous   without crystalline structure

analytical approach   consideration of complex rheological behavior in terms of the interaction of a number of simple prototypes
méthode analytique (f)
analytische Betrachtungsweise (f)
Аналитический метод рассмотрения (m)

anelasticity   delayed elasticity, especially of the linear viscoelastic solid; to be distinguished from inelasticity (see *Anelasticity*, Section II)

angle of extinction   in flow birefrigence and in Couette flow the angle between the direction of greatest extinction of light and the direction of polarization; $[L°]$; χ
angle d'extinction (m)
Auslöschungswinkel (m)
Угол угасания (m)

angle of friction   the angle whose tangent is equal to the coefficient of friction; $[L°]$; ϕ

angle de frottement (m)
Reibungswinkel (m)
Угол трения (m)

angle of isocline   the complement of the angle of extinction; $[L°]$; ϕ

angle of repose   the maximal natural angle at which a pile of granular material will lie without moving
angle de talus naturel (m)
Böschungswinkel (m)
Угол сползання (m); угол естественного откоса (m)

angular velocity   the time derivative of an angle; $[T^{-1}]$; $\omega, \dot{\Omega}, \dot{\theta}$
vitesse angulaire (f)
Winkelgeschwindigkeit (f)
Угловая скорость (f)

annealing   the stabilization of the internal structure of a solid substance, such as glass or metal, by protracted heating followed by slow cooling
recuit (m)
Ausglühen (n)
Отжиг (m)

anomalous viscosity   a viscosity that does not remain constant when the rate of shear is changed
viscosité anormale (f)
variable Viskosität (f)
Аномальная вязкость (f)

anti-thixotropy   comparatively slow fall, on standing of the sample, of a consistency that was gained as a result of shearing

apparent viscosity   a coefficient calculated from empirical data as if Newton's law held, when the coefficient is not a constant; $[ML^{-1}T^{-1}]$; η'
viscosité apparente (f)
scheinbare Viskosität (f)
Кажущаяся вязкость (f)

attenuation coefficient   in a damped oscillation the reciprocal of the distance within which the amplitude falls off to a fraction of $1/e$; $[L^{-1}]$; α
coefficient d'amortissement (m)
Dämpfungskoeffizient (m)
Коэффициент затухания (m)

axial force   a force acting in the direction of the axis of a prism or a cylinder; P_l, N
force axiale (f)
Axialkraft (f)
Осевая сила (f)

bar   the unit of traction; measured in dynes per square centimeter $[ML^{-1}T^{-2}]$

Barus effect ≡ *melt swell*   the swelling of an extrudate on its emergence from a tube or an orifice

Bauschinger effect   a lowering of the yield point when the deformation is reversed

bending stress   the normal stress in the cross section of a bent beam; $[ML^{-1}T^{-2}]$; σ_B
contrainte de flexion (f)
Biegungsspannung (f)
Напряжение изгиба (n)

Bingham body   a Newtonian fluid with yield value, whose structural formula is $H(StV|N)$; B

biorheology   the rheology either of living materials or of materials from living sources studied for their biological interest

body   an assessable property of the consistency of a material
corps (m)
Körper (m)
Тело (n)

body force ≡ *mass force*   a force proportional to the mass of a body, such as gravity
force de masse (f)
Massenkraft (f)
Массовая сила (f)

Boltzmann's superposition principle   If a deformation is the sum of various deformations, each of which may be an arbitrary function of time, the stress at any time will be the sum of the stresses that would have been caused by each of the deformations separately

boundary condition   the tractions or displacements prescribed on the boundary of a body

condition à la limite (f)
Randbedingung (f)
Граничное условие (n)

breakage ≡ *rupture*   the appearance of a surface of discontinuity in the course of the deformation of a body (see *Breakage, rupture, fracture*, Section II)
rupture (f)
Bruch (m)
Разрушение (n);   разрыв (m)

breakage by separation ≡ *cleavage breakage* ≡ *fracture*   a breakage occurring in a plane perpendicular to the greatest principal strain
rupture par décohésion (f)
Trennungsbruch (m)
Хрупкое разрушение (n);   разрыв (m)

breaking energy ≡ *breaking resilience*   energy consumed per unit area of stress surface; R_b
énergie de rupture (f)
Bruch-Energie (f)
Энергия разрушения (f)

breaking resilience ≡ *breaking energy*, q.v.

breaking strength   see *strength*

brittle   tending to break under the condition of minimal previous plastic deformation
fragile
spröde
Хрупкий

bulk modulus ≡ *modulus of volume expansion*, q.v.

bulk viscosity   the viscosity of volume flow; ζ
viscosité volumique (f)
Volumenviskosität (f)
Объемная вязкость, вязкость нзменения объема (f)

Burgers body   an ideal material, whose structural formula is K–M; Bu

Cauchy strain   a strain whose principal component is an extension divided by the original length
déformation relative de Cauchy (f)
Cauchy-Verzerrung (f)
Деформация по Коши (f)

cavitation the appearance of cavities or bubbles in a liquid subjected to high stresses

churning an increase in consistency, on shearing of a material, which is not lost by subsequent resting

baratage (m); rhéoépaississement non reversible (m)

Butterung (f)

Упрочение при сдвиге (n)

circle of stress ≡ *Mohr's circle*, q.v.

cleavage controlled opening (extending) of a wedge-shaped kerb in a specimen

cleavage breakage ≡ *breakage cleavage*, q.v.

coacervation the formation of new liquid phases from destabilized, solvated macromolecules or their aggregates

coefficient of cross viscosity the second-order viscosity coefficient of a Reiner–Rivlin fluid; $[ML^{-1}]$; η_c

coefficient de viscosité transversale (m)

Koeffizient der Querviskosität (m)

Коэффициент поперечной вязкости (m)

coefficient of fluidity reciprocal of the coefficient of viscosity; measured in rhe's; $[M^{-1}LT]$; φ

coefficient de fluidité (m)

Fluiditätskoeffizient (m)

Коэффициент текучести (m)

coefficient of friction the ratio of the frictional resistance between two bodies in contact to the pressure force normal to the surface of contact; $[L°]$

coefficient de frottement (m)

Reibungskoeffizient (m)

Коэффициент трения (m)

coefficient of viscosity ≡ *viscosity coefficient* the ratio of the shearing stress to the rate of shear in laminar motion; $[ML^{-1}T^{-1}]$; η

coefficient de viscosité (m)

Viskositätskoeffizient (m)

Коэффициент вязкости (m)

coefficient of viscous traction ≡ *Trouton's coefficient* the ratio of the tensile stress to the rate of extension; $[ML^{-1}T^{-1}]$; λ_T

coefficient de viscosité axial (m)

Viskositätskoeffizient bei Dehnung (m)

Коэффициент вязкости при растяжении (m)

cohesion the resistance of solid and liquid materials to cavitation and to external forces tending to separate parts of the body

cohesive-energy density cohesive energy per unit volume (latent heat of evaporation per unit volume); energy changes encountered in the division of matter, often equal to twice the change in surface energy; CED

cohesive-energy density parameter square root of cohesive-energy density

cold flow ≡ *creep*, q.v.

écoulement à froid (m)

Kriechen (n)

Хладотекучесть (f)

complex compliance a complex quantity, of which the storage compliance is the real part and the loss compliance the imaginary part; J^* (in shear), B^* (in bulk); D^* (in extension)

compliance complexe (f)

komplexe Nachgiebigkeit (f)

Комплексная податливость (f)

complex modulus the complex sum of moduli, of which the storage modulus is the real part and the loss modulus the imaginary part; G^* (in shear), K^* (in bulk); E^* (in extension)

complex viscosity the complex modulus divided by the frequency in an experiment with sinusoidal time dependence; η^*

compliance ration of a strain to its corresponding stress; $[M^{-1}LT^2]$

compliance (f)

Nachgiebigkeit (f)

Податливость (f)

compressibility the reciprocal of the bulk modulus; $[M^{-1}LT^2]$; κ

compressibilité (f)

Kompressibilität (f)

Сжимаемость (f)

consistency an assessable the property of a material by which it resists permanent change of shape, defined by the complete stress-flow relation

consistency curve a curve of rate of shear versus shearing stress in a deformation

constitutive equation ≡ *rheological equation*, q.v.

Couette flow the rotational shearing flow in the annulus between two coaxial cylinders
écoulement de Couette (m)
Couette-Strömung (f)
Течение по Куэтту (n)

crazing the spontaneous formation of cracks or lower density regions, following volume changes in the drying out of colloidal materials
craquelure (f)
Schrumpfreissen (n)
Растрескивание (n)

creaming the rising to the surface of particles of the dispersed phase of dispersion
crémage (m)
Aufrahmung (f)
Отстаивание (n)

creep ≡ *cold flow* the deformation of a stressed solid proceeding in time (see *primary c., secondary c.*, and *tertiary c.*)
fluage (m)
Kriechen (n)
Ползучесть (f)

creep function the deformation as a function of time, when a unit stress is applied instantaneously at zero time and kept constant thereafter; $J(t)$ (in shear), $B(t)$ (in bulk), $D(t)$ (in extension)
fonction de fluage (f)
Nachgiebigkeit bei Kriechen (f)
Податливость ползучести (f)

cross elasticity in simple shear an elastic response in the direction of the trace components of the strain tensor
élasticité transversale (f)
Quer-Elastizität (f)
Поперечная упругость (f)

cross stresses normal stresses associated with simple shear
contraintes transversales (f)
Querspannung (f)
Поперечные напряжения (n)

cross viscosity in simple shear a viscous response in the directions of the trace components of the flow tensor
viscosité transversale (f)
Quer Viskosität (f)
Поперечная вязкость (f)

crowding factor ≡ *self-crowding factor* in a modified Arrhenius equation a factor that allows for the reduction in volume of the continuous phase of a suspension of high concentration (see *crowding factor, self crowding actor*, Section II)
facteur d'encombrement (m); facteur d'auto-contraction (m)
Selbstfüllungseffekt (m)
Фактор самонаполнения (m)

cubical dilatation ≡ *volume dilatation*, q.v.

cyboma an ordered structure of molecules forming a temporary association during the flow of a liquid

cybotaxis the formation of cybomas in liquids

damped oscillation an oscillation whose amplitude diminishes with time
oscillation amortie (f)
gedämpfte Schwingung (f)
Затухающее колебание (n)

damping viscosity a viscosity that delays elastic response

dashpot a model for a viscous element consisting of a piston sliding in a cylinder filled with a true liquid
amortisseur (m)
Dämpfer (m); Dämpfungselement (m)
Поршень (m)

deformation change of shape, volume, or both, of a body (see *relative deformation*); [$L°$]; D_{lm}, d_{lm}

deformational nonlinearity in a rheological equation when the kinematic argument is not a linear function of the gradients

non-linéarité de déformation (f)
Nicht-Linearität der Deformation (f)
Нелинейность деформационного процесса (f)

delayed elasticity damped elastic response
élasticité retardée (f); élasticité différée (f)
verzögerte Elastizität (f)
Задержанная упругость (f)

deviator a symmetrical tensor of zero trace; $\overset{\circ}{t}$

differential viscosity the derivative of stress
with respect to the rate of deformation
in a consistency curve; $[ML^{-1}T^{-1}]$;
η_Δ
viscosité différentielle (f)
Ableitungsviskosität (f); differentielle Viskosität (f)
Дифференциальная вязкость (f)

dilatancy an increase in volume caused by
shear (see *Shear thinning, shear thickening,* Section II)

dispersion usually a dilute two-phase or multiphase system

displacement the vector defining the change
of position of a point; $[L]$; $\mathbf{u}$, u_l; u_x,
u_y, u_z
déplacement (m)
Verschiebung (f)
Перемещение, смещение (n)

displacement gradient the derivative of a displacement of a particle in relation to
its coordinates; $[L^\circ]$; Γ_{lm}, γ_{lm}
gradient de déplacement (m)
Verschiebungsgradient (m)
Градиент смещения $u \sim u$ перемещения (m)

dissipation coefficient the exponent of time
in an equation relating stress, strain,
and time by means of powers; $[L^\circ]$;
k

dissipation of energy the conversion of
applied or stored mechanical energy
into heat

distortion change of shape of a body at
constant volume
distorsion (f)
Formänderung (f)
Искажение, формоизменение (n)

ductility the extent of relative plastic extension at fracture
ductilité (f)
Zähigkeit (f)
Предельная пластическая растяжнмость (f)

dynamic friction a friction occurring when
there is a finite motion between two
contiguous surfaces of bodies
frottement dynamique (m)
gleitende Reibung (f)
Динамическое трение (n)

dynamic viscosity the real component of the
complex viscosity (see *Dynamic viscosity,* Section II)

efflux viscometer a viscometer that discharges the test material through an
orifice
viscosimètre à écoulement (m)
Ausfluss-Viskometer (n)
Вискозиметр на истечение (m)

elastic after-effect ≣ *after effect* ≡ *after-working* the delayed recovery from an
elastic strain
recouvrance élastique déférée (f)
elastische Nachwirkung (f)
Упругое последействие разгрузки (n)

elastic fore-effect the delayed formation of
an elastic strain
déformation élastique déférée (f)
elastischer Verzögerungseffekt (m)
Упругое последействие нагрузки (n)

elastic hysteresis the appearance of a loop in
a closed stress–strain cycle
hystérésis élastique (m)
elastische Hysteresis (f)
Упругий гистерезис (m)

elasticity the property of a material by virtue
of which, after deformation and upon
removal of stress, it tends to recover
part or all of its original size, shape, or
both

elastic limit the limit of the purely elastic
part of a stress–strain curve
limite d'élasticité (f)
Elastizitätsgrenze (f)

elastic liquid a liquid showing elastic properties on shearing at a finite rate

liquide élastique (m)
elastische Flüssigkeit (f)
Упругая жидкость (f)

elastic modulus ≡ *modulus of elasticity*, q.v.

elasticoviscosity   viscous flow accompanied by elastic behavior

elastoplastic system   a plastic system showing some elasticity
système élastoplastique (m)
elastoplastischer Körper (m)
Упруго-пластическая система (f)

electroviscous effect   the increase in viscosity, produced in various ways, resulting from electric charges on dispersed particles
effet électro-visqueux (m)
elektroviskoser Effekt (m)
Электро-вязкость (f)

elongation   increase in length; [L]; Δl
allongement (m)
Verlängerung (f)
Удлинение (n)

emulsion   the dispersion of a liquid in a liquid

endurance limit ≡ *fatigue limit*

epibolic stress   the part of a stress that disappears in relaxation
partie transitoire de la contrainte (f)
epibolische Spannung (f)
Эпиболическое напряжение: релаксирующее напряжение (n)

epsilon potential   the potential on a colloidal particle in a liquid, measured by treating the particle itself as an electrode; ε

equation of state   an equation that relates parameters defining a system
équation d'état (f)
Zustandsgleichung (f)
Уравнение состояния (f)

Euclid solid   an ideal, completely rigid, material

extension ≡ *linear dilatation*   the elongation per unit length; [L°]
extension (f); allongement unitaire (m)
Dehnung (f)
Относительное удлинание (n)

fatigue   a decrease in the breaking strength of a material induced by the application of stress cycles
fatigue (f)
Ermüdung (f)
Усталость (f)

fatigue limit ≡ *endurance limit*   onset of rapid deterioration through fatigue

firmoviscosity ≡ *viscoelasticity*, q.v.

flow   a nonrecoverable deformation proceeding in time; [T⁻¹]; f_{lm}
écoulement (m)
Fliessen (n); Fluss (m)
Течение (n)

flow birefringence   birefringence observed usually in a dispersed system, only when it is flowing
birefringence d'écoulement (f)
Strömungsdoppelbrechung (f)
Двупреломление в потоке (n)

flow condition   an equation defining the yield value
condition d'écoulement (f)
Fliessbedingung (f)
Условие течения (n)

flow elasticity   the ability of a liquid to recover part of the deformation undergone during flowing
élasticité d'écoulement (f)
Fliesselastizität (f)
Упругость в потоке (f)

fluid   a liquid or a gas
fluide (m)
fluid (n)
Текучая среда (f)

foam   a coherent system of open or closed cells filled with gas

fracture ≡ *breakage by separation*   (see also *Breakage, rupture, fracture*, Section II)
rupture (f)
Bruch (m)
Разрыв (m); излом (m); разрушение (n)

friction   the resistance to the relative motions of two contiguous bodies, acting at the surfaces of contact

frottement (m)
Reibung (R)

frozen stresses and strains localized stresses and strains within a body in the absence of external forces

gel a dispersion containing jellylike particles

generalized Newtonian liquid a viscous liquid, having tensorial linearity, whose viscosity is not a constant; sometimes called non-Newtonian liquid, q.v.
liquide de Newton généralise (m)
verallgemeinerte Newtonsche Flüssigkeit (f)
Обобщенная ньютоновская жидкость (f)

hardness the resistance of the surface of a body to penetration
dureté (f)
Härte (f)
Твердость (f)

hemorheology the study of the rheological properties of cellular and plasmatic components of blood in macroscopic, microscopic, and submicroscopic dimensions and of the vessel structure with which blood comes into direct contact

Hencky flow condition the occurrence of plastic yielding when the resilience of a material has been reached
condition d'écoulement de Hencky (f)
Hencky–Fliessbedingung (f)
Условия течения по Генки (n)

Hencky strain ≡ *logarithmic strain* ≡ *natural strain* a measure of strain equal to the logarithm of the stretch
déformation de Hencky (f); déformation logarithmique (f)
Hencky–Verzerrung (f)
Деформация по Генки (f)

hesion attractive forces exerted by matter as in adhesion and cohesion

high elasticity elastic deformations too large to be sustained by stretching of the atomic structure; often, in common parlance, elastic strains greater than about 5%
haute élasticité (f)

Hochelastizität (f)
Высоко-эластичность (f)

homogeneous independent of the coordinates

Hooke body ≡ *Hookean solid* an ideal elastic solid for which the shearing stress is proportional to the shear, straining and recovery taking place near sonic speeds; H

Hookean solid ≡ *Hooke body*, q.v.

Hydrostatic stress ≡ isotropic stress

hyperelasticity in classical theory, finite elastic strain defined by a strain–energy function

hypoelasticity a theory in which stressing is determined by currently existing stress and strain

hysteresis return of a function to its origin by a different path

ideally elastic behavior time-independent elasticity
comportement élastique idéal (m)
ideal-elastisches Verhalten (n)
Идеально упругое поведение (n)

inelastic nonelastic
non-élastique; inélastique
unelastisch
Неупругий

infinitesimal deformation a relative deformation that is so small that orders of it higher than the first can be neglected as less than experimental error

inherent viscosity the logarithm of the relative viscosity divided by the weight concentration (see *Inherent viscosity*, Section II); $[M^{-1}L^3]$
viscosité inhérente (f)
logarithmische Viskositätszahl (f)
Приведенная вязкость или логарифмическая приведенная вязкость (f)

initial stress external stress present before the start of an imposed deformation
contrainte initiale (f)
Vorspannung (f)
Начальное напряжение (n)

instantaneous deformation a deformation that proceeds without damping
déformation instantanée (f)
instantane Deformation (f)
Мгновенная деформация (f)

instantaneous set the nonrecoverable deformation that appears immediately on loading
déformation permanente instantanée (f)
augenblickliche permanente Deformation (f)
Мгновенная остаточная деформация (f)

integrative approach consideration of complex rheological behavior as intermediate between that of a number of simple prototypes
méthode intégrative (f)
integrative Betrachtungsweise (f)
Обобщающее приближнние (n)

interaction coefficient coefficient of second-order and higher-order terms in a power expansion of the specific viscosity as a function of the concentration
coefficient d'interaction (m)
Wechselwirkungskoeffizient (m)
Коэффициент взаимодействия (m)

internal friction ≡ *plastic resistance*, q.v. (see also *solid viscosity*)

intrinsic viscosity ≡ *limiting inherent viscosity* ≡ *viscosity number* reduced viscosity extrapolated to zero concentration, specific-viscosity volume; $[L^3 M^{-1}]$; $[\eta]$
viscosité intrinsèque (f)
Viskositätszahl (f)
Характеристическая вязкость (f)

inviscid ≡ nonviscous

isotropic having the same properties in all directions of the solid angle

isotropic stress ≡ *hydrostatic stress*

Jeffreys body an ideal liquid, whose structural formula is N|M : J

jelly a gel which in the solid phase forms a continuous network throughout the liquid
gelée (f)

Gallerte (f)
Студень (m)

katastatic stress the part of the stress that does not disappear in relaxation
contrainte limité (a l'équilibre) (f)
katastatische Spannung (f)
Катастатическое напряжение-нерелаксирующее напряжение (n)

Kelvin body ≡ *Voigt body* an ideal solid, whose structural formula is N|H : K

kinematic viscosity the ratio of viscosity to density, measured in stokes, in c.g.s. units; $[L^2 T^{-1}]$; v

kinetic friction friction between two solid bodies in contact and in relative motion
frottement de glissement (m); frottement dynamique (m)
Gleitreibung (f)
Кинетическое трение (n)

Kronecker's delta a tensor of second rank, of which the diagonal components are all equal to unity and the other components all vanish; δ_{lm}
symbole de Kronecker (m)
Kroneckers delta (n)
Дельта-функция Кронекера (f)

Lamé's constant ≡ *Lamé's modulus* the parameter of the volume dilatation term in the classical stress–strain relation of elasticity; $[ML^{-1}T^{-2}]$; λ_L

laminar flow a flow taking place in laminae that move as wholes relative to each other
écoulement laminaire (m)
Laminarströmung (f)
ламинарное течение (n); ламинарный поток (m)

Lethersich body an ideal liquid whose structural formula is N–K : L

limiting inherent viscosity ≡ *intrinsic viscosity*, q.v.

limit of proportionality the limit of the straight part of a stress–strain curve or of a stress–strain rate curve
limite de proportionalité (f)
Proportionalitätsgrenze (f)
Предел пропорциональности (m)

linear dilatation ≡ *linear extension*, q.v.

linear phenomena a phenomena in which the experimental variables involved appear only as zero or first powers (see *Linear phenomena, nonlinear phenomena*, Section II)
phénomène linéaire (m)
lineares Verhalten (n)
Линейное явление (n)

liquid a material, of definite volume at zero isotropic pressure, that deforms indefinitely under any finite shearing stress, however small
liquide (m)
Flüssigkeit (f)
Жидкость (f)

liquid crystal a liquid whose molecules show a degree of order but maintain some mobility relative to each other

liquid limit ≡ *sticky point* ≡ *upper plastic limit* the minimal moisture content at which soil will barely flow under a standard stress
limite supérieure de plasticité (f)
obere Plastizitätsgrenze (f)
Предельная влажность (f)

load an external force acting on a body
charge (f)
Last (f)
Нагрузка (f)

logarithmic strain ≡ *natural strain* ≡ *Hencky strain*, q.v.

loss angle the phase lag between stress and strain under sinusoidal stress in the absence of inertial forces
angle de perte (m)
Verlustwinkel (m)
Угол потерь (m)

loss compliance the ratio of the amplitude of that part of the strain which has a phase lag of 90° with a sinusoidal stress of angular frequency ω, to the stress; J'', B'', D''
partie imaginaire de la compliance complexe (f)
Verlustkomplianz (f)
Податливость потерь (f)

loss modulus the ratio of the amplitude of that part of the stress which has a phase lag of 90° with a sinusoidal strain of angular frequency ω, to the strain; G'', K'', E''
partie imaginaire du module complexe (f)
Verlustmodul (m)
Модуль потерь (m)

low elasticity describes elastic deformations which are small enough to be caused by the straining of the atomic structure; often in common parlance, elastic strains less than about 1%
élasticité des petites déformations (f)
niedrige Elastizität (f)
Слабая деформируемость (f)

lower plastic limit ≡ *plastic limit* the minimal moisture content that will permit deformation of a small soil sample without rupture
limite intérieure de plasticité (f)
untere Plastizitätsgrenze (f)
Нижняя граница пластичности (f)

lyophil a material that as a dispersed phase attracts to itself molecules of the continuous phase

lyophobe a material that is not lyophilic (see *lyophil*)

macrorheology the rheology in which no account is taken of the microstructure of materials

mass force ≡ *body force*, q.v.

Maxwell body ≡ *Maxwell liquid* an ideal material whose structural formula is N–H: M
liquide de Maxwell (m)
Maxwellscher Körper (m)
Тело Максвеля (n)

Maxwell liquid ≡ *Maxwell body*, q.v.

melt swell ≡ *Barus effect*, q.v.

memory function stress as a function not only of the deformation at a given time but also of the preceding deformation history
fonction mémoire (f)

Nachwirkungsfunktion (f)

Функция памяти (f)

metarheology subjects in science that border on rheology

micromeritics the study of the behavior, especially of the flow, of powders
écoulement des poudres (m)
Pulver-Rheologie (f)
Микромеритика (f)

microrheology the rheology in which account is taken of the microstructure of materials

microstress a local stress concentration
microtension (f)
Micro-spannung (f)
Локальное напряжение (n)

modulus of cross elasticity the second-order modulus in elastic tensorial non-linearity; $[ML^{-1}T^{-2}]$; μ_c
module d'élasticité transversale (m)
Modul der Querelastizität (m)
Модуль поперечной упругости (m)

modulus of elasticity ≡ *elastic modulus* the ratio of a stress to its corresponding elastic strain; $[ML^{-1}T^{-2}]$; μ, G, E, K
module d'élasticité (m)
Elastizitätsmodul (m)
Модуль упругости (m)

modulus of rigidity ≡ *shear modulus* the ratio of a shearing stress to its corresponding shearing strain; $[ML^{-1}T^{-2}]$; μ, G
module de rigidité (m);   module de cisaillement (m)
Schubmodul (m)
Модуль жесткости, модуль сдвига (m)

modulus of volume expansion ≡ *bulk modulus* an elastic modulus that is the ratio of the isotropic stress to the relative change of volume; $[ML^{-1}T^{-2}]$; K
module de compressibilité (m)
Kompressionsmodul (m)
Объемный модуль (m)

Mohr's circle ≡ *circle of stress* the circle, in rectangular coordinates σ and t, repre-senting geometrically the relation be-tween σ and t and the position of the cross section perpéndicular to the traction

natural strain ≡ *logarithmic strain* ≡ *Hencky strain*, q.v.

necking the thinning of a tensile specimen at specific points
striction (f)
Einschnürung (f)
Образование шейки (n)

nematic pertaining to a form of liquid crystal whose molecules lie with their long axes parallel to one another but are not arranged in layers

Newtonian liquid an ideal liquid, devoid of shear elasticity, for which the shearing stress is proportional to the rate of shear
liquide Newtonien (m)
Newtonsche Flüssigkeit (f)
Жидкость Ньютона, Ньютоновская жид-кость (f)

nonlinear phenomena phenomena that are not linear; they may be parametrical, deformational, or tensorial (see *Linear phenomena, nonlinear phenomena*, Sec-tion II)
phénomènes non-linéaires (m)
nicht-lineare Erscheinungen (f)
Нелинейние явление (n)

non-Newtonian liquid a viscous liquid that exhibits shear elasticity or is of ten-sorial nonlinearity, or both (see *gen-eralized Newtonian liquid*)

normal component the component of a vector at right angles to the surface element to which it applies

normal stress the normal component of traction; σ
contrainte normale (f)
Normalspannung (f)
Нормальное напряжение (n)

Ostwald curve the consistency curve of an inelastic shear thinning, non-New-tonian liquid approaching a straight line, which extrapolates to the origin at high rates of shear

parametrical nonlinearity in a rheological equation the condition that the parameters are not constants but depend upon the arguments of the equation
non-linéarité paramétrique (f)
parametrische Nicht–Linearität (f)
Параметрическая нелинейность (f)

paste concentrated dispersion with a yield value

peeling test a test in which the adhesion of a film or tape or of an adhesive material on a tape is measured by pulling the tape away from the surface to which it adheres
essai de pelage (m)
Abreisstest (m)
Испытание на отслаивание (n)

permanent deformation ≡ *set* an unrecoverable deformation
déformation permanente (f)
bleibende Deformation (f)
Остаточная деформация (f)

phenomenological rheology the rheology that treats the material as a continuum which is characterized by experimental rheological parameters, without consideration of molecular structure
rhéologie phénoménologique (f)
phänomenologische Rheologie (f)
Феноменологическая реология (f)

plane strain strain in two dimensions
déformation plane (f)
ebene Verzerrung (f)
Двумерная деформация, плоскостная деформация (f)

plane stress stress in two dimensions
contrainte plane (f)
ebene Spannung (f)
Плоскостное напряжение, двумерное напряжение (n)

plasticity the capacity to be moulded but to retain shape under finite forces such as gravity; showing flow above a yield value

plastic limits ≡ *lower liquid limit sticky point* ≡ *upper plastic limit*, q.v.

plastic mobility the reciprocal of plastic viscosity; [$M^{-1}LT$]
mobilité (f)
Mobilitätskoeffizient (m)
Подвижность (f)

plastic resistance ≡ *internal friction*

plastic strength see *strength*

plastic viscosity the viscous resistance of a plastic material in flow; [$ML^{-1}T^{-1}$]; η_{pl}
viscosité plastique (f)
plastische Viskosität (f)
Пластическая вязкость (f)

plug flow the movement of a material as a solid plug along a tube
écoulement en bouchon (m); écoulement à noyau (m)
Propfen–Fliessen (n)
Стержневой поток (m)

poise the unit of *viscosity*, q.v.

poiseuille 10^5 poises

Poisson's ratio the ratio of the transversal contraction to the axial elongation of a rod in simple tension; [L^0]; v
rapport de Poisson (m)
Poissonsches Verhältnis (n)
Коэффициент Пуассона (m)

Poynting–Thomson body ≡ *standard linear solid* an ideal solid, whose structural formula is H–K; PTh

Prandtl body an ideal solid, whose structural formula is StV–H; P

pressure negative tension; [$ML^{-1}T^{-2}$]; P
pression (f)
Druck (m)
Сжатие, давление (n)

primary creep decelerating creep
fluage primaire (m); fluage ralenti (m)
primäres Kriechen (n)
Неустановившаяся ползучесть (f)

principal axes axes of a tensor along which there are no tangential components

principal direction the directions with respect to which only the normal components of a symmetrical tensor are different

from zero and the tangential components vanish; i, j, k
directions principales (f)
Hauptrichtungen (f)
Главные направления (n)

pseudofinite strain a finite elastic deformation of a body when the elemental
strains are infinitesimal and large deformation is due to an accumulation
of elemental rotations
déplacement relatif fini avec déformations
locales infinitésimales (m)
pseudo-endliche Verzerrung (f)
Псевдоконечная деформация (f)

pseudoplastic a material that shows no yield
value but whose viscosity falls progressively with rising stress, such that
the consistency curve approaches a
straight line that has a finite intercept
on the stress axis

psychorheology the rheology that attempts to
relate the results of assessing rheological properties by handling or other
subjective observation to quantitative
measurements of physical properties

pure shear a shear without rotation
rate the changing with time of a variable,
indicated by Newton's dot (see *Rate*,
Section II)
vitesse (f)
Zeit-Ableitung (f)
Скорость (f)

recovery the re-establishment of an earlier
condition, especially of strain
recouvrance (f)
Erholung (f)
Восстановление (n)

redress elastic strain recovery of a material,
induced by the application of such
stimuli as heat or ultrasonic radiation
recouvrance stimulée (f)
stimulierte elastische Erholung (f)
Стимулированное восстановление (n):
редресс (m)

reduced viscosity the ratio of the specific
viscosity to the concentration of the
dispersion, the latter usually being

given in grams per deciliter or grams
per milliliter; $[M^{-1}L^3]$; η_{red}
viscosité réduite (f)
reduzierte Viskosität (f)
Приведенная вязкость (f)

Reiner–Rivlin fluid an inelastic viscous fluid
with tensorial nonlinearity

relative deformation deformation relative to
the original dimensions; $[L^\circ]$; D, d

relative viscosity the ratio of the viscosity of
a solution to that of its solvent or of
the viscosity of a dispersion to that of
its continuous phase; $[L^\circ]$; η_{rel}

relaxation the gradual reduction of stress
while strain is maintained or persists

relaxation function the stress as a function
of time when a unit deformation is
applied stepwise at zero time and
maintained constant thereafter

relaxation time the time taken for the internal
stress to drop to $1/e$ of its initial value
under constant strain; $[T]$; T_{rel}
temps de relaxation (m)
Relaxationszeit (f)
Время релаксации (n)

replicability implies that a test cannot be
repeated satisfactorily on a single
sample but that there is agreement
between tests on replicate samples
réplicabilité (f)
Wiederholbarkeit (f)
Повторяемость (f)

reproducibility implies that a test can be
carried out more than once on the
same sample with the same result (often
used as synonym for replicability)
reproductibilité (f)
Reproduzierbarkeit (f)
Воспроизводимость (f)

residual deformation a deformation un-
recovered after a given time

residual stress a stress that does not vanish
after an unloading
contrainte résiduel (f)
Nachspannung (f)
Остаточное напряжение (n)

resilience the maximum ability of a body to store energy elastically per unit volume; $[ML^{-1}T^{-2}]$; *R*
résilience (f)
elastische Formveränderungsarbeit (f)
Резильянс (дпредельный) (m)

retardation time the time taken by the strain to fall to $1/e$ of its original value in an elastic after-effect; $[T]$; T_{ret}
temps de retard (m)
Retardationszeit (f)
Время запаздывания (n); время ретардации (n)

rheochor a parameter defined as the product of the molecular weight of a liquid and the one-eighth power of its viscosity, divided by the density; it enables molecular volumes of different liquids to be compared at the same viscosity; *R*

rheodichroism in a two-phase system a differential absorption of light parallel and perpendicular to the direction of the flow producing it

rheoencephalograph an instrument for measuring the flow of blood in the brain by passing small electric currents through the skull

rheogram an automatically written record of the deformation of a test piece

rheological equation ≡ *rheological equation of state* ≡ *constitutive equation* an equation between stress, deformation, and time, and in which the parameters define rheological properties
équation rhéologique (f)
rheologische Gleichung (f)
Реологическое уравнение (n)

rheology the study of the deformation of materials, including flow

rheomalaxis on the shearing of a material a loss of consistency which is not regained on subsequent resting, heating, etc. (see *Rheomalaxis*, Section II)
ramollissement permanent (m)
permanente Scherverdünnung (f)
Необратимое сдвиговое разрушение (n)

rheometer an instrument for measuring rheological properties

rheopexy the solidification of a thixotropic system by gentle and regular movements; see *shear thickening* (see also *Shear thinning*, *shear thickening* and *Thixotropy*, *rheopexy*, Section II)

rigidity the property of the absence of deformation under any stress; also the resistance to deformation
rigidité (f)
Starrheit (f)
жесткость (f)

rotational velocity the velocity of a point at unit distance from the center of rotation of a rotating body; $[T^{-1}]$; ω
vitesse de rotation (f)
Rotationsgeschwindigkeit (f)
Скорость вращения (f)

rupture ≡ *breakage*, q.v. (see also *Breakage*, *rupture*, *fracture*, Section II)

St. Venant body an ideal material that is plastically deformed without viscous resistance when the stress exceeds a yield stress; StV

Schofield–Scott-Blair body an ideal solid, whose structural formula is B–K; SchScB

Schwedoff body an ideal material, whose structural formula is H–(StV M); Schw

secondary creep steady creep
fluage secondaire (m); fluage établi (m)
sekundäres Kriechen (n)
Установившийся крип (m)

self-crowding factor ≡ *crowding factor*, q.v. (see also *Crowding factor*, *self-crowding factor*, Section II)

set ≡ *permanent deformation*, q.v.

shear the change of angle between any two material lines in a deforming body; $[L°]$; γ
cisaillement (m); glissement (m)
Schub (m)
Сдвиг (m)

shear breakage   a breakage occurring in the plane of maximal shearing
rupture par glissement (f)
Verschiebungsbruch (m)
Сдвиговое разрушение (n)

shear hardening   an increase of consistency, the result of shearing deformation
écrouissage par distortion (m)
Schubverfestigung (f)
Сдвиговое упрочнение (n)

shearing strain   the gradient of a shearing displacement; [L°]; γ
deformation de cisaillement (f); glissement (m)
Schubverzerrung (f)
Деформация сдвига (f)

shearing stress ≡ *tangential stress*   the tangential component of a traction; $[ML^{-1}T^{-2}]$; τ
contrainte tangentielle (f); cission (f)
Schubspannung (f)
Сдвиговое напряжение (n)

shear modulus ≡ *modulus of rigidity*, q.v.

shear softening   a decrease of consistency, the result of shearing deformation
rhéofluidification (f)
Schuberweichung (f)
Сдвиговое размягчение (n)

shear thickening   a univalued increase of the viscosity or consistency with increasing rate of shear; sometimes called *rheopexy*, q.v. (see also *Shear thinning, shear thickening* and *Thixotropy, rheopexy*, Section II)
rhéo-épaississement (m); épaississement par cisaillement (m)
Scherverdickung (f)
Сдвиговое загущение (n)

shear thinning   a univalued reduction of the viscosity or consistency with increasing rate of shear (see *Shear thinning, shear thickening* and *Thixotropy, rheopexy*, Section II)
rhéo-fluidification (f); fluidification par cisaillement (f)
Scherverdünnung (f)
Сдвиговое разжижение (n)

shortness   a tendency to tear easily
friabilité (m)
Kürze (f)
Хрупкость, ломкость, слабая деформируемость (f)

shrinkage   a nonelastic reduction in volume
retrait (m)
Schwinden (n)
Усадка (f)

sigma effect   the phenomenon that the ratio of shear rate to shearing stress at the wall of a material flowing in a capillary tube is not proportional to the fourth power of the radius, excluding cases in which it is due to plug flow or slip at the wall

simple shear   a laminar deformation parallel to a stationary plane

slider   a model of a plastic element showing friction
patin-à-friction (m)
Reibungselement (n)
Ползун (m)

slippage   the discontinuous displacement of two bodies along contiguous planes
glissement (m)
Gleitung (f)
Скольжение, проскальзывание (n)

smectic   pertaining to a form of liquid crystal whose molecules lie with their long axes parallel to one another and are arranged in layers that are free to slide over one another

solid   a body that permanently retains its deformed shape under shearing stress
solide (m)
Festkörper (m)
Твердое тело (n)

solid viscosity   the viscous resistance in a solid; sometimes called *internal friction*, q.v.; η_s
viscosité solide (f)
Viskosität des Festkörpers (f)
Вязкость твердого тела (f)

solvation   attachment of part of the liquid continuous phase to the dispersed particles in a dispersion

specific viscosity   the ratio of the increase in viscosity of a dispersion over the viscosity of its continuous phase to the viscosity of the continuous phase; $[L°]$; η_{sp}
viscosité spécifique (f)
spezifische Viskosität (f)
Удельная вязкость (f)

spinability   the capacity of a liquid for being drawn into threads out of all proportion to its viscosity, presumably due to work hardening or shear thickening
filabilité (f)
Spinnbarkeit (f)
Прядомость (f)

spontaneous elastic recovery   the elastic recovery of strain, not involving redress
recouvrance élastique spontanée (f)
spontane elastische Erholung (f)
Самопроизвольное упругое восстановление (n)

standard linear solid ≡ *Poynting-Thompson body*, q.v.

static fatigue   the reduction in the strength or in the real modulus of a complex deformation during creep or under static load
fatigue statique (f)
statische Ermüdung (f)
Статическая усталость (f)

static friction   a tangential force preventing relative finite motion between the contiguous surfaces of two bodies
frottement statique (m)
Haftreibung (f)
Статическое трение (n)

stick point   the temperature of a polymer or of a melting block at which it becomes sticky
point d'adhésivité (m)
Klebepunkt (m)
Точка прилипания (f)

sticky point

stiffness   the variable ratio of the tensile stress to the extension in simple tension (see *stiffness*, Section II); $[ML^{-1}T^{-2}]$; S

raideur (f)
Steifigkeit (f)
Сопротивляемость растяжению (f)

stimulated elastic restoration ≡ *redress*, q.v.

stokes   the unit of *kinematic viscosity*, q.v.

storage compliance   the ratio of the amplitude of that part of the strain which is in phase with the sinusoidal stress of angular frequency and of unit amplitude, to the stress; J', B', D'
partie réelle de la complaisance complexe (f)
speiche Komplianz (f); speiche Nachgiebigkeit (f)
Упругая податливость (f)

storage modulus   the ratio of the amplitude of that part of the stress which is in phase with the sinusoidal strain of angular frequency ω and of unit amplitude, to the strain; G', K', E'
partie réelle du module complexe (f)
Speichermodul (m)
Модуль накопления (m)

stored-energy function ≡ *strain-energy function*, q.v.

strain   the relative deformation or only its recoverable part (see *Stress, strain*, Section II); $[L°]$; ε_{lm}, e_{lm}
déformation relative (f)
Verzerrung (f)
Относительная деформация (f)

strain energy   the energy per unit volume stored in a strained body; $[ML^{-1}T^{-2}]$
énergie de déformation (f)
Verzerrungsenergie (f)
Энергия деформации, деформирования(f)

strain energy function ≡ *stored-energy function*   a scalar potential for the stress tensor, specified in the classical theory of finite elastic strain
fonction d'énergie de déformation (m)
Verzerrungsenergie-Funktion (f)
Функция энергии деформации, (f); упругий потенциал (m)

strain work   the recoverable part of the stress work
travail de déformation réversible (m)
Verzerrungsarbeit (f)
Обратимая часть работы деформирования (f)

streamline a curve whose tangents are formed by the velocity vectors of flow at a given instant
ligne de courant (f)
Stromlinie (f)
Линия тока (f)

strength the resistance to plastic flow (plastic strength, the yield value) or to breaking (breaking strength); measured as a stress
résistance (f)
Festigkeit (f)
Прочность (f)

stress the aggregate of all tractions over the solid angle of the directions; see *traction* (see also *Stress, strain*, Section II); $[ML^{-1}T^{-2}]$; s_{lm}
contrainte (f)
Spannung (f)
Напряжение (n)

stress corrosion a decline with time of the static strength of a specimen under stress

stress work the work performed by the stresses in the deformation of a body; $[ML^{-3}T^{-2}]$; w_s
travail de contrainte (m)
Spannungsarbeit (f)
Работа напряжений (f)

stretch the ratio of the extended length to the initial length in elongation; $[L°]$; λ
rapport d'extension (m)
Streckung (f)
Относительное удлинение (n)

structural formula a formula representing a mechanical rheological model, in which a horizontal line signifies coupling in series, a vertical line signifies coupling in parallel, and abbreviations stand for the materials
formule analogique (f)
Strukturformel (f)
Модельная формула (f)

structural turbulence a turbulence induced through the breakdown of the internal structure of a liquid
turbulence de structure (f)
Strukturturbulenz (f)
Структурная турбулентность (f)

structural viscosity the property of an Ostwald curve (see *Shear thinning, shear thickening*, Section II)
viscosité de structure (f)
Strukturviskosität (f)
Структурная вязкость (f)

superfluidity the extremely high fluidity of certain inert elements, at temperatures approaching absolute zero (e.g., liquid helium II)
suprafluidité (f)
Superfluidität (f)
Сверхтекучесть (f)

suppleness the inverse of hardness
soupelesse (f)
Weichheit (f)

suspension a dispersion of a solid in a liquid
suspension (f)
Suspension (f)
Суспензия (f)

syneresis spontaneous expulsion with time of a liquid from gelled or solid colloidal systems

tack the force required to part at a particular speed two surfaces, or two surfaces separated by material, of a specified thickness, behaving as a liquid, after a short time of contact

tactoid a material that shows patches of oriented colloidal particles, usually eye-shaped, within a liquid

tangential stress ≡ *shearing stress*, q.v.

tearing the pulling apart of a material by the application of a force on the outside of a V-shaped kerb
action de déchirer (f)
Zerreissen (n)
Разрывание (n)

tectonites materials that form tubes when rolled into rod shapes

telescopic flow laminar flow in the direction

of the axis of concentric tubes of circular cross section

tension a traction acting in the direction external normal to the surface element on which it acts
traction normale (f); tension (f)
Zug (m)
Натяжение, напряжение растяжения (n)

tensorial nonlinearity when the rheological equation contains the second power of the kinematical argument

tertiary creep accelerating creep
fluage tertiaire (m); fluage accéléré (m)
tertiäres Kriechen (n)
Стадия ускоряющейся ползучести (f)

thermoplastic having the property of softening upon repeated application of heat and of hardening upon the subsequent application of cold
matière thermoplastique (f)
thermoplastischer Kunststoff (f)
Термопластическое вещество (n)

thermosetting plastic a material having the property of hardening upon the application of sufficient heat and of not softening upon the further application of heat
matière thermodurcissable (f)
hitzehärtbarer Kunststoff (m)
Термореактивное вещество (n)

thixotropy an isothermal and comparatively slow recovery, on standing, of a material, of a consistency lost through shearing (see *Shear thinning, shear thickening* and *Thixotropy, rheopexy,* Section II)

torque the moment of loads producing torsion; $[ML^2T^2]$; M_t
moment de torsion (m)
Torsionsmoment (n)
Момент кручения (m)

torsion the laminar deformation of a prism, in which cross sections are relatively rotated
torsion (f)
Torsion (f)
Кручение (n)

toughness the capacity of a material for absorbing energy before breaking
ténacité (f)
Zähigkeit (f)
Тягучесть, ударная вязкость (f)

traction the ratio of the force to the area of a surface element on which the force acts; sometimes called *stress* (q.v.) but imprecisely, actually being the vectorial component of the stress
traction (f); vecteur contraint (m)
Spannungsvektor (m)
Вектор напряжения (m)

trajectory the line that a moving point describes

Trouton–Rankine body an ideal liquid, whose structural formula is N–PTh; TR

Trouton's coefficient ≡ *coefficient of viscous traction*, q.v.

turbulence a condition at which eddies occur in a flowing material

twist the angle of torsion
angle de torsion (m)
Verdrehungswinkel (m)
Угол закручивания, угол кручения (m)

velocity gradient the derivative of the velocity of a particle in relation to its coordinates; $[T^{-1}]$; $\dot{\gamma}_{lm}$
gradient de vitesse (m)
Geschwindigkeits-Gradient (m)
Градиент скорости (m)

viscoelasticity ≡ *firmoviscosity* elasticity accompanied by viscous resistance (see *Viscoelasticity,* Section II)
visco-élasticité (f)
Viskoelastizität (f)
Вязко-упругость (f)

viscosity the resistance to a deformation, increasing with the rate of deformation; measured in poises, in c.g.s. units; $[ML^{-1}T^{-1}]$; η

viscosity coefficient ≡ *coefficient of viscosity*, q.v.

viscosity number ≡ limiting inherent viscosity
    ≡ intrinsic viscosity, q.v.

Voigt body ≡ Kelvin body, q.v.

volume dilatation ≡ cubical dilatation the
    relative increase in volume of an ele-
    ment of a body; [L°]; D_v, d_v, ε_v, e_v
dilatation en volume (f)
Volumdehnung (f)
Объемное расширение (n)

volume elasticity the elastic response to a
    change of volume
élasticité de volume (f)
Volumelastizität (f)
Объемная упругость (f)

volume flow a continuous change of volume,
    proceeding in time under isotropic
    stresses; f_v
écoulement de volume (m)
Volumfliessen (n)
Объемное течение (n)

volume viscosity the viscous resistance to
    volume flow; $[ML^{-1}T^{-1}]$; ζ
viscosité de volume (f)
Volumviskosität (f)
Объемная вязкость (f)

voluminosity constant the factor that allows
    for the solvation of dispersed particles
    and accounts for a change in the
    numerical factor 2.5.in Einstein's equa-
    tion of viscosity versus concentration
    of spheres; [L°]
coefficient de solvatation (m)
Solvatationskoeffizient (m)
Константа объемности (f)

Weissenberg effect an elasticity cross effect
    in which tractions in the directions
    normal to the direction of flow are
    equal

work hardening a rise in yield value induced
    by the plastic deformation of a material
écrouissage (m)
Kalthärtung (f);   Werkstoffverfestigung (f)
Механическое упрочнение (n)

work softening a fall in yield value induced
    by the plastic deformation of a material
ramollissement mécanique (m)
Werkstofferweichung (f)
Механическое размягчение (n)

yield point the point on the stress–strain
    curve at which yielding starts abruptly
seuil d'écoulement (m)
Fliessgrenze (f)
Предел текучести (m)

yield strain the strain corresponding to the
    yield point; [L°]; θ
déformation d'écoulement (f)
Fliessdeformation (f)
Деформация, отвечающая пределу теку-
    чести (f)

yield stress the stress corresponding to the
    yield point; $[ML^{-1}T^{-2}]$; ϑ
contrainte d'écoulement (f)
Fliessspannung (f)
Предельное напряжение (n)

yield value ≡ yield stress, q.v.

yielding the beginning of flow of a plastic
    material

Young's modulus the constant ratio of the
    tensile stress to the extension in simple
    tension; $[ML^{-1}T^{-2}]$; E

zeta potential the potential on a particle
    dispersed in a liquid, measured by
    electrophoresis

IV. Symbols

It is hardly necessary to emphasize the desirability or even the necessity of a unified standard system of symbols. Symbols are the words in a mathematical language, and it is very tiring for the reader of scientific articles to have to learn a new language for every article he reads.

In the following compilation proposed we have kept as far as possible to existing standards, such as those given in *Symbols, Units and Nomenclature in Physics, Document U.I.P. 9* (S.U.N.61-44), published by the International Union of Pure and Applied Physics, 1961, and *Letter Symbols, Signs and Abbreviations* (Part I, General), *British Standard 1991*, Part 1, 1954.

The limited number of readily available alphabets and fonts renders it impossible to reserve any symbol for the presentation of only one quantity for all cases.

Most symbols are printed in italic letters; some Roman letters are used, such as "e," the basis of natural logarithms and the abbreviations for rheological bodies. Roman numerals *I, II, III* are used for the principal invariants. Sanserif light face capitals M, L, and T are used for dimensional units. Greek letters are used for angles, material parameters, and dummy indices. Wherever an author desires to avoid the use of subscripts to emphasize the tensor character of a quantity, he may use boldface letters. Vectors may be indicated by an arrow on top of the symbol. Truesdell and Toupin[14] have introduced the Hebrew letters א, ב, and ג for scalar coefficients in expressions for isotropic functions, and we have adopted this use. It may be mentioned that the symbol *nabla*, ∇, introduced by Hamilton is also Hebrew, meaning "harp," although the symbol is not a Hebrew letter. We have not used block letters or German for standard symbols, and these are therefore available for uses not covered in our list.

Symbols for mathematical operations and constants have usually not been included, nor have chemical symbols or abbreviations of any words, including abbreviations of the names of units.

Newton's dots are to be used for all material time derivatives; for example, $\dot{x}_l$ for velocity and $\ddot{x}_l$ for acceleration. Such symbols are not shown in the list, except for that for "power."

We have used the letters *l, m, n, o* for tensor indices. These are often indicated by *i, j,* but we applied these letters (and also *k*) for principal directions.

The use of *u, v, w* for the components of velocity should be discontinued.

---

[14] C. Truesdell and R. Toupin, The classical field theories. *In* "Handbuch der Physik" (S. Flügge, ed.), Vol. 3, Pt. I. Springer, Berlin, 1960.

Symbol	Meaning
A	Area
A	As superscript represents "of Almansi" (e.g., D^A)
$\vec{a}, a_l$	Acceleration
a, b, c	Initial coordinates
a_l	Initial coordinates
B	Bingham body
B_l	Body force
B^*	Complex compliance in bulk
B'	Storage compliance in bulk
B''	Loss compliance in bulk
Bu	Burgers body
b	Breadth
C	Constant
C	As superscript represents "of Cauchy" (e.g., D^C)
$\vec{c}$	Cross direction
c_v	Concentration by volume
D	Deformation
D_{lm}	Finite deformation tensor
D_v	Volume dilatation
D^*	Complex compliance in extension
D'	Storage compliance in extension
D''	Loss compliance in extension
d	Thickness
d_{lm}	Infinitesimal deformation tensor
d_v	Volume dilatation
E	Young's modulus
E	Energy
E^*	Complex modulus in extension
E'	Storage modulus in extension
E''	Loss modulus in extension
E_k	Kinetic energy
e	Basis of natural logarithms
e_{lm}	Finite strain tensor
e_v	Volumetric strain
f_{lm}	Flow tensor
f_v	Volume flow
G	Shear modulus (also μ)
G	As superscript represents "of Green" (e.g., D^G)
G^*	Complex modulus in shear
G'	Storage modulus in shear
G''	Loss modulus in shear
g	Acceleration of gravity
g_{lm}	Metric tensor
H	As superscript represents "of Hencky" (e.g., D^H)

Symbol	Meaning
H	Hooke body
h	Height
I_{lm}	Tensor of inertia
i, j, k	Principal directions
J	Jeffreys body
J^*	Complex modulus in bulk
J'	Storage modulus in bulk
J''	Loss modulus in bulk
K	Bulk modulus
K	Kelvin body
k	As subscript represents "kinetic" (e.g., E_k, kinetic energy)
k	Dissipation coefficient
L	Unit and dimension of length
L	Lethersich body
l	Length
l	As subscript represents "longitudinal" (e.g., e_1, longitudinal strain); also represents "liquid" (e.g., μ_1, shear modulus of a liquid)
l, m, n, o	Tensor indices
M	Moment of force
M	Unit and dimension of mass
M	Maxwell liquid
m	Mass
M_t	Torque
(m)	As subscript represents "mean" (e.g., $s_{(m)}$, mean stress); alternative suggestion, $\bar{s}$
N	Newtonian liquid, also axial force
$\vec{n}$	Normal direction
(n)	As subscript represents "related to normal n" (e.g., $s_{(n)}$, traction related to normal n)
o, (o)	As subscript represents "original" (e.g., l_o, original length)
P	Prandtl body
P, P_l	Force, longitudinal force
PTh	Poynting–Thompson body
p	Isotropic pressure
pl	As subscript represents "plastic" (e.g., η_{pl})
$p_{(m)}$	Mean pressure
Q	Volume of flow in time t
R	Large radius
R	Rheochor
R_b	Breaking energy
R	Resilience

Symbol	Meaning	Symbol	Meaning
r	Radius or radial distance as that, for example, from the center of a tube	δ	Cohesive energy density parameter
		δ	Modulus of dilatancy
		δ_{lm}	Kronecker's delta
r	Cylindrical coordinate	ε_{lm}	Infinitesimal strain
S	Stiffness	ε_v	Volume dilatation
S	As superscript represents "of Swainger" (e.g., D^S)	ζ	Volume viscosity
		ζ	Zeta potential
StV	St. Venant body	η	Viscosity, shear viscosity
SchScB	Schofield–Scott-Blair body	η'	Apparent viscosity
Schw	Schwedoff body	η_s	Solid viscosity
s	As subscript represents "solid" (e.g., η_s)	η_Δ	Differential viscosity
		η_{sp}	Specific viscosity
s_{lm}	Stress	$[\eta]$	Intrinsic viscosity
$s_{(m)}$	Mean stress	η_c	Coefficient of cross viscosity
$\vec{s}_{(n)}$	Traction on surface element with external normal n	η_{pl}	Plastic viscosity, plastic mobility
		η_{red}	Reduced viscosity
T	Unit and dimension time	η_{rel}	Relative viscosity
T	Period	θ	Second cylindrical coordinate
TR	Trouton–Rankine body	θ	Yield strain
T_{rel}	Relaxation time	ϑ	Yield stress
T_{ret}	Retardation time	κ	Compressibility
t	Time	λ	Stretch
$\vec{t}$	Tangential direction	λ_L	Lamé's modulus
t_{lm}	tensor of second rank	λ_T	Trouton's coefficient
$\overset{\circ}{t}_{lm}$	Deviator of tensor t_{lm}	μ	Shear modulus (also G)
$\vec{u}, u_1$	Direction of displacement, longitudinal displacement	μ_c	Modulus of cross elasticity
		ν	Poisson's ratio
V	Volume	ν	Kinematic viscosity
v	As subscript represents "of volume" (e.g., e_v, volumetric strain)	ξ	Bulk viscosity
		ρ	Density
$\vec{v}, v_1$	Velocity	Σ	"the sum of..."
W	Work	σ	Normal traction
$\dot{W}$	Power	σ_B	Bending stress
w	Work per unit volume	τ	Tangential traction
w_d	Dissipated work per unit volume	ϕ	Free-energy density, also angle of friction: and angle of isoeline
w_e	Strain work per unit volume		
w_s	Stress work per unit volume	ϕ	Coefficient of fluidity
x_1	Cartesian coordinates	χ	Angle of extinction
x, y, z	Cartesian coordinates	ψ	Bound-energy density
α	Attenuation coefficient	Ω	Angle
α, β, γ	Summation indices	ω	Angular velocity, angular frequency
Γ_{lm}	Finite displacement gradient	ω_{lm}	Tensor of infinitesimal rotation
γ	Shear	$\aleph, \beth, \gimel$	(Aleph, bet, gimmel) scalar coefficients in isotropic functions
γ_{lm}	Infinitesimal displacement gradient		
Δ	Difference	—	Coupled in series (mechanical models of rheological equations)
Δl	Elongation		
Δs	Distance	|	Coupled in parallel (mechanical models of rheological equations)
$\Delta \vec{s}$	Directed segment		

Symbol	Meaning	Symbol	Meaning
	As superscript indicates "deviation" (e.g., $\overset{\circ}{D}_{lm}$, deviator of finite deformation tensor)	I, II, III	First, second, and third invariant
		∇	Nabla, the operator

ACKNOWLEDGMENTS

We are indebted to Dr. H. Weiss, Dr. F. Schwarzl, Professor A. A. Trapenznikov, Mr. H. Brooks, and others for helping us with French, German, and Russian translations, and to the Editor also for extensive discussions.

We owe much to too many rheological colleagues who have given us valuable advice to thank them all by name. We ask them to accept our general expression of gratitude. We have also, of course, made free use of the subject indexes of the earlier volumes of the present series.

Sources

International Union of Pure and Applied Physics, "Symbols, Units . . ." 1961.
International Union of Pure and Applied Chemistry, *J. Polymer Sci.* **8**, 957 (1952).
British Standards Institute. "Letters, Symbols etc. . . ." 1954.
British Standards Institute. Glossary of Terms used in the Plastics Industry, 1951.
J. M. Burgers and G. W. Scott Blair, Report on the Principles of Rheological Nomenclature. Joint Committee on Rheology, International Council of Scientific Unions, 1949.
H. Leaderman, Society of Rheology, Committee on Nomenclature Reports 1–4 (1956, 1957).
Proposals for Rheological Nomenclature and Definitions, Nederlandse Reologsche Vereniging, 1956.
"Monographs on the Rheological Behaviour of Natural and Synthetic Products." North Holland Publ., Amsterdam, 1952–1956.

AUTHOR INDEX

Numbers in parentheses are reference numbers and are included to assist in locating references in which authors' names are not mentioned in the text.

A

Abramson, H., 425, 457(12)
Abu, H., 425, 450(10), 457(10)
Adams, N., 392, 393(85), 394(85), 397(85), 402(85)
Aggarwal, S. L., 252, 268
Aihara, A., 11
Albert, C. G., 429
Alexander, P., 281
Alfrey, Jr., T., 293, 294(11), 296(11), 297(11), 299(11), 300(11), 301(11), 440
Allan, R. S., 87, 108(4), 111(4), 112, 114(4), 130, 131(63), 134, 162, 165, 168(4), 177 (124), 178(124), 179, 180, 183(124), 193(4), 214(4)
Allen, V. R., 40, 64
Allis, J. W., 321, 329(24)
Allnat, A. R., 70
Alves, G. E., 429
Anczurowski, E., 93, 95(23), 96(24), 101, 102(24), 185, 186(35, 137), 190(137), 193(35), 195(137), 196, 197, 198(23), 200(23), 201, 202(137), 203(137), 206(35, 137), 207(35), 208(35), 213(35), 214(137)
Anderson, J. A., 457, 458
Andrade, E. N. C., 272
Andrews, M., 61
Arlov, A. P., 141(85), 144
Arnold, J. E., 457, 458
Aschenbrenner, M., 425, 457(9)
Atabek, H. B., 227
Attinger, E. O., 235
Auer, P. L., 261, 324, 343(40), 344(40), 345
Auerbech, P., 272

B

Baer, E., 277
Bagley, E. B., 268, 278
Ballman, R. L., 269
Barber, E. M., 271
Barker, J. A., 79
Barlow, A. J., 56
Barnum, E. R., 46
Barrall, II, E. M., 320, 329
Bartok, W., 92, 94, 96(27), 101(22, 27), 105, 115(42), 116(42), 117(42), 118, 119(42), 120, 122, 144(48), 162(27, 48), 164, 167(27), 168, 174(27), 177(48), 178(48), 183(48)
Bastow, S. H., 326, 340(43), 341(43)
Bateman, H., 370, 371(55)
Batschinsky, A., 34
Bauer, W. H., 433, 439, 440, 457(88), 459
Bayliss, L. E., 147, 233, 235(87)
Beazley, K. M., 456, 459(124)
Becherer, G., 331, 333(54)
Beech, D. G., 457, 458
Beevers, R. B., 321
Belcher, H. V., 281
Belkin, I. M., 446
Bellemans, A., 3
Belner, R. J., 209
Benbow, J. J., 278, 370
Benis, A. M., 239, 242(211), 243(211)
Berg, D., 320, 336(12)
Bergen, J. T., 296, 304, 312
Bergmann, K., 10
Bernal, J. D., 320
Berry, G. C., 56
Best, R. J., 60

489

Ginn, R. F., 395
Glasstone, S., 23, 257, 341
Glushenkova, V. R., 324
Goel, S., 275
Gogos, C. G., 457, 459
Goldman, A. J., 152, 153(113)
Goldschmid, O., 147, 148(97a), 215(97a)
Goldsmith, H. L., 87, 88, 96, 99, 100, 101(31),
  102, 148, 149, 150(11), 154(11), 155, 156,
  157(11), 159(11), 167, 168(128), 169, 174(128),
  176(131), 177, 180(128), 181(11, 128), 182,
  183, 184(128), 185(131), 186(105, 131), 187
  (131), 188(131), 189, 190(131), 191(131),
  192(131), 196(11), 197, 207(31), 210(31a),
  211(131), 216, 217(31a), 218(31a, 38, 165),
  219(31a), 220(31a), 223(31a), 224(15, 168),
  226(168), 227(15), 228(15), 229(183), 230(15,
  31a), 231(31a), 232(15), 235, 236, 237(205),
  240(121, 138), 242(31a), 244, 245
Goldstein, S., 221, 370
Gollis, M. N., 32
Golub, M. A., 255, 263(19)
Gonzales, R. N., 429
Goodeve, C. F., 433, 434, 457(81), 465
Goodman, P., 281
Gore, W. L., 302
Goren, S., 240
Gotze, W., 457, 458
Grassie, N., 281
Gratch, S., 64
Gray, G. W., 318, 320, 333(1), 339(1), 343(1)
Gray, J. P., 289
Green, H., 147, 209(89), 428, 435, 437, 459(35)
Greenblatt, J. H., 429, 454(68), 455(68), 457(68),
  459(68)
Greensmith, H. W., 397, 398(94), 399(94),
  400(94)
Greenwood, N. N., 24
Gregerson, M. I., 233, 456
Griest, E. M., 78
Griffith, R. M., 302, 303(18), 306(18)
Groom, A. C., 242, 243(233)
Gross, B., 361
Grossman, P. U. A., 368
Grotz, L. C., 313, 314
Grubb, W. T., 32
Grunberg, L., 1, 58(5), 68, 78(5), 263
Guest, M. M., 239, 243(217)
Gunderson, R. C., 42
Guthknecht, R., 457

Guttman, F., 55
Guzman, G. M., 255

H

Haberman, W. L., 154
Hagerty, W. W., 263
Hahn, P. F., 242
Hahn, S. J., 440, 457, 458
Hallam, H. E., 18
Haller, W., 452
Ham, T. H., 234
Hamilton, C. W., 278
Hamilton, W. F., 242
Hanks, R. W., 275
Happel, J., 151, 153, 154, 156, 216(109)
Harkins, W. D., 124
Harper, H. R., 308
Harrington, R. E., 141
Harris, F. E., 130, 131
Harris, J. W., 233, 234
Hart, A. W., 42
Hartland, A., 66
Hartley, G. S., 429, 455
Haszeldine, R. N., 34
Hatschek, E., 147, 272
Hatton, R. E., 43
Hauser, E. A., 428
Hawksley, P. G. W., 428, 452
Haycock, P. J., 123
Hayes, J. W., 386, 387(74)
Haynes, R. H., 210, 242(152)
Heckert, W. W., 427, 452
Helders, F. E., 322, 345
Heller, J. P., 193
Heller, W., 457
Henning, T., 457
Henry, J. E., 307
Hermans, J. J., 430
Hermans, Jr., J., 321, 322, 323, 324(27), 325,
  326(27)
Hershey, A. D., 141
Herzfeld, K. F., 66
Herzog, R. O., 344
Hess, K., 256, 263(25), 264(25), 268(25)
Hewlett, V. A., 216
Hiby, J. W., 192
Higasi, K., 10
Higginbotham, G. H., 147, 210(93), 211(93),
  242(93)

494   AUTHOR INDEX

Hildebrand, J. H., 7
Hiller, K. H., 457
Hind, S. R., 457, 458
Hlynka, I., 457, 458
Hochmuth, R. M., 244
Hoeppler, F., 464
Hoffman, H., 457, 458
Hollomon, J. H., 463
Horn, E., 457, 458
Horowitz, H. H., 252
Howells, E. R., 278
Howlett, K. E., 59
Hsia, G., 243
Huber, T., 101
Hudson, D. R., 457, 459
Hudy, J. A., 457, 458
Huppler, J. D., 360, 386, 387(75), 388(75), 401 (75)
Hurd, C. B., 32, 47
Hurd, R., 457, 458
Hutton, J. F., 278, 279

I

Immergut, E. H., 261, 264
Ippen, A. T., 276

J

James, D. M., 457, 458
James, R. E., 216
Jane, R. S., 272
Janusek, J., 457, 458
Jeffery, G. B., 87, 89, 91, 94, 96, 98, 101, 135, 137(2), 141, 195, 211, 212, 213, 218(2), 268, 344
Jeffrey, R. C., 155, 210(120), 216(120), 219(120), 220(120), 230
Jeffreys, H., 465
Jeffries, H. J., 457, 458
Jelley, E. E., 321, 323(16)
Jellinek, H. H. G., 281
Jenkins, C. F., 428
Jepson, C. H., 291, 292(9), 302(9), 303(9)
Jessen, F. W., 457
Jobling, A., 79, 350, 369(15), 391, 393(15), 394(15), 395(15), 428, 442, 447, 448, 452
Johnson Dix, F., 210
Johnson, D. L., 277

Johnson, J. F., 40, 252, 260, 263(39), 268, 272, 281, 320, 328, 329, 331(47), 333(48), 336, 338(70), 339(70), 340(70)
Johnson, M. M., 275
Johnson, W. R., 141, 281
Jolly, R. E., 252
Joly, M., 456
Jones, A. D., 428, 453, 454(29), 457(29)
Jones, R. H., 259, 262(34), 266(34)
Jonsson, I., 243
Jordan, D. O., 322
Juliusberger, F., 428, 452(39), 457(39, 40), 465

K

Kahanovitz, A., 276
Kalejs, J., 123
Kaneko, M., 324, 344(41), 345(41)
Kanig, G., 72
Kaplun, S., 221
Karnis, A., 99, 102, 148(31a, 38, 39), 155(31a), 158, 176, 177, 185(131), 186(131), 187, 188, 189(39, 131), 190, 191, 192, 210(31a), 211 (131), 216(31a, 38), 217(31a), 218, 219(31a), 220(31a), 223(31a), 230(31a), 231, 242(31a)
Kapur, C., 243, 275
Kast, W., 331, 333(54)
Katchalsky, A., 429, 456(48)
Katz, D., 429, 457(60)
Kearsley, E., 408
Keller, J. B., 221, 222
Kelley, E. L., 233, 263, 270(74), 272
Khanna, S. K., 457, 459
Khailenk, L. V., 263
Kim, K. Y., 271, 368, 457, 458
Kirkwood, J. G., 15, 16, 18, 261, 262, 324, 343, 344, 345, 405
Klage, G., 12
Klaus, E. E., 281
Kleinschmidt, R. V., 49
Klimov, K. I., 457, 459
Knobloch, P., 12
Komura, K., 457, 459
Kotaka, T., 404, 405
Kotaki, T., 345
Kramer, H., 11
Krasucki, Z., 87, 96(8), 130(8), 133
Kratky, O., 457, 458
Kremann, R., 1

O

Oka, S., 271, 295, 350, 369(7), 373(7), 384(7), 390(7)
O'Konski, C. T., 87, 130, 131, 132
Oldroyd, J. D., 124, 350, 359, 369(26), 371(6)
Oldshue, J. Y., 286, 289(2)
Oliver, D. R., 147, 148, 155(101), 210(93, 101), 211(93), 216(101), 217, 220(101), 223(101), 242(93)
Ormsby, W. C., 429
Osborne, D. G., 457, 458
Oseen, C. W., 220
Osthoff, J. C., 32
Ostwald, W., 255, 321, 336, 338, 339, 340, 342, 343(13, 69), 345(69), 425, 452
Overbeek, J. T. G., 450
Oyanagi, Y., 40

P

Palade, G., 34
Pallett, W. F. O., 457, 459
Palmer, A. A., 243
Panchenkov, G. M., 80
Pao, Y. H., 267
Park, M. G., 276, 277(121)
Parry, H. L., 76
Paslay, P. R., 451
Passaglia, E., 264
Pearson, G. P., 459
Pearson, J. R. A., 221
Penther, C. J., 446
Penzin, G., 252
Perkinson, G., 263, 265(70)
Peter, S., 329, 330, 331, 332, 429
Peterfi, T., 424, 425
Peteril, E., 457, 458
Peterlin, A., 255, 261, 262, 266, 267, 280, 281(145), 429, 452, 454(44, 45), 455(44), 456(44), 459(44, 109)
Peters, H., 329, 330, 331, 332
Phibbs, R. H., 235
Philippoff, W., 256, 263, 264, 267, 268, 269, 270(75), 278, 368, 395
Pierce, P., 438, 457(85)
Piercy, J. E., 26
Pierotti, G. J., 72
Pigford, R. L., 429

Pimentel, G. C., 67
Pipkin, A. C., 408
Pirquet, A., 428, 457(40), 465
Plazek, D. J., 56
Plock, R. J., 343, 344(80)
Plymale, C. E., 307
Pockter, A., 450
Pohl, H. A., 282, 283(158), 457, 459
Poiseuille, J. M. L., 147
Pople, J., 18
Porod, G., 45
Porter, R. S., 40, 252, 260, 263(39), 268, 272, 281, 320, 328, 329, 331(47), 333(48), 336, 338(70), 339(70), 340(70)
Powell, B. D., 341, 342(78)
Powell, R. E., 40, 71, 77, 255
Powles, J. G., 13, 14, 66
Prager, S., 405
Prescott, J., 139
Price, C. C., 141, 281
Prideaux, E. B. R., 76
Prigogine, I., 3, 5(6), 70
Prothero, J. W., 224, 245
Proudman, I., 221
Pryce-Jones, J., 426, 429, 430, 431, 452, 454(71, 77), 455, 457(51, 71, 76, 77, 78), 459(51, 71, 77, 78)
Puddington, I. E., 341, 342(78), 429, 454, 459(69, 114)

Q

Quan, C., 429, 454(45), 456(45)

R

Raasch, J. K., 93, 150(25), 153, 162, 169, 170(25), 184(114), 344
Rabinowitsch, B., 270
Rae, D., 444
Rajagopal, E. S., 130
Rakov, A. V., 15
Ram, A., 260, 263, 265(41, 70, 71, 72), 270(63), 272, 274(63), 281, 282(148)
Rand, R. P., 239, 240
Rawitzer, W., 424, 425, 426(8)
Ray, T. C., 18
Rea, D. R., 316
Redlich, O., 72

SUBJECT INDEX

See also Chapter 9 on Rheological Terminology. In the index, F signifies figure and T signifies table.

502

W

Wall effect(s), 150, 189, 209, 242
  higher-order calculations, 154
  neutrally buoyant spheres, 153
  sphere settling, 220
  suspended spheres, 223
Weight, effect in spinning, 413
Weissenberg effect, 307, 380, 455, *see*
  also Normal stress effect
  suspension of flexible fibers, 214
Weissenberg fluid, 408
Weissenberg number, 279
Weissenberg relation, 394
  suspensions of dumbbells, pearl neck-
    laces, rigid rods, 405
Wetting, particle mixing, 307
White blood corpuscles, *see* White cell

White cell, 232
  motion, 242
Windup roll, 416
  cross section, 417
  take-up roll, 415
W-L-F (Williams, Landel, Ferry) equation, 55, 64
Wood chips, pipeline transportation, 231
Wood pulp fibers, 144

Y

Yield stress, 354
Yield value, 427

Z

Zinc oxide, 459T